"十三五"国家重点图书出版物出版规划

经典建筑理论书系

加州大学伯克利分校环境结构中心系列

建筑模式语言

——城镇·建筑·构造

（下册）

A Pattern Language

—Towns · Buildings · Construction

［美］ C. 亚历山大　 S. 伊希卡娃　 M. 西尔沃斯坦

　　　 M. 雅各布逊　 I. 菲克斯达尔 - 金　 S. 安格尔　　著

王听度　 周序鸿　译

李道增　 高亦兰

关肇邺　 刘鸿滨　审校

知识产权出版社

全国百佳图书出版单位

—北 京—

A Pattern Language was originally published in English in 1977. This translation is published by arrangement with Oxford University Press. Intellectual Property Publishing House Co., Ltd. is solely responsible for this translation from the original work and Oxford University Press shall have no liability for any errors, omissions or inaccuracies or ambiguities in such translation or for any losses caused by reliance thereon.

图书在版编目（CIP）数据

建筑模式语言：城镇·建筑·构造 /（美）C. 亚历山大等著；王昕度,周序鸿译 . —北京：知识产权出版社，2022.7
（经典建筑理论书系）
书名原文：A Pattern Language：Towns · Buildings · Construction
ISBN 978-7-5130-7406-3

Ⅰ .①建⋯　Ⅱ .① C⋯②王⋯③周⋯　Ⅲ .①建筑设计—研究　Ⅳ .① TU2

中国版本图书馆 CIP 数据核字（2021）第 013803 号

责任编辑：李　潇　刘　嚚　　　　　责任校对：谷　洋
封面设计：红石榴文化 · 王英磊　　　责任印制：刘译文

经典建筑理论书系

建筑模式语言——城镇·建筑·构造（下册）

A Pattern Language—Towns · Buildings · Construction

［美］　C. 亚历山大　S. 伊希卡娃　M. 西尔沃斯坦
　　　　M. 雅各布逊　I. 菲克斯达尔－金　S. 安格尔　著

王昕度　周序鸿　译

李道增　高亦兰
　　　　　　　审校
关肇邺　刘鸿滨

出版发行：知识产权出版社 有限责任公司	网　　址：http://www.ipph.cn
社　　址：北京市海淀区气象路 50 号院	邮　　编：100081
责编电话：010-82000860 转 8119	责编邮箱：liuhe@cnipr.com
发行电话：010-82000860 转 8101	发行传真：010-82000893/82005070
印　　刷：三河市国英印务有限公司	经　　销：新华书店、各大网上书店及相关专业网点
开　　本：880mm×1230mm　1/32	总 印 张：69.5
版　　次：2022 年 7 月第 1 版	印　　次：2022 年 7 月第 1 次印刷
总 字 数：1730 千字	总 定 价：268.00 元（上、下册）
ISBN 978-7-5130-7406-3	
京权图字：01-2016-8195	

目　录

CONTENTS

CONTENTS
目　录

CONTENTS

目　录

目 录
CONTENTS

CONTENTS
目 录

CONTENTS
目 录

BUILDINGS

建筑

用于规划城镇或社区的综合性模式业已完成。现在我们开始讨论新的一部分模式语言，它规定土地上的建筑群和个体建筑三维空间的形状。这是一些可以"设计"或"建造"的模式，即规定个体建筑物和各建筑物之间的空间的模式。我们在这些地方使用的模式是个人或少数人能够支配的，是他们能够立即建造起来的。这种情况在本书还是首次出现。

৪০৪৩

我们假定，根据"本模式语言概要"中的规定，你已经编排好模式的程序。现在我们就要按部就班地进行设计。

1. 基本规程如下：按照程序的顺序逐一采用这些模式，使这些模式与工地及你个人的才能结合在一起，产生属于自己的模型。

2. 工作必须安排在工程准备动工的现场；在打算改造的房间内部；在即将盖起楼房的地段；以及诸如此类的地方。只要有可能，要同房屋盖好后那些实际的房屋使用者一起工作。如你本人使用，那再好不过了。但最主要的是，在工地工作和生活，**让工地告诉你它的秘密**。

3. 还应记住，模型是随着你完成该程序而日臻完善的。开始时很松散、杂乱，逐渐变得复杂、精致、多样化和完美起来。不要匆忙走过这个过程。你的模型需要处处与这些模式和工地条件相适应，式样不能过多。事实上，

This completes the global patterns which define a town or a community. We now start that part of the language which gives shape to groups of buildings, and individual buildings, on the land, in three dimensions. These are the patterns which can be "designed" or "built" —the patterns which define the individual buildings and the space between buildings, where we are dealing for the first time with patterns that are under the control of individuals or small groups of individuals, who are able to build the patterns all at once.

ଧୟଓଷ

We assume that, based on the instructions in "Summary of the Language," you have already constructed a sequence of patterns. We shall now go through a step-by-step procedure for building this sequence into a design.

1. The basic instruction is this: Take the patterns in the order of the sequence, one by one, and let the form grow from the fusion of these patterns, the site, and your own instincts.

2. It is essential to work on the site, where the project is to be built; inside the room that is to be remodeled; on the land where the building is to go up; and so forth. And as far as possible, work with the people that are actually going to use the place when it is finished: if you are the user, all the better. But, above all, work on the site, stay on the site, *let the site tell you its secrets.*

3. Remember too, that the form will grow gradually as you go through the sequence, beginning as something very loose and amorphous, gradually becoming more and more complicated, more refined and more differentiated, more finished. Don't rush this process. Don't give the form more order than it

随着每一个模式体现在设计中,你会看到一个逐渐连贯的整体。

4. 一次取一个模式。打开本书,翻到第一个模式,再读它一遍。这一模式描述其他模式如何影响它或者被它所影响。到目前为止,这一信息的用处还只是有助于你将面前的这一模式作为一个整体加以想象。

5. 现在,请你设想一下,在这个有自己特点的工地上该如何建立起这个模式。请站在该地基上,闭上眼睛想象一下,假如你已经了解的模式突然间出现在那儿,情况会是怎样。一旦你心目中有了它可能具有的形象,就可以在这块地段上来回走走,用脚步量出大致的面积,用绳子和厚纸板把墙标记出来,然后在地上立桩或放置碎石标出重要的角落。

6. 先把这个模式想好,然后再着手考虑下一个。这就是说,你得把这个模式作为"整体"来看待,竭力想象这个整体,使它全面而完整,然后才开始去创作其他的模式。

7. 本语言的程序可以保证你不需要作大的改动,不会使你取消原先的设想。不仅如此,随着你越来越多地采用这些模式,精益求精,改动就越少,最终使你的设计完美无缺。

8. 由于是一次一个模式进行设计,因此在你从一个模式过渡到另一个模式的时候,必须使设计尽可能地容易变动。当你一个接一个地利用这些模式时,就需要不断调整你的设计以适应新模式的要求。重要的是,要灵活机动,留有余地,不使设计不必要地过分固定,不必害怕改动。

needs to meet the patterns and the conditions of the site,each step of the way.In effect,as you build each pattern into the design,you will experience a single gestalt that is gradually becoming more and more coherent.

4. Take one pattern at a time.Open the page to the first one and read it again.The pattern statement describes the ways in which other patterns either influence this pattern,or are influenced by it.*For now*,this information is useful only in so far as it helps you to envision *the one pattern before you,as a whole.*

5. Now,try to imagine how,on your particular site,you can establish this pattern.Stand on the site with your eyes closed.Imagine how things might be,if the pattern,as you have understood it,had suddenly sprung up there overnight.Once you have an image of how it might be,walk about the site,pacing out approximate areas,marking the walls,using string and cardboard,and putting stakes in the ground,or loose stones,to mark the important corners.

6. Complete your thought about this pattern,before you go on to the next one.This means you must treat the pattern as an "entity";and try to conceive of this entity,entire and whole,before you start creating any other patterns.

7. The sequence of the language will guarantee that you will not have to make enormous changes which cancel out your earlier decisions.Instead,the changes you make will get smaller and smaller,as you build in more and more patterns,like a series of progressive refinements,until you finally have a complete design.

8. Since you are building up your design,one pattern at a time,it is essential to keep your design as fluid as possible,while

可视需要对设计进行修改，只要坚持原先的模式规定的基本关系和特性不变。你会看到，保持这些主要的内容不变，在设计中作些小的修改仍是可能的。每当你采用一个新模式时，要重新调整总体设计，以便使它同你正在制作的模式协调。

9. 在你设想如何建立一个模式的同时，请考虑同它列在一起的其他模式。有些比它大，有些比它小。对于比它大的，你可以设想它们会在某天怎样出现在你工作的地区，并要问问自己，你现在正在建造的模式如何能够有助于修改或形成这些更大的模式。

10. 对于那些较小的模式，在你进行构想时，务必以后把它们也包括在内。在建造这些模式时，如果你能大致确定，将如何把这些模式建在主要模式之中，那是很有裨益的。

11. 从一开始就应算好面积，以便使你的建筑费用始终合情合理，不超过你实际的支付能力。我们见过许多这样的事情：人们想方设法设计自己的或别人的住宅，但后来却大失所望，因为最终的费用太高，于是他们不得不改弦更张，重起炉灶。

要做到这一点，需先确定预算，然后采用一种合理的单方造价方案，将该预算落实到每平方英尺的建筑物中去。为了便于讨论，假定你的建房预算为30000美元。在建筑业经纪人的帮助下，算一算，哪一种平方英尺造价对你的房屋建造方案最合适。例如，在1976年的加利福尼亚，有一座与本语言最后一部分中的模式相一致的、比较朴素的住宅，每平方英尺的造价大约为28美

you go from pattern to pattern.As you use the patterns,one after another,you will find that you keep needing to adjust your design to accommodate new patterns.It is important that you do this in a loose and relaxed way,without getting the design more fixed than necessary,and without being afraid to make changes. The design can change as it needs to,so long as you maintain the essential relationships and characteristics which earlier patterns have prescribed.You will see that it is possible to keep these essentials constant,and still make minor changes in the design.As you include each new pattern,you readjust the total gestalt of your design,to bring it into line with the pattern you are working on.

9. While you are imagining how to establish one pattern, consider the other patterns listed with it.Some are larger.Some are smaller.For the larger ones,try to see how they can one day be present in the areas you are working on,and ask yourself how the pattern you are now building can contribute to the repair or formation of these larger patterns.

10. For the smaller ones,make sure that your conception of the main pattern will allow you to make these smaller patterns within it later.It will probably be helpful if you try to decide roughly how you are going to build these smaller patterns in,when you come to them.

11. Keep track of the area from the very beginning so that you are always reasonably close to something you can actually afford.We have had many experiences in which people try to design their own houses,or other buildings,and then get discouraged because the final cost is too high,and they have to go back and change it.

元。如果你希望装修要更豪华些，造价就要高一点。若你的建房预算有 36000 美元，就可以盖一幢大约 1300ft^2 的房屋。

12. 现在，在整个设计过程中，都要把这 1300ft^2 的数字记在心上。如果你想盖两层楼，地面面积应保持 650ft^2。如果仅仅使用楼上的部分面积，地面面积可达 800 ～ 900ft^2。如果你决定建造相当精致的有围合的户外小空间、墙、棚架，那就要减少室内面积以补偿这些室外费用——可能需要把室内面积降到 1100 ～ 1200ft^2。每当用一种模式进一步丰富你的建筑布局时，请记住这个总面积，这样，就绝不会使你的预算超支。

13. 最后，在工地上用砖、木棒或桩做出固定这种模式所需要的点和线。最好不要设计在纸上；就算是再复杂的建筑物也要设法把记号做在工地上。

实际设计过程中更详细的规程和具体事例请参阅《建筑的永恒之道》（*The Timeless Way of Building*）第 20 章、第 21 章和第 22 章。

第一组模式有助于对建筑群的整体布局、建筑物的高度和数目、该地区的各种入口、主要停车场以及经过建筑群体的通路进行设计。

95. 建筑群体

96. 楼层数

97. 有遮挡的停车场

98. 内部交通领域

99. 主要建筑

100. 步行街

To do this,decide on a budget,and use a reasonable average square foot cost,to translate this budget into square feet of construction.Say for the sake of argument,that you have a budget of $30000 for construction.With help from builders,find out what kind of square foot cost is reasonable for the kind of building you are making.For example,in 1976,in California,a reasonable house,compatible with the patterns in the last part of the language,can be built for some $28/square foot.If you want expensive finishes,it will be more.With $36,000 for construction,this will give you some 1300 square feet.

12. Now,throughout the design process,keep this 1300 square-foot figure in mind.If you go to two stories,keep the ground area to 650 square feet.If you use only part of the upstairs volume,the ground floor can go as high as 800 or 900 square feet.If you decide to build rather elaborate outdoor rooms,walls,trellises,reduce the indoor area to make up for these outdoor costs—perhaps down to 1100 or 1200.And,each time you use a pattern to differentiate the layout of your building further,keep this total area in mind,so that you do not,ever,allow yourself to go beyond your budget.

13. Finally,make the essential points and lines which are needed to fix the pattern,on the site with bricks,or sticks or stakes.Try not to design on paper;even in the case of complicated buildings find a way to make your marks on the site.

More detailed instructions,and detailed examples of the design process in action,are given in chapters 20,21,and 22 of *The Timeless Way of Building*.

The first group of patterns helps to lay out the overall arrangement of a group of buildings:the height and number of these buildings,the entrances to the site,main parking areas,and lines of movement through the complex;

95. BUILDING COMPLEX

96. NUMBER OF STORIES

97. SHIELDED PARKING

98. CIRCULATION REALMS

99. MAIN BUILDING

100. PEDESTRIAN STREET

101. BUILDING THOROUGHFARE

102. FAMILY OF ENTRANCES

103. SMALL PARKING LOTS

模式95　建筑群体**

95   BUILDING COMPLEX**

...this pattern,the first of the 130 patterns which deal specifically with buildings,is the bottleneck through which all languages pass from the social layouts of the earlier patterns to the smaller ones which define individual spaces.

Assume that you have decided to build a certain building. The social groups or institutions which the building is meant to house are given—partly by the facts peculiar to your own case,and partly,perhaps,by earlier patterns.Now this pattern and the next one—NUMBER OF STORIES(96),give you the basis of the building's layout on the site.This pattern shows you roughly how to break the building into parts.NUMBER OF STORIES helps you decide how high to make each part. Obviously,the two patterns must be used together.

❧❧❧

A building cannot be a human building unless it is a complex of still smaller buildings or smaller parts which manifest its own internal social facts.

A building is a visible,concrete manifestation of a social group or social institution.And since every social institution has smaller groups and institutions within it,a human building will always reveal itself,not as a monolith,but as a complex of these smaller institutions,made manifest and concrete too.

A family has couples and groups within it;a factory has teams of workers;a town hall has divisions,departments within the large divisions,and working groups within these

……此模式是专门研究建筑物的 130 个模式中的第一个。它是一个瓶颈口，所有的模式语言，从之前那些具有社会性质的模式到限定属于个体空间的较小模式都得通过它。

假定你已经决定承建某一幢楼房。你知道有哪些社会组织或机构准备搬进这幢楼房——一部分是根据你个人情况所特有的事实，而另一部分也许是根据前面那些模式。本模式及下一个模式——**楼层数**（96），提供了在工地上建筑物布局的基础。本模式大体表明，如何将建筑物分割成几部分。**楼层数**（96）可以帮你决定，每一部分应盖多高。显然，这两个模式必须一起使用。

⊗⊙⊗

一幢建筑应该是一个建筑群体，它由一些较小建筑或较小部分所组成，通过它们表现其内部社会功能，不然的话，该建筑就会毫无生气。

建筑物是社会集团或社会机构的外部的、具体的表现。由于每一个社会机构其内部有着更小的团体和机构，那么生机蓬勃的建筑物绝不会是整体式的，而是表现为由这些小机构组成的群体，可以使身处外部的人一目了然。

家庭内部有夫妇和其他成员；工厂有一批工人；市政厅设有各个局，这些大的局内有处和科，各科又有工作小组。能够把一个机构中的分支及其之间的联系表现出来的建筑物是生机蓬勃的建筑物——因为它让我们根据人们相互结合的方式来生活。相反，那种整体式建筑物正是因为无视自身社会结构的实际情况，它便缺少生活气息，只能使人们的生活勉强去适应它。

departments.A building which shows these subdivisions and articulations in its fabric is a human building—because it lets us live according to the way that people group themselves.By contrast, any monolithic building is denying the facts of its own social structure,and in denying these facts it is asserting other facts of a less human kind and forcing people to adapt their lives to them instead.

We have tried to make this feeling more precise by means of the following conjecture:the more monolithic a building is,and the less differentiated,the more it presents itself as an inhuman,mechanical factory.And when human organizations are housed in enormous,undifferentiated buildings,people stop identifying with the staff who work there as personalities and think only of the institution as an impersonal monolith,staffed by personnel.In short, the more monolithic the building is,the more it prevents people from being personal,and from making human contact with the other people in the building.

The strongest evidence for this conjecture that we have found to date comes from a survey of visitors to public service buildings in Vancouver,British Columbia.(*Preliminary Program for Massing Studies,Document 5:Visitor Survey*, Environmental Analysis Group,Vancouver,B.C.,August 1970.)Two kinds of public service buildings were studied—old,three story buildings and huge modern office buildings.The reactions of visitors to the small building differed from the reactions of visitors to the large buildings in an extraordinary way.The people going to the small buildings most often mentioned friendly and competent staff as the important factor in their satisfaction with the service. In many cases the visitors were able to give names and describe the people with whom they had done business.Visitors to the huge

我们一直试图借助以下推测来阐明这一感受：建筑物越是整体式、缺少变化，它越表现得不合情理、机械呆板。当社会机构设在巨大的、千篇一律的建筑物里时，人们不再把自己当作有个性的工作人员，仅把该机构看作由工作人员组成的无个性的团体。简言之，建筑物越是整体式，它越使人失去个性，越使建筑物中人与人之间不愿来往。

　　迄今为止，我们所发现的这一设想的最强有力的证据，来自对加拿大的不列颠哥伦比亚省温哥华市的公共服务楼的参观者的调查。(*Preliminary Program for Massing Studies*, *Document 5*: *Visitor Survey*, Environmental Analysis Group, Vancouver, B.C., August 1970.) 被调查的是两类公共服务楼。其中一类是旧的三层楼房，另一类是现代化的办公大楼。参观者对小楼房的反应同对大办公楼的反应截然不同。去小楼参观过的人屡屡提及态度友好、办事干练的工作人员，这是他们对服务感到满意的一个重要原因。在许多场合，参观者能够说出同他们打交道的人的名字，并能描述出这些人的特征。可是，友好气氛和办公人员干练之类的话很少能从大办公楼的参观者口中听到。绝大多数的参观者只是说他们对"建筑物漂亮的外表和设备"感到满意。

　　参观者感觉到整体式建筑物缺乏个性。他们不能首先考虑他们想去看的人以及人们之间的关系，而只能把注意力集中在建筑物本身及其特点上。工作人员都是千人一面，彼此难分，态度冷漠，因此参观者很少注意他们的为人——友好还是不友好，能干还是不能干。

　　从这一调查中我们也能了解到，在大建筑物里参观的人会时常抱怨，这种建筑物给他们留下的是笼统的印象，说不出它的好坏。参观者对较小的建筑物就没有这种怨言。看来，这种整体式建筑物会使人产生一种心神不安的感觉。

office buildings, on the other hand,mentioned friendliness and staff competence rather infrequently.The great majority of these visitors found their satisfaction in "good physical appearance, and equipment."

In the monoliths,the visitors' experience is depersonalized. They stop thinking primarily of the people they are going to see and the quality of the relationship and focus instead on the building itself and its features.The staff becomes "person nel," interchangeable,and indifferent,and the visitors pay little attention to them as people—friendly or unfriendly,competent or incompetent.

We learn also from this study that in the large buildings visitors complained frequently about the "general atmosphere" of the building,without naming specific problems.There were no such complaints among the visitors to the smaller buildings.It is as if the monoliths induce a kind of free-floating anxiety in people:the environment "feels wrong," but it is hard to give a reason.It may be that the cause of the uneasiness is so simple— the place is too big,it is difficult to grasp,the people are like bees in a hive—that people are embarrassed to say it outright. ("If it is as simple as that, I must be wrong—after all, there are so many of these buildings.")

However it is,we take this evidence to indicate deep disaffection from the human environment in the huge, undifferentiated office buildings.The buildings impress themselves upon us as things:objects,commodities;they make us forget the people inside,as people; yet when we use these buildings we complain vaguely about the "general atmosphere."

It seems then that the degree to which a building is broken

环境让人"感到不对头",但却说不出所以然来。心神不安的理由可能太简单:这儿地方太大,让人捉摸不透,人们像蜂窝里的蜜蜂——以致大家都不好意思直截了当地把它说出来。("如果道理这么简单,那我一定错了——这样的建筑物毕竟多得很呐。")

不管怎样,我们认为这一证据反映了人们对呆板的大办公楼里生活环境的不满。这些建筑给我们的印象是:物品和商品。它们使我们忘记里面住的是人,但当我们使用这样的楼房时,我们只是模糊地对"总体气氛"感到不满。

那么看来,如何把一幢建筑物分解成明显的几部分,的确会影响建筑物里人们之间的关系。如果,出于心理学上的考虑,必须把一幢建筑物分解为若干部分,看来不可能找到比我们所提出的更自然的分解法了。这就是说,各种机构、团体、子团体、下属机构,都在这个建筑实体的具体关联中显现出来,因为只有当这个建筑是一个**建筑群体**时,人们才得以充分认识该建筑物中的人的身份。

哥特式大教堂虽然是很大的建筑,却堪称建筑群体的范例。它的各个部分,尖塔、走廊、中殿、高坛、西门都精确地反映了教徒、唱诗班、特别弥撒仪式等各社会集团的需要。

当然,非洲的茅屋群也富有生活气息,因为它是建筑群体,而不是一幢大建筑物。

对于密度很高的建筑群体来说,要使其中的各部分都有自己的特色,最容易的办法是由一些临街的狭窄楼房组合成为一个建筑群体,每一幢楼房都有各自的内部楼梯。这是乔治亚式连排房或纽约褐砂石房屋的基本结构。

into visible parts *does* affect the human relations among people in the building.And if a building must,for psychological reasons,be broken into parts,it seems impossible to find any more natural way of breaking it down,than the one we have suggested.Namely,that the various institutions,groups,subgr oups,activities,are visible in the concrete articulation of the physical building,on the grounds that people will only be fully able to identify with people in the building,when the building is a building *complex*.

A gothic cathedral—though an immense building—is an example of a building complex.Its various parts, the spire, the aisle, the nave, the chancel, the west gate, are a precise reflection of the social groups—the congregation,the choir,the special mass, and so forth.

And,of course,a group of huts in Africa,is human too,because it too is a complex of buildings, not one huge building by itself.

For a complex of buildings at high density,the easiest way of all,of making its human parts identifiable,is to build it up from narrow fronted buildings,each with its own internal stair.This is the basic structure of a Georgian terrace,or the brownstones of New York.

Therefore:

Never build large monolithic buildings. Whenever possible translate your building program into a building complex,whose parts manifest the actual social facts of the situation. At low densities, a building complex may take the form of a collection of small buildings connected by arcades, paths, bridges, shared gardens, and walls.

At higher densities, a single building can be treated as a

因此：

绝对不要建造大的整体式建筑物。只要有可能，把你的建筑计划改为建筑群体，使其各部分都能表现实际的社会内容。在密度低的情况下，一个建筑群体可采用小建筑群的形式，各建筑物之间用拱廊、小道、桥、共用花园及墙连接起来。

在密度高的地方，只要把单个的建筑物其重要部分突出出来并使之各具特色，但仍是这个三维结构的一部分，就可把该建筑物看作一个建筑群体。

甚至一幢小建筑，比方一幢住宅，也能被当作一个"建筑群体"，因为这幢房屋有一部分多半会高于其翼楼和与其相连的平房。

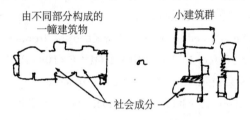

由不同部分构成的一幢建筑物　　小建筑群

社会成分

&

在密度很高的情况下，沿着步行街，将建筑物盖成高耸的狭长形，每幢三层或四层，鳞次栉比，每一幢楼都有其自己的室内或室外楼梯。只要有可能，一定要分步骤盖这些楼房，一次盖一幢，这样才有时间考虑如何使每一幢楼同与它相邻的那一幢能够做到彼此映衬。使正面保持 25 ～ 30ft 的高度。**狭长形住宅**（109）、**建筑物正面**（122）；**主入口**（110），也许还有连接邻楼的部分**拱廊**（119）。

把群体内各建筑物排列妥当，使之形成内部交通领域——**内部交通领域**（98）；在群体内盖一幢主建筑——

building complex,if its important parts are picked out and made identifiable while still part of one three-dimensional fabric.

Even a small building, a house for example, can be conceived as a "building complex"—perhaps part of it is higher than the rest with wings and an adjoining cotage.

৪০৩৪

At the highest densities,3 or 4 stories,and along pedestrian streets,break the buildings into narrow,tall separate buildings,side by side,with common walls,each with its own internal or external stair.As far as possible insist that they be built piecemeal,one at a time,so that each one has time to be adapted to its neighbor.Keep the frontage as low as 25 or 30 feet.LONG THIN HOUSE (109),BUILDING FRONTS(122);MAIN ENTRANCE(110)and perhaps a part of an ARCADE(119)which connects to next door buildings.

Arrange the buildings in the complex to form realms of movement—CIRCULATION REALMS(98);build one building from the collection as a main building—the natural center of the site—MAIN BUILDING(99);place individual buildings where the land is least beautiful,least healthy—SITE REPAIR(104);and put them to the north of their respective open space to keep the gardens sunny—SOUTH-FACING OUTDOORS(105);subdivide them further,into narrow wings,no more than 25 or 30 feet across—WINGS OF LIGHT(107).For details of construction,start with STRUCTURE FOLLOWS SOCIAL SPACES(205)...

该地段的自然中心——**主建筑**（99）；将个体建筑盖在最难看、最脏乱的地方——**基地修整**（104）；使它们坐落在各自所在空地的北边以保持花园朝阳——**朝南的户外空间**（105）；再进一步把它们细分成狭窄形的翼楼，宽度不超过 25ft 或 30ft——**有天然采光的翼楼**（107）。关于构造施工的具体问题，参阅**结构服从社会空间的需要**（205）及之后各模式……

模式96　楼层数*

　　……现在假定，你已大致了解了建筑群体各部分如何连接——**建筑群体**（95），它们有多大。再假定，你已经选好了基地。为了确保能在这个基地范围内盖成一幢实用的楼房，你得确定各部分盖多少层。每一部分的高度必须有限制，**不高于四层楼**（21）。除此之外，它还取决于该地基的面积，以及每一部分所需要的建筑面积。

96 NUMBER OF STORIES*

...assume now,that you know roughly how the parts of the building complex are to be articulated—BUILDING OOMPLEX(95),and how large they are.Assume,also,that you have a site.In order to be sure that your building complex is workable within the limits of the site,you must decide how many stories its different parts will have.The height of each part must be constrained by the FOUR-STORY LIMIT(21).Beyond that,it depends on the area of your site,and the floor area which each part needs.

୫୦୯ଓ

Within the four-story height limit,just exactly how high should your buildings be?

To keep them small in scale,for human reasons,and to keep the costs down,they should be as low as possible.But to make the best use of land and to form a continuous fabric with surrounding buildings,they should perhaps be two or three or four stories instead of one.In this pattern we give rules for striking the balance.

*Rule 1:Set a four-story height limit on the site.*This rule comes directly from FOUR-STORY LIMIT(21)and the reasons for establishing this limit are described there.

*Rule 2:For any given site,do not let the ground area covered by buildings exceed 50 per cent of the site.*This rule requires that for any given site,where it belongs to a single household or a corporation,or whether it is a part of a larger site

你盖的楼房应不高于四层楼，确切地讲，它该有多高?

考虑到人的因素，建筑物的规模要小，同时还希望可以降低造价，楼房应尽可能盖得低一点。但为了合理地利用土地，并同周围的房屋联系起来考虑，它们多半应该是两层、三层或四层，而不是一层。在本模式中我们提出维持这种平衡的一些规则。

规则 1：规定工地上盖楼高度不超过四层。这一规则直接来自**不高于四层楼**（21），确定这一限度的原因已在彼处说过了。

规则 2：对于任何一个选定的基地，被建筑物覆盖的地面面积不应超过该地基的 50%。这一规则要求，对于任何一个选定的基地，不管它是属于独家使用还是由数家联合使用，也不管它是否包括有多幢建筑物的更大基地的一部分，至少该基地的一半要留作空地。这是地面覆盖的限度，只有在这个限度内才能进行合理的总体规划。因此这条规则决定了可以在一个选定的基地上盖某一层数楼房的最大建筑面积。建筑面积对基地面积的比率（FAR，Floor Area Ratio，土地容积率），一楼不能超过 0.5，二楼为 1.0，三楼为 1.5，四楼为 2.0。

如果你打算建造的总楼面面积加上该基地上已建成楼面面积大于该基地本身面积的两倍，那就超过了这个限度。在这种情况下，我们建议缩减建筑设计，少建造一些空间；也许可以把工程一部分盖在其他基地上。

规则 3：别让你的楼房高度同周围大部分房屋的高度过于悬殊。有一个大体上行得通的办法：楼房同周围建筑物不要相差一层楼以上的高度。总的来说，彼此相邻房屋的高度应大致相同。

which contains several buildings,at least half of the site is left as open space.This is the limit of ground coverage within which reasonable site planning can take place.The rule therefore determines the maximum floor area that can be built with any given number of stories on a given site.The ratio of indoor area to site area(FAR—for floor area ratio)cannot thus exceed 0.5 in a single story building,1.0 in a two story,1.5 in a three story and 2.0 in a four story building.

If the total floor area you intend to build plus the built floor area that exists on the site is more than twice the area of the site itself,then you are exceeding this limit.In this case,we advise that you cut back your program;build less space;perhaps build some of your project on another site.

Rule 3:Do not let the height of your building(s)vary too much from the predominant height of surrounding buildings. A rule of thumb:do not let your buildings deviate more than one story from surrounding buildings.On the whole,adjacent buildings should be roughly the same height.

I live in a small one-story garden cottage at the back of a large house in Berkeley.All around the cottage there are two-story houses,some as close as thirty feet.I thought when I moved in,that a garden cottage would be secluded and I would have some private outdoor space.But instead I feel that I'm living in a goldfish bowlx—every one of the second-story windows around me looks right down into my living room,or into my garden.The garden outside is useless,and I don't sit near the window.

Therefore:

First, decide how many square feet of built space you need, and divide by the area of the site to get the floor area ratio.Then

违反常规
Breaking the rule of thumb

　　我住在伯克利一幢大楼后面的一座小的单层花园别墅里。别墅周围都是两层楼房，有些离我只有 30ft。当我搬进来的时候曾想，一座花园别墅一定会十分幽静，我会拥有某种私密的户外小空间。但事与愿违，我感觉自己是住在一个金鱼缸里——周围二层楼里的每个人都能从窗户看到我的居室或我的花园。室外的花园成了废物，而且我还不能倚窗而坐。

　　因此：

　　首先，确定你需要多少平方英尺的建筑空间，将此除以基地的面积，就得到该基地的土地容积率。然后根据下表中的土地容积率和周围楼房的高度选择你的建筑物的高度。建筑物的覆盖面绝不可超过土地面积的 50%。

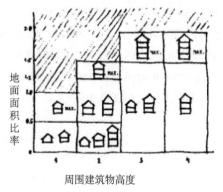

地面面积比率

周围建筑物高度

choose the height of your buildings according to the floor area ratio and the height of the surrounding buildings from the following table.In no case build on more than 50 per cent of the land.

8003

Once you have the number of stories and the area of each part clear,decide which building or which part of the building will be the MAIN BUILDING(99).Vary the number of floors within the building—CASCADE OF ROOFS(116). Place the buildings on the site,with special reverence for the land,and trees,and sun—SITE REPAIR(104),SOUTH FACING OUTDOORS(105),TREE PLACES(171).In your calculations,remember that the effective area of the top story will be no more than threequarters of the area of lower floors if it is in the roof,according to SHELTERING ROOF(117).

If the density is so high all around,that it is quite impossible to leave 50 per cent of the site open(as might be true in central London or New York), then cover the ground floor completely,but devote at least 50 per cent of the upper floors to open gardens—ROOF GARDEN(118).

Give each story a different ceiling height-bottom story biggest,top story smallest—and vary the column spacings accordingly—FINAL COLUMN DISTRIBUTION(213).The same building system applies,whether there are 1,2,3 or 4 stories—STRUCTURE FOLLOWS SOCIAL SPACES(205)...

8003

一旦楼层数和每一部分的面积都已确定，你就可以决定哪幢楼房或楼房的哪部分作为**主要建筑**（99）。使楼房内部有层数变化——**重叠交错的屋顶**（116）。在工地盖房屋时要特别爱护土地、树木和阳光——**基地修整**（104）、**朝南的户外空间**（105）、**树荫空间**（171）。根据**带阁楼的坡屋顶**（117），在规划时请记住，最高层若在坡屋顶之内，其有效面积不能超过较低层面积的 3/4。

如周围建筑密度很高，完全不可能留出 50％的空地来（伦敦中心区或纽约可能就是这样），那么一层的楼面面积就可以全部利用上，但要把上层楼面面积至少留出 50％用作露天花园——**屋顶花园**（118）。

每一层都有不同的天花板高度——底层最高，顶层最低——并相应改变柱间距——**柱的最后分布**（213）。不管有一层楼、二层楼、三层楼或四层楼，都可以应用相同的建筑系统——**结构服从社会空间的需要**（205）……

模式97 有遮挡的停车场*

　　……我们提出过的许多模式都不赞成过多使用小汽车。我们希望，采用这些模式会逐步地做到完全不需要大停车场和停车设施——**地方交通区**（11）、**停车场不超过用地的**9％（22）。然而，遗憾的是，在某些情况下，大面积停车场仍然被需要。现实情况既然如此，这样的停车场就要及早布置，务必使它不致破坏**建筑群体**（95）的完整性。

⊱⊰

　　停满汽车的大停车场或车库是没有生活气息的死建筑——没有人愿意看见它们或者在它们旁边行走。但与此同时，如果你开车，停车场的入口基本上是建筑物的主要入口——它非让人看见不可。

　　在**不超过用地的**9％（22）中，我们已经规定了一个周边地区的停车总数的上限。在**小停车场**（103）中，我们提出了地面上停车场的最佳面积及分布。但在某些情况下，还需要建造较大的停车场或停车设施。只要建造起来的停车场和停车设施不污染其周围的土地，环境就可以不受它们的影响。

　　这是一个简单的生物学原理。例如，在人体中有废料，这些废料是人体工作方式的一部分，显然它们得占有一个位置。但胃和结肠的构造起到保护其他内脏器官的作用，使之不会受到这些废料所携带的毒素的损害。

97   SHIELDED PARKING*

...many patterns we have given discourage dependence on the use of cars; we hope that these patterns will gradually get rid, altogether, of the need for large parking lots and parking structures—LOCAL TRANSPORT AREAS(11), NINE PER CENT PARKING(22). However, in certain cases, unfortunately, large areas of parking are still necessary.Whenever this is so, this parking must be placed very early, to be sure that it does not destroy the BUILDING COMPLEX(95)altogether.

❧❧❧

Large parking structures full of cars are inhuman and dead buildings—no one wants to see them or walk by them. At the same time,if you are driving,the entrance to a parking structure is essentially the main entrance to the building—and it needs to be visible.

In NINE PER CENT PARKING(22) ,we have already defined an upper limit on the total amount of parking in a neighborhood.In SMALL PARKING LOTS(103)we give the best size and the distribution of the lots when they are on the ground.But in certain eases it is still necessary to build larger parking lots or parking structures.The environment can tolerate these larger lots and structures,provided that they are built so that they do not pollute the land around them.

This is a simple biological principle.In the human body,for example,there are waste products;the waste products are part of the way the body works, and obviously they must have a place.

在城市中的情况也是如此。在现阶段，城市规划要求有一定数量的停车场；目前来看这是不可避免的。但建造的停车场必须有遮挡。可作遮挡的有商店、房屋、长满青草的土堆或任何其他建筑物——任何东西，只要使周围的地方看不见停车设施内部和车辆就行。在地面的一层，遮挡特别重要。商店起很大作用，因为商店能立即招来路人。又因为停车的需要是同商业发展息息相关的，从经济角度看，也应多开设商店。

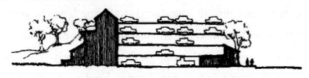

一座有遮挡的停车场
A shielded parking structure

当然，房屋本身也可以起相同的作用。在巴黎，许多极富魅力的漂亮的公寓房子分布在庭院的四周，小汽车可以停在院子里面，远离街道。汽车不多，因此对这些住宅来说，汽车没有破坏它们的庭院；而街道则根本不停靠汽车。

除了停车场需要遮挡之外，司机也需要能够很快认出停车场来——并知晓这个停车场如何通往他要去的楼房。在楼房附近停车的人，他们最常发出的抱怨，不是停车地点太远，而是不知道该到哪儿去找停车场，并如何顺利地把车开回到那个大楼。

由此可见：

1. 专供来访者使用的停车场，必须在通往停车场的方向设置明显的标志，即使停车场整个都是有遮挡的。坐车过来的人要找的是建筑物，而不是停车场。停车场的入口要作为重要的入口——大门——做出标志，这样，在寻找建筑物的时候，人们自然就能看到它。停车场入口处的位置应

But the stomach and colon are built in such a way as to shield the other internal organs from the poisons carried by the wastes.

The same is true in a city.At this moment in history the city requires a certain limited amount of parking;and for the time being there is no getting away from that.But the parking must be built in such a way that it is shielded—by shops,houses,hills of grassy earth,walls,or any other buildings of any kind—anything,so long as the interior of the parking structure and the cars are not visible from the surrounding land.On ground level,the shield is especially critical.Shops are useful since they generate their own pedestrian scale immediately.And since the need for parking often goes hand in hand with commercial development,shops are often very feasible economically.

And of course,the houses themselves can serve the same function.In Paris,many of the most channing and beautiful apartment houses are arranged around courtyards,which permit parking inside,away from the street.There are few enough cars,so that they don't destroy the courtyard,for the houses;and the street is left free of parked cars entirely.

Along with the need to shield parking structures there is the equally pressing need on the part of a driver to be able to spot the parking structure quickly—and see how it is connected to the building he is headed for.One of the most frequent complaints about the parking near a building is not that it is too far away,but that you don't know where you can go to find a parking spot and still be sure of how to get back into the building.

This means that

1. Parking, which is specifically for the use of visitors, must be clearly marked from the directions of approach,even

选择恰当，使来访者在看到建筑物的主要入口时，差不多同时也能找到它。

2. 在停放汽车时，你必须能看到停车场的出口，由这个出口你可以走到大楼。这会使你去寻找距离最近的停车处，而不必为寻找出口跑来跑去。

因此：

将所有大的停车场或停车棚设在某些自然形成的遮挡后面，使人不能从外面看到汽车和停车设施。用以遮挡汽车的屏障可以是建筑物、连成一片的房屋，或者是建有房屋的土丘、便道或商店等。

停车场的入口成为使用该停车场的建筑物的天然门道，使得来访者能从停车场的入口很容易看见他要去的那个建筑物的主要入口处。

屏蔽　　　　停车场围屏

停车场大门

✿❀✿❀

关于遮挡，请参阅**丘状住宅**（39）、**住宅与其他建筑间杂**（48）、**个体商店**（87）、**室外楼梯**（158）、**回廊**（166）。使停车场可以实现有效遮挡的一种最廉价的办法是利用帆布篷——帆布的颜色可以多种多样：这样，篷下的光线就会绚丽多彩——**帆布顶篷**（244）。在你把车开往停车场的地方和你步行走出停车场的地方，都要能清楚地看到建筑物的主要入口——**内部交通领域**（98）、**各种入口**（102）、**主入口**（110）。在封闭式的停车结构中，利用强烈的日光作天然方向，它会告诉人们，往哪个方向走可以离开停车

though the structure as a whole is shielded.The person who is coming by car will be looking for the building,not the parking lot.The entrance to parking must be marked as an important entrance—a gate—so that you can see it automatically, in the process of looking for the building.And it must be placed so that you find it about the same time that you see the building's main entrance.

2.While you are parking your car you must be able to see the exit from the parking area which will lead you into the building.This will let you search for the closest spots,and will mean that you don't have to walk around searching for the exit.

Therefore.

Put all large parking lots, or parking garages, behind some kind of natural wall, so that the cars and parking structures cannot be seen from outside. The wall which surrounds the cars may be a building,connected houses, or housing hills, earth berms, or shops.

Make the entrance to the parking lot a natural gateway to the buildings which it serves, and place it so that you can easily see the main entrance to the building from the entrance to th parking.

⁎

For shields see HOUSING HILL(39),HOUSING IN BETWEEN(48),INDIVIDUALLY OWNED SHOPS(87),OPEN STAIRS(158),GALLERY SURROUND(166).One of the cheapest ways of all to shield a parking lot is with canvas awnings—the canvas can be many colors:underneath,the light is beautiful—CANVAS ROOFS(244).Make certain that the major entrances of buildings are quite clearly visible from the

库——**明暗交织**（135）；最后，关于承重结构、工程和构造施工，请参阅**结构服从社会空间的需要**（205）及之后各模式……

place where you drive into parking lots,and from the places where you leave the parking lots on foot—CIRCULATION REALMS(98),FAMILY OF ENTRANCES(102),MAIN ENTRANCES(110).In covered parking struc-tures,use a huge shaft of daylight as a natural direction which tells people where to walk to leave the parking—TAPESTRY OF LIGHT AND DARK(135);and finally,for the load-bearing structure,engineering,and construction,begin with STRUCTURE FOLLOWS SOCLAL SPACES (205)...

模式98 内部交通领域**

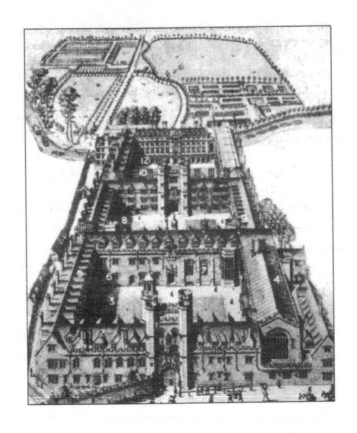

　　……一旦你已经基本上想好了，打算盖多少楼房——**建筑群体**（95），准备把它们盖多高——**楼层数**（96），你就可以大致确定，它们应该有怎样一种布局，才能使人容易看到通向它们的引道，并使人们在这条引道上走起来舒适方便。本模式阐明布局的一般原理。

98   CIRCULATION REALMS**

...once you have some rough idea how many buildings you are going to build—BUILDING COMPLEX(95), and how high they are to be—NUMBER OF STORIES(96), you can work out roughly what kind of layout they should have to make the access to them clear and comfortable.This pattern defines the overall philosophy of layout.

৪০০৪

In many modern building complexes the problem of disorientation is acute. People have no idea where they are, and they experience considerable mental stress as a result.

...the terror of being lost comes from the necessity that a mobile organism be oriented in its surroundings.Jaccard quotes an incident of native Africans who became disoriented.They were stricken with panic and plunged wildly into the bush.Witkin tells of an experienced pilot who lost his orientation to the vertical,and who described it as the most terrifying experience in his life.Many other writers in describing the phenomenon of temporary disorientation in the modern city,speak of the accompanying emotions of distress.(Kevin Lynch, *The Image of the City*, Cambridge,Mass.:MIT Press, 1960, p.125.)

It is easiest to state the circulation problem for the case of a complete stranger who has to find his way around the complex of buildings.Imagine yourself as the stranger,looking for a particular address,within the building.From your point of view,the building is easy to grasp if someone can explain the position of this address to you,in a way you can remember easily,and carry in your head while you are looking for it.To

ഽഽരു

在许多现代建筑群体内，方向不清的问题很尖锐。人们弄不清楚他们自己所处的方位，结果他们的精神压力非常大。

……迷路产生的恐惧感其根源在于，一个运动着的有机体需要在其周围环境中有一个方向。杰卡特曾引述过一些非洲当地居民的迷路事件。他们为恐怖所袭击而奔进了丛林。威特金讲过，一位有经验的飞行员在空中迷失了方向，不知自己的高度，这位飞行员把这说成是他一生中最恐怖的经历。许多其他作者描写在现代化城市中暂时迷路的现象时，都讲到随之而来的苦恼心情。(Kevin Lynch, *The Image of the City*, Cambridge, Mass.: MIT Press, 1960, p.125.)

对于一个在大楼周围四处找路的陌生人，将该楼的通道标识清楚是很容易的。请把你自己想象为一个陌生人，在这样的大楼里找某一家人的住址。在你看来，如果有人能够用你很容易记住的办法向你说清楚这个住址的位置，这个大楼就变得容易认知了，你在找它的时候心中就有数了。总而言之，一个人必须能够用一句话向不认识路的其他人说清楚大楼内的任何一个特定住址。譬如说："从大门一直往前走，顺着大路进到第二道小门，就是带蓝格的那个小门——你就能找到我家的门了。"

初看起来，好像只有陌生人才会感到困惑——因为熟悉一幢建筑的人总能够找到他周围的道路，不管这道路设计得多么糟糕。然而，心理学理论表明，通道的布局杂乱无章，对熟悉该幢建筑的人产生的影响几乎同对陌生人一样严重。我们可以设想，一个人每次要到某个目的地去，他头脑里总得带着类似地图或说明之类的东西。于是问题出现了：他得花多少时间想着这张地图和目的地呢？如果他花很多时间注意路标，盘算接下来该往哪走，他的时间

put this in its most pungent form:*a person must be able to explain any given address within the building,to any other person,who does not know his way around,in one sentence.*For instance, "Come straight through the main gate,down the main path and turn into the second little gate,the small one with the blue grillwork—you can't miss my door."

At first sight,it might seem that the problem is only important for strangers—since a person who is familiar with a building can find his way around no matter how badly it is organized.However,psychological theory suggests that the effect of badly laid out circulation has almost as bad an effect on a person who knows a building,as it does on a stranger. We may assume that every time a person goes toward some destination,he must carry some form of map or instruction in his mind.The question arises:How much of the time does he have to be consciously thinking about this map and his destination? If he spends a great deal of time looking out for landmarks,thinking about where to go next,then his time is entirely occupied,and leaves him little time for the process of reflection,tranquil contemplation,and thought.

We conclude that any environment which requires that a person pay attention to it constantly is as bad for a person who knows it,as for a stranger.A good environment is one which is easy to understand,without conscious attention.

What makes an environment easy to understand? What makes an environment confusing? Let us imagine that a person is going to a particular address within a building.Call this address A.The person who is looking for A does not go directly toward A—unless it happens to be visible from the point where he starts.Instead,he sets his journey up to form a series of

就全被占满了，没有工夫去沉思、默想和考虑问题。

我们可以由此得出结论，任何一种环境，只要它需要人们不断注意它，对陌生人来说很糟糕，对于熟人亦然。一个好的环境是不需人们劳心费神就能一清二楚的。

怎样才能使环境易于为人了解呢？是什么使环境令人晕头转向呢？让我们设想一下，某人到一个大楼里面找一个特定地址。设此地址为 A。去找 A 的人如果不是在他出发的地方碰巧可以看到 A，就不能直接找到 A。他走这一趟得经过好几个步骤，每一个步骤都只是到达暂时的中间目的地，又是下一个步骤的出发点。譬如说，先经过大门，接着到了左边的第二个院子，然后到了院子里右边的拱廊，再下一步走过第三道门。这些步骤就是这个人头脑里的那张地图。如果能轻易绘制出这样一张地图，那你就很容易找到大楼周围的通道。如果无法轻易绘制出这张地图，那你找起路来就难了。

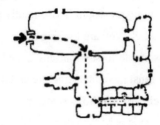

你头脑中的地图是这样起作用的
The way the map in your mind works

因为地图可以帮你弄清区域的大小范围，它很有用（在上例中首先是区域即建筑本身，其次是院落，再次是拱廊，最后就是房间，即目的地）。地图给你指出最大区域的入口，从该入口到下一个大区域的入口等。你一次作出一个判断，而你所作的每一次判断都缩小了建筑中尚待寻找的范围，直到最后把范围缩小到你要寻找的特定地址为止。

看来有理由这么说：任何一张可清晰绘出一个建筑群

steps,in which each step is a kind of temporary intermediate goal,and a taking off point for the next step.For example:First go through the gate,then to the second courtyard on the left,then to the right-hand arcade of the courtyard,and then through the third door.This sequence is a kind of map which the person has in his head.If it is always easy to construct such a map,it is easy to find your way around the building.If it is not easy,it is hard to find your way around.

A map works because it identifies a nested system of realms(in the case of our example the realms are first,the building itself,then the courtyard,then the arcade,then the room itself,the destination).The map guides you to the entrance of the largest realm,and from there to the entrance of the next largest realm,and so on.You make one decision at a time,and each decision you make narrows down the extent of the building which remains to be explored,until you finally narrow it down to the particular address you are looking for.

It seems reasonable to say that any useful map through a building complex must have this structure,and that any building complex in which you cannot create maps of this kind is confusing to be in.This is borne out by intuition.Consider these two examples;each has a system of realms which allows you to make such maps very easily.

An Oxford college.Here the college is made up of courts,each court has a collection of rooms called a "staircase" opening off it,and the individual suites of rooms open off these staircases. The realms are:College,Courts,Staircases,Rooms.

Manhattan.Here the city is made up of major areas,each major area has certain central streets and arteries.The realms are:Manhattan,Districts,Realms defined by the avenues,and

的实用地图必定具有这种结构，而任何你不能绘出这类地图来的建筑物，会使你住在里面感到晕头转向。这可以凭直感来证实。看看下面两个实例；每一个例子都有你能够轻松画出地图来的区域格局。

牛津大学的某个学院。这个学院由许多庭院组成，每个庭院有一系列与它相通的被称为"楼梯间"的房间，而每套房间则通向楼梯间。这些区域包括学院、庭院、楼梯间、房间。

曼哈顿。纽约市划分为几个大区，每个大区都有几条中心街道和主干道。这些区划是：曼哈顿、市区、由南北街道所形成的区划和由东西横街及单个建筑所形成的区段。因为曼哈顿的区域划分得很清楚，东西街道形成的区段从属于南北街道形成的区段，所以曼哈顿可称得上泾渭分明了。

我们的结论是：为了使人清楚，一个建筑群体必须遵循以下三条规则：

1. 建筑群体内各区域的布局能够为人所识别，这些区域中最大的一个是整个群体。

2. 每个区域都有一个主要的流通空间，它直接从入口处通向该区域。

3. 任何一个区域的入口都直接通向另一个比它更大的区域的流通空间。

最后要强调，这些区域每一级都应该有相应的名字；而这又要求这些区域规划整齐，做到名副其实，使人了解冠有那个名字的区域由何处起始又终止于何处。这些区域不一定非像前面两例中所提到的区域那样整齐不可。但它们必须具备充分的心理学意义上的内容和实质，才能使它们在人们的心里起到区域划分的作用。

因此：

巨大的建筑物和小建筑群体的布局应能做到：人们经

Realms defined by cross streets and individual buildings. Manhattan is clear because the districts are so well defined,and the realms defined by the streets are subordinate to the realms defined by the avenues.

We conclude that in order to be clear,a building complex must follow three rules:

1.It is possible to identify a nested system of realms in the complex,the first and largest of these realms being the entire complex.

2.Each realm has a main circulation space,which opens directly from the entrances to that realm.

3.The entrances to any realm open directly off the circulation space of the next larger realm above it.

We emphasize finally,that these realms at every level must have *names*;and this requires,in turn,that they be well enough defined physically,so that they can in fact be named,and so that one knows where the realm of that name starts,and where it stops.The realms do not have to be as precise as in the two examples we have given.But they must have enough psychological substance and existence so that they can honestly work as realms in somebody's mind.

Therefore:

Lay out very large buildings and collections of small buildings so that one reaches a given point inside by passing through a sequence of realms,each marked by a gateway and becoming smaller and smaller,as one passes from each one,through a gateway,to the next.Choose the realms so that each one can be easily named,so that you can tell a person where to go,simply by telling him which realms to go through.

过一系列区域，每一区域都有一个入口，随着人们通过入口,从一个区域走到另一个区域时,这些入口一个比一个小,最后到达里面某一个地点。规划这些区域时，务必使每一个区域都便于命名。这样，你只要告诉一个人经过哪几个区域就可以告诉他往哪里去。

将整个区域第一批入口，即最大的一些入口，作为大门口——**主门道**（53）；使从大门进去的主要区域成为步行街或公共用地——**公共用地**（67）、**步行街**（100）；然后规划有单个建筑物的较小区域、庭院和大的有顶街道——**主要建筑**（99）、**有顶街道**（101）、**外部空间的层次**（114）、**有生气的庭院**（115）；用稍小一些的但仍然十分显眼的大门作为这些较小区域入口的标志——**各种入口**（102）、**主入口**（110）。使小路的布局同**小路和标志物**（120）相一致……

ജ‌ൽ

Treat the first entrances to the whole system of circulation realms, the very largest ones, as gateways—MAIN GATEWAYS(53); make the major realms, which open off the gateways, pedestrian streets or connon land—COMMON LAND(67), PEDESTRIAN STREET (100); then, make minor realms with individual buildings, and courtyards, and major indoor streets—MAIN BUILDING(99), BUILDING THOROUGHFARE(101), HIERARCHY OF OPEN SPACE(114), COURTYARDS WHICH LIVE(115); and mark the entrance to these minor realms with minor entrances that still stand out quite clearly—FAMILY OF ENTRANCES(102), MAIN ENTRANCE(110).Make the layout of paths consonant with PATHS AND GOALS(120)...

模式99 主要建筑*

……一旦你已基本弄清，在一个**建筑群体**（95）内人们如何走动，建筑物大致有多高——**楼层数**（96），就该设法找出该建筑群体的中心，使之有助于完善**内部交通领域**（98）。

৩৩৫৪

一个建筑群体没有中心，犹如一个人没有脑袋一样。

99 MAIN BUILDING*

...once you have decided more or less how people will move around within the BUILDING COMPLEX(95),and roughly how high the buildings will be—NUMBER OF STORIES(96)—it is time to try and find the natural heart or center of the building complex,to help complete its CIRCULATION REALMS(98).

৪০৫৪

A complex of buildings with no center is like a man without a head.

In circulation realms we have explained how people understand their surroundings and orient themselves in their surroundings by making mental maps.Such a map needs a point of reference:some point in the complex of buildings,which is very obvious,and so placed,that it is possible to refer all the other paths and buildings to it.A main building,which is also the functional soul of the complex,is the most likely candidate for this reference point.Without a main building,there is very little chance of any natural points of reference being strong enough to act as an organizer for one's mental map.

Furthermore,from the point of view of the group of users—the workers or the inhabitants—the sense of community and connection is heightened when one building or a part of one building is singled out and treated as a main building,common to all,the heart of the institution.Some examples:the meeting hall among a collection of government buildings;a guild hall in

在内部交通领域中我们已经解释了，人们如何了解他们所处的环境及绘制心中的地图，在周围环境中为他们自己找到道路。这样一张地图需要一个参照点：即建筑群体中的某个点，它应该很显眼，它的位置使人可以参照它来找到所有其他的路径和建筑物。一幢主要建筑，由于实际上也起着建筑群体的灵魂作用，最有可能充当这个参照点。若没有一个主要建筑，任何自然参照点都不大可能目标十分突出，足以构成人们心中的地图。

此外，从用户（工作人员或居民）的角度看，如果一幢楼房或楼房的一部分目标突出，大家都把它看作主要建筑，看作这个机构的中心，就会增强互相交往和联系的意识。可举数例如下：在政府大楼群中的会议厅；同业公会的礼堂；公寓住宅的厨房和家庭室；公园中的转椅游乐场；宗教圣地的寺庙；保健中心的游泳馆；办公楼的工作间。

要十分谨慎地选择实际上起楼群灵魂作用的建筑物作为主要建筑。不然的话，某些无关的功能就会主宰建筑群体。纽约联合国大厦设计的失误原因就在于此。联合国大会是这一机构的心脏和灵魂，却比秘书处矮了一截。这一机构深受文牍主义之害。(See the excellent series of articles by kewis Mumford, discussing the U.N.buildings in *From the Ground Up*, Harvest Books, 1956, pp.20~70.)

因此：

对于任何一个建筑群，都要确定其中哪幢建筑物发挥最主要的作用——哪幢建筑物和人们的关系最大，是该建筑群的灵魂。然后就把这幢建筑物作为主要建筑，使之处于中心位置，应高于别的建筑。

即使一个建筑群体十分密集，只是一幢单一的建筑物，也要把它的主要部分盖得比别的部分高一些和突出一些，这样人们一眼就能看到最重要的那一部分。

a work community;the kitchen and family room in a communal household;the merry-goround in a park;a temple on sacred ground;the swimming pavilion in a health center;the workshop in an office.

Great care must be taken to pick that function which is actually the soul of the group,in human terms,for the main building.Otherwise,some irrelevant set of functions will dominate the building complex.The United Nations complex in New York fails for just this reason.The General Assembly,the heart and soul of the institution,is dwarfed by the bureaucratic Secretariat.And,indeed,this institution has suffered from the red-tape mentality.(See the excellent series of articles by Lewis Mumford,discussing the U.N.buildings in *From the Ground Up*, Harvest Books,1956,pp.20-70.)

Therefore:

For any collection of buildings,decide which building in the group houses the most essential function—which building is the soul of the group,as a human institution.Then form this building as the main building,with a central position,higher roof.

Even if the building complex is so dense that it is a single building,build the main part of it higher and more prominent than the rest,so that the eye goes immediately to the part which is the most important.

⊗⊗⊗

Build all the main paths tangent to the main building,in arcades or glazed corridors,with a direct view into its main functions—COMMON AREAS AT THE HEART(129).

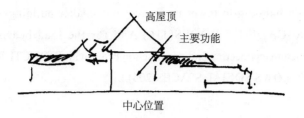

　　把同主要建筑连接的所有主要走道建成拱廊或装有玻璃的走廊，能直接看到它的主要功能——**中心公用区**（129）。赋予主要建筑以高屋顶，较小建筑以较低屋顶，使屋顶实现重叠交错——**重叠交错的屋顶**（116）。关于承重结构、工程以及构造施工，请参阅**结构服从社会空间的需要**（205）及其之后各模式……

Make the roof cascade down from the high roof over the main building to lower roofs over the smaller buildings— CASCADE OF ROOFS(116).And for the load bearing structure,engineering,and construction,begin with STRUCTURE FOLLOWS SOCIAL SPACES(205)...

模式100　步行街**

100 PEDESTRIAN STREET**

...the earlier patterns—PROMENADE(31),SHOPPING STREET(32)and NETWORK OF PATHS AND CARS(52),all call for dense pedestrian streets;ROW HOUSES(38),HOUSING HILL(39),UNIVERSITY AS A MARKETPLACE(43),MARKET OF MANY SHOPS(46),all do the same;and within the BUILDING COMPLEX(95),CIRCULATION REALMS(98) calls for the same.As you build a pedestrian street,make sure you place it so that it helps to generate a NETWORK OF PATHS AND CARS(52),RAISED WALKS(55),and CIRCULATION REALMS(98)in the town around it.

❧❧❧

The simple social intercourse created when people rub shoulders in public is one of the most essential kinds of social"glue"in society.

In today's society this situation,and therefore this glue,is largely missing.It is missing in large part because so much of the actual process of movement is now taking place in indoor corridors and lobbies,instead of outdoors.This happens partly because the cars have taken over streets,and made them uninhabitable,and partly because the corridors,which have been built in response,encourage the same process.But it is doubly damaging in its effect.

It is damaging because it robs the streets of people.Most of the moving about which people do is indoors—hence lost to the street;the street becomes abandoned and dangerous.

And it is damaging because the indoor lobbies and corridors

……前面那些模式——**散步场所**（31）、**商业街**（32）以及**小路网络和汽车**（52），都要求有密集的步行街；**联排式住宅**（38）、**丘状住宅**（39）、**像市场一样的开放大学**（43）、**综合商场**（46），都起着与步行街相同的作用；在**建筑群体**（95）内部，**内部交通领域**（98）也有同样的要求。在建造一条步行街时，务必使它有助于在周围的城市产生**小路网络和汽车**（52）、**高出路面的便道**（55）及**内部交通领域**（98）。

∞○逢

人们在公共场所摩肩接踵造成的简单的社会交往是社会上最基本的相互接触方式之一。

在当今社会，这种状况以及由这种状况引起的相互接触，大部分正在消失。消失的原因是，现今许多活动实际上发生在室内走廊和大厅里，而不在户外。这种情况之所以发生，部分由于汽车取代了街道而使街道上没有人行走；部分由于相应建造起来的走廊起着同样的作用。但其结果具有双重的破坏性。

它的破坏性表现在街道没有行人。人们大都在室内走动——因而街上无人；街道被废弃而变得危险了。

同时，它的破坏性在于，室内大厅和走廊大部分时间是死寂的。之所以这样，一部分原因是，室内空间不像户外空间那样具有公共性质；另一部分原因在于，在多层建筑物中，每一条走廊行人密度都比户外街道低。因此，人走过这些地方会感到不愉快，甚至提心吊胆；大楼里面的人不能进行社交或从社交中得到好处。

为了重新建立起公共活动中的社交关系，只要有可能，各房间、办公室、部门、楼房之间的活动实际上都应该安排在户外，在有顶街道、拱廊、小路、大街上进行，这些

are most often dead.This happens partly because indoor space is not as public as outdoor space;and partly because,in a multi-story building each corridor carries a lower density of traffic than a public outdoor street.It is therefore unpleasant,even unnerving,to move through them;people in them are in no state to generate,or benefit from,social intercourse.

To recreate the social intercourse of public movement,as far as possible,the movement between rooms,offices,departme nts,buildings,must actually be outdoors,on sheltered walks,arca des,paths,streets,which are truly public and separate from cars. Individual wings,small buildings,departments must as often as possible have their own entrances—so that the number of entrances onto the street increases and life comes back to the street.

In short,the solution to these two problems we have mentioned—the streets infected by cars and the bland corridors—is the pedestrian street.Pedestrian streets are both places to walk along(from car,bus,or train to one's destination) and places to pass through (between apartments,shops,offices,s ervices,classes).

To function properly,pedestrian streets need two special properties.First,of course, no cars;but frequent crossings by streets with traffic,see NETWORK OF PATHS AND CARS(52):deliveries and other activities which make it essential to bring cars and trucks onto the pedestrian street must be arranged at the early hours of the morning,when the streets are deserted.Second,the buildings along pedestrian streets must be planned in a way which as nearly as possible eliminates indoor staircases,corridors,and lobbies,and leaves most circulation outdoors.This creates a street lined with

都应该成为真正的公共设施并同机动车道分开。单独的翼楼、小建筑物、部门，都应尽可能地有自己的入口——这样，通向街道的出入口数目增加，从而使街道重新恢复活力。

简言之，要解决我们提到过的两个问题——受汽车影响的街道和空荡荡的走廊，办法是建造步行街。步行街既是行走的地方（从小汽车、公共汽车或火车下来走到目的地），也是在公寓楼、商店、办公楼、服务设施、教室之间通过的地方。

为了做到名副其实，步行街需要具备两个特性。首先，当然是没有汽车，但又要同交通频繁的街道不断有交叉，**请参阅小路网络和汽车**（52）：需要把大型或小型机动车开到步行街来的送货和其他活动必须安排在大清早，也就是在街道尚无行人的时候。其次，在设计步行街的建筑物时，要尽可能消除室内楼梯、走廊和大厅，将大部分通道设在户外。这样就能使街道两旁布满楼梯，楼上的办公室和房间可以直接通往街道，还形成许许多多的入口，使街道生机盎然。

最后，应该指出，最舒适的步行街看来应该是那些宽度不超过周围建筑物高度的街道（参阅 "Vercle free zones in city centers," *International Brief* #16，U.S.Department of Housing and Urban Development，Office of International Affairs，June 1972）。

大约正方形……或甚至窄一些
About square...or even narrower

因此：

设计建筑物时要考虑它们能够形成步行街，街上有许

stairs,which lead from all upstairs offices and rooms directly to the street,and many many entrances,which help to increase the life of the street.

Finally it should be noted that the pedestrian streets which seem most comfortable are the ones where the width of the street does not exceed the height of the surrounding buildings. (See "Vehicle free zones in city centers," *International Brief* # 16,U.S.Department of Housing and Urban Development,Office of International Affairs,June 1972).

Therefore.

Arrange buildings so that they form pedestrian streets with many entrances and open stairs directly from the upper storys to the street,so that even movement between rooms is outdoors,not just movement between buildings.

ßæĜß

The street absolutely will not work unless its total area is small enough to be well filled by the pedestrians in it—PEDESTRIAN DENSITY(123).Make frequent entrances and open stairs along the street,instead of building indoor corridors,to bring the people out;and give these entrances a family resemblance so one sees them as a system—FAMILY OF ENTRANCES(102),OPEN STAIRS(158);give people indoor and outdoor spaces which look on the street—PRIVATE TERRACE ON THE STREET(140),STREET WINDOWS(164),OPENING TO THE STREET (165),GALLERY SURROUND(166),SIX-FOOT BALCONY(167);and shape the street to make a space of it— ARCADE(119),PATH SHAPE(121)...

多入口和使人可以直接从楼里走向街道的室外楼梯，这样，不仅楼和楼之间的活动，甚至房间和房间之间的活动也都可以在户外进行。

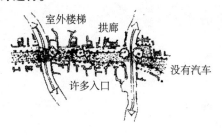

室外楼梯　拱廊　　没有汽车　许多入口

౸৩

除非街道总面积很小，街上行人总能够熙熙攘攘，否则，它就绝对不能成为步行街——**行人密度**（123）。沿街频频出现入口和室外楼梯，少造室内走廊，这样人们就会走到外面来；同时还要将彼此间外形相似的入口编成一组。让人们把它们看作一个系统——**各种入口**（102）、**室外楼梯**（158）；给人们以面向街道的室内和室外空间——**私家的沿街露台**（140）、**临街窗户**（164）、**向街道的开敞**（165）、**回廊**（166）、**六英尺深的阳台**（167）；形成步行街时使其能够营造出一种空间——**拱廊**（119）、**小路的形状**（121）……

模式101 有顶街道

101   BUILDING THOROUGHFARE

...if the building complex is built at high density,then at least part of the circulation cannot be made of outdoor PEDESTRIAN STREETS(100)because the buildings cover too much of the land;in this case,the main spines of the CIRCULATION REALMS(98)must take the form of building thoroughfares similar to pedestrian streets,but partly or wholly inside the buildings.Building thoroughfares replace the terrible corridors which destroy so much of modem building,and help to generate the indoor layout of a BUILDING COMPLEX(95)

❧✲❧

When a public building complex cannot be completely served by outdoor pedestrian streets, a new form of indoor street, quite different from the conventional corridor, is needed.

The problem arises under two conditions.

1.*Cold weather*. In very cold climates to have all circulation outdoors inhibits social communication instead of helping it.Of course,a street can be roofed,particularly with a glass roof.But as soon as it becomes enclosed,it has a different social ecology and begins to function differently.

2.*High density*. When a building complex is so tightly packed on the site that there is no reasonable space for outdoor streets because the entire building complex is a continuous two,three,or four story building,it becomes necessary to think of major thoroughfares in different terms.

To solve the problems posed by these conditions, streets must be replaced by indoor thoroughfares or corridors.But

……如果建筑群密度很高，那么至少有部分流通不能依靠室外**步行街**（100），因为这些建筑物占地面积太大。在这种情况下，**内部交通领域**（98）的主干道必须建成有顶街道的形式，它类似步行街，但部分或全部位于建筑物内部。有顶街道取代糟蹋现代建筑的令人讨厌的走廊，并有助于搞好**建筑群体**（95）的室内布局。

శుౖౖౖౖ

当一个公共建筑群体的流通问题不能完全依靠步行街来解决时，就需要有一种新型的、完全不同于传统走廊的室内街道。

室内街道
An indoor street

这个问题是在下面两种情况下提出来的。

1. 冷天气。在气候非常寒冷的地方，如让所有的流通都在户外进行，势必会引起社交的断绝而绝不能促进社交活动。当然，街道也可以有顶棚，特别是玻璃顶棚。但只要街道是封闭的，它就具有了不同的社会生态学的内容而开始起不同的作用。

2. 高密度。如果一个建筑群体的布局非常紧凑，以致没有给户外的街道以合理的空间，因为整个建筑群体是一个连片的二层楼、三层楼或四层楼建筑，这时有必要考虑

the moment we put them indoors and under cover,they begin to suffer from entirely new problems,which are caused by the fact that they get sterilized by their isolation.First,they become removed from the public realm,and are often deserted.People hardly ever feel free to linger in public corridors when they are off the street.And second,the corridors become so unfriendly that nothing ever happens there.They are designed for scuttling people through,but not for staying in.

In order to solve these new problems, created when we try to put a street indoors, the indoor streets—or building thoroughfares—need five specific characteristics.

1.*Shortcut*

Public places are meant to invite free loitering.The public places in community buildings(city halls,community centers,public libraries)especially need this quality,because when people feel free to hang around they will necessarily get acquainted with what goes on in the building and may begin to use it.

But people rarely feel free to stay in these places without an Official Reason.Goffman describes this situation as follows:

Being present in a public place without an orientation to apparent goals outside the situation is sometimes called lolling,when position is fixed,and loitering,when some movement is entailed.Either can be deemed sufficiently improper to merit legal action.On many of our city streets,especially at certain hours,the police will question anyone who appears to be doing nothing and ask him to "move along." (In London,a recent court ruling established that an individual has a right to walk on the street but no legal right merely to stand on it.) In Chicago,an individual in the uniform of a hobo can loll on "the stem," but once off this preserve he is required to look as if he were intent on getting to some business destination.Similarly,some mental

不同样式的主要通道。

为了解决由这些情况产生的问题，街道必须由室内街道或走廊来代替。但一旦我们把街道设进楼内并给它们加上顶棚，它们就面临新的问题，这就是它们因被隔离开而无人行走。首先，这些街道脱离了公共环境而时常闲置。人们离开大街，在公共走廊漫步绝不会感到自由自在。其次，走廊不会让人感觉可亲，那里总是无人光顾。设计走廊，目的只是让人匆匆而过，而不是让人停留。

为了解决这些我们想在室内建造街道所产生的新问题，室内街道——或称为有顶街道——需要具备五个特征。

1. 捷径

公共场所本来就是可以让人闲逛的。社区建筑物中的公共场所（市政厅、社区中心、公共图书馆）尤其需要具备这种性质，因为当人们有空闲逛时，他们必定会去熟悉建筑物内的情况并可能开始利用它。

但人们如果没有冠冕堂皇的理由，就会不好意思停留在那些地方。戈夫曼将这种情形描述如下：

出现在某个公共场合而不向该场合外一个明显的目标走去，如不挪动位置，有时叫闲坐着；如稍微动一下，叫作闲逛。无论是哪一种，都很难称为合法行为。在我们许多城市街道上，特别在某些时刻，警察会盘问任何一个看起来无所事事的人，然后要求他"走开"。（在伦敦，新近一条法规规定，人们有权在街上行走，但无权停留）。在芝加哥，穿着体面制服的成功人士可以在大街上游荡；但脱了体面制服，警察就要求他留神，似乎他在一心一意寻找另一个职业。同样，某些精神病人被收容，是由于警察发现他们在下班高峰时段踯躅街头，漫无目的地游荡。（Erving Goffman, *Behavior in Public Places*, New York: Free Press, 1963, p.56.）

要使一个公共场所真正发挥其作用，它必须有助于反

patients owe their commitment to the fact that the police found them wandering on the streets at off hours without any apparent destination or purpose in mind.(Erving Goffman,*Behaviorin Public Places*,New York:Free Press,1963,p.56.)

If a public space is to be really useful it must somehow help to counter the antiloitering tendency in modern society. Specifically,we have observed these problems:

a. A person will not use a public place if he has to make a special motion toward it,a motion which indicates the intention to use the facility "officially."

b. If people are asked to state their reason for being in a place (for example,by a receptionist or clerk)they won't use it freely.

c. Entering a public space through doors,corridors,changes of level,and so on,tends to keep away people who are not entering with a specific goal in mind.

Places which overcome these problems,like the Galleria in Milan,all have a common characteristic:they all have public thoroughfares which slice through them,lined with places to stop and loiter and watch the scene.

2.*Width*

An indoor street needs to be wide enough for people to feel comfortable walking or stopping along the way.Informal experiments help to determine how much space people need when they pass others.Since the likelihood of three people passing three people is not high,we consider as a maximum two people passing two people,or three people passing one person. Each person takes about two feet; there needs to be about one foot between two groups which pass,so that they do not feel crowded;and people usually walk at least one foot away from

击现代社会中反游逛的倾向。特别是，我们注意到如下的问题：

a. 如果一个人必须要特意去到一个公共场所，以表明他"正式"确实在使用这一设施，那他是不会去利用它的。

b. 如果有人（例如接待人员或职员）盘问你，要你说出待在一个地方的理由，你就不能随意待在这个地方。

c. 进入公共场所需要通过的大门、走廊、台阶等，势必将没有特定目的而闲逛的人拒之门外。

凡是解决了这些问题的地方，如意大利米兰的大市场，都有共同的特点：它们都有一些公共街道，从这些地方穿插过去，街道两旁有可供人停留、徜徉和观光的处所。

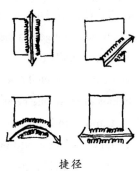

捷径
Shortcuts

2. 宽度

室内街道需要有足够的宽度，让人们在街道上或走或停都会感到舒适。简单的实验可有助于确定人们跟他人迎面走过时需要多大的空间。因为三个人在行走，并同时迎面穿过三个人的这种可能性不太大，我们就把两个人和两个人对穿或三个人和一个人对穿作为最大人流。每个人约占2ft；两组人对穿需要有1ft的间隔，这样他们才不会感到拥挤；还有，人们通常行走时至少离墙1ft。因此，街道宽

the wall.The street width,therefore,should be at least 11 feet.

Our experiments also indicate that a person seated or standing at the edge of a street feels uncomfortable if anyone passes closer than five feet.Thus,in places in the street where seats,activities,entrances,and counters are placed,the street should widen to about 16 feet(one-sided)or 20 feet(two-sided).

3. *Height*

Ceiling heights should also feel comfortable for people walking or standing along an indoor street.According to CEILING HEIGHT VARIETY(190),the height of any space should be equal to the appropriate horizontal social distances between people for the given situation—the higher the ceiling,the more distant people seem from each other.

Edward Hall,in The Hidden Dimension,suggests that a comfortable distance between strangers is the distance at which you cannot distinguish the details of their facial features.He gives this distance as being between 12 and 16 feet.Thus,the ceiling height in an indoor street should be at least in that range.

Where people sit and stand talking to each other,the appropriate social distance is more intimate.Hall gives it a dimension of four to seven feet.Thus,the ceiling in activity and "edge" places should be seven feet.

This suggests,for a large indoor street,a ceiling that is high in the middle and low at the edges.In the middle,where people are passing through and are more anonymous,the ceiling may be 12 to 20 feet high,or even higher,according to the scale of the passage.Along the edges of the thoroughfare,where people are invited to stop and become slightly more engaged in the life of the building,the ceilings may be lower.Here are three sections through an indoor street which have this property.

度（应至少为 11ft）。我们的实验也表明，一个人在街边无论是坐还是站，如果在 5ft 距离内有行人经过，他会感到不舒服。因此，在街道上有坐位、活动、入口以及柜台的地方，街道宽度应达到 16ft（单面）或 20ft（双面）。

3. 高度

天花板高度也应使在室内街道上行走或站立的人感到舒适。根据**天花板高度变化**（190），任何一个空间的高度都应同一定场合人们之间合适的水平社交距离相等——天花板越高，看起来人们彼此之间的距离越大。

爱德华·霍尔在《看不见的尺寸》（*The Hidden Dimension*）中指出，陌生人之间的合适的距离应当是使人彼此看不清面貌细部的距离。他认为这一距离是 12 ～ 16ft。这样，室内街道的天花板高度应至少在这个间距内。

当人们坐着或站着讲话时，合适的社交距离要近一些。霍尔认为该距离应是 4 ～ 7ft。因此，天花板在活动场所和"边缘"应为 7ft。

这表明，对于大的室内街道来说，天花板应该中间高两边低。在中间，彼此陌生的人在这里经过，根据过道的宽度，天花板可能高达 12 ～ 20ft，甚至更高。在室内街道的边缘，人们常去歇脚，他们在这里较多地参与楼内的生活，那么天花板可以低一些。以下是具有这一性质的室内街道的三个部分。

室内街道剖面
Cross sections of an indoor street

4. *Wide entrance*

As far as possible,the indoor street should be a continuation of the circulation outside the building.To this end,the path into the building should be as continuous as possible,and the entrance quite wide—more a gateway than a door.An entrance that is 15 feet wide begins to have this character.

5. *Involvements along the edge*

To invite the free loitering described above under *Shortcut*,the street needs a continuum of various "involvements" along its edge.

Rooms next to the street should have windows opening onto the street.We know it is unpleasant to walk down a corridor lined with blank walls.Not only do you lose the sense of where you are but you get the feeling that all the life in the building is on the other side of the walls,and you feel cut off from it.We guess that this contact with the public is not objectionable for the workers,so long as it is not too extreme,that is,as long as the workplace is protected either by distance or by a partial wall.

The corridor should be lined with seats and places to stop,such as newspaper,magazine,and candy stands,bulletin boards,exhibits,and displays.

Where there are entrances and counters of offices and services off the corridor,they should project into the corridor. Like activities,entrances and counters create places in the corridor,and should be combined with seats and other places to stop.In most public service buildings these counters and entrances are usually set back from corridors which makes them hard to see,and emphasizes the difference between the corridor as a place for passing through,and the office as a place where things happen.The problems can be solved if the entrances and counters project into the corridor and become part of it.

4. 宽入口

室内街道应尽可能成为建筑物外面通道的延续。为此目的，进入楼内的走道应尽可能地连续，入口要很宽——应该是大门，而不是房门。15ft 宽的大门开始具有这种性质了。

5. 街边的有关设施

为了使上述"捷径"中所描写的自由游逛得以实现，街道需要有一个沿其边缘有各种有关设施的连续统一体。

临街房间应有开向街道的窗户。我们知道，沿着两边墙上没有窗户的走廊行走是不愉快的。你不仅不知道你所在的位置，而且会觉得生活在这个大楼里如同置身墙外，你感到自己和大楼天各一方。我们相信，楼内的工作人员是不会反对同公众保持这种接触的，只要不是太过分；也就是说，只要工作空间被一段距离或矮墙所隔断就行。

走廊两旁应设置坐椅并有诸如报刊亭、糖果摊、布告牌、展览室、橱窗等设施。

凡是走廊旁有办公室和服务处的入口和柜台的地方，它们应向走廊伸出。入口和柜台，和其他活动一样，也给走廊创造一些空间，它们应同坐椅和其他引人留步的场所结合起来。在大多数公共服务类建筑内，这些柜台和入口通常位于从走廊凹进的地方，这不仅使人难以看见它们，而且突出了作为过道的走廊和作为处理事务的办公室之间的不同。如果入口和柜台伸向走廊而构成走廊的一部分，这些问题是可以解决的。

因此：

凡因建筑物密集或气候寒冷致使街道干线设在室内的地方，应将这些街道建成有顶街道。每一条街道的位置都要起捷径作用，尽可能同室外的公共街道连接起来，并有很宽的敞开的入口。街道两边应有窗户、坐位、柜台和

Therefore:

Wherever density or climate force the main lines of circulation indoors,build them as building thoroughfares. Place each thoroughfare in a position where it functions as a shortcut,as continuous as possible with the public street outside,with wide open entrances.And line its edges with windows,places to sit,counters,and entrances which project out into the hall and expose the buildings'main functions to the public.Make it wider than a normal corridor—at least 11 feet wide and more usually,15 to 20 feet wide;give it a high ceiling,at least 15 feet,with a glazed roof if possible and low places along the edge.If the street is several stories high,then the walkways along the edges,on the different stories,can be used to form the low places.

∞∞∞

Treat the thoroughfare as much like a PEDESTRIAN STREET(100)as possible，with OPEN STAIRS(158)coming into it from upper storys.Place entrances，reception points，and seats to form the pockets of activity under the lower ceilings at the edges—FAMILY OF ENTANCES (102),ACTIVITY POCKETS(124),RECEPTION WELCOMES YOU(149), WINDOW PLACE(180),CEILING HEIGHT VARIETY(190),and give these places strong natural light—TAPESTRY OF LIGHT AND DARK(135). Make a connection to adjacent rooms with INTERIOR WINDOWS(194)and SOLID DOORS WITH GLASS(237). To give the building thoroughfare the proper sense of liveliness, calculate its overall size according to PEDESTRIAN DENSITY(123)...

伸向大厅并向公众显示建筑物主要功能的入口。使这种街道的宽度超过普通走廊——至少11ft宽，更为常见的是15～20ft宽；使之具有高天花板，至少15ft高，如有可能，屋顶装上玻璃，街道两边降低高度。如该街道有几层楼高，在各层楼两边的人行道就可以用来形成高度较低的活动场所。

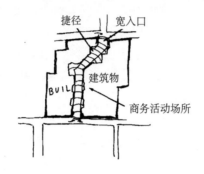

捷径　宽入口

建筑物

商务活动场所

BUIL

⊰⊱

　　尽可能把有顶街道当作**步行街**（100），有**室外楼梯**（158）将它同上面楼层连接。在两旁低天花板下设置入口、接待点和坐位以形成袋形活动场地——**各种入口**（102）、**袋形活动场地**（124）、**宾至如归**（149）、**窗前空间**（180）、**天花板高度变化**（190），使这些地方可以拥有很强的自然光——**明暗交织**（135）。用**内窗**（194）和**镶玻璃板门**（237）把相邻房间连接起来。为了使有顶街道给人以生动活泼的感觉，根据**行人密度**（123），计算一下它的整体尺寸……

模式102 各种入口*

　　……本模式是**内部交通领域**（98）的补充。内部交通
领域描述了在一幢大建筑物或建筑群体内一系列的区域，
进入每一个区域都有一个主入口即大门，离开每个区域有
许多较小的门道、大门和通路。本模式描述这些较小出入
口之间的关系。

∞✕∾

　　当一个人初到一幢办公大楼、服务大楼或联合车间，
抑或是到了一组住宅楼群时，除非整幢建筑的布局呈现在
他的眼前，使他能一眼就看到所要去的那个地方的入口，
不然，他会感到无所适从。

102 FAMILY OF ENTRANCES*

...this pattern is an embellishment of CIRCULATION REALMS(98).CIRCULATION REALMS portrayed a series of realms,in a large building or a building complex,with a major entrance or gateway into each realm and a collection of minor doorways,gates,and openings off each realm.This pattern applies to the relationship between these "minor" entrances.

⊗⊗⊘⊗

When a person arrives in a complex of offices or services or workshops,or in a group of related houses,there is a good chance he will experience confusion unless the whole collection is laid out before him,so that he can see the entrance of the place where he is going.

In our work at the Center we have encountered and defined several versions of this pattern.To make the general problem clear,we shall go through these cases and then draw out the general rule.

1. In our multi-service center project we called this pattern Overview of Services.We found that people could find their way around and see exactly what the building had to offer,if the various services were laid out in a horseshoe, directly visible from the threshold of the building.See *A Pattern Language Which Generates Multi-Service Centers*, pp.123-126.

2. Another version of the pattern,called Reception Nodes, was used for mental health clinics.In these cases we specified one clearly defined main entrance,with main reception clearly

在本书的中间部分我们遇到并描写过本模式的几种样式。为了阐明这一普遍性问题，我们将探讨以下情况，然后总结出一般规律。

　　1. 在我们的多项服务中心方案中，我们将此模式称作"服务项目一览"。我们发现，只要各种服务项目以马蹄形展开，使人进入大楼一目了然，人们就能够找到他们周围的道路，并确切了解大楼所提供的服务。参阅 *A Pattern Language Which Generates Multi-Service Centers*，pp.123~126。

服务项目一览
Overview of services

　　2. 本模式的另一形式叫作"接待点"，曾用于精神类疾病诊所。在这些情况下，我们指定一个已明确的主入口，在此主入口内可清楚地看到主接待室，然后从主接待室看到所有"下一个"接待室，这样，一个可能受惊或晕头转向的病人问一下接待员就会找到自己的路——他能顺着这条路被领到下一个看得见的接待室。

接待点
Reception nodes

　　3. 在重建伯克利市政厅大厦的工程中，我们利用的是该模式的另一种形式。在室内街道内部，每一服务处入口的样子都很相似——每个入口都略微向街道方向凸出，这样人们就很容易在各入口之间找到周围的道路。

　　4. 我们已经把这一模式用于构成住宅团组的房屋。在某案例中，本模式把不同房屋的入口排列成彼此看得见的一组入口；再使它们具有相似的外形。

各种入口
Family of entrances

　　在所有这些情况下，存在着相同的重要问题。一个人在有好几个入口的地方寻找其中的一个，而此人又不认得路，需要有某种能找到他要找的大门的简便方法。如它是"蓝

visible inside this main entrance and each "next" point of reception then visible from the previous one,so that a patient who might be frightened or confused could find his way about by asking receptionists—and could always be directed to the next, visible receptionist down the line.

3. In our project for re-building the Berkeley City Hall complex,we used another version of the pattern.Within the indoor streets,the entrance to each service was made in a similar way—each one bulged out slightly into the street,so that people could easily find their way around among the resulting family of entrances.

4. We have also applied the pattern to houses which are laid out to form a cluster.In one example the pattern drew different house entrances together to make a mutually visible collection of them,and again gave each of them a similar shape.

in all these cases, the same central problem exists. A person who is looking for one of several entrances,and doesn't know his way around,needs to have some simple way of identifying the one entrance he wants.It can be identified as "the blue one," "the one with the mimosa bush outside," "the one with a big 18 on it," or "the last one on the right,after you get round the corner," but in every case the identification of "the one..." can only make sense if the entire collection of possible entrances can first be seen and understood as a collection.Then it is possible to pick one particular entrance out,without conscious effort.

色的那个",是"门外有含羞草丛的那个",是"门上写着很大的 18 的那个",或是"你走到拐角后,往右最后的那一个大门",但无论在哪种情况下,只有首先能看见所有这些大门并认出它们是一组入口时,你才能利用"……那个"来找路。这时你才有可能不花费太大力气把有特殊记号的那个入口找出来。

因此:

将入口设置成一组。这就是说:

1. 它们形成一组,使人一目了然,从每一个大门都可以看见所有其他的大门。

2. 它们都大体相似,如所有的入口都有门廊,或所有的入口两边都有墙,或所有的入口门道都差不多。

各种入口

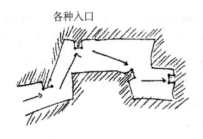

&⨯&

具体地讲,要使入口醒目、突出——**主入口**(110);当入口通往住宅等私人建筑时,在公共街道和住宅内部之间应营造出一种过渡空间——**入口的过渡空间**(112);并使入口本身形成一个空间,该空间跨着墙,因而是一个半里半外的凸出建筑物,空间有顶,可遮风挡雨——**入口空间**(130)。如果它是由室内街道进入公共办公楼的入口,可以在入口空间辟出一个接待室——**宾至如归**(149)……

Therefore:

Lay out the entrances to form a family.This means:

1. They form a group,are visible together,and each is visible from all the others.

2.They are all broadly similar,for instance all porches,or all gates in a wall,or all marked by a similar kind of doorway.

৪০৫৪

In detail,make the entrances bold and easy to see—MAIN ENTRANCE(110);when they lead into private domains,houses and the like,make a transition in between the public street and the inside—ENTRANCE TRANSITION(112);and shape the entrance itself as a room,which straddles the wall,and is thus both inside and outside as a projecting volume,covered and protected from the rain and sun—ENTRANCE ROOM(130). If it is an entrance from an indoor street into a public office,make reception part of the entrance room—RECEPTION WELCOMES YOU(149)...

模式103 小停车场*

     ……因为小停车场是某种形式的门道——是你停车和进入步行区的地方——这一模式有助于完善**商业街**（32）、**住宅团组**（37）、**工作社区**（41）、**绿茵街道**（51）、**主门道**（53）、**内部交通领域**（98），以及任何需要很方便地停放少量车辆的地方。但最重要的，如使用得当，这一模式配合**有遮挡的停车场**（97），会有助于逐步形成**停车场不超过用地的9%**（22）。

103   SMALL PARKING LOTS*

...since a small parking lot is a kind of gateway—the place where you leave your car, and enter a pedestrian realm—this pattern helps to complete SHOPPING STREETS(32), HOUSE CLUSTER(37), WORK COMMUNITY(41), GREEN STREETS(51), MAIN GATEWAYS(53), CIRCULATION REALMS(98), and any other areas which need small and convenient amounts of parking.But above all, if it is used correctly, this pattern, together with SHIELDED PARKING(97), will help to generate NINE PER CENT PARKING(22)gradually, by increments.

ജ്ഞ

Vast parking lots wreck the land for people.

In NINE PER CENT PARKING(22),we have suggested that the fabric of society is threatened by the mere existence of cars,if areas for parked cars take up more than 9 or 10 per cent of the land in a community.

We now face a second problem.Even when parked cars occupy less than 9 per cent of the land,they can still be distributed in two entirely different ways.They can be concentrated in a few huge parking lots;or they can be scattered in many tiny parking lots.The tiny parking lots are far better for the environment than the large ones,even when their total areas are the same.

Large parking lots have a way of taking over the landscape, creating unpleasant places, and having a depressing effect on the open space around them.They make people feel dominated by cars; they separate people from the pleasure

大型停车场让我们损失土地。

在**停车场不超过用地的 9%**（22）中，我们指出，如果停放汽车的场地占据一个社区土地的 9% 或 10% 以上，那么汽车的存在足以威胁社会组织。

我们现在面临第二个问题。即使当停放的汽车占据不到 9% 的土地，它们仍然可能以两种截然不同的方式分布。它们可以集中在几个大停车场；它们也可以分散在许多小停车场。小停车场即使占地总面积同大停车场一样多，对于保护环境来说也远比后者有效。

大停车场破坏风景，使得这片地方无法令人赏心悦目，其破坏作用还波及周围的空地。这些大停车场让人们感觉汽车在主宰一切；它们让人不觉得接近自己的汽车是一种愉快和便利；而且，如果这种停车场大到可能会发生无法预测的交通事故时，它们对儿童来说是危险的，因为小孩子难免会在停车场玩耍。

人无立锥之地
The destruction of human scale

这些问题的产生，根本原因在于汽车的体量比人大得多。大型停车场虽然适合于汽车停放，却给人造成许多不便。这些大型停车场太宽，需有过多的铺装地面，却没有可供人逗留的地方。事实上，我们注意到，人们在走过大型停车场时，总是加快脚步尽快离开。

BUILDINGS
建　筑
985

and convenience of being near their cars;and, if they are large enough to contain unpredictable traffic, they are dangerous for children, since children inevitably play in parking lots.

The problems stem essentially from the fact that a car is so much bigger than a person.Large parking lots,suited for the cars,have all the wrong properties for people.They are too wide;they contain too much pavement;they have no place to linger.In fact,we have noticed that people speed up when they are walking through large parking lots to get out of them as fast as possible.

It is hard to pin down the exact size at which parking lots become too big.Our observations suggest that parking lots for four cars are still essentially pedestrian and human in character;that lots for six cars are acceptable;but that any area near a parking lot which holds eight cars is already clearly identifiable as "car dominated territory."

This may be connected with the well-known perceptual facts about the number seven.A collection of less than five to seven objects can be grasped as one thing,and the objects in it can be grasped as individuals.A collection of more than five to seven things is perceived as "many things." (See G.Miller, "The Magical Number Seven,Plus or Minus Two:Some Limits on Our Capacity for Processing Information," in D.Beardslee and M.Wertheimer,eds., *Readings in Perception*, New York, 1958, esp.p.103.)It may be true that the impression of a "sea of cars" first comes into being with about seven cars.

Therefore:

Make parking lots small,serving no more than ffie to seven cars,each lot surrounded by garden walls, hedges, fences, slopes, and trees, so that from outside the cars are

很难准确控制停车场的面积而不使之过大。据我们观察，停放 4 辆小汽车的停车场基本上对人无碍；停放 6 辆小汽车的停车场尚属可行；但在停放 8 辆小汽车的停车场附近的任何地方都会使人明显感到这里是"汽车王国"了。

这可能跟 7 这个使人敏感的数字有关。一组少于 5 ～ 7 个的物体可能被看作一件事物，其中各物体可能被看作个别的东西。多于 5 ～ 7 个事物的集合体令人产生"许多东西"的感觉。（参阅 G.Miller, "The Magical Number Seven, Plus or Minus Two : Some Limits on Our Capacity for Processing Information", in D.Beardslee and M.Wertheimer, eds., *Readings in Perception*, New York, 1958, esp.p.103.）情况可能真是那样，"车海"的印象首先是从大约 7 辆汽车产生的。

小停车场可以随意停放汽车
The small lots can be quite loosely placed

因此：

停车场建得小一点，可停放不超过 5 ～ 7 辆小汽车。每个停车场周围有花园围墙、矮树丛、篱笆、土坡和树木，人们从外面看不到汽车。这些停车场之间的间隔不少于 100ft。

almost invisible. Space these small lots so that they are at least 100 feet apart.

Place entrances and exits of the parking lots in such a way that they fit naturally into the pattern of pedestrian movement and lead directly,without confusion,to the major entrances to individual buildings—CIRCULATION REALMS(98).Shield even these quite modest parking lots with garden walls,and trees, and fences,so that they help to generate the space around them—POSITIVE OUTDOOR SPACE (106), TREE PLACES(171), GARDEN WALLS(173)...

fix the position of individual buildings on the site,within the complex,one by one,according to the nature of the site,the trees,the sun:this is one of the most important moments in the language;

104. SITE REPAIR

105. SOUTH FACING OUTDOORS

106. POSITIVE OUTDOOR SPACE

107. WINGS OF LIGHT

108. CONNECTED BUILDINGS

109. LONG THIN HOUSE

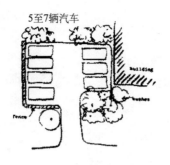

5至7辆汽车

&✕☙

　　停车场出入口的设置自然地适应行人的活动方式，并直接无误地通向各个建筑物的主入口——**内部交通领域**（98）。对这些不大的停车场也应以花园围墙、树木和篱笆遮挡，以便使它们有助于形成停车场周围的空间——**户外正空间**（106）、**树荫空间**（171）、**花园围墙**（173）……

　　把建筑群内各建筑物的位置根据地段、树木和阳光的特点，逐一确定下来：这是本模式语言的最重要阶段之一。

　　104. 基地修整
　　105. 朝南的户外空间
　　106. 户外正空间
　　107. 有天然采光的翼楼
　　108. 鳞次栉比的建筑
　　109. 狭长形住宅

模式104　基地修整**

　　……建筑群体的大体情况已在**建筑群体**（95）、**楼层数**（96）、**内部交通领域**（98）中确定下来。以下几个模式以及本语言中全部其他模式都是关于如何设计单个建筑物及其环境的。本模式阐明首先应该采取的行动——修整基地

104   SITE REPAIR**

...the most general aspects of a building complex are established in BUILDING COMPLEX(95),NUMBER OF STORIES(96),and CIRCULATION REALMS(98).The patterns which follow,and all remaining patterns in the language,concern the design of one single building and its surroundings.This pattern explains the very first action you must take—the process of repairing the site.Since it tends to identify very particular small areas of any site as promising areas of development,it is greatly supported by BUILDING COMPLEX(95)which breaks buildings into smaller parts,and therefore makes it possible to tuck them into different corners of the site in the best places.

ଞୠଓଷ

Buildings must always be built on those parts of the land which are in the worst condition,not the best.

This idea is indeed very simple.But it is the exact opposite of what usually happens;and it takes enormous will power to follow it through.

What usually happens when someone thinks of building on a piece of land?He looks for the best site—where the grass is most beautiful,the trees most healthy,the slope of the land most even,the view most lovely,the soil most fertile—and that is just where he decides to put his house.The same thing happens whether the piece of land is large or small.On a small lot in a town the building goes in the sunniest corner,wherever it is most pleasant.On a hundred acres in the country,the buildings go on the most pleasant hillside.

的工作。由于本模式认为，任一基地的非常独特的一小块地方都是可以建造房屋的，因而它完全符合**建筑群体**（95）的要求，因为本模式主张把建筑物分割成小部分，从而有可能把它们恰如其分地安排在基地的不同角落。

🙰🙰🙰

房屋一定要盖在条件最差而不是最好的地方。

这个想法确实很简单。但通常发生的情况却恰好相反；将这一思想贯彻始终需要很大的意志力。

当一个人考虑要在一片土地上盖房子的时候，通常他是怎么做的呢？他寻找最好的基地——那儿芳草芬菲、树木繁茂、土地平整、风光秀丽、土壤肥沃——这就是他决定建房的地方。不管那片土地是大是小，这样的选择标准是不会改变的。在城市的小块土地上，房子总要盖在阳光最充沛的一隅，那儿才令人心旷神怡。在农村的百亩土地上，房子要盖在风景秀丽的山边。

这只不过是人的本性；对于缺乏土地生态学知识的人而言，这样做看来合情合理。如果想盖一幢房屋，"……把它盖在尽可能好的地方。"

但现在请考虑一下，可供使用的土地中有 3/4 都是不那么美好的。由于人们总是把房屋盖在那 1/4 的富饶的土地上，另外那 3/4 的土地，本来生态环境就差，自然就更无人问津了。逐渐地，这些地方变得越来越荒凉。那垃圾成堆的黑暗阴湿的角落，那满是臭水的沼泽地，那没有植被的干旱多石的山坡，有谁会去光顾呢？

不仅如此。当我们选择把房屋建造在湖光山色中时，已经存在的美景——每年春天开遍草地的番红花，有蜥蜴躺着晒太阳的洒满阳光的石堆，人们喜爱漫步的铺着碎石的小径——要被破坏殆尽了。当在景物美好的地方施工时，随着建筑工程的进展，无数美景也将被毁于一旦。

It is only human nature;and,for a person who lacks a total view of the ecology of the land,it seems the most obvious and sensible thing to do.If you are going to build a building, "...build it in the best possible place."

But think now of the three-quarters of the available land which are not quite so nice.Since people always build on the one-quarter which is healthiest,the other three-quarters,already less healthy ecologically,become neglected.Gradually,they become less and less healthy.Who is ever going to do anything on that corner of the lot which is dark and dank,where the garbage accumulates,or that part of the land which is a stagnant swamp, or the dry,stony hillside,where no plants are growing?

Not only that.When we build on the best parts of the land,those beauties which are there already—the crocuses that break through the lawn each spring,the sunny pile of stones where lizards sun themselves,the favorite gravel path,which we love walking on—it is always these things which get lost in the shuffle.When the construction starts on the parts of the land which are already healthy,innumerable beauties are wiped out with every act of building.

People always say to themselves,well,of course,we can always start another garden,build another trellis,put in another gravel path,put new crocuses in the new lawn,and the lizards will find some other pile of stones.*But it just is not so.*These simple things take years to grow—it isn't all that easy to create them,just by wanting to.And every time we disturb one of these precious details,it may take twenty years,a lifetime even,before some comparable details grow again from our small daily acts.

If we always build on that part of the land which is most healthy,we can be virtually certain that a great deal of the

人们会自我解嘲说，得了，我们当然可以随时造出另一个花园，搭起另一个花棚，砌出另一条碎石路，在新的草坪上种上新的番红花，蜥蜴又会找到另外的石堆。可是**这谈何容易**。这些看似简单的东西要长出来得花费几年工夫——把它们造出来并非轻而易举、想要就要。我们每一次的破坏，要花费 20 年甚至一辈子，才能用我们日常点滴的劳动创造出同样美好、宝贵的景物。

如果我们总是在肥美的土地上建造房屋，那其他大量的土地无疑将永远贫瘠。如果我们想要使所有的地方都秀丽肥沃，我们必须采取相反的行动。我们必须把每盖一幢新房子看作一次在一块布料上修补裂缝的机会；每一次盖房都给我们机会把环境中最丑最糟的一部分变得美好起来——至于对那些本来就是优美的环境不必再加青睐。事实上，我们倒应该**严格约束自己别去触动它们**，这样我们就能把精力花在需要我们注意的地方。这就是基地整理的原则。

事实是，目前的社区建设几乎从不按这个模式去做：人人都会告诉你，某幢新建筑和某条新马路破坏了他们心爱的去处。《旧金山纪事报》（1973 年 2 月 6 日）一则标题为"愤怒的男孩推倒房屋"的报道正是这样一个令人感到震惊的典型案例：

两个 13 岁的男孩，由于一排郊区新房盖在他们猎兔场的中间而怒不可遏——在承认用偷来的推土机铲平其中的一幢之后被捕。

据 Washoe 县行政司法部门报告，这两个男孩开动在 Reno 北边约 4miles 处的建筑工地上的推土机，然后用这个庞然大物推压一幢房子达 4 次之多，此事发生在星期五夜间。

这幢行将完工的牧场主住房在昨天早上工人赶到时已成一片废墟。据该幢建筑承包人估算，损失达 7800 美元。其中一个男孩告诉当局说，这幢住房和附近其他几幢破坏了"我们心爱的野兔猎场"。

这两名男孩以暴力破坏罪被起诉。

关于基地修整的想法在前文刚刚提及。它研究如何减少

land will always be less than healthy.If we want the land to be healthy all over—all of it—then we must do the opposite. We must treat every new act of building as an opportunity to mend some rent in the existing cloth;each act of building gives us the chance to make one of the ugliest and least healthy parts of the environment more healthy—as for those parts which are already healthy and beautiful—they of course need no attention.And in fact,we must discipline ourselves most strictly to *leave them alone*, so that our energy actually goes to the places which need it.This is the principle of site repair.

The fact is,that current development hardly ever does well by this pattern:everyone has a story about how some new building or road destroyed a place dear to them.The following news article from the San Francisco Chronicle(February 6,1973)headlined "Angry Boys Bulldoze House" struck us as the perfect case:

Two 13-year old boys—enraged over a swath of suburban homes being built in the midst of their rabbit-hunting turf—were arrested after they admitted flattening one of the homes with a purloined bulldozer.

According to the Washoe County sheriff's office,the youths started up a bulldozer used at the construction site about four miles north of Reno,then plowed the sturdy vehicle through one of the homes four times late last Friday night.

The ranch-style house—which was nearly completed-was a shambles when workmen arrived yesterday morning.Damage was estimated at $7800 by the contractor.One of the boys told authorities the home along with several others nearby was ruining a "favorite rabbit-hunting preserve."

The two boys were booked on charges of felonious destruction.

破坏的问题。但大部分有才华的传统建筑师总有办法利用建筑形式，不仅避免了破坏，而且改善了自然景色。这种态度同我们当前的建筑观点真是有天壤之别。现在，那种能帮助我们选定房屋所在基地以改善风景的观念根本就不存在了。

因此：

绝不能把房屋盖在风景优美的地方。实际上，应该反其道行之。把基地及基地上建筑物视为单一的活动的生态系统。放过那些人们珍爱、风景优美、环境舒适、土地肥沃的地区，而把新房建造在目前最不招人喜欢的那部分基地上。

需要修整
的地区

使其原封不
动的地区

❦❧

最重要的，要保持树木原封不动，在树木周围盖房子要特别小心——**树荫空间**（171）；使空地朝向房屋南边——**朝南的户外空间**（105）；通常在形成空间时应设法使每一处空间都理所当然地成为正空间——**户外正空间**（106）。利用**梯形台地**（169）修整斜坡，如果它们需要修整；并要使户外空间尽可能保持其自然风貌——**花园野趣**（172）。如有必要，可使房屋退缩到偏僻的角落以保存一株古藤，一丛你喜爱的灌木，一片可爱的绿草——**有天然采光的翼楼**（107）、**狭长形住宅**（109）……

The idea of site repair is just a beginning.It deals with the problem of how to minimize damage. But the most talented of traditional builders have always been able to use built form, not only to avoid damage, but also to improve the natural landscape. This attitude is so profoundly different from our current view of building, that concepts which will help us decide how to place buildings to *improve* the landscape don't even exist yet.

Therefore.

On no account place buildings in the places which are most beautiful. In fact, do the opposite. Consider the site and its buildings as a single living ecosystem. Leave those areas that are the most precious,beautiful,comfortable,and healthy as they are, and build new structures in those parts of the site which are least pleasant now.

❧❦

Above all,leave trees intact and build around them with great care—TREE PLACES(171);keep open spaces open to the south of buildings,for the sun—SOUTH FACING OUTDOORS(105);try,generally,to shape space in such a way that each place becomes positive,in its own right—POSITIVE OUTDOOR SPACE(106).Repair slopes if they need it with TERRACED SLOPE(169),and leave the outdoors in its natural state as much as possible—GARDEN GROWING WILD(172). If necessary,push and shove the building into odd corners to preserve the beauty of an old vine,a bush you love,a patch of lovely grass—WINGS OF LIGHT(107),LONG THIN HOUSE(109)...

模式105　朝南的户外空间**

105 SOUTH FACING OUTDOORS**

...within the general ideas of location which SITE REPAIR (104) creates,this pattern governs the fundamental placing of the building and the open space around it wth respect to sun.

❧❦❧

People use open space if it is sunny, and do not use it if it isn't, in all but desert climates.

This is perhaps the most important single fact about a building. If the building is placed right, the building and its gardens will be happy places full of activity and laughter. If it is done wrong, then all the attention in the world, and the most beautiful details, will not prevent it from being a silent gloomy place. Thousands of acres of open space in every city are wasted because they are north of buildings and never get the sun. This is true for public buildings, and it is true for private houses. The recently built Bank of America building in San Francisco—a giant building built by a major firm of architects—has its plaza on the north side. At lunchtime, the plaza is empty, and people eat their sandwiches in the street, on the south side where the sun is.

Just so for small private houses.The shape and orientation of lots common in most developments force houses to be surrounded by open space which no one will ever use because it isn't in the sun.

A survey of a residential block in Berkeley, California, confirms this problem dramatically.Along Webster Street—an

……在不违背**基地修整**（104）中关于选择基地的普遍原则的情况下，本模式规定房屋及其周围空地相对于太阳的基本位置。

⊰⊱

如果空地朝阳，人们就加利用；不朝阳，人们不加利用，在除沙漠以外的所有气候条件下的情况都是如此。

这或许是关于建筑物的最重要的一个事实。如果房屋坐落在合适的位置，房屋及其花园中会充满活力和欢声笑语。如果朝向不好，即使百般装饰、各种陈设，也难免死气沉沉，暗淡无光。在每个城市里，成千上万亩的空地白白废弃，因为它们在建筑物北边，终日不见阳光。公共建筑如此，私人住宅亦然。最近在旧金山落成的美国银行大楼——由一个大型建筑公司承建的大厦——有一个广场在北边。吃午饭时，这个广场上空无一人，人们都聚焦在有阳光的南面街道上吃他们的三明治。

朝北的户外空间
North facing outdoors

小的私人住宅亦是如此。在大部分居民住宅区中普遍存在的地基的形状和方位迫使房屋处于空地包围之中，这些空地无人利用，因为它们没有阳光。

对加利福尼亚州伯克利市一个住宅区的一份调查有力地肯定了这个问题。沿着韦伯斯特大街——一条东西向大街——受到访问的 20 个人中有 18 个人说，他们仅利用院子里向阳的那一部分。这些人中有半数住在街北——这些

eastwest street—18 of 20 persons interviewed said they used only the sunny part of their yards.Half of these were people living on the north side of the street—*these people did not use their backyards at all*, but would sit in the front yard,beside the sidewalk,to be in the south sun.The north-facing back yards were used primarily for storing junk.Not one of the persons interviewed indicated preference for a shady yard.

The survey also gave credence to the idea that sunny areas won't be used if there is a deep band of shade up against the house,through which you must pass to get to the sun.Four north facing backyards were large enough to be sunny toward the rear.In only one of these yards was the sunny area reported as being used—in just the one where it was possible to get to the sun without passing through a deep band of shade.

Although the idea of south-facing open space is simple,it has great consequences, and there will have to be major changes in land use to make it come right. For example, residential neighborhoods would have to be organized quite differently from the way they are laid out today. Private lots would have to be longer north to south,with the houses on the north side.

Note that this pattern was developed in the San Francisco Bay Area. Of course, its significance varies as latitude and climate change. In Eugene, Oregon, for example, with a rather rainy climate, at about 50°latitude, the pattern is even more essential:the south faces of the buildings are the most valuable outdoor spaces on sunny days.In desert climates, the pattern is less important; people will want to stay in outdoor spaces that have a balance of sun and shade. But remember that in one way or another, this pattern is absolutely fundamental.

人根本不使用他们的后院，而是常常坐在前院人行道旁，这样可以享受南面的阳光。朝北的后院主要用来储存旧货。被访问的人中无一人说喜欢背阴面的院子。

受人喜爱的朝南户外空间
Favorite outdoor places to the south

调查还使人相信这样一个想法：如果有一道很宽的阴影笼罩着这幢房屋，你得经过这道阴影才能到阳光中来，那么向阳地段也不会有人利用。四户人家其朝北的后院都很大，其后部能照进阳光。调查表明，这些院子中只有一户利用了有阳光的地方——正是那个有可能不经过一道很深的阴影就能到阳光中去的院子。

尽管空地朝南的想法很简单，但其结果影响很大，必须对土地使用的习惯做出巨大改变才能使它实现。例如，住户的规划和今天它们的分布状态很不相同。私人住宅南北方向应长一些，房子应坐落在北边。

重新规划街区以得到阳光
Blocks reorganized to catch the sun

请注意，本模式是在旧金山湾区发展出来的。当然，其效果的显著程度随纬度和气候而变化。例如，在俄勒冈的尤金，由于雨量较多，而且在大约50°的纬度上，这个模式甚至更加重要：建筑物南面在阳光和煦的日子里是最宝贵的户外空间。在沙漠气候里，这一模式则次要一些；人们想要待在半阴半阳的户外空间里。但要记住，无论如何这都是一个极为重要的模式。

Therefore:

Always place buildings to the north of the outdoor spaces that go with them,and keep the outdoor spaces to the south. Never leave a deep band of shade between the building and the sunny part of the outdoors.

୨୦୦ଓ

Let HALF-HIDDEN GARDEN(111)influence the position of the outdoors too. Make the outdoor spaces positive—POSITIVE OUTDOOR SPACE(106)—and break the building into narrow wings—WINGS OF LIGHT(107). Keep the most important rooms to the south of these wings—INDOOR SUNLIGHT(128); and keep storage,parking,etc,to the north—NORTH FACE(162). When the building is more developed, you can concentrate on the special sunny areas where the outdoors and building meet, and make definite places there, where people can sit in the sun—SUNNY PLACE(161)...

因此：

应始终使建筑物坐落在同它相连的户外空间的北边，而使户外空间朝南。绝对不要让建筑物和户外空间有阳光的部分之间留下一道很宽的阴影。

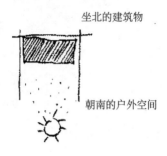

坐北的建筑物

朝南的户外空间

⁙⁙⁙

使**半隐蔽花园**（111）也对户外空间产生影响。使户外空间成为正空间——**户外正空间**（106）——将建筑物分散成狭窄的翼楼——**有天然采光的翼楼**（107）。使一些最重要的房间朝向翼楼的南面——**室内阳光**（128）；而使储藏室、汽车房等朝北——**背阴面**（162）。当建筑物的规模扩大时，要把注意力集中在建筑物同户外空间交界处某些阳光充沛的地方，使它们成为人们可以坐着晒太阳的去处——**有阳光的地方**（161）……

模式106　户外正空间**

　　……在形成**朝南的户外空间**（105）时，你必须既选择盖房屋的地方，也选择户外空间。你不能顾此而失彼。本模式告诉你户外空间的几何形状；下一个模式——**有天然采光的翼楼**（107）——作为本模式的补充，告诉你室内空间的形状。

⍣⍣⍣

　　户外空间若仅仅是建筑物之间"留下的"空地，通常是不能利用的。

106 POSITIVE OUTDOOR SPACE**

...in making SOUTH FACING OUTDOORS(105) you must both choose the place to build,and also choose the place for the outdoors.You cannot shape the one without the other.This pattern gives you the geometric character of the outdoors;the next one—WINGS OF LIGHT (107)—gives you the complementary shape of the indoors.

⚜

Outdoor spaces which are merely "left over" between buildings will, in generaI, not be used.

There gte two fundamentally different kinds of outdoor space:negative space and positive space. Outdoor space is negative when it is shapeless, the residue left behind when buildings—which gte generally viewed as positive—are placed on the land.An outdoor space is positive when it has a distinct and definite shape,as definite as the shape of a room,and when its shape is as important as the shapes of the buildings which surround it.These two kinds of space have entirely different plan geometries,which may be most easily distinguished by their figure-ground reversal.

If you look at the plan of an environment where outdoor spaces are negative, you see the buildings as figure,and the outdoor space as ground.There is no reversal. It is impossible to see the outdoor space as figure, and the buildings as ground. If you look at the plan of an environment where outdoor spaces are positive, you may see the buildings as figure,and outdoor

有两种根本不同的户外空间："负空间"和"正空间"。当建筑物（它通常被认为是"正"的）建造起来以后，余下来的形状不规则的户外空间是"负空间"，即不成形的空间。当户外空间有明显而固定的形状时（就像房间的形状那样固定），当它的形状同周围建筑物的形状同样重要时，这样的户外空间是"正空间"。这两种空间有着截然不同的平面几何图形，根据它们的黑色块和白底是否可逆可以轻而易举地将它们区分开来。

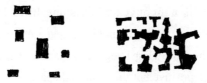

造成残留负空间的建筑物及造成户外正空间的建筑物
Buildings that create negative, leftover space...
Buildings that create positive outdoor space

如果你研究上面这张户外空间为"负空间"的环境的平面图，你要把建筑物看作黑色块，把户外空间看作白底。不能把它们颠倒过来。不可能把户外空间看作黑色块，而把建筑物当作白底。如果你研究一个室外空间为"正空间"的环境平面图，你可把建筑物看作黑色块，把户外空间看作白底——而且，你也可以把户外空间看作以建筑物作白底反衬出来的黑色块。在这个图形上黑色块和白底是可逆的。

确定户外空间是"正"还是"负"的另一方法是看它们的围合程度和凸出程度。

在数学中，当连接空间内任意两点的线完全处于内部空间时，该空间是凸空间。当连接两点的某些线至少部分露在这个空间之外时，该空间是非凸空间。根据这一定义，以下不规则的四方形为凸空间，因而是"正空间"；但L形空间是不凸出空间或非"正空间"，因为连接两个端点的

spaces as ground—and, you may also see the outdoor spaces as figure against the ground of the buildings.The plans have figure-ground reversal.

Another way of defining the difference between "positive" and "negative" outdoor spaces is by their degree of enclosure and their degree of convexity.

In mathematics,a space is convex when a line joining any two points inside the space itself lies totally inside the space.It is nonconvex,when some lines joining two points lie at least partly outside the space.According to this definition,the following irregular squarish space is convex and therefore positive;but the L-shaped space is not convex or positive,because the line joining its two end points cuts across the corner and therefore goes outside the space.

Positive spaces are partly enclosed,at least to the extent that their areas seem bounded(even though they are not,in fact,because there are always paths leading out,even whole sides open),and the "virtual" area which seems to exist is *convex*.Negative spaces are so poorly defined that you cannot really tell where their boundaries are,and to the extent that you can tell,the shapes are *nonconvex*.

Now, what is the functional relevance of the distinction between "positive" and "negative" outdoor spaces.We put forward the following hypothesis. *People feel comfortable in spaces which are "positive"and use these spaces;people feel relatively uncomfortable in spaces which are"negative"and such spaces tend to remain unused.*

The case for this hypothesis has been most fully argued by Camillo Sitte, in *City Planning According to Artistic Principles* (republished by Random House in 1965).Sitte has analyzed

线对直越过它们的角经过空间外部。

"正空间"是部分围合的，至少它的面积看来是有界线的（虽然实际上没有界线，因为总有道路通向外界，甚至整个边界是敞开的），因而看来存在的"真正的"面积是凸面积。"负空间"则很难明确其定义，它的边界在哪里你实在无法弄清，但你能辨别出来，它的形状是不凸出的。

凸出的和不凸出的
Convex and nonconvex

这个空间是可以感觉出来的：它是清楚的——这是一个场地……而且是凸出的
This space can be felt:it is distinct—a place...and it is convex

这个空间是模糊的、不定形的、"虚无的"
This space is vague, amorphous, "nothing"

那么在实用方面，"正"的和"负"的户外空间有何区别呢？我们提出以下假设。人们在"正空间"感到舒适而利用这些空间；人们在"负空间"感到不怎么舒适，因此这些空间无人利用。

这一假说成立的理由在卡米洛·西特（Camillo Sitte）的《根据艺术原则进行城市规划》（*City Planning According to Artistic Principle*）一书（1965 年由 Random House 再版）中得到淋漓尽致的说明。西特分析了大量的欧洲城市广场，将那些看来有人光顾、富有生气的广场跟那些一无是处的——加以鉴别，力图解释那些有生命力的广场成功的秘诀。

a very large number of European city squares, distinguishing those which seem used and lively from those which don't, trying to account for the success of the lively squares.He shows, with example after example, that the successful ones—those which are greatly used and enjoyed—have two properties.On the one hand, they are partly enclosed; on the other hand, they are also open to one another, so that aach one leads into the next.

The fact that people feel more comfortable in a space which is at least partly enclosed is hard to explain.To begin with, it is obviously not always true.For example, people feel very comfortable indeed on an open beach,or on a rolling plain,where there may be no enclosure at all.But in the smaller outdoor spaces-gardens, parks, walks, plazas-enclosure does, for some reason, seem to create a feeling of security.

It seems likely that the need for enclosure goes back to our most primitive instincts.For example,when a person looks for a place to sit down outdoors,he rarely chooses to sit exposed in the middle of an open space—he usually looks for a tree to put his back against;a hollow in the ground,a natural cleft which will partly enclose and shelter him.Our studies of people's space needs in workplaces show a similar phenomenon. To be comfortable, a person wants a certain amount of enclosure around him and his work—but not too much—see WORKSPACE ENCLOSURE(183). Clare Cooper has found the same thing in her study of parks:people seek areas which are partially enclosed and partly open—not too open,not too enclosed(Clare Cooper, Open Space Study, *San Francisco Urban Design Study*, San Francisco City Planning Dept.,1969).

他举了一个又一个的例子来证明，成功的广场——被广泛使用和喜爱的那些——具有两大特点：一方面它们局部围合；另一方面它们也相互开放，彼此沟通。

人们在至少局部围合的空间感到比较舒适这个事实很难解释。首先，这个规律很明显并不是绝对的。例如，人们在开阔的海滨或在绵延起伏的原野上也确实感到非常舒服，而这些地方可是完全无遮无挡的。但在较小的户外空间——花园、公园、散步场所、城市广场——由于某种原因，围合看来却会让人产生一种安全感。

户外正空间的四个例子
Four examples of positive outdoor space

看来对围合的需要很可能出于人的非常原始的本能。例如，当一个人在户外要找一个地方坐下，他不大会选择户外空间正中间的某个位置去坐，——他通常会找一棵树把背靠在那儿；或找一个地洞让天然缝隙部分可以包覆和遮挡他。我们对人们在工作场所中的空间需要所作的研究发现了类似的现象。为了舒服一些，一个人要求他和他的工作区周围有一定距离的围隔——但也不要太多——参阅**工作空间的围隔**（183）。克莱尔·库珀在她对公园的研究中发现了同样的情况：人们寻求部分围合部分开放的地方——别太开放，也别太封闭。(Clare Cooper, Open Soace Study, *San Francisco Urban Design study*, San Francisco City

Most often,positive outdoor space is created at the same time that other patterns are created.The following photograph shows one of the few places in the world where a considerable amount of building had no other purpose whatsoever except to create a positive outdoor space.It somehow underlines the pattern's urgency.

When open space is negative, for example, L-shaped—it is always possible to place small buildings, or building projections, or walls in such a way as to break the space into positive pieces.

And when an existing open space is too enclosed,it may be possible to break a hole through the building to open the space up.

Therefore:

Make all the outdoor spaces which surround and lie between your buildings positive. Give each one some degree of enclosure;surround each space with wings of buildings, trees, hedges, fences, arcades,and trellised walks, until it becomes an entity with a positive quality and does not spill out indefinitely aroud corners.

Place WINGS OF LIGHT(107)to form the spaces. Use open trellised walks, walls, and trees to close off spaces which are too exposed—TREE PLACES(171), GARDEN WALL(173), TRELLISED WALK (174); but make sure that every space is always open to some larger space,so that it is not too enclosed—HIERARCHY OF OPEN SPACE(114). Use BUILDING FRONTS(122)to help create the shape of space. Complete the positive character of the outdoors by

Planning Dept.，1969)

最为常见的情况是，户外正空间是在创造其他模式的同时创造出来的。下面这张照片所显示的这个地方是把建筑物的很大一部分专门用于营造户外"正空间"，这样的地方在全世界也是不多见的。它在一定程度上强调了本模式在应用上的迫切性。

位于（法国）南锡的广场
The square at Nancy

如果空地是负空间，例如是 L 形的——还有可能通过设置小的建筑物或建筑的伸出部分或墙垣将该空间分割成"正"的小空间。

把这个……转化为这个
Transform this...to this

而当现有的户外空间过于封闭时，可将建筑物打开一个缺口，使空间开敞。

把这个……转化为这个
Transform this...to this

making places all around the edge of buildings,and so make the outdoors as much a focus of attention as the buildings— BUILDING EDGE(160).Apply this pattern to COURTYARDS WHICH LIVE(115),ROOF GARDENS(118),PATH SHAPE(121),OUTDOOR ROOM(163),GARDEN GROWING WILD(172).

因此：

**使你的建筑物周围或建筑物之间的全部户外空间成为
"正空间"。使每一处空间都有某种程度的围合；在每一处
空间周围都设置翼楼、树木、矮墙、篱笆、拱廊和棚下小径，
直到它成为一个整体，不会从各个拐角处无限地向外流通。**

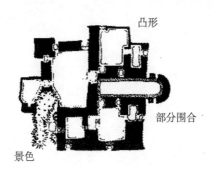

凸形

部分围合

景色

&cx;

通过建造**有天然采光的翼楼**（107）来形成这样的
空间。利用敞开的棚下小径、墙和树木以围合太暴露的
空间——**树荫空间**（171）、**花园围墙**（173）、**棚下小径**
（174）；但一定要使每一空间始终朝向某个更大的空间开
放，这样它才不至于太封闭——**外部空间的层次**（114）。
利用**建筑物正面**（122）使之有助于形成空间形状。在建筑
物边缘的周围形成空间以完善户外正空间，并使之同建筑
物一样令人瞩目——**建筑物边缘**（160）。将本模式用于**有
生气的庭院**（115）、**屋顶花园**（118）、**小路的形状**（121）、
有围合的户外小空间（163）、**花园野趣**（172）。

模式107　有天然采光的翼楼**

　　……在这一阶段，从**朝南的户外空间**（105）和**户外正空间**（106）中，你已经选好建筑物在基地上的大概位置。在仔细布置建筑物内部之前，有必要更具体地确定屋顶和建筑物的造型。为了做到这一点，请重温你对建筑物基本的社会成分所作出的决定。在某些情况下，你会根据个别情形作出这些决定；在其他情况下，可能利用基本的社会性模式来确定这些基本实体——**家庭**（75）、**小家庭住宅**（76）、**夫妻住宅**（77）、**单人住宅**（78）、**自治工作间和办公室**（80）、**小服务行业**（81）、**办公室之间的联系**（82）、

107 WINGS OF LIGHT**

...at this stage, you have a rough position for the building or buildings on the site from SOUTH FACING OUTDOORS(105) and POSITIVE OUTDOOR SPACE(106). Before you lay out the interior of the building in detail, it is necessary to define the shapes of roofs and buildings in rather more detail. To do this, go back to the decisions you have already made about the basic social components of the building. In some cases, you will have made these decisions according to the individual case; in other cases you may have used the fundamental social patterns to define the basic entities—THE FAMILY (75), HOUSE FOR A SMALL FAMILY(76), HOUSE FOR A COUPLE(77), HOUSE FOR ONE PERSON(78), SELF-GOVERNING WORKSHOPS AND OFFICES (80), SMALL SERVICES WITHOUT RED TAPE(81), OFFICE CONNECTIONS (82), MASTER AND APPRENTICES(83), INDIVIDUALLY OWNED SHOPS (87). Now it is time to start giving the building a more definite shape based on these social groupings. Start by realizing that the building needn't be a massive hulk, but may be broken into wings.

❧❦

Modern buildings are often shaped with no concern for natural light—they depend almost entirely on artificial light. But buildings which displace natural light as the major source of illumination are not fit places to spend the day.

师徒情谊（83）、**个体商店**（87）。现在，应该根据这些社会组合开始给建筑物以更加确定的形状。请根据这样的认知来开始你的工作：楼房不需要成为庞然大物，反而可以分解为多个小翼楼。

∞∞

现代建筑物的造型往往不考虑自然光——它们几乎完全依赖人造光。但不采用自然光作为照明主要来源的建筑物是不适于白天居住的。

一座古怪的大楼——对室内阳光毫不在意
A monster building—no concern for daylight inside

这一简单说明，如果加以认真对待，将会在建筑造型上引发一场革命。目前，人们已习惯性地认为，用人造光源照明的室内空间是可以利用的；因而建筑物形状各异，进深不同。

如果认为天然光是室内空间必不可少的——而不是可有可无的——特点，那么建筑物的进深不能超过 20 ~ 25ft，因为距离窗户超过大约 12ft 或 15ft 的建筑物内的任何一点都得不到充足的阳光。

以后，在**两面采光**（159）中，我们还会有争论，甚至

This simple statement, if taken seriously, will make a revolution in the shape of buildings. At present, people take for granted that it is possible to use indoor space which is lit by artificial light; and buildings therefore take on all kinds of shapes and depths.

If we treat the presence of natural light as an *essential*—not optional-feature of indoor space, then no building could ever be more than 20-25 feet deep, since no point in a building which is more than about 12 or 15 feet from a window, can get good natural light.

Later on, in LIGHT ON TWO SIDES(159), We shall argue, even more sharply, that every room where people can feel comfortable must have not merely one window, but two, on different sides. This adds even further structure to the building shape: it requires not only that the building be no more than 25 feet deep, but also that its outer walls are continually broken up by corners and reentrant corners to give every room two outside walls.

The present pattern, which requires that buildings be made up of long and narrow wings, lays the groundwork for the later pattern. Unless the building is first conceived as being made of long, thin wings, there is no possible way of introducing LIGHT ON TWO SIDES(159), in its complete form, later in the process. Therefore, we first build up the argument for this pattern, based on the human requirements for natural light, and later, in LIGHT ON TWO SIDES (159), we shall be concerned with the organization of windows within a particular room.

There are two reasons for believing that people must have buildings lit essentially by sun.

更加激烈，那就是使人感到舒服的房间必须不止有一个窗户，而应有两个窗户，在不同的面上。这就需要给建筑物的形状再增添一些结构：它不仅要求建筑物的进深不超过25ft，而且要求其外墙接连为角落和凹角所断开而使每个房间都有两面外墙。

本模式要求建筑物由长而狭窄的翼楼所构成，为后面的模式奠定基础。除非首先把建筑物设计为长而狭窄的翼楼，否则，在以后的建筑过程中无法完整地引入**两面采光**（159）。因此我们必须首先讨论这一以人对自然光的需要为依据的模式，然后，在**两面采光**（159）中，我们会讨论某一特定房间内部的窗户布置。

我们认为，人们的住宅必须基本上采用自然光，其理由有二：

首先，在全世界，人们都在反对无窗建筑；在没有阳光的地方工作的人总有颇多怨言。阿莫斯·拉波波特分析了人们说过的话，认定人们住在有窗的房间会比住在无窗的房间心情要好。(Amos Rapoport, "Some Consumer Comments on a Designed Environment," *Arena*, January, 1967, pp.176~178.) 爱德华·霍尔讲了一个故事：一个人在没有窗户的办公室工作了一段时间，老是说"这儿挺好，这儿挺好。"但后来却遽然离去。霍尔说："问题是这样深刻、这样严重，以致这个人甚至连谈论它都感到不能忍受，因为一谈论它就会闸门打开，停都停不住。"

其次，人们确实需要阳光的证据越来越多，因为阳光在某种程度上起着维持人体生理节奏的决定性作用，而白天光线的变化，虽然有着明显的不同，但就上述意义来讲，不会引起人体保持其与环境关系的根本变化。（例见R.G.Hopkinson, *Architectural Physics*: *Lighting*, Department of Scientific&Industrial Research, Building Research Station, HMSO, London, 1963, pp.116~117.）如果真是这样，太

First,all over the world,people are rebelling against window-less buildings;people complain when they have to work in places without daylight.By analyzing words they use, Rapoport has shown that people are in a more positive frame of mind in rooms with windows than in rooms without windows.(Amos Rapoport, "Some Consumer Comments on a Designed Environment," *Arena*, January, 1967, pp.176-178.) Edward Hall tells the story of a man who worked in a windowless office for some time, all the time saying that it was "just fine, just fine," and then abruptly quit.Hall says, "The issue was so deep, and so serious, that this man could not even bear to discuss it,since just discussing it would have opened the floodgates."

Second, there is a growing body of evidence which suggests that man actually *needs* daylight,since the cycle of daylight somehow plays a vital role in the maintenance of the body's circadian rhythms,and that the change of light during the day, though apparently variable, is in this sense a fundamental constant by which the human body maintains its relationship to the environment.(See,for instance,R.G.Hopkinson, *Architectural Physics:Lighting*, Department of Scientific & Industrial Research,Building Research Station,HMSO,London, 1963,pp.116-117.)If this is true,then too much artificial light actually creates a rift between a person and his surroundings and upsets the human physiology.

Many people will agree with these arguments.Indeed,the arguments merely express precisely what all of us know already:that it is much more pleasant to be in a building lit by daylight than in one which is not.But the trouble is that many of the buildings which are built without daylight are built

多的人工光线实际上造成人与生活环境的割裂，因而扰乱了人的生理机能。

许多人会赞同这些论点的。的确，这些论点仅仅准确地表达了我们已经了解的情况：住在有阳光照明的房子里比住在阳光照不进的房子里要愉快得多。但问题在于，许多建筑物没有阳光是由于房屋太密集造成的。房屋盖得鳞次栉比，有人认为，为了达到高密度，有必要牺牲阳光。莱斯利·马丁和莱昂内尔·马奇对阐明这个问题作过很大贡献。(Leslie Martin and Lionel March, *Land Use and Built Form*, Cambridge Research, Cambridge Univer sity, April 1966.) 利用建筑物地面面积与总的基地面积之比作为判定密度的标准，而把建筑物的进深之半作为判定光线的照射条件，他们比较了建筑物和空地的三种不同的布局，并将其命名为 S_0、S_1 及 S_2。

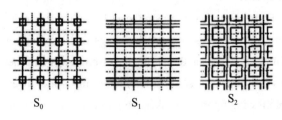

S_0 S_1 S_2

三种建筑类型
Three building types

在这三种布局中，S_2 是建筑物以狭长的翼楼包围户外空间的，它在密度不变的情况下阳光条件最好。它也为一定水平的阳光提供最高密度。

有另一种批评意见经常指责这一模式。说该模式由于要建造狭窄而无规则的翼楼，它增加了建筑物的外围，从而使建筑物的造价大为提高。造价高低差别有多大？下列数字取自斯基德莫尔·欧文和梅里尔在 BOP（建筑优选法）计划中使用的标准办公楼的造价分析。这些数字表明一幢

that way because of density.They are built compact,in the belief that it is necessary to sacrifice daylight in order to reach high densities.

Lionel March and Leslie Martin have made a major contribution to this discussion.(Leslie Martin and Lionel March, *Land Use and Built Form*, Cambridge Research, Cambridge University, April 1966.) Using the ratio of built floor area to total site area as a measure of density and the semi-depth of the building as a measure of daylight conditions, they have compared three different arrangements of building and open space,which they call S_0, S_1, and S_2.

Of the three arrangements, S_2, in which buildings surround the outdoors with thin wings,gives the best daylight conditions for a fixed density. It also gives the highest density for a fixed level of daylight.

There is another criticism that is often leveled against this pattern.Since it tends to create buildings which are narrow and rambling, it increases the perimeter of buildings and therefore raises building cost substantially.How big is the difference? The following figures are taken from a cost analysis of standard office buildings used by Skidmore Owings and Merrill, in the program BOP(Building Optimization).These figures illustrate costs for a typical floor of an office building and are based on costs of 21 dollars per square foot for the structure, floors, finishes, mechanical, and so on, not including exterior wall, and a cost of 110 dollars per running foot for the perimeter wall. (Costs are for 1969.)

办公楼的标准层的价格，所依据的是用于结构、地面、装修、机械等（不包括外墙）每平方英尺 21 美元的造价以及外围边墙的每一延英尺 110 美元的造价。（1969 年造价）。

面积/ft²	形状/ft	外围造价/美元	每方英尺外围造价/美元	每方英尺总造价/美元
15000	120×125	54000	3.6	24.6
15000	100×150	55000	3.7	24.7
15000	75×200	60500	4.0	25.0
15000	60×250	68000	4.5	25.5
15000	50×300	77000	5.1	26.1

延长外围对建筑物造价所增无几

从上表我们可以看出，至少在这一情况下，加长外围对建筑物的造价所增无几。最狭长的建筑物比完全正方形建筑物的造价只多 6%。我们认为这个情况是相当典型的，那种认为正方形的集中的建筑形式可以节省造价的说法实属夸大其词。

现在假定，本模式同密度问题和外围造价问题没有矛盾，于是我们必须确定，一幢建筑物可能有多宽，仍基本上可用自然光照明。

首先，我们假定，建筑物内任一点的每平方英尺照明度都应不少于 20lm。这是标准走廊内的照明度水平，它正好低于阅读所要求的水平。其次，我们假定，一个空间，如果它的光线 50% 以上来自自然光，那么它看起来是由"自然光"照明的：这就是说，甚至离窗最远的地方也必须获得来自自然光的至少每平方英尺 10lm 的照明度。

现在我们看一下被霍普金森和凯详细分析过的一个房间。这是一间教室，进深 18ft，面宽 24ft，一面全是窗户，高出地面 3ft。墙有 40% 的反射率——这是相当标准的数值。在自然光处于正常状态时，距窗 15ft 的课桌从自然光获得每平方英尺 10lm——这是最小值。但这已经是一个采光相

Area (Sq.Ft.)	Shape(Ft.)	Perimeter Cost($)	Perimeter Cost Per Sq.Ft.($)	Total Cost Per Sq.Ft.($)
15, 000	120×125	54, 000	3.6	24.6
15, 000	100×150	55, 000	3.7	24.7
15, 000	75×200	60, 500	4.0	25.0
15, 000	60×250	68, 000	4.5	25.5
15, 000	50×300	77, 000	5.1	26.1

The extra perimeter adds little to building costs.

We see then,that at least in this one case,the cost of the extra perimeter adds very little to the cost of the building.The narrowest building costs only 6 per cent more than the squarest. We believe this case is fairly typical and that the cost savings to be achieved by square and compact building forms have been greatly exaggerated.

Now,assuming that this pattern is compatible with the problems of density and perimeter cost,we must decide how wide a building can be,and still be essentially lit by the sun.

We assume,first of all,that no point in the building should have less than 20 lumens per square foot of illumination.This is the level found in a typical corridor and is just below the level required for reading.We assume,second,that a place will only seem "naturally" lit,if more than 50 per cent of its light comes from the sky:that is,even the points furthest from the windows must be getting at least 10 lumens per square foot of their illumination from the sky.

Let us now look at a room analyzed in detail by Hopkinson and Kay.The room,a classroom,is 18 feet deep,24 feet wide,with a window all along one side starting three feet above the floor.Walls have a reflectance of 40 per cent— a fairly typical value.With a standard sky,the desks 15 feet from the window are just getting 10 lumens per square foot

当好的房间了（R.G.Hopkinson and J.G.Kay，*The Lighting of Buildings*，New York：Praeger，1969，p.108）。

那么，很难想象许多进深在 15ft 以上的房间能够达到我们的标准。确实，本书中许多模式都趋向于减小窗的面积——**俯视外界生活之窗（192）、借景的门窗（221）、深窗洞（223）、小窗格（239）**，因此在许多情况下房间不应超过12ft深——只有墙很亮或天花很高时可超过12ft。因此，我们得出结论，一个建筑物的翼楼要真地成为"有天然采光的翼楼"，其宽度必须约为 25ft——绝不能宽于 30ft——翼楼内的房间都是"同一进深"。当建筑物的宽度大于此数时，势必要采用人工照明。

一幢宽度必须很大的建筑物——例如，一座大厅——如果屋顶另有天窗的话，仍然能得到适当的自然光。

因此：

把每幢建筑物都分成若干翼楼，使它们大致同建筑物内部最重要的自然形成的社会群体相适应。使每一幢翼楼都成为长条形，尽可能地狭窄——绝不要使其宽度超过 25ft。

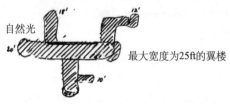

自然光

最大宽度为25ft的翼楼

✿❀✿❀

利用翼楼来营造出户外空间，使之像庭院和房间那样，有一定形状——**户外正空间（106）**；只要有可能，将各翼楼同周围已有的建筑物连接起来，以便使该建筑物在一个长而不规则的连续结构中占有自己的位置——**鳞次栉比的建筑（108）**。在进一步开始认真考虑和确定单个房间时，利用翼楼使每一房间有**两面采光（159）**所提供的阳光。

from the sky—our minimum.Yet this is a rather well lit room. R.G.Hopkinson and J.G.Kay,*The Lighting of Buildings*, New York:Praeger,1969,p.108).

It is hard to imagine then,that many rooms more than 15 feet deep will meet our standards.Indeed,many patterns in this book will tend to reduce the window area—WINDOWS OVERLOOKING LIFE (192),NATURAL DOORS AND WINDOYVS(221), DEEP REVEALS(223), SMALL PANES(239), so that in many eases rooms should be no more than 12 feet deep—more only if the walls are very light or the ceilings very high.We conclude,therefore,that a building wing that is truly a "wing of light" must be about 25 feet wide—never wider than 30 feet—with the interior rooms "one deep" along the wing. When buildings are wider than this,artificial light,of necessity, takes over.

A building which simply has to be wide—a large hall for example—can have the proper level of natural light if there are extra clerestory windows in the roof.

Therefore:

Arrange each building so that it breaks down into wings which correspond,approximately,to the most important natural social groups within the building.Make each wing long and as narrow as you can—never more than 25 feet wide.

❧❦

Use the wings to form outdoor areas which have a definite shape, like courts and rooms—POSITIVE OUTDOOR SPACE(106); connect the wings, whenever possible, to the existing buildings round about so that the building takes its place within a long and rambling continuous fabric—CONNECTED BUILDINGS(108).When you

每一幢翼楼最好都能有自己的屋顶，这样，所有的翼楼在一起就形成一个巨大的重叠交错的屋顶——**重叠交错的屋顶**（116）；如果翼楼内有不同的住宅、工作房间或一系列主要房间，则要建造从一面、从一个拱廊或回廊通向这些房间或一组房间的通路，而不要从中心走廊直接通进去——**拱廊**（119）、**短过道**（132）。关于翼楼的承重结构，参阅**结构服从社会空间的需要**（205）及之后各模式……

get further down and start defining individual rooms,make use of the daylight which the wings provide by giving each room LIGHT ON TWO SIDES(159).

Give each wing its own roof in such a way that all the wings together form a great cascade of roofs—CASCADE OF ROOFS(116); if the wing contains various houses,or workgroups, or a sequence of major rooms, build access to these rooms and groups of rooms from one side, from an arcade, or gallery, not from a central corridor—ARCADES(119), SHORT PASSAGES(132).For the load bearing structure of the wings, begin with STRUCTURE FOLLOWS SOCIAL SPACES(205)...

模式108 鳞次栉比的建筑*

……本模式有助于完善**建筑群体**（95）、**有天然采光的翼楼**（107）和**户外正空间**（106）。特别是，它通过清除建筑物之间的所有无用的地方来创造户外正空间。当你把每幢建筑物同相邻的建筑物连接在一起时会发现，户外空间几乎自然而然地变成"正空间"。

108   CONNECTED BUILDINGS*

...this pattern helps to complete BUILDING COMPLEX(95), WINGS OF LIGHT(107), and POSITIVE OUTDOOR SPACE(106). It helps to create positive outdoor space, especially, by eliminating all the wasted areas between buildings.As you connect each building to the next you will find that you make the outdoor space positive, almost instinctively.

❧❦

Isolated buildings are symptoms of a disconnected sick society.

Even in medium and high density areas where buildings are very close to each other and where there are strong reasons to connect them in a single fabric,people still insist on building isolated structures,with little bits of useless space around them.

Indeed,in our time,isolated,free-standing buildings are so common,that we have learned to take them for granted,without realizing that all the psycho-social disintegration of society is embodied in the fact of their existence.

It is easiest to understand this at the emotional level.The house, in dreams, most often means the self or person of the dreamer.A town of disconnected buildings,in a dream,would be a picture of society,made up of disconnected,isolated,selves. And the real towns which have this form,like dreams,embody just this meaning:they perpetuate the arrogant assumption that people stand alone and exist independently of one another.

When buildings are isolated and free standing,it is of course not necessary for the people who own them,use

80○03

孤立的建筑物是互不联系的病态社会的表征。

甚至在建筑物互相十分靠近，而且有充分理由把它们连接成一片的中等和高密度地区，人们还是坚持盖孤立的建筑物，周围只有一点无用的小空间。

这些建筑物装成似乎互相独立——这一
装扮使它们周围的空间完全不能利用
These buildings pretend to be independent of one another—
and this pretense leads to useless space around them

的确，在我们所处的这个时代，孤立的、自成一统的建筑物是如此普遍，以致我们已经见怪不怪了。我们没有认识到，社会上一切心理上的社会性崩溃都孕育于存在这一现象的事实之中。

从感情的角度来看，这是很容易理解的。在梦中，房子常意味着做梦者本人。由互不关联的建筑物所构成的城市，在梦中会是由互不联系的、孤立的个人组成的社会的图画。而具有这种形式的真实的城市，正如梦境，恰好体现这层意义：它们使人们独立存在、互不依靠的这一孤傲的思想成为永恒。

当建筑物互相分离、各自独立时，它们的所有人、住户和维修者当然完全无须相互来往。相反，在各建筑物实际上相依相靠的城市中，彼此相邻这一基本事实迫使人们

them,and repair them to interact with one another at all.By contrast,in a town where buildings lean against each other physically,the sheer fact of their adjacency forces people to confront their neighbors,forces them to solve the myriad of little problems which occur between them,forces them to learn how to adapt to other people's foibles,forces them to learn how to adapt to the realities outside them,which are greater,and more impenetrable than they are.

Not only is it true that connected buildings have these healthy consequences and that isolated buildings have unhealthy ones.It seems very likely—though we have no evidence to prove it—that,in fact,isolated buildings have become so popular,so automatic,so taken for granted in our time,because people seek refuge from the need to confront their neighbors,refuge from the need to work out common problems. In this sense,the isolated buildings are not only symptoms of withdrawal,but they also perpetuate and nurture the sickness.

If this is so,it is literally not too much to say that *in those parts of town where densities are relatively high*, isolated buildings, and the laws which create and enforce them,are undermining the fabric of society as forcibly and as persistently as any other social evil of our time.

By contrast,Sitte gives a beautiful discussion,with many examples,of the normal way that buildings were connected in ancient times:

The result is indeed astonishing,since from amongst 255 churches:

41 have one side attached to other buildings

96 have two sides attached to other buildings

110 have three sides attached to other buildings

必须面对他们的左邻右舍，迫使他们必须去解决他们之间发生的无数小问题，迫使他们必须去学习如何适应其他人的小缺点，迫使他们去学习如何适应他们周围的现实世界，这样的现实世界比起他们自己来更加重大，更加难以理解。

鳞次栉比的建筑物相较于孤立的建筑物利多弊少，这是千真万确的。但情况不仅如此。事实上，在我们所处的这个时代，孤立的建筑物已经如此普遍、如此自然、如此理所当然，因为人们想避开跟邻居打交道的麻烦，不想去解决一些共同的问题。事情看来就是这样，虽然我们尚无证据来证明它。在这个意义上，孤立的建筑物不仅是逃避现实的表现，而且它们还使这种病态成为痼疾并且任其发展。

如果情况果真如此，那么，**在城镇中人口密度比较高的那些地方**，孤立的建筑物，以及帮助建造这些建筑物并使其合法化的法律，如同当代其他公害一样，都在强烈而持久地损害社会结构，我们这么说实在并不过分。

作为比较，西特以丰富的例证透彻地说明了，古代建筑物鳞次栉比的情况是司空见惯的：

结果的确惊人，因为在 255 个教堂中：

41 个一面同另外的建筑物相连

96 个两面相邻于其他建筑物

110 个三面与其他建筑物比邻

2 个四面都有其他建筑物阻挡

6 个无依无靠

一共 255 个教堂；只有 6 个是孤立的。

对当时的罗马来说，教堂从来不是独自耸立的建筑物，这可以认为是一条规律。事实上，对整个意大利来说情况也差不多如此。可以清楚地看出，今人的态度与彼时罗马所采取的十分完整而又显然是精心安排的布局完全背道而驰。我们可能认为，一个新教堂除了盖在建筑基地的中心使其四周有空地之外，就不能坐落别处。但这样选址绝无优点，而只有缺点。这对建筑物毫无好处，

2 have four sides obstructed by other buildings

6 are free standing

255 churches in all; only 6 free-standing.

Regarding Rome then,it can be taken as a rule that churches were never erected as free-standing structures.Almost the same is true,in fact,for the whole of Italy.As is becoming clear,our modern attitude runs precisely contrary to this well-integrated and obviously thoughtout procedure.We do not seem to think it possible that a new church can be located anywhere except in the middle of its building lot,so that there is space all around it.But this placement offers only disadvantages and not a single advantage.It is the least favorable for building,since its effect is not concentrated anywhere but is scattered all about it.Such an exposed building will always appear like a cake on a serving-platter.To start with,any lifelike organic integration with the site is ruled out...

It is really a foolish fad, this craze for isolating buildings... (Camillo Sitte, City Planning According to Artistic Principles, New York:Random House, 1965, pp. 25-31.)

Therefore:

Connect your building up, wherever possible, to the existing buildings round about.Do not keep set backs between buildings;instead,try to form new buildings as continuations of the older buildings.

⟐⟐⟐

Connect buildings with arcades,and outdoor rooms, and courtyards where they cannot be connected physically, wall to wall—COURTYARDS WHICH LIVE (115), ARCADES(119), OUTDOOR ROOMS(163)...

因为建筑物的效能不集中在任何地方，而是分散在其周围。这样一种暴露式建筑看去总像盛在一只大的浅盘子里的一块蛋糕。首先，它不可能同基地构成有生气的有机整体。

这种把建筑物孤立起来的浪潮，实在是一种愚蠢的时髦……
（Camillo Sitte，*City Planning According to Artistic Principles*，NewYork：Random House，1965，pp.25~31.）

鳞次栉比建筑物的结构
A fabric of connected buildings

因此：

只要有可能，就应将房屋同周围现有的建筑物连成一片。不要在建筑物之间设置出一段距离；相反，要使新房屋成为老建筑的延续。

连接

❧❧

在建筑物之间不能用墙相连的地方，用拱廊、有围合的户外小空间和庭院把它们连接起来——**有生气的庭院**（115）、**拱廊**（119）、**有围合的户外小空间**（163）……

109   LONG THIN HOUSE*

...for a very small house or office the pattern of WINGS OF LIGHT (106)is almost automatically solved—no one would imagine that the house should be more than 25 feet wide.But in such a house or office there are strong reasons to make the building even longer and thinner still.This pattern was originally formulated by Christie Coffin.

❧❧

The shape of a building has a great effect on the relative degrees of privacy and overcrowding in it, and this in turn has a critical effect on people's comfort and well being.

There is widespread evidence to show that overcrowding in small dwellings causes psychological and social damage.(For example,Wiliam C.Loring, "Housing Characteristics and Social Disorganization," *Social Problems*, January 1956;Chombart de Lauwe,*Famille et Habitation*,Editions du Centre National de la Recherche Scientifique,Paris,1959;Bernard Lander,*Towards an Understanding of Juvenile Delinquency*, New York:Columbia University Press,1954.)Everyone seems to be on top of everyone else.Everything seems to be too near everything else.Privacy for individuals or couples is almost impossible.

It would be simple to solve these problems by providing more space—but space is expensive,and it is usually impossible to buy more than a certain very limited amount of it.So the question is: *For a given fixed area,which shape will create the*

模式109 狭长形住宅*

……对于一幢很小的房屋或办公楼来说，有**天然采光的翼楼**（107）这个模式差不多自然而然地解决了——没有人会认为这样的房屋其宽度应超过25ft。但在这样的房屋或办公楼里面，有充分的理由使建筑物变得更加狭长一些。这一模式最初是由克里斯蒂·科芬制定出来的。

めOCぴ

建筑物的形状对于其内部清静还是嘈杂在某种程度上起着很大作用，而这一点又对人们是否感觉舒适和安宁起着决定性的作用。

有许多证据证明，住宅小、人口多会造成心理上和社会性上的损害。（例见 Wiliam C.Loring，"Housing Characteristics and Social Disorganization，" *Social Problems*，January 1956；Chombart de Lauwe，*Famille et Habitation*，Editions du Centre National de la Recherche Scientifique，Paris，1959；Bernard Lander，*Towards an Understanding of Juvenile Delinquency*，New York：Columbia University Press，1954.）每个人看来都紧挨着其他的人。每样东西看来都太接近其他东西。个人或夫妻几乎不可能有个私密地方。

提供更多的空间就能轻松解决这些问题——但地皮昂贵，除了很有限的一点土地之外，通常不可能去购买更多的地皮。因此，问题在于：对于某一固定的土地面积，哪种房屋形状会营造出最大的空间感？

对这一问题有一种数学解答。

拥挤的感觉多半是建筑物内部点与点之间的平均距离造成的。在一幢小房子里这些距离很小——结果在屋内不

greatest feeling of spaciousness?

There is a mathematical answer to this question.

The feeling of overcrowding is largely created by the mean point-to-point distances inside a building.In a small house these distances are small—as a result it is not possible to walk far inside the house nor to get away from annoying disturbances;and it is hard to get away from noise sources,even when they are in other rooms.

To reduce this effect the building should have a shape for which the mean point-to-point distance is high.(For any given shape,we may compute the mean or average distance between two randomly chosen points within the shape).The mean point-to-point distance is low in compact shapes like circles and squares,and high in those distended shapes like long thin rectangles,and branched shapes,and tall narrow towers. These shapes increase the separation between places inside the building and therefore increase the relative privacy which people are able to get within a given area.

Of course,in practice there are limits on the long— thinness of a building.If it is too long and thin,the cost of walls becomes prohibitive,the cost of heating is too high,and the plan is not useful.But this is still no reason to settle only for box-like forms.

A small building can actually be much narrower than people imagine.It can certainly be much narrower than the 25ft width proposed in WINGS OF LIGHT(107).We have seen successful buildings as narrow as 12 feet wide—indeed,Richard Neutra's own house in Los Angeles is even less.

可能走得很远，也不可能躲开烦人的打扰；因此，即使这些噪声在其他房间产生，也很难摆脱噪声源。

为了减少这种影响，建筑物应该具有这样的形状，能使点与点之间的平均距离变长。（对于任何一种形状的建筑物，我们都可以计算该建筑物内部任意选择的两点之间的平均距离。）在诸如圆形和正方形这样的密集形中点到点的平均距离是短的，在狭窄的长方形、枝叉形以及高而窄的塔形那类伸长的形状中，点到点的平均距离较长。这些形状增加了建筑物内部各空间之间的距离，因而相对增强了人们在一定的区域内能获得的私密感。

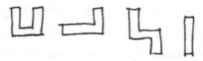

扩大各个点之间距离的建筑物
Buildings which increase the distance between points

当然，实际做起来，建筑物的狭长形状是有一定限度的。如果太长太窄，墙的造价不允许，供暖费用太高，而且不好做平面布置。但这并非只能采用盒子形建筑物的理由。

一幢小楼房其实可以比人们想象的要狭窄得多。它的宽度相较于**有天然采光的翼楼**（107）中提出的 25ft 的宽度当然要狭窄得多。我们看到过 12ft 宽的成功的狭长形住宅。理查德·纽特拉本人在洛杉矶的住宅就比这还要狭窄。

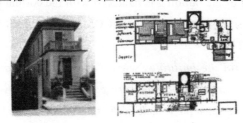

狭长形住宅
Long thin houses

And a long thin house can also be a tower,or a pair of towers,connected at ground level.Towers,like floors can be much narrower than people realize.A building which is 12 feet square,and three stories high,with an exterior stair,makes a wonderful house.The rooms are so far apart,psychologically,that you feel as if you are in a mansion.

Therefore:

In small buildings,don't cluster all the rooms together around each other;instead string out the rooms one after another,so that distance between each room is as great as it can be.You can do this horizontally—so that the plan becomes a thin,long rectangle;or you can do it vertically—so that the building becomes a tall narrow tower.In either case,the building can be surprisingly narrow and still work—8,10,and 12 feet are all quite possible.

❧✤❧

Use the long thin plan to help shape outdoor space on the site—POSITIVE OUTDOOR SPACE(106);the long perimeter of the building sets the stage for INTIMACY GRADIENT(127)and for the CASCADE OF ROOFS (116). Make certain that the privacy which is achieved with the thinness of the building is balanced with the cormnunality at the cross—roads of the house—COMMON AREAS AT THE HEART(129)...

within the buildings'wings,lay out the entrances,the gardens,courtyards,roofs and terraces:shape both the volume of the buildings and the volume of the space between the buildings at the same time—remembering that indoor space and outdoor space,like yin and yang,must always get their shape together.

一幢狭长形住宅也可以是塔形建筑，或底层是相连的双塔形建筑。塔形建筑如同楼层一样可以比人们所知道的要狭窄得多。一幢住宅，面积 12ft 见方，高三层，带室外楼梯，是一幢很精致的住宅。房间彼此远远分开，在心理上，你会感到犹如置身于深宅大院。

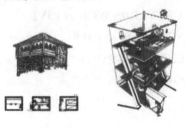

俄罗斯式塔形建筑
A Russian tower

因此：

在小建筑物里，不要把所有的房间都集中在一起，一个套一个；而应把房间一字排开，一个挨着一个，这样才能使每个房间之间的距离尽可能地大。你可以将其做成水平的——使图面呈狭长的长方形；也可以做成垂直的——这样，建筑物就成为高耸、狭长的塔楼。无论在哪种情况下，建筑物都可以成为惊人的狭长形，但仍能很好地使用——8ft、10ft 和 12ft 都可以。

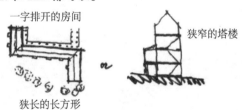

一字排开的房间

狭窄的塔楼

狭长的长方形

❧❦

利用狭长形建筑物以有助于形成基地上的户外空间——**户外正空间**（106）；建筑物的长边便于造成**私密性**

层次（127）和**重叠交错的屋顶**（116）。一定要让因建筑物的外形狭长而得到的私密性同住宅中中心区的公共性取得平衡——**中心公用区**（129）……

在建筑物的各翼楼之间，布置入口、花园、庭院、屋顶和露台：既形成建筑物的体量，同时也形成建筑物之间的空间的体量——记住：室内空间和户外空间，如同阴和阳，务必同时形成。

110. 主入口

111. 半隐蔽花园

112. 入口的过渡空间

113. 与车位的联系

114. 外部空间的层次

115. 有生气的庭院

116. 重叠交错的屋顶

117. 带阁楼的坡屋顶

118. 屋顶花园

模式110　主入口**

......你已经大致确定了房屋在基地上的位置——**基地修整**（104）、**朝南的户外空间**（105）、**有天然采光的翼楼**（107）。你也考虑到了建筑群内的主要通道和通向建筑物的交通线路——**内部交通领域**（98）、**各种入口**（102）。现在该确定建筑物的入口了。

110   MAIN ENTRANCE**

...you have a rough position for your building on the site—
SITE REPAIR(104), SOUTH FACING OUTDOORS(105),
WINGS OF LIGHT (107).You also have an idea of the
major circulation in the building complex and the lines of
approach which lead toward the building—CIRCULATION
REALMS(98), FAMILY OF ENTRANCES(102). Now it is
time to fix the entrance of the building.

❧☙

**Placing the main entrance(or main entrances)is perhaps
the single most important step you take during the evolution of
a building plan.**

The position of main entrances controls the layout of the
building.It controls movement to and from the building,and
all the other decisions about layout flow from this decision.
When the entrances are placed correctly,the layout of the
building unfolds naturally and simply;when the entrances are
badly placed,the rest of the building never seems quite right.
It is therefore vital that the position of the main entrance(or
entrances)be made early and correctly.

The functional problem which guides the placing of main
entrances is simple. *The entrance must be placed in such a
way that people who approach the building see the entrance
or some hint of where the entrance is, as soon as they see the
building itself.* This makes it possible for them to orient their
movements toward the entrance as soon as they start moving

选定一个主入口（或几处主入口）的位置也许是在规划该幢建筑的过程中你应采取的最重要的一步。

　　主入口的位置决定建筑物的布局，它决定了进出建筑物的活动，而其他一切关于建筑物布局的决定都来自这个决定。只要把入口安排妥当了，就能自然而轻松地展开建筑物的布局；如果入口的位置选择不当，建筑物的其他部分看起来总不会合适。因此把主入口（或几处主入口）的位置及早而妥善地选好是至关重要的。

　　从使用角度决定主入口的位置是简单的。**选择入口位置时应考虑要能够让走近建筑物的人一看见建筑物就能看到入口或看到有关入口在哪里的提示。这样他们有可能在开始朝建筑物走去的时候，就能找到入口的方向，而不至于必须改变方向或改变他们走近建筑物的计划。**

　　功能问题是显而易见的，但不应过高估计它给建筑物带来的好处。我们已经有过多次这样的经验：这个问题不解决，合适的位置尚未选定，工程便摆脱不了僵局。相反，主入口的位置一旦选定，而且使人感到选得恰到好处，那时其他决策自然迎刃而解。对于独户住宅、住宅团组、小的公共建筑物、大的公共建筑群，情况都是如此。显然，不管建筑物规模大小，这个模式总是用得着的。

　　让我们再详细探讨一下这个功能问题。在一个建筑物周围或建筑群范围内寻找合适的入口是叫人心烦的。只有当你知道入口在哪里的时候，才不会费心去想它。这是很自然的事——你边走边想着心中的事情，看着引你关注的东西——你不必转来转去以集中注意力来找路。但许多建筑物的入口确实很难找；从这点来看，它们很不"自然"。

　　解决这个问题有两个步骤：首先，主入口的位置必须选得恰当；其次，它们的外观必须让人一眼就能看清楚。

toward the building,without having to change direction or change their plan of how they will approach the building.

The functional problem is rather obvious,but it is hard to overestimate the contribution it makes to a good building.We have had the experience over and again,that until this question is settled and an appropriate position chosen,a project is at a stalemate.And conversely,once the main entrances have been located and they can be felt to be in the right position,then other decisions begin to come naturally.This is true for single houses,house clusters,small public buildings,large complexes of public buildings.Apparently,the pattern is basic,no matter the scale of the building.

Let us look into the functional question in more detail. Everyone finds it annoying to search around a building,or a precinct of buildings,looking for the proper entrance.When you know just where the entrance is,you don't have to bother thinking about it.It's automatic—you walk in,thinking about whatever's on your mind,looking at whatever catches your eye—you are not forced to pay attention to the environment simply to get around.Yet the entrances to many buildings are hard to find;they are not "automatic" in this sense.

There are two steps to solving the problem.First,the main entrances must be placed correctly.Second,they must be shaped so they are clearly visible.

1. *Position*

Consciously or unconsciously, a person walking works out his path some distance ahead, so as to take the shortest path. (See Tyrus Porter, *A Study of Path Choosing Behavior*, thesis, University of California, Berkeley, 1964).If the entrance is not visible when the building itself becomes visible, he cannot

1. 位置

走路的人总要自觉或不自觉地选择面前的道路，以便能抄近道。(参阅 Tyrus Porter, *A Study of Path Choosing Behavior*, thesis, University of California, Berkeley, 1964) 如果只见楼房不见入口，他不知该如何走。为了走对路，他一定要一见到楼房就能很快找到入口。

为了其他的理由，你也必须首先找到入口。如果你不得不绕着楼房转一大圈才能进去，很可能你走进去之后还得往回走，沿原路返回。这不仅让人感到心烦，甚至可能会怀疑自己的路是不是走对了，是不是从正确的入口进去的。这个情况很难用数字来确定；但我们提出，每隔大约50ft，应有一入口。绕行50ft，没有人会对此感到心烦；如果比这远得多，那就要烦人了。

因此，安排入口位置的第一步是考虑靠近该基地的交通干线。把入口的位置设在人们一看到建筑物就能看到入口的地方；沿着建筑物走到入口的路程不应超过50ft。

入口位置
Entrance position

2. 形状

朝建筑物走来的人需要清楚地看到入口。但许多走近建筑物的人是沿着建筑物的正面跟建筑物平行行走的。他们走近建筑物的角度是锐角。从这个角度来看，许多入口是看不清楚的。沿锐角走近建筑物能看到入口的条件是：

work out his path.To be able to work out his path,he must be able to see the entrance early,as soon as he sees the building.

And for other reasons too,the entrance needs to be the first thing that you come to.If you have to walk a long distance along the building before you can enter,the chances are high that you will have to turn back after entering,and walk back in the direction you came from.This is not only annoying,but you may even begin to wonder whether you are going the right way and whether you haven't perhaps even missed the proper entrance.It is hard to pin this down numerically,but we suggest a threshold of some 50 feet.No one is bothered by a detour of 50 feet;if it gets much longer,it begins to be annoying.

Therefore,the first step in placing the entrances is to consider the main lines of approach to the site.Locate entrances so that,once the building(s)come into view,the entrance,too,comes into view;and the path toward the entrance is not more than 50 feet along the building.

2. *Shape*

A person approaching a building needs to see the entrance clearly.Yet many of the people approaching the building are walking along the front of the building and parallel to it.Their angle of approach is acute.From this angle,many entrances are hardly visible.An entrance will be visible from an acute angled approach if:

a.The entrance sticks out beyond the building line.

b.The building is higher around the entrance,and this height is visible along the approach.

a. 入口突出于建筑物边线之外。

b. 建筑物在入口处高于别处，而这一高度在朝建筑物走来的路上是看得见的。

当然，入口处的相对色彩、入口处周围的明暗对比以及有无线脚和装饰，这些也都会起作用的。但最重要的是，入口处要同其周围环境有显著区别。

因此：

将建筑物主入口的位置规定在从接近入口的主要街道能立即看到的地方，并使之具有突出在建筑物正面的显而易见的形状。

入口形状
Entrance shape

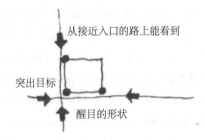

从接近入口的路上能看到

突出目标

醒目的形状

❧◉◈◈

可能的话使一组入口彼此相似，以便它们在街上或建筑群里尽可能醒目地突出出来——**各种入口**（102）；在入口处突出的那一部分营造出一个空间，使之有足够的面积成为一个舒适、明亮和美观的场所——**入口空间**（130），使街道和该入口空间之间的小路经过一系列光线、高度和风景的转换——**入口的过渡空间**（112）。务必使入口同停车场有适当的联系——**有遮挡的停车场**（97）、**与车位的联系**（113）……

And of course,the relative color of the entrance,the light and shade immediately around it,the presence of mouldings and ornaments,may all play a part too.But above all,it is important that the entrance be strongly differentiated from its immediate surroundings.

Therefore:

Place the main entrance of the building at a point where it can be seen immediately from the main avenues of approach and give it a bold,visible shape which stands out in front of the buildng.

❧⋙⋘☙

If possible,make the entrance one of a family of similar entrances,so that they all stand out as visibly as possible within the street or building complex—FAMILY OF ENTRANCES(102);build that part of the entrance which sticks out,as a room,large enough to be a pleasant,light,and beautiful place—ENTRANCE ROOM(130)and bring the path between the street and this entrance room through a series of transitions of light and level and view—ENTRANCE TRANSITION(112).Make sure that the entrance has the proper relationship to parking—SHIELDED PARKING(97),CAR CONNECTION(113)...

模式111　半隐蔽花园*

……本模式有助于形成**住宅团组**（37）、**联排式住宅**（38）、**工作社区**（41）、**自己的家**（79）和**建筑群体**（95）的基本布局，因为它影响建筑物及其花园的相对位置。由于它影响建筑物的位置以及花园的形状和位置，它也可用以造成**朝南的户外空间**（105），并有助于全面进行**基地修整**（104）。

❧❧❧

如果花园离街道太近，人们不愿使用它，因为它不够私密。但如果花园离街道太远，它也不会被使用，因为它太闭塞了。

首先考虑一下你熟悉的前花园。它们通常是装饰性的，人们在这里植草种花。可是人们常去那儿坐坐吗？除了当人们专门想要看看街道那些特别的时刻，前花园只不过用来装点门面。半私密性的家庭之间的聚会、朋友们的宴饮、跟孩子玩球、在草地上躺着休息——这些活动所需要的保护比起典型的前花园所能提供的要更多一些。

而后花园其实也不能解决这个问题。那些完全闭塞的、完全"在后面"的后花园如此远离街道，人们在那儿也不会感到舒服。后花园常常离街道很远，你听不到人们到屋子里来的声音；你看不到任何更大、更开阔的空间，你感觉不到他人的存在——唯有这个封闭的、孤立的、围着篱笆的一家人的天地。儿童们的情绪表达则会自然而然和直觉得多，他们会使我们看清问题。他们很少在纯粹的后花园里游戏；他们更多地喜欢位于房屋侧面的院子和花园，这些院子或花园具有几分私密，又跟街道有些相通。

111   HALF-HIDDEN GARDEN*

...this pattern helps to form the fundamental layout
of HOUSE CLUSTERS(37),ROW HOUSES(38),WORK
COMMUNITY(41),YOUR OWN HOME(79),and BUILDING
COMPLEX(95),because it influences the relative position of the
buildings and their gardens.Since it affects the position of the
buildings,and the shape and position of the gardens,it can also
be used to help create SOUTH FACING OUTDOORS(105)and
to help the general process of SITE REPAIR(104).

❧❧❧

**If a garden is too close to the street, people won't use it
because it isn't private enough. But if it is too far from the
street, then it won't be used either,because it is too isolated.**

Start by thinking about the front gardens which you know.
They are often decorative,lawns,flowers.But how often are
people sitting there? Except at those special moments,when
people want specifically to be watching the street,the front
garden is nothing but a decoration.The half-private family
groups,drinks with friends,playing ball with the children,lying
in the grass—these need more protection than the typical front
garden can create.

And the back gardens do not really solve the problem either.
Those back gardens which are entirely isolated, entirely "in
back" —are so remote from the street, that people often don't
feel comfortable there either. Often the back garden is so
remote from the street,that you can't hear people coming to
the house;you have no sense of any larger,more open space,no

那么，看起来适合作花园的地方既非在屋前，也非完全在屋后。花园需要有一定程度的僻静，但也要有跟街道和门口有某种细微的联系。这个平衡只能在这样的情况下达到，即花园的位置半前半后——简而言之，在边上，隔以围墙，以免过于与街道相通；但又要通过小径、大门、拱廊、棚架而充分开敞，使人在园中仍能瞥见街头，看得见大门或通往大门的路径。

所有这些都要求创新通常关于"地基"的观念。地基通常沿街很窄，而且很深。但要造成半隐蔽花园，地基沿街部分必须长，而且要浅，这样，每幢住宅的边旁都可以有花园。在这里我们列出一些住宅和半隐蔽花园的范例。

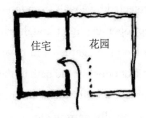

半隐蔽花园的范例
Archetype of a half-hidden garden

实现这种想法的办法很多。其中有一种特别令人感兴趣，那是我们在一幢充当我们办公楼的老住宅中所做的实验。

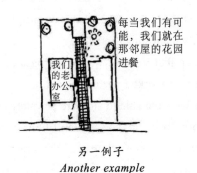

另一例子
Another example

sense of other people—only the enclosed,isolated,fenced—in world of one family.Children,so much more spontaneous and intuitive,give us a view in microcosm.How rarely they play in the full back garden;how much more often they prefer these side yards and gardens which have some privacy,yet also some exposure to the street.

It seems then,that the proper place for a garden is neither in front,nor fully behind.The garden needs a certain degree of privacy,yet also wants some kind of tenuous connection to the street and entrance.This balance can only be created in a situation where the garden is half in front,half in back—in a word,at the side,protected by a wall from too great an exposure to the street;and yet open enough,through paths,gates,arcades, trellises,so that people in the garden still have a glimpse of the street,a view of the front door or the path to the front door.

All this requires a revolution in the normal conception of a "lot." Lots are usually narrow along the street and deep. But to create half-hidden gardens,.the lots must be long along the street,and shallow,so that each house can have a garden at its side.This gives the following archetype for house and half-hidden.

There are many ways of developing this idea.One version we experienced in an old house where we once had our offices was particularly interesting.

The garden that *we* used was to the back,but behind the nextdoor house.It worked perfectly as a half-hidden garden for our house.We were able to sit there privately and have our lunch,and work on warm days,and still be in touch with the main entrance and even a glimpse of the street.But our own back garden was entirely hidden—and we never used it.

我们用过的花园在屋后，但在邻屋的后面。它对我们的房子来说，恰好起到半隐蔽花园的作用。在天气暖和的时候，我们可以安静地坐在那儿就餐或工作，但仍然跟大门口保持联系，甚至能看到大街。但我们自己的后花园是完全隐蔽的——因此我们从来不用它。

因此：

不要把花园完全置于住宅的前面，也不要使它完全在后面。而要使它处于某种中间位置，可以与房子并排，这个位置应半隐半露于街道。

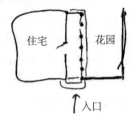

* * *

如果可能，利用本模式去影响住宅地基的形状，使它们沿街部分尽可能类似两个正方形；给花园周围砌上围墙，把通向住宅的入口设在住宅和花园之间，这样，人在花园里可以悠闲自得，但仍未与尘世隔绝，并可随时知悉有人在往这房子走来——**主入口**（110）、**花园围墙**（173）；使花园富于野趣——**花园野趣**（172），使通道经过街道和房屋之间过渡的主要地段，或在它旁边绕行——**入口的过渡空间**（112）。半隐蔽花园可以是**有生气的庭院**（115）、**屋顶花园**（118）或**私家的沿街露台**（140）……

Therefore:

Do not place the garden fully in front of the house,nor fully to the back.Instead,place it in some kind of half-way position,side-by-side with the house,in a position which is half-hidden from the street,and half-exposed.

୫୬ଔଓ

If possible,use this pattern to influence the shape of house lots too,and make them as near double squares along the street as possible;build a partial wall around the garden,and locate the entrance to the house between the house and the garden,so that people in the garden can be private,yet still aware of the street,and aware of anybody coming up to the house—MAIN ENTRANCE(110),GARDEN WALL(173);allow the garden to grow wild—GARDEN GROWING WILD (172),and make the passage through,or alongside it,a major part of the transition between street and house—ENTRANCE TRANSITION (112).Half-hidden gardens may be COURTYARDS WHICH LIVE(115),ROOF GARDENS(118),or a PRIVATE TERRACE ON THE STREET(140)...

模式112 入口的过渡空间**

112 ENTRANCE TRANSITION**

...whatever kind of building or building complex you are making,you have a rough position for its major entrances—the gateways to the site from MAIN GATEWAYS(53);the entrances to individual buildings from FAMILY OF ENTRANCES (102),MAIN ENTRANCE(110).In every case,the entrances create a transition between the "outside" —the public world—and some less public inner world.If you have HALF-HIDDEN GARDENS(111)the gardens help to intensify the beauty of the transition.This pattern now elaborates and reinforces the transition which entrances and gardens generate.

❦

Buildings,and especially houses,with a graceful transition between the street and the inside,are more tranquil than those which open directly off the street.

The experience of entering a building influences the way you feel inside the building.If the transition is too abrupt there is no feeling of arrival,and the inside of the building fails to be an inner sanctum.

The following argument may help to explain it.While people are on the street,they adopt a style of "street behavior." When they come into a house they naturally want to get rid of this street behavior and settle down completely into the more intimate spirit appropriate to a house.But it seems likely that they cannot do this unless there is a transition from one to the other which helps them to lose the street behavior.

……无论你建造哪种建筑物和建筑群，总要给主要的一些入口确定一个大致的位置——从**主门道**（53）中确定该建筑群的各个大门；从**各种入口**（102）和**主入口**（110）中确定单个建筑物的入口。在每种情况下，这些入口形成了"外界"——公共环境——和某些较少公共性质的内部环境之间的过渡。如果你有**半隐蔽花园**（111），这花园可用以美化这种过渡。本模式详细描述并促成由入口和花园而形成的过渡空间。

✺

建筑物，特别是住宅，在它们内部和街道之间如有一个优美的过渡，比起径直通向大街的房屋来要宁静得多。

你是怎样走进建筑物的会影响你走进建筑物以后的感觉。如这个过渡太过突兀，你会感到似乎还没有进屋，而屋内也不能成为内室。

突如其来的入口，没有过渡
An abrupt entrance—no transition

下面的论点可能有助于解释这一现象。当人们在街上行走时，他们是一种"街道行为"。当他们走进屋子之后，他们自然而然地会摒弃这种街道行为而完全安静下来，进入适合于住宅的比较亲切的气氛里。但除非有一个能促使他们摒弃街道行为的由此及彼的过渡，不然看起来他们有可能做不到这一点。这个过渡实际上必须在人们能够完全放松之前消除街道所特有的那种拥挤、紧张和"疏远"的因素。

The transition must,in effect,destroy the momentum of the closedness,tension and "distance" which are appropriate to street behavior,before people can relax completely.

Evidence comes from the report by Robert Weiss and Serge Bouterline,*Fairs,Exhibits,Pavilions,and their Audiences*,Cambridge,Mass.,1962.The authors noticed that many exhibits failed to "hold" people;people drifted in and then drifted out again within a very short time.However,in one exhibit people had to cross a huge,deep-pile,bright orange carpet on the way in.In this case,though the exhibit was no better than other exhibits,people stayed.The authors concluded that people were,in general,under the influence of their own "street and crowd behavior," and that while under this influence could not relax enough to make contact with the exhibits.But the bright carpet presented them with such a strong contrast as they walked in,that it broke the effect of their outside behavior,in effect "wiped them clean," with the result that they could then get absorbed in the exhibit.

Michael Christiano,while a student at the University of California,made the following experiment.He showed people photographs and drawings of house entrances with varying degrees of transition and then asked them which of these had the most "houseness." He found that the more changes and transitions a house entrance has,the more it seems to be "houselike." And the entrancewhich was judged most houselike of all is one which is approached by a long open sheltered gallery from which there is a view into the distance.

There is another argument which helps to explain the importance of the transition:people want their house,and especially the entrance,to be a private domain.If the front door

证据来自罗伯特·韦斯和塞奇·博特莱因的报告。(Robert Weiss and Serge Bouterline, *Fairs*, *Exhibits*, *Pavilions*, *and their Audiences*, Cambridge, Mass., 1962.) 作者们注意到了，许多展览馆没有能够"吸引住"观众；人们随着大人流走了进来，过不一会儿又随大人流走出去了。然而，在一个展览厅里人们走进来的时候得经过一块大的、长绒毛的、鲜橘色的地毯。在这种情况下，虽然该展厅并不比别的展厅出色，人们却停留下来了。作者认为，一般地说，人们受到他们自身"街道和人群行为"的影响，在这种影响下无法做到很放松地去参观展品。但这块鲜艳的地毯在他们走进来的时候提供了一个很强烈的对照，它把他们所受外界行为的影响打破了，实际上"把他们擦洗干净了"，结果他们就有可能为展览所吸引。

迈克尔·克里斯蒂安诺在加利福尼亚大学求学时，曾作过如下的实验。他让人们看有不同程度过渡的房屋入口的照片和图画，问他们其中哪些最有"住宅味儿"。他发现一个住宅入口的变化和过渡越多，看起来越"像住宅"。被大家公认为最像住宅的那一幢房屋的入口有一条长长的、有顶且敞开的走廊，从这条走廊可以看到远处。

还有一个论点有助于解释过渡空间的重要性：人们要使自己的住宅特别是入口具有私密的领域感。如果前门往后退一点儿，在门和街道之间有一个过渡空间，这种领域感就能很好地建立起来。这就说明，为什么人们不愿意放弃前门草坪，尽管并不"使用它"。西里尔·伯德发现，一个住宅区 90% 的居民说，他们的大约 20ft 深的前花园大小完全合适，甚至有的说太小——但仅有 15% 的人将它作为休憩处所。("Reactions to Radburn: A Study of Radburn Type Housing, in Hemel Hempstead," RIBA final thesis, 1960.)

迄今我们谈论的主要是住宅。但我们相信，本模式也适用于各式各样的入口。它当然适用于包括公寓在内的所

is set back,and there is a transition space between it and the street,this domain is well established.This would explain why people are often unwilling to go without a front lawn,even though they do not "use it." Cyril Bird found that 90 per cent of the inhabitants of a housing project said their front gardens,which were some 20 feet deep,were just right or even too small—yet only 15 per cent of them ever used the gardens as a place to sit.("Reactions to Radburn:A Study of Radburn Type Housing,in Hemel Hempstead," RIBA final thesis,1960.)

So far we have spoken mainly about houses.But we believe this pattern applies to a wide variety of entrances. It certainly applies to all dwellings including apartments— even though it is usually missing from apartments today.It also applies to those public buildings which thrive on a sense of seclusion from the world:a clinic,a jewelry store,a church,a public library.It does not apply to public buildings or any buildings which thrive on the fact of being continuous with the public world.

Here are four examples of successful entrance transitions.

As you see from these examples,it is possible to make the transition itself in many different physical ways.In some cases,for example,it may be just inside the front door—a kind of entry court,leading to another door or opening that is more definitely inside.In another case,the transition may be formed by a bend in the path that takes you through a gate and brushes past the fuchsia on the way to the door.Or again,you might create a transition by changing the texture of the path,so that you step off the sidewalk onto a gravel path and then up a step or two and under a trellis.

In all these cases,what matters most is that the transition

有住宅——尽管现今的公寓楼通常不采用这种模式。它也适用于不大抛头露面的那些公共建筑物：诊所、珠宝店、教堂、公共图书馆。它不适用于某些与公共环境相衔接的公共建筑物或任何其他建筑物。

以下是入口过渡的四个成功的例子。

每个入口都利用不同方法将各种成分组合起来以形成过渡
Each creates the transition with a different combination of elements

从这些例子中可以看到，有可能以许多不同的方法来形成过渡。例如，在一些情况下，它可以安排在前门内——一种入口的庭院，从它过渡到另外一个属于内宅的门户。在另一种情况下，建造一条通过大门的曲径和在通往门口的道路上遍植树丛以形成过渡。再则，过渡也可以用改变路径的质地来形成，这样，你可以从人行道走到卵石路上，再登上一二级台阶，然后在棚架下穿行。

在所有这些情况下，最重要的是要有过渡，使它成为存在于建筑物内外之间的实际空间，使景物、音响、光线和你行走的地面在你经过这个地方时都起了变化。这是实体的变化——最重要的是景物的变化——它在你的心中形成了心理上的过渡。

exists,as an actual physical place,between the outside and the inside, and that the view,and sounds,and light,and surface which you walk on change as you pass through this place.It is the physical changes—and above all the change of view—which creates the psychological transition in your mind.

Therefore:

Make a transition space between the street and the front door.Bring the path which connects street and entrance through this transition space,and mark it with a change of light,a change of sound,a change of direction,a change of surface,a change of level,perhaps by gateways which make a change of enclosure,and above all with a change of view.

୫୦୧୫

Emphasize the momentary view which marks the transition by a glimpse of a distant place—ZEN VIEW(134);perhaps make a gateway or a simple garden gate to mark the entrance—GARDEN WALL (173);and emphasize the change of light—TAPESTRY OF LIGHT AND DARK(135),TRELLISED WALK(174).The transition runs right up to the front door,up to the ENTRANCE ROOM(130),and marks the beginning of the INTIMACY GRADIENT(127)...

因此：

要在街道和前门之间形成一个过渡空间。使连接街道和入口的路径经过这个过渡空间，利用光线的变化、音响的变化、方向的变化、表面的变化、高度的变化，也许还要通过能变化封闭程度的大门，但最重要的是利用景物的变化，把这个空间标识出来。

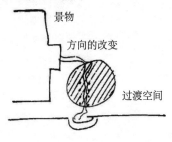

特别注意，使远望时有一个标志过渡的景物的瞬息变化——**禅宗观景**（134）；多半要造一个大门口或一个简朴的花园门以标识入口——**花园围墙**（173）；要强调光线的变化——**明暗交织**（135）、**棚下小径**（174）。过渡要一直到达前门，到达**入口空间**（130），并标志**私密性层次**（127）的开始……

模式113　与车位的联系

　　……一旦你确定了建筑物的入口并使之有明确的过渡——**主入口**（110）、**入口的过渡空间**（112）——有必要解决一个人如何坐汽车驶近建筑物的问题。当然，在步行区这是不适用的；但通常小轿车本身必须在建筑物附近有一个车房；既然如此，车房的选址和性能是至关重要的。

113   CAR CONNECTION

...once you have the entrance of the building fixed and its transition clear—MAIN ENTRANCE(110),ENTRANCE TRANSITION(112)—it is necessary to work out how a person can approach the building by car.Of course,in a pedestrian precinct this will not apply;but generally the car itself must have a housing somewhere near the building;and when this is so,its place and character are critical.

☙❧

The process of arriving in a house,and leaving it,is fundamental to our daily lives;and very often it involves a car. But the place where cars connect to houses,far from being important and beautiful,is often off to one side and neglected.

This neglect can wreck havoc with the circulation in the house,especially in those houses with the traditional "front door and back door" relationship.Both family and visitors tend,more and more,to come and go by car.Since people always try to use the door nearest the car(see Vere Hole,et al., "Studies of 800 Houses in Conventional and Radburn Layouts," Building Research Station, Garston, Herts, England, 1966), the entrance nearest the parking spot always becomes the "main" entrance,even if it was not planned that way.

If this entrance is a "back" door,then the back of the house becomes less a sanctuary for the family and perhaps the housewife feels uncomfortable about guests traipsing through. On the other hand,if this entrance is a formal "front" door,it is not really appropriate for family and good friends.In

在我们日常生活中，进出住宅的活动是基本需要；而且常常要用汽车。但汽车停靠在房屋旁边的那块地方却很不受重视也很不美观，它只是宅旁一隅，经常被人们所忽视。

这种忽视的态度势必危及住宅的进出通道，尤以维持传统的"前门和后门"关系的住宅为甚。一家人和外来访问者来来往往越来越多地依靠汽车。由于人们总想利用和汽车最靠近的那个门（参见 Vere Hole, et al., "Studies of 800 Houses in Conventional and Radburn Iayouts," Building Research Station, Garston, Herts, England, 1966）, 距离停车地点最近的入口就成了"主要"入口，即使原来不是这么规划的。

汽车入口成了主入口——不考虑原来规划
The car entrance becomes the main entrance—regardless of the plan

如果这个入口是"后门"，那么住宅的后部对于家庭来说就不是一个安静处所，对客人们的进进出出主妇会感到不方便。反过来，如果这个入口是正式的"前门"，它对家里人和好友又太不合适了。在拉德伯恩，后门面对停车场，前门面对供人们步行的草地。对于有小汽车的家庭来说，因为后门在停车的地方，人们出入总爱走后门，但来客人"理应"让他们走前门。

Radburn,the back doors face the parking lot,and the front doors face a pedestrian green.For families with cars,the back door,being on the car side,dominates exit and entry,yet visitors are "supposed" to come to the front door.

In order to ensure that both the kitchen and formal living room are conveniently located with respect to cars and that each space maintains its integrity in terms of use and privacy,there must be one and only one primary entrance into the house,and the kitchen and living room must be both directly accessible from this entrance.We do not mean that a house needs to have only one entrance.There is no reason why a house cannot have several entrances—indeed there are good reasons why it probably should have more than one.Secondary entrances,like patio and garden doors and teenager's private entrances,are very important.But they should never be placed so that they are in between the main entrance and the natural place to arrive by car—otherwise，they will compete with the main entrance and,again,confuse the way the house plan works.

Finally,it is essential to make something of the space which connects the house and the car,to make it a positive space—a space which supports the experience of coming and going.Essentially this means making a room out of the place for the car,the path from the ear door to the house,and the front door.It may be achieved with columns,low walls,the edge of the house,plants,a trellised walk,a place to sit.This is the place we call the CAR CONNECTION(113).A proper car connection is a place where people can walk together,lean,say goodbye;perhaps it is integrated with the structure and form of the house.

An ancient inn,built in the days of coach and horses,has a layout which treats the coach as a fundamental part of the

为了保证厨房和正式的起居室与停放汽车的地方之间有一个方便的相对位置，又要使它们各自的空间保持完整以利于使用和安静，住宅必须有一个也只能有一个主入口，而且厨房和起居室都必须直接与此入口相通。我们并不认为一幢住宅只需有一个入口。没有什么理由能表明一幢住宅不能有若干个入口——倒是有充分的理由表明一幢住宅或许应该有一个以上的入口。次要入口，如天井或花园小门和供青少年们专用的入口，也很重要。但这些门都不要设置在主入口与汽车开过来的必经之地之间——否则，它们就会跟主入口发生竞争，结果又会使住宅的功能变得混乱。

　　最后，务必在能够把住宅和汽车联系起来的空间造些东西，使此空间成为"正空间"——使人们来去方便的空间。这主要是说，可以利用这个空间盖一间汽车房，修一条从汽车门通往住宅和前门的走道。利用柱、矮墙、房屋边缘、植物、棚下小径、休息场地都能达到这个目的。我们将此称为**与车位的联系**（113）。一个与车位联系得合适的地方可供人并肩行走、倚靠、话别；它多半也可以与住宅的结构和外形形成一个整体。

　　一个乘坐马车和骑马的时代修建的古代客栈，是将马车当作环境的一个基本部分并把客栈和马车之间的联系看作是客栈的重要部分来进行布局的——这成为客栈的特点。飞机场、游艇码头、马厩、火车站都是这样。但由于某种原因，尽管汽车对现代化住宅里的生活方式如此重要，房子与汽车之间发生联系的那个地方却几乎从未被认真地用来当作理所应当的美观而重要的地方。

environment and makes the connection between the two a significant part of thc inn——so much so that it gives the inn its character.Airports,boathouses,stables,railway stations,all do the same.But for some reason,even though the car is so important to the way of life in a modem house,the place where car and house meet is almost never treated seriously as a beautiful and significant place in its own right.

Therefore:

Place the parking place for the car and the main entrance,in such a relation to each other,that the shortest route from the parked car into the house,both to the kitchen and to the living rooms,is always through the main entrance.Make the parking place for the car into an actual room which makes a positive and graceful place where the car stands,not just a gap in the terrain.

∞≫≪∞

Place both kitchen and main common living room just inside the main entrance—INTIMACY GRADIENT(127),COMMON AREAS AT THE HEART(129);treat the place for the car as if it were an actual outdoor room—OUTDOOR ROOM(163).If it is enclosed,build the enclosure according to STRUCTURE FOLLOWS SOCIAL SPACES(205);and make the path between this room and the front door a beautiful path,preferably the same as the one used by people who come on foot— ENTRANCE TRANSITION(112),ARCADES(119),PATHS AND GOALS(120),RAISED FLOWERS(245).If you can,put the car connection on the north face of the building—NORTH FACE(162)...

因此：

把停放汽车的地方和主要入口的相互关系作这样的安排：即从停下的汽车里出来，走到屋内厨房和起居室的最短路程总要经过主要入口。要让停车处真正成为一个有使用价值而美观的地方，这是一个车位，而不只是地上的一个缺口。

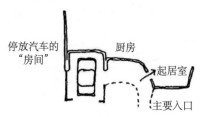

使厨房和主要的共用起居室靠近主要入口——**私密性层次**（127）、**中心公用区**（129）；把停车的地方看作真正有围合的户外小空间——**有围合的户外小空间**（163）。如果它是有围合的，根据**结构服从社会空间的需要**（205）将它加以围合；使这个空间和前门之间的路径成为优美的小路，最好和步行来的人所走的路径一样美——**入口的过渡空间**（112）、**拱廊**（119）、**小路和标志物**（120）、**高花台**（245）。如有可能，将与车位的联系安排在建筑物的北边——**背阴面**（162）……

模式114 外部空间的层次*

　　……主要户外空间的性质在**基地修整**（104）、**朝南的户外空间**（105）和**户外正空间**（106）中已经谈过。但你可以优化它们，完善它们的性能。办法是：务必做到从每一空间向外看，总能看到其他某个更大的空间，使所有的空间组合在一起形成层次。

৪০৫৪

　　在户外，人们总要想办法寻找一个使自己的背部得到保护的处所，同时能眺望到紧邻他们前方这个空间之外的某个更大的开口。

114　HIERARCHY OF OPEN SPACE*

...the main outdoor spaces are given their character by SITE REPAIR (104),SOUTH FACING OUTDOORS(105) and POSITIVE OUTDOOR SPACE (106).But you can refine them,and complete their character by making certain that every space always has a view out into some other larger one,and that all the spaces work together to form hierarchies.

※※※

Outdoors,people always try to find a spot where they can have their backs protected,looking out toward some larger opening,beyond the space immediately in front of them.

In short,people do not sit facing brick walls—they place themselves toward the view or toward whatever there is in the distance that comes nearest to a view.

Simple as this observation is,there is almost no more basic statement to make about the way people place themselves in space.And this observation has enormous implications for the spaces in which people can feel comfortable.Essentially,it means that any place where people can feel comfortable has

1. A back.

2. A view into a larger space.

In order to understand the implications of this pattern,let us look at the three major cases where it applies.

简而言之,人们不会面对砖墙坐着——他们面对的是景物或是最容易看见的远方某样物体。

这一观察结果虽然简单,但关于人们在空间中如何自我定位这一方面却没有比这更重要的说法了。而且这一观察所得对于使人们如何在户外空间感到舒适有很大的启发意义。它表明,任何使人们感到舒适的地方基本上需要具备:

1. 靠背。

2. 能看到更大的空间。

为了了解这一模式所包含的意义,让我们看一下它应用的三种主要情况。

在很小的户外空间,在私人花园里,本模式表明,要以角落作靠背,在此设置坐椅往外看着花园。如果安排得恰当,这个角落是舒适的,而绝不会给人以幽禁的感觉。

坐位和花园
Seat and garden

规模稍稍大一点,在露台或某种户外空间和更大一些的户外空间(街道和广场)之间有这种联系。本模式具备这一规模的最普通的形式是前门廊,它形成一定的封闭和靠背,与公共街道相分隔。

在规模最大时,本模式表明,使广场和草坪在一端向大的街景开敞。具备这种规模时,广场本身起某种一个人可以依靠的背景作用,以此为背景此人可以向外眺望更大的远景。

露台和街道或广场
Terrace and street or square

因此:

无论你正在营造什么样的空间——不管是花园、露台、街道、公园、公共户外空间,还是庭院,一定要做到两件事。首先,要至少营造出另一个较小的空间,使它看到该空间

In the very smallest of outdoor spaces,in private gardens, this pattern tells you to make a corner of the space as a "back" with a seat,looking out on the garden.If it is rightly made,this corner will be snug,but not at all claustrophobic.

Slightly larger in scale,there is the connection between a terrace or an outdoor room of some kind and a larger open space,the street or a square.The most common form of the pattern at this scale is the front stoop,which forms a definite enclosure and a back,off the public street.

At the largest scale,this pattern tells you to open up public squares and greens,at one end,to great vistas.At this scale,the square itself acts as a kind of back which a person can occupy,and from which he can look out upon an even larger expanse.

Therefore:

Whatever space you are shaping—whether it is a garden, terrace, street, park, public outdoor room, or courtyard, make sure of two things. First, make at least one smaller space, which looks into it and forms a natural back for it.Second,place it,and its openings,so that it looks into at least one larger space.

When you have done this,every outdoor space will have a natural"back";and every person who takes up the natural position,with his back to this"back,"will be looking out toward some larger distant view.

的内部，并构成该空间的天然背景。其次，选好该空间及其通路的位置，从该空间这里能向外看到至少一个更大的空间的内部。

完成了此项工作，每一个户外空间都会有一个天然的"背景"；每一个占据这个天然位置的人，背靠着这个"背景"，可以往外看到某个更大的远景。

广场和街景
Square and vista

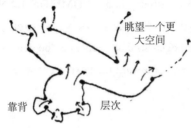

例如：园中坐椅面向花园——**园中坐椅**（176）、**半隐蔽花园**（111）；袋形活动场地面向公共广场——**袋形活动场地**（124）、**小广场**（61）；花园面向区内道路——**私家的沿街露台**（140）、**区内弯曲的道路**（49）；道路面向田野——**绿茵街道**（51）、**近宅绿地**（60）；田野面向景物开阔的乡村——**公共用地**（67）、**乡村**（7）。务必使每一层次排列分明，这样，人们在它里面向外看到邻近一个更大的空间，才会感到舒适……

For example:garden seats open to gardens—GARDEN SEAT (176),HALF-HIDDEN GARDEN(106);activity pockets open to public squares—ACTIVITY POCKETS(124),SMALL PUBLIC SQUARE(61);gardens open to local roads—PRIVATE TERRACE ON THE STREET(140),LOOPED LOCAL ROAD(49),roads open to fields—GREEN STREETS (51),ACCESSIBLE GREENS(60);fields open to the countryside,on a great vista—COMMON LAND(67),THE COUNTRYSIDE(7).Make certain that each piece of the hierarchy is arranged so that people can be comfortably settled within it,oriented out toward the next larger space...

模式115　有生气的庭院＊＊

115 COURTYARD WHICH LIVE**

...within the general scheme of outdoor spaces,made positive according to the patterns POSITIVE OUTDOOR SPACE(106)and HIERARCHY OF OPEN SPACE(114),it is necessary to pay special attention to those smallest ones,less than 30 or 40 feet across—the courtyards—because it is especially easy to make them in such a way that they do not live.

❧

The courtyards built in modern buildings are very often dead.They are intended to be private open spaces for people to use—but they end up unused,full of gravel and abstract sculptures.

There seem to be three distinct ways in which these courtyards fail.

1. *There is too little ambiguity between indoors and outdoors.*If the walls,sliding doors,doors which lead from the indoors to the outdoors,are too abrupt,then there is no opportunity for a person to find himself half way between the two—and then,on the impulse of a second,to drift toward the outside.People need an ambiguous in-between realm-a porch,or a veranda,which they naturally pass onto often,as part of their ordinary life within the house,so that they can drift naturally to the outside.

2. *There are not enough doors into the courtyard.*If there is just one door,then the courtyard never lies between two activities inside the house;and so people are never passing through it,and enlivening it,while they go about their daily

……根据**户外正空间**（106）和**外部空间的层次**（114）对户外空间进行整体规划时，必须要特别注意那些宽度小于 30 ～ 40ft 的小的空间——庭院，因为它们极易被弄得毫无生气。

∞∞

现代建筑物中的庭院往往死气沉沉。原来是打算把它们做成私密的户外空间供人利用的——但结果毫无用处，空有一堆碎石和令人不解的雕塑品。

死气沉沉的庭院
Dead courtyard

看来造成这些庭院的失败有三方面显著的原因。

1. **室内外之间的界线太分明。**如果从室内过渡到室外的墙、推拉门和其他的门都是十分突兀的，一个人就不可能使自己处于室内外之间的地方——于是，在一秒钟内，人就到了户外。人们需要一个模糊不清的、介于两种空间之间的地带——门廊或游廊，他们自然会常常去这个地方，这是他们在住宅内生活中的常事，这样他们就能很自然地步行到户外去了。

2. **通往庭院的门太少。**如果只有一扇门，这个庭院绝对无法起到沟通其内部两种活动的作用；因此人们在日常生活中进出绝不会经过这里而造成此处毫无生气。为了克

business.To overcome this,the courtyard should have doors on at least two opposite sides,so that it becomes a meeting point for different activities,provides access to them,provides overflow from them,and provides the cross-circulation between them.

3. *They are too enclosed.*Courtyards which are pleasant to be in always seem to have "loopholes" which allow you to see beyond them into some larger,further space.The courtyard should never be perfectly enclosed by the rooms which surround it,but should give at least a glimpse of some other space beyond.

Here are several examples of courtyards,large and small,from various parts of the world,which are alive.

Each one is partly open to the activity of the building that surrounds it and yet still private.A person passing through the courtyard arid children running by can all be glimpsed and felt,but they are not disruptive.Again,notice that all these courtyards have strong connections to other spaces.The photographs do not tell the whole story;but still,you can see that the courtyards look out,along paths,through the buildings,to larger spaces.And most spectacular,notice the many different positions that one can take up in each courtyard,depending on mood and climate.There are covered places,places in the sun,places spotted with filtered light,places to lie on the ground,places where a person can sleep.The edge and the corners of the courtyards are ambiguous and richly textured;in some places the walls of the buildings open,and connect the courtyard with the inside of the building,directly.

服这一点，至少在庭院的相对两边都应该有门，这样它才能成为人们各种活动的会合点，为人们提供通道，疏导人流，并使人们得以交流。

3. 它们太闭塞。使人待着舒服的庭院看来得有"观察孔"，它使你能够看到外面的某个更大更远的空间。庭院不应完全被其周围的空间包围，而至少应该能看到它外面的其他某个空间。

下面这几个庭院来自世界各地，有大有小，都充满生气。

有生气的庭院
Courtyards which live

每个庭院都有一部分是向着周围建筑物的活动开敞，但仍然是幽静的。经过庭院的人或从庭院跑过去的儿童都能让人看到或感觉得到，但他们都不破坏庭院的幽静。还请注意，所有这些庭院同其他空间都是互通的。这些照片不能说明全部情况；但你还是可以看见，从庭院向外沿着小径，通过建筑物，可以看到更大的空间。最突出的是，你会注意到，人们因心境和天气的不同在每个庭院里占据的位置也很不相同。有的地方有顶棚，有的地方照得到阳光，有的地方阳光透过绿荫斑斑点点地照射进来，有的地方人们可以躺在地上，有的地方可供人睡觉。庭院的边缘和角落都没有明显的界线，而且被装饰得丰富多采；在某些地

Therefore:

Place every courtyard in such a way that there is a view out of it to some larger open space;place it so that at least two or three doors open from the building into it and so that the natural paths which connect these doors pass across the courtyard.And,at one edge,beside a door,make a roofed veranda or a porch,which is continuous with both the inside and the courtyard.

�នន✧

Build the porch according to the patterns for ARCADE(119), GALLERY SURROUND(166), and SIX-FOOT BALCONY (167);make sure that it is in the sun—SUNNY PLACE (161);build the view out according to the HIERARCHY OF OPEN SPACE(114)and ZEN VIEW (134);make the courtyard like an OUTDOOR ROOM(163)and a GARDEN WALL(173) for more enclosure;make the height of the eaves around any courtyard of even height;if there are gable ends,hip them to make the roof edge level—ROOF LAYOUT(209);put SOMETHING ROUGHLY IN THE MIDDLE(126)...

方建筑物的墙是开敞的，将庭院同建筑物内部直接连接起来。

因此：

对每个庭院都可以这样布置：在庭院就可以看到某个更大的外部空间，使建筑物至少有两三道门向庭院开出，还要使连接这些门的自然路径穿过庭院。而在门的一边盖一有顶棚的游廊或门廊，可以使建筑物内部同庭院连接起来。

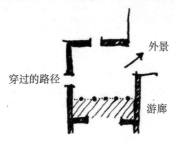

根据**拱廊**（119）、**回廊**（166）、**六英尺深的阳台**（167）等模式来建造门廊；务必使门廊能够照得着太阳——**有阳光的地方**（161）；根据**外部空间的层次**（114）和**禅宗观景**（134）来设计如何观看外部景物；把庭院盖成类似**有围合的户外小空间**（163）和**花园围墙**（173），使它有更多的围合；使庭院周围屋檐的高度保持一致；如有山墙的尖顶，把它们作成坡屋顶，使屋顶的边缘平整——**屋顶布置**（209）；使**空间中心有景物**（126）……

模式116　重叠交错的屋顶*

116 CASCADE OF ROOFS*

...this pattern helps complete the BUILDING COMPLEX(95), NUMBER OF STORIES(96),MAIN BUILDING(99),and WINGS OF LIGHT(107),and it can also be used to help create these patterns.If you are designing a building from scratch,these larger patterns have already helped you to decide how high your buildings are;and they have given you a rough layout,in wings,with an idea of what spaces there are going to be in each floor of the wings.Now we come to the stage where it is necessary to visualize the building as a volume and,therefore,above all else,as a system of roofs.

❧❧

Few buildings will be structurally and socially intact, unless the floors step down toward the ends of wings,and unless the roof,accordingly,forms a cascade.

This is a strange pattern.Several problems,from entirely different spheres,point in the same direction;but there is no obvious common bond which binds these different problems to one another—we have not succeeded in seizing the single kernel which forms the pivot of the pattern.

Let us observe,first,that many beautiful buildings have the form of a cascade:a tumbling arrangement of wings and lower wings and smaller rooms and sheds,often with a single highest center.Hagia Sophia,the Norwegian stave churches,and Palladio's villas are imposing and magnificent examples.Simple houses,small informal building complexes,and even clusters of

……本模式有助于完善**建筑群体**（95）、**楼层数**（96）、**主要建筑**（99）和**有天然采光的翼楼**（107），也可以用来帮助创造这些模式。如果你开始着手草拟一个建筑设计，这些更大的模式已经帮你决定了你的建筑物有多高；而且它们已经明示出了布置翼楼的大概情况，还提出了翼楼每一层应具有的空间。现在我们开始这样一个阶段，在这个阶段有必要把建筑看成一个体量，因而最重要的是，把它看成是一个屋顶的体系。

⁑⁂⁑

除非建筑物层层往翼楼端部降低，而且除非屋顶相应地形成重叠交错，不然，很少有建筑物在结构上和社会功能上是完整的。

这是一个怪模式。几个问题，来自完全不同的方面，却指向同一个方向，但没有显而易见的共同点能把这些不同的问题彼此结合起来——我们未能抓住一个构成本模式枢轴的中心。

首先，让我们观察一下。许多漂亮的房屋都有重叠交错的形式：从翼楼、较低的翼楼、更小的空间和小储藏室层层下降，通常只有一个位置最高的中心。圣索菲亚教堂、挪威的鳞板教堂和帕拉第奥的别墅都属于庄严而宏伟的建筑之列。简易的住宅、小的非正式的建筑群，以至茅屋群，则比较朴素。

圣索菲亚教堂
Hagia sophia

mud huts are more modest ones.

What is it that makes the cascading character of these buildings so sound and so appropriate?

First of all,there is a social meaning in this form.The largest gathering places with the highest ceilings are in the middle because they are the social centers of activities;smaller groups of people,individual rooms,and alcoves fall naturally around the edges.

Second,there is a structural meaning in the form.Buildings tend to be of materials that are strong in compression; compressive strength is cheaper than tensile strength or strength in bending.Any building which stands in pure compression will tend toward the overall outline of an inverted catenary—ROOF LAYOUT(209).When a building does take this form,each outlying space acts to buttress the higher spaces.The building is stable in just the same way that a pile of earth,which has assumed the line of least resistance,is also stable.

And third,there is a practical consideration.We shall explain that ROOF GARDENS(118),wherever they occur,should not be over the top floor,but always on the same level as the rooms they serve.This means,naturally,that the building tends to get lower toward the edges since the roof gardens step down from the top toward the outer edge of the ground floor.

Why do these three apparently different problems lead to the same pattern?We don't know.But we suspect that there is some deeper essence behind the apparent coincidence.We leave the pattern intact in the hope that someone else will understand its meaning.

Finally,a note on the application of the pattern.One must

是什么使得这些建筑物的重叠交错如此合情合理和恰到好处呢？

首先，这种形式带有社会意义。天花最高的最大集会场所位于当中，因为它们是交往活动的中心；人数较少的团体、单个的房间、凹室，很自然地坐落于四周。

其次，这一形式有着结构上的意义。建筑用的材料要求抗压强度高；具有抗压强度的材料比抗张强度或抗弯强度的材料便宜。凡以纯抗压结构形成的建筑其整体外形必趋向于一条倒过来的垂链线——**屋顶布置**（209）。当一幢建筑物真正采用这种形式时，每一远离中心的空间都起着支撑较高空间的作用。这样的建筑物是稳定的，与外形阻力最小的土堆其结构也是最稳的是相同的原理。

再次，还有实用方面的考虑。我们将说明，**屋顶花园**（118）不管设在什么地方，都不应该高于最高一层，而始终应该与花园中的房间处于同一水平。这自然意味着，建筑物的趋势是往边缘降低的，因为屋顶花园从高处向第一层楼的外部边缘渐趋下降的。

为什么这三个显然不同的问题会归于同一模式呢？原因我们不清楚。但我们推测，在这明显的巧合背后会有着某种更深刻的本质。我们和盘托出这一模式，是希望会有其他人理解它的意义。

最后，要对本模式的使用作些注解。在设计大型建筑物时，人们应该注意，重叠交错的设计与**有天然采光的翼楼**（107）并不互相矛盾。如果把重叠交错设

弗兰克·赖特的草图
A sketch of Frank Lloyd wright's

计为金字塔形，同时建筑物又很大，建筑物的中心部分就进不来阳光。反之，将重叠交错与有天然采光的翼楼适当

take care,in laying out large buildings,to make the cascade compatible with WINGS OF LIGHT(107).If you conceive of the cascade as pyramidal and the building is large,the middle section of the building will be cut off from daylight.Instead,the proper synthesis of cascades and wings of light will generate a building that tumbles down along relatively narrow wings,the wings turning corners and becoming lower where they will.

Therefore:

Visualize the whole building,or building complex, as a system of roofs.

Place the largest,highest,and widest roofs over those parts of the building which are most significant:when you come to lay the roofs out in detail,you will be able to make all lesser roofs cascade off these large roofs and form a stable self-buttressing system,which is congruent with the hierarchy of social spaes underneath the roofs.

೮೦Ӂ೦ಟ

Make the roofs a combination of steeply pitched or domed, and flat shapes—SHELTERING ROOF(117), ROOF GARDEN(118).Prepare to place small rooms at the outside and ends of wings,and large rooms in the middle—CEILING HEIGHT VARIETY(190).Later,once the plan of the building is more exactly defined,you can lay out the roofs exactly to fit the cascade to individual rooms;and at that stage the cascade will begin to have a structural effect of great importance—STRUCTURE FOLLOWS SOCIAL SPACES(205),ROOF LAYOUT(209)...

地结合起来，生成的建筑物会沿着比较狭窄的翼楼产生跌落效果，这些翼楼在拐角处会降低高度。

因此：

把整个建筑物或建筑群看作是屋顶的系统。把最大、最高和最宽的屋顶盖在建筑物最重要的部分：

社会整体　　相应的屋顶

重叠交错

中心最高

开始具体布置屋顶的时候，你就能够把所有较小的屋顶与这些大的屋顶进行重叠交错设计而形成一个稳定的自持系统，这个系统同屋顶下的社会空间的层次相一致。

              ഇൽയ

将高耸的坡屋顶或圆形屋顶与平屋顶结合起来——**带阁楼的坡屋顶**（117）、**屋顶花园**（118）。在翼楼外面和各端准备盖小房间，在中心盖大房间——**天花板高度变化**（190）。然后，当建筑物的平面图更精确地被制定出来时，可以布置屋顶使之恰好重叠交错于各个空间；而在此阶段重叠交错会开始具有十分重要的结构上的效果——**结构服从社会空间的需要**（205）、**屋顶布置**（209）……

模式117 带阁楼的坡屋顶**

……在有天然采光的翼楼（107）之上，在**重叠交错的屋顶**（116）之内，重叠交错的屋顶的某些部分是水平的，某些则呈陡坡状或筒拱状。本模式讨论陡坡状或筒拱状屋顶的特征；下一个模式将讨论那些水平屋顶的性质。

୧୦୪ଓ

屋顶在我们的生活中起着十分重要的作用。最原始的建筑物只不过是个屋顶而已。如果屋顶隐而不露，如果人们在建筑物周围感觉不到屋顶的存在，或者不能利用它，那么他们就会缺少基本的安居感。

117   SHELTERING ROOF**

...over the WINGS OF LIGHT(107),within the overall
CASCADE OF ROOFS(116),some parts of the cascade are flat
and some are steeply pitched or vaulted.This pattern gives the
character of those parts which are steeply pitched or vaulted;the
next one gives the character of those which must be flat.

❧✿❧

**The roof plays a primal role in our lives.The most primitive
buildings are nothing but a roof.If the roof is hidden,if its
presence cannot be felt around the building,or if it cannot be
used,then people will lack a fundamental sense of shelter.**

This sheltering function cannot he created by a pitched
roof,or large roof,which is merely added to the top of an
existing structure.The roof itself only shelters if it contains,em
braces,covers,surrounds the process of living.This means very
simply,that the roof must not only be large and visible,but it
must also include living quarters *within* its volume,not only
underneath it.

Compare the following examples.They show clearly how
different roofs are,when they have living quarters within them
and when they don't.

The difference between these two houses comes largely from
the fact that in one the roof is an integral part of the volume of
the building,while in the other it is no more than a cap that has
been set down on top of the building.In the first case,where the
building conveys an enormous sense of shelter,it is impossible

如果一个坡屋顶（大屋顶）只是凭空被添加到现有结构的顶部，它就起不到庇护作用。只有当屋顶能包容、拢合、遮盖、围挡住生活的纷繁过程时，它才具有庇护功能。不言而喻，屋顶不仅应该大且醒目，而且还必须在自己的体量内部而不仅在其下部有住人空间。

　　试比较下面两个实例。它们清楚地表明两种屋顶是多么不同：一种是屋顶内部有居室的，另一种是屋顶内部没有居室的。

一个屋顶内部是可以住人的，另一个屋顶是另加上去的
One roof lived in, the other stuck on

　　上列两幢房屋之所以不同，多半是由于其中的一幢房屋的屋顶是该建筑体量的不可或缺的组成部分，而另一幢房屋的屋顶只不过是戴在它上面的一顶帽子而已。在第一种情况下，建筑物给人一种极其强烈的庇护感，人们不可能在其正面画一条水平线将它的屋顶与住人的部分分隔开来。但在第二种情况下，屋顶分离得如此明显，以至这样一条线就不画自明了。

　　我们认为，屋顶的几何形状和它们提供心理庇护的能力之间的这种联系是可以加以验证的：首先，有证据表明，无论是儿童或是成年人都十分自然地倾向于带阁楼的坡屋顶，仿佛这种屋顶具有典范性质。例如阿莫斯·拉波波特对此就作过如下论述：

　　"……'屋顶'是家的象征，正如俗语所云：'人人头上有屋顶'。屋顶的重要性已为若干研究报告所强调。其中一份研究报告颇为强调房屋的形象，即象征。它认为坡屋顶是庇护的象征，而平屋顶则不然，因此就象征性而言，它是难以被接受的。关于这

to draw a horizontal line across the facade of the building and separate the roof from the inhabited parts of the building.But in the second case,the roof is so separate and distinct a thing,that such a line almost draws itself.

We believe that this connection between the geometry of roofs,and their capacity to provide psychological shelter,can be put on empirical grounds:first,there is a kind of evidence which shows that both children and adults naturally incline toward the sheltering roofs,almost as if they had archetypal properties.For example,here is Amos Rapoport on the subject:

... "roof" is a symbol of home,as in the phrase "a roof over one's head," and its importance has been stressed in a number of studies.In one study,the importance of images—i.e., symbol—sfor house form is stressed,and the pitched roof is said to be symbolic of shelter while the flat roof is not,and is therefore unacceptable on symbolic grounds.Another study of this subject shows the importance of these aspects in the choice of house form in England,and also shows that the pitched,tile roof is a symbol of security.It is considered,and even shown in a building—society advertisement,as an umbrella,and the houses directly reflect this view.(Amos Rapoport,*House Form and Culture*,Englewood Cliffs,N.J.:Prentice-Hall,1969,P.134.)

George Rand has drawn a similar point from his research. Rand finds that people are extremely conservative about their images of home and shelter.Despite 50 years of the flat roofs of the "modem movement," people still find the simple pitched roof the most powerful symbol of shelter.(George Rand, "Children's Images of Houses:A Prolegomena to the Study of Why People Still Want Pitched Roofs," *Environmental Design:Research and Practice*,Proceedings of the EDRA 3/AR 8)Conference,University of California at Los Angeles,William

一题目的另一份研究报告表明，英国人在选择房屋造型时对屋顶是否带有阁楼一向很重视；该报告还表明，瓦铺的坡屋顶是安全的象征。人们把这种屋顶视为保护伞，甚至在建筑协会的广告中还加以图示，这些房屋直接反映了这种观点。"（Amos Rapoport, *House Form and Culture*, Englewood Cliffs, N.J. : Prentice-Hall, 1969，p.134.）

乔治·兰德在他的研究工作中也得出相似的论点。兰德发现，人们在对待家和房屋的形象方面是极为保守的。尽管旨在推广平屋顶的"现代运动"已有50年的历史，人们仍然认为，简单的坡屋顶是房屋最有力的象征。（George Rand, "Children's Images of Houses : A Prolegomena to the Study of Why People Still Want Pitched Roofs，"*Environmental Design : Research and Practice*, Proceedings of the EDRA3/AR 8 Conference，Universitv of California at LosAngeles，William J.Mitchell，ed.，January 1972，pp.6-9-2~6-9-10.）

法国精神病学者加勒戈里对儿童进行了如下的观察：

"在南锡，有人要求住公寓的儿童们画一幢住房。这些儿童出生在水泥板块砌成的公寓里，这些公寓多么像搭在孤零零的小山头上的纸糊房子，茕茕孑立。但儿童们却无一例外地画了一间有两扇窗的小农舍，一缕炊烟从屋顶的烟囱里袅袅升起"。（M.Gregoire, "The Child in the HighRise，"*Ekistics*，May 1971，pp.331~333.）

类似这样的证据有人认为不足为凭，说这是文化诱导的结果。但还有第二种更为明显的证据，所依据的是使屋顶的特征和庇护感之间的联系更加一目了然这一简单的事实。在下面的叙述中我们要解释，为了营造出一种庇护气氛，屋顶应该具有哪些几何特征。

1. 屋顶上下的空间必须是有用的空间，人们每天与之接触的空间。完整的庇护感来自如下的事实：屋顶在覆盖

J.Mitchell,ed.,January 1972,pp.6-9-2 to 6-9-10.)

And the French psychiatrist,Menie Gregoire,makes the following observation about children:

At Nancy the children from the apartments were asked to draw a house.These children had been born in these apartment slabs which stand up like a house of cards upon an isolated hill.Without exception they each drew a small cottage with two windows and smoke curling up from a chimney on the roof.(M.Gregoire, "The Child in the HighRise," *Ekistics*,May 1971,pp.331-333.)

Such evidence as this can perhaps be dismissed on the grounds that it is culturally induced.But there is a second kind of evidence,more obvious,which lies in the simple fact of making the connection between the features of a roof and the feeling of shelter completely clear.In the passage which follows,we explain the geometric features which a roof must have in order to create an atmosphere of shelter.

1.The space under or on the roof must be useful space,space that people come into contact with daily.The whole feeling of shelter comes from the fact that the roof surrounds people at the same time that it covers them.You can imagine this taking either of the following forms.In both cases,the rooms under the roof are actually surrounded by the roof.

2.Seen from afar,the roof of the building must be made to form a massive part of the building.When you see the building,you see the roof.This is perhaps the most dramatic feature of a strong,sheltering roof.

What constitutes the charm to the eye of the old—fashioned country harn but its immense roof—a slope of gray shingle exposed to the weather like the side of a hill,and by its amplitude suggesting a bounty that warms the heart.Many of the old farmhouses,too,were

着人们的同时也包围着人们。你可以用下列两种形状来想象这一点。在这两种情况下，屋顶下面的住人空间实际上是被屋顶围住的。

两种屋顶的剖面图
Two roof sections

2. 从远处看，房屋必须有一个体量巨大的屋顶。当人们看见房屋的同时也看见了屋顶。这也许是一个结实牢固的、带阁楼的坡屋顶的最显著的特点。

"老式的农村谷仓之所以令人赏心悦目，就是因为它有大屋顶——它那露明的灰色鱼鳞板瓦斜面，犹如小山的斜坡；它的宏大规模表现出暖人心房的博爱精神。许多旧式农舍也有异曲同工之妙，它们宽大宏伟，在景色迷蒙的远方首先映入眼帘的是这些农舍的巨大的斜屋顶。它们覆盖着家家户户，如同母鸡庇护着它的小雏，呈现出一幅幅质朴的家庭气氛的感人图景。"（John Burroughs，*Signs and Seasons*，New York 4：Houghton Mifflin，1914，p.252.）

3. 带阁楼的屋顶必须使人能触摸到它——从外部触摸到它。如果这是坡屋顶或筒拱屋顶，它的某一部分必须向地面低垂，这部分就在小路经过的地方，当你走过小路时，很自然地会去触摸它的边缘。

你能触摸到的屋檐
Roof edges you can touch

modelled on the same generous scale,and at a distance little was visible but their great sloping roofs.They covered their inmates as a hen covereth her brood,and are touching pictures of the domestic spirit in its simpler forms.(John Burroughs,*Signs and Seasons*,New York:Houghton Mifflin,1914,p.252.)

3.And a sheltering roof must be placed so that one can touch it—touch it from outside.If it is pitched or vaulted,some part of the roof must come down low to the ground,just in a place where there is a path,so that it becomes a natural thing to touch the roof edge as you pass it.

Therefore:

Slope the roof or make a vault of it,make its entire surface visible,and bring the eaves of the roof down low,as low as 6′0″or 6′6″at places like the entrance, where people pause. Build thetop story of each wing right into the roof, so that the roof does not only cover it, but actually surrounds it.

ಬಾಜಚಿ

Get the exact shape of the cross section from ROOF VAULTS (220);use the space inside the top of the sloped roof for BULK STORAGE(145);where the roof comes down low,perhaps make it continuous with an ARCADE(119)or GALLERY SURROUND(166).Build the roof flat,not sloped,only where people can get out to it to use it as a garden—ROOF GARDENS(118);where rooms are built into the roof, make windows in the roof—DORMER WINDOWS(231).If the building plan is complex,get the exact way that different sloped roofs meet from ROOF LAYOUT(209)...

因此：

把屋顶建成斜坡状或筒拱状，使人能看见它的全部外表，并使屋檐低垂，比如在入口处——人们往往要停留的地方——要下垂到6'0"或6'6"。使每一翼楼的顶层直入屋顶之内，使得屋顶不仅覆盖着它，而且实际上包围着它。

不是这样 → 这样 →

❦

从**拱式屋顶**（220）获得截面的准确形状；利用坡屋顶顶端内部的空间作为**大储藏室**（145）；在屋顶低垂的地方，可能要将它同**拱廊**（119）或**回廊**（166）连接起来。只有在人们能到屋顶上去并将它作为花园时才把屋顶建成平屋顶而不是坡屋顶——**屋顶花园**（118）；在屋顶之内有居室的地方，在其顶上要开窗户——**老虎窗**（231）。如果建筑规划非常复杂，要根据**屋顶布置**（209）取得不同的斜屋顶重叠交错的准确方法……

模式118　屋顶花园*

118　ROOF GARDEN*

...in between the sloping roofs created by SHELTERING ROOF(117),the roofs are flat where people can walk out on them.This pattern describes the best position for these roof gardens and specifies their character.If they are correctly placed,they will most often form the ends of WINGS OF LIGHT(107)at different stories and will,therefore,automatically help to complete the overall CASCADE OF ROOFS(116).

❧❧❧

A vast part of the earth's surface,in a town,consists of roofs.Couple this with the fact that the total area of a town which can be exposed to the sun is finite,and you will realize that it is natural,and indeed essential,to make roofs which take advantage of the sun and air.

However,as we know from SHELTERING ROOF(117) and ROOF VAULTS(220),the flat shape is quite unnatural for roofs from psychological,structural,and climatic points of view. It is therefore sensible to use a flat roof only where the roof will actually become a garden or an outdoor room;to make as many of these "useful" roofs as possible;but to make all other roofs,which cannot be used,the sloping,vaulted,shell—like structures specified by SHELTERING ROOF(117)and ROOF VAULT(220).

Here is a rule of thumb:if possible,make at least one small roof garden in every building,more if you are sure people will actually use them.Make the remaining roofs steep roofs.Since,as we shall see,the roof gardens which work are almost always at the same level as some indoor rooms,this means that at least

……在由**带阁楼的坡屋顶**（117）所产生的各坡屋顶之间，人们可以走出来到那儿去的屋顶是平屋顶。本模式描写这些屋顶花园的最佳位置并规定它们的性质。如果它们的位置选得合适，常常会在各层楼形成**有天然采光的翼楼**（107）的两端，因而，自然有助于全面完善**重叠交错的屋顶**（116）。

※

在城市，地面大部分为屋顶所覆盖。由此联想到，整个城市地面得到的阳光是很有限的，于是你就会认识到，建造能利用阳光和新鲜空气的屋顶是理所当然和确有必要的。

然而，我们从**带阁楼的坡屋顶**（117）和**拱式屋顶**（220）中得知，平屋顶从心理学、结构和气候的观点看都是很不合理的屋顶。因此，只有在屋顶实际上能用做花园或户外房间的地方，利用平屋顶才是合理的；要尽可能多地建造这些"有用的"屋顶，但要把所有其他不能利用的屋顶建成为**带阁楼的坡屋顶**（117）和**拱式屋顶**（220）所确定的斜坡状、拱状和壳状结构。

此处提供一个经验法则：如有可能，在每一幢建筑物里都至少建造一个小型的屋顶花园，如果你确信人们会真正利用它们，还可以多造几个。把其余的屋顶都建成坡屋顶。因为，我们将看到，能派上用场的屋顶花园几乎都是同某些内部空间处于同一高度，这就是说，至少建筑物屋顶的某些部分将始终是坡屋顶。于是，我们期望，本模式将产生一种屋顶景观，那就是几乎在每幢建筑物里都有屋顶花园和坡屋顶交织在一起。

现在我们根据平屋顶本身的条件简略地谈一下这种屋顶。平的屋顶花园总是在干燥、温暖的气候条件下流行的，在那些地方的屋顶花园可以住人。在那些处于地中海气候

some part of the building's roofs will always be steep.We shall expect,then,that this pattern will generate a roof landscape in which roof gardens and steep roofs are mixed in almost every building.

We now consider the flat roof,briefly,on its own terms.Flat roof gardens have always been prevalent in dry,warm climates,where they can be made into livable environments.In the dense parts of towns in Mediterranean climates,nearly every roof is habitable:they are full of green,private screens,with lovely views,places to cook out and eat and sleep.And even in temperate climates they are beautiful. They can be designed as rooms without ceilings,places that are protected from the wind,but open to the sky.

However,the flat roofs that have become architectural fads during the last 40 years are quite another matter.Gray gravel covered asphalt structures,these flat roofs are very rarely useful places;they are not gardens;and taken as a whole,they do not meet the psychological requirements that we have outlined in SHELTERING ROOF(117).To make the flat parts of roofs truly useful,and compatible with the need for sloping roofs,it seems necessary to build flat roof gardens off the indoor parts of the buildings.In other words,do not make them the highest part of the roof;let the highest parts of the roof slope;and make it possible to walk out to the roof garden from an interior room,without climbing special stairs.We have found that roof gardens that have this relationship are used far more intensely than those rooftops which must be reached by climbing stairs.The explanation is obvious:it is far more comfortable to walk straight out onto a roof and feel the comfort of part of the building behind and to one side of you,then it is to climb up to a place you cannot see.

Therefore:

Make parts of almost every roof system usable as roof gardens.Make these parts flat,perhaps terraced for planting,with

的城市中住房拥挤的地方，差不多每家屋顶都可以住人：这些屋顶绿叶婆娑，私密静僻，有秀丽的风光，有在户外烹调、就餐和露宿的地方。即使在气候不十分燥热的地方，屋顶花园也是很宜人的。它们可以被设计成没有顶棚的房间，能够避风，却又是露天。

然而，平屋顶成为最近40年来建筑学上的时尚却另当别论。灰色的碎石铺就的柏油结构，这些平屋顶有用的地方很少；它们不是花园；作为整体，它们不能满足我们在**带阁楼的坡屋顶**（117）中阐述过的那种心理上的要求。为了使屋顶的平顶部分真正有用，使居住者对它的需要同对坡屋顶一样，看来有必要在建筑物的室内部分以外造平坦的屋顶花园。换言之，不要把它们盖在屋顶的最高部位；让最高部位用坡屋顶；要使居住者可以从内部空间走出来到屋顶花园而又毋须爬专门的楼梯。我们发现，具有这种关系的屋顶花园比起那些要靠爬楼梯才能到达的屋顶来，使用率要高很多。这很容易解释：从室内走出来就到屋顶上，看到你的背后和一旁是房子的一部分会使你感到安心，这比你爬高到一个你看不见的地方要舒服很多。

因此：

几乎每幢楼房的一部分屋顶都可以用来做屋顶花园。把这些屋顶建成平屋顶，多半要筑坛以种植花木，并有私密空间可以坐下来休息和躺下睡觉。在各层楼都可以建造屋顶花园，而且人们都可以从楼内某个地方直接走到屋顶花园。

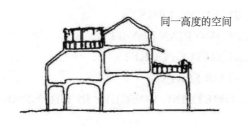

同一高度的空间

places to sit and sleep,private places.Place the roof gardens at various stories,and always make it possible to walk directly ont onto the roof garden from some lived-in part of the building.

తుల్తు

Remember to try and put the roof gardens at the open ends of WINGS OF LIGHT(107)so as not to take the daylight away from lower stories.Some roof gardens may be like balconys or galleries or terraces—PRIVATE TERRACE ON THE STREET(140),GALLERY SURROUND (166),SIX-FOOT BALCONY(167).In any case,place the roof garden so that it is sheltered from the wind—SUNNY PLACE(161),and give part of the roof some extra kind of shelter—perhaps a canvas awning—so that people can stay on the roof but keep out of the hot sun—CANVAS ROOFS(244).Treat each individual garden much the way as any other garden,with flowers,vegetables,outdoor rooms,canvas awnings,climbing plants—OUTDOOR ROOMS(163),VEGETABLE GARDEN(177),RAISED FLOWERS(245),CLIMBING PLANTS (246)...

when the major parts of buildings and the outdoor areas have been given their rough shape,it is the right time to give more detailed attention to the paths and squares between the buildings.

119. ARCADES

120. PATHS AND GOALS

121. PATH SHAPE

122. BUILDING FRONTS

123. PEDESTRIAN DENSITY

124. ACTIVITY POCKETS

125. STAIR SEATS

126. SOMETHING ROUGHLY IN THE MIDDLE

记住尽量把屋顶花园建造在**有天然采光的翼楼**（107）敞开的两端，这样才不致夺走照到底层的阳光。某些屋顶花园可以像阳台、回廊或露台——**私家的沿街露台**（140）、**回廊**（166）、**六英尺深的阳台**（167）。在任何一种情况下，都要把屋顶花园建造在背风处——**有阳光的地方**（161），还要对屋顶的一部分另外加以遮挡——多半可用帆布篷——以使人们能够待在屋顶上而又可以避开炎热的太阳——**帆布顶篷**（244）。把每一个花园都布置成同任何其他花园一样，养花种菜；造起有围合的户外小空间和搭起帆布篷；种植攀援植物——**有围合的户外小空间**（163）、**菜园**（177）、**高花台**（245）、**攀援植物**（246）……

当建筑物的主要部分和户外空间已经大体成形时，就该更仔细地注意建筑物之间的小路和场地了。

119. 拱廊

120. 小路和标志物

121. 小路的形状

122. 建筑物正面

123. 行人密度

124. 袋形活动场地

125. 能坐的台阶

126. 空间中心有景物

模式119　拱廊**

......**重叠交错的屋顶**（116）可以通过拱廊来实现。沿
着建筑物的小路，建筑物之间的短道，**步行街**（100），**鳞
次栉比的建筑**（108）间的走道以及**内部交通领域**（98）的
各部分都最好做成拱廊。这是本语言中最美的模式之一；
它和少数其他几个模式一样，全面影响建筑物的性质。

৪০০৪

　　拱廊——建筑物边缘上有遮盖的过道，部分在里面，
部分露在外面——在人和建筑物相互作用方面起着至关重
要的作用。

119　ARCADES**

...the CASCADE OF ROOFS(116)may be completed by arcades.Paths along the building,short paths between buildings,PEDESTRIAN STREET(100),paths between CONNECTED BUILDINGS(108),and parts of CIRCULATION REALMS(98)are all best as arcades.This is one of the most beautiful patterns in the language;it affects the total character of buildings as few other patterns do.

৪৩৫৪

Arcades—covered walkways at the edge of buildings, which are partly inside, partly outside—play a vital role in the way that people interact with buildings.

Buildings are often much more unfriendly than they need to be.They do not create the possibility of a connection with the public world outside.They do not genuinely invite the public in;they operate essentially as private territory for the people who are inside.

The problem lies in the fact that there are no strong connections between the territorial world within the building and the purely public world outside.There are no realms between the two kinds of spaces which are ambiguously a part of each—places that are both characteristic of the territory inside and,simultaneously,part of the public world.

The classic solution to this problem is the arcade:arcades create an ambiguous territory between the public world and the private world,and so make buildings friendly.But they need the

建筑物应该是令人感到亲切的，但事实却远非如此。它们没有提供与外部公共环境联系的可能性。它们并不真正接待公众入内；它们基本上是住在里面的人的私有领地。

问题就在于在建筑物内部这块领地与外界纯粹的公共环境之间没有强有力的联系。在这两种空间之间没有这样一种地方，它不明确属于哪个空间——这样的地方既具有领地内部的特征，同时也是公共环境的一部分。

解决这个问题的经典方法是采用拱廊：拱廊形成了公共环境和私密环境之间的界线模糊的地带，因而使建筑物变得亲切。但它们要成功，需要具备以下的性质。

1. 要使它们具有公共性，通往建筑物的公共小路本身必须成为部分在建筑物里面的地段；而且这个地段必须包含建筑物内部的性质。

如果通过建筑物和在建筑物外沿的主要走道的确是公共的，受到房屋的附加物遮挡，成为一个低的拱廊，有开口通往建筑物——许多门和窗以及半敞开的墙——那么人们就被吸引到这个建筑物里来了；这个行为表明，人们感到自己跟这个建筑物或多或少有了联系。也许他们会观望，然后走进去，问一个问题。

2. 为了把这个地方确定为与公共环境有一定距离的领域，一定要使人感到它是建筑物内部的附加物，因此应该把它遮挡住。

拱廊是造成这样一个领域的最简单的办法，而且也最美观。拱廊建造在建筑物同公共环境相连的地方；它们向公众开放，但又有部分嵌在建筑物内部，深度至少7ft。

3. 如果顶盖的边缘太高，拱廊起不了作用。要使拱廊顶盖的边缘低垂。

following properties to be successful.

1. To make them public,the public path to the building must itself become a *place* that is partly inside the building;and this place must contain the character of the inside.

If the major paths through and beside the buildings are genuinely public,covered by an extension of the building,a low arcade,with openings into the building—many doors and windows and half-open walls—then people are drawn into the building;the action is on display,they feel tangentially a part of it.Perhaps they will watch,step inside,and ask a question.

2. To establish this place as a territory which is also *apart* from the public world,it must be felt as an extension of the building interior and therefore covered.

The arcade is the most simple and beautiful way of making such a territory.Arcades run along the building,where it meets the public world;they are open to the public,yet set partly into the building and at least seven feet deep.

3. Arcades don't work if the edges of the ceiling are too high.Keep the edges of the arcade ceilings low.

4. In certain cases,the effect of the arcade can be increased if the paths open to the public pass right through the building. This is especially effective io those places where the building wings are narrow—then the passage through the building need be no more than 25 feet long.It is very beautiful if these "tunnels" connect arcades on both sides of the wing. The importance of these arcades which pass right through a building,depends on the same functional effects as those described in BUILDING THOROUGHFARE(101).

顶盖的边缘太高
The edges of the ceiling are too high

4. 在某些情况下，如果通往公共环境的走道径直穿过
建筑物，拱廊的作用可以得到加强。这在建筑物翼楼狭长
的那些地方效果特别显著——这时穿过建筑物的过道的长
度不要超过 25ft。如果这些"隧道"将翼楼两旁的拱廊连
接起来，那会是十分美观的。这些直接穿过建筑物的拱廊
所起的作用是否重要，要看实际效果，这同我们在**有顶街
道**（101）中描述过的情况相同。

穿过建筑物的拱廊
Arcades which pass through buildings

在世界上那些采用本模式的地方，有好几英里长的相
连或半相连的拱廊和有顶的过道绕过并穿过城市的公共地
带。这些有顶的空间于是就成为城市的许多非正式的贸易
场所。鲁道夫斯基认为这样的地方确实"代替了古代的市
场"。他的著作《为人群服务的街道》有许多地方论及拱廊
和拱廊空间公私难分的绝妙性质：

"我们从未曾做到把街道变成绿洲而不是沙漠。在那些街道
尚未变质为公路和停车场的国家，采取了一系列的措施使街道适
合于人的需要；那儿有藤架和篷幕（即篷幕横跨街道），帐篷式结
构物或永久性的屋顶。一切都是东方特色，或是有着东方传统的
西方国家（如西班牙）的特色。这极为精致的街道天幕，是一种
市民们团结一致的表现——也可称为博爱主义的表现——就是拱

In those parts of the world where this pattern has taken hold,there are miles of linked and half-linked arcades and covered walks passing by and through the public parts of the town.This covered space then becomes the setting for much of the informal business of the city.Indeed,Rudofsky claims that such space "takes the place of the ancient forum." A good deal of his book,*Streets for People*,is concerned with the arcade and the marvelous ambiguities of its space:

It simply never occurs to us to make streets into oases rather than deserts.In countries where their function has not yet deteriorated into highways and parking lots,a number of arrangements make streets fit for humans;pergole and awnings(that is,awnings spread across a street),tentlike structures,or permanent roofs.All are characteristic of the Orient,or countries with an oriental heritage,like Spain. The most refined street coverings,a tangible expression of civic solidarity—or,should one say,of philanthropy—are arcades.Unknown and unappreciated in our latitudes,the function of this singularly ingratiating feature goes far beyond providing shelter against the elements or protecting pedestrians from traffic hazards.Apart from lending unity to the streetscape,they often take the place of the ancient forums.Throughout Europe,North Africa,and Asia,arcades are a common sight because they also have been incorporated into "formal" architecture.Bologna's streets,to cite but one example,areaccompanied by nearly twenty miles of portici.(Bernard Rudofsky,*Streets forPeople*,**New York:Doubleday**,1969,p.13.)

Therefore:

Wherever paths run along the edge of buildings,build arcades,and use the arcades,above all,to connect up the buildings to one another,so that a person can walk from place to place under the cover of the arcades.

廊。虽然在处于我们这样纬度的人们不了解也不欣赏它那奇异的造福于人的特点，它的作用远非只是为路人提供躲避风雨的处所或保护路人使之免受交通事故的伤害。除了使街景和谐之外，它们还时常起到古代的市场的作用。遍及欧洲、北美洲和亚洲，拱廊随处可见，因为在"正式"的建筑学中已经有它们的一席之地。仅举一例，意大利博洛尼亚的街道差不多有 20miles 的回廊。"

(Bernard Rudofsky, *Streets for People*, New York : Doubleday, 1969, p.13.)

因此：

凡是小路沿着建筑物边缘经过的地方，都应建造拱廊，最重要的是，利用拱廊将该建筑物与另一建筑物连接起来，这样，人们可以在拱廊的屋顶下从一个地方走到另一个地方。

简单而美观
Simple and beautiful

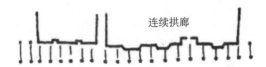

连续拱廊

☙〇☙

把拱廊造得低一点——**天花板高度变化**（190）；使拱廊的屋顶尽量地低——**带阁楼的坡屋顶**（117）；把柱做得粗一点以便供人倚靠——**柱旁空间**（226）；使柱之间的开度窄而低——**低门道**（224）、**柱的连接**（227）——或者把它们做成拱形，或者制作深梁或用花格子结构——这样，在内部才有围隔的感觉——**建筑物边缘**（160）、**半敞开墙**（193）。关于构造施工的问题请参阅**结构服从社会空间的需要**（205）和**加厚外墙**（211）……

BUILDINGS
建 筑
1127

Keep the arcade low—CEILING HEIGHT VARIETY(190); bring the roof of the arcade as low as possible—SHELTERING ROOF(117);make the columns thick enough to lean against—COLUMN PLACE(226);and make the openings between columns narrow and low—LOW DOOR-WAY(224),COLUMN CONNECTION(227)—either by arching them or by making deep beams or with lattice work—so that the inside feels enclosed—BUILDING EDGE(160),HALF-OPEN WALL(193).For construction see STRUCTURE FOLLOWS SOCIAL SPACES(205)and THICKENING THE OUTER WALLS(211)...

模式120 小路和标志物*

……当建筑物、拱廊和户外空间通过**建筑群体**（95）、**有天然采光的翼楼**（107）、**户外正空间**（106）、**拱廊**（119）大致确定了下来——就该把注意力放在建筑物之间的小路上了。本模式对这些小路的形状进行了规定，并且也有助于赋予**公共性的程度**（36）、**小路网络和汽车**（52）及**内部交通领域**（98）以更具体的形式。

120   PATHS AND GOALS*

...once buildings and arcades and open spaces have been roughly fixed by BUILDING COMPLEX(95),WINGS OF LIGHT(107),POSITIVE OUTDOOR SPACE(106), ARCADES(119)—it is time to pay attention to the paths which run between the buildings.This pattern shapes these paths and also helps to give more detailed form to DEGREES OF PUBLICNESS(36),NETWORK OF PATHS AND CARS(52),and CIRCULATION REALMS(98).

ഇൻൽ

The layout of paths will seem right and comfortable only when it is compatible with the process of walking.And the process of walking is far more subtle than one might imagine.

Essentially there arc three complementary processes:

1. As you walk along you scan the landscape for intermediate destinations—the furthest points along the path which you can see.You try,more or less,to walk in a straight line toward these points.This naturally has the effect that you will cut corners and take "diagonal" paths,since these are the ones which often form straight lines between your present position and the point which you are making for.

2. These intermediate destinations keep changing.The further you walk,the more you can see around the corner.If you always walk straight toward this furthest point and the furthest point keeps changing,you will actually move in a slow curve,like a missile tracking a moving target.

&20C3&

小路的布置只有在适宜人们步行时才会让人觉得合适和舒服。而步行的过程比我们想象的要微妙得多。

基本上有三个互相依赖的过程。

1. 在步行时,你会通过观看景物来寻找中间目的地——即你所能看到的小路上的最远点。你总想走一条直道以到达目的地。这自然会导致这样一种结果:你会抄近路走"对角线",因为这些近路常常在你现在的位置和你要去的目标之间形成直线。

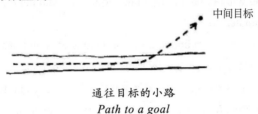

中间目标

通往目标的小路
Path to a goal

2. 这些中间目标一直在不断变换。你越往前走,在拐弯处你会看见越多。如果你总是直线地走向这个最远的点,而这个最远的点又不断变化,你实际上是在走一条缓慢的弯道,如同导弹追踪一个移动的目标。

一系列标志物
Series of goals

3. 既然你在行走时不愿不断改变方向,也不愿把全部时间花在反复考虑你走路的最佳方向上,你就要利用这样的方法来行走:选择一个临时"标志物"——某一个看得很清楚的路标——它总是在你要去的方向上,然后直线向着它行走 100yd;然后,当你走近时,选择另一个新的目标,

3. Since you do not want to keep changing direction while you walk and do not want to spend your whole time recalculating your best direction of travel,you arrange your walking process in such a way that you pick a temporary "goal" —some clearly visible landmark—which is more or less in the direction you want to take and then walk in a straight line toward it for a hundred yards,then,as you get close,pick another new goal,once more a hundred yards further on,and walk toward it...You do this so that in between,you cantalk,think,daydream,smell thespring,without having to think about your walking direction every minute.

In the diagram above a person begins at A and heads for point E.Along the way,his intermediate goals are points B,C,and D.Since he is trying to walk in a roughly straight line toward E,his intermediate goal changes from B to C,as soon as C is visible;and from C to D,as soon as D is visible.

The proper arrangements of paths is one with enough intermediate goals,to make this process workable.If there aren't enough intermediate goals,the process of walking becomes more difficult, and consumes unnecessary emotional energy.

Therefore:

To lay ont paths,first place goals at natural points of interest.Then connect the goals to one another to form the paths.The paths may be straight,or gently curving between goals;their paving should swell around the goal.The goals should never be more than a few hundred feet apart.

再往前 100yd，向着这个目标走去⋯⋯这样做，就能使你在行走中间可以谈话、思考、遐想、呼吸春天的空气，而无须时时想着你的行走方向。

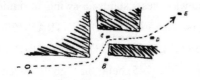

实际的路径
The actual path

在上图中，一个人从 A 出发往 E 走。在一路上，他的中间目标有 B、C 和 D。由于他想大致以直线往 E 走，一旦看见 C，他的中间目标就从 B 变为 C；当看见 D 的时候，目标又从 C 变为 D 了。

把小路安排恰当，在该小路上设置足够的中间标志物，以便这种行走方法可行。如果没有足够的中间标志物，行走过程变得比较困难，还要空劳神思。

因此：

布置好小路，首先在引人注目的各个自然点上把标志物设置好。然后把这些标志物相互连接起来形成小路。这些小路可以是直的，或者在各标志物之间略有弯曲；铺路的时候应使标志物周围稍微隆起。这些标志物彼此间相距不应超过几百英尺。

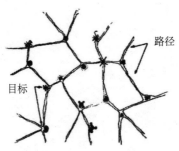

路径

目标

All the ordinary things in the outdoors—trees,fountains, entrances,gateways,seats,statues,a swing,an outdoor room— can be the goals.See FAMILY OF ENTRANCES(102),MAIN ENTRANCE (110),TREE PLACES(171),SEAT SPOTS(241), RAISED FLOWERS(245);build the "goals" according to the rules of SOMETHING ROUGHLY IN THE MIDDLE(126); and shape the paths according to PATH SHAPE (121).To pave the paths use PAVING WITH CRACKS BETWEEN THE STONES(247)...

　　户外一切平常的东西——树木、喷泉、入口、大门、坐椅、塑像、一架秋千、一个有围合的户外小空间——都可以设为标志物。参阅**各种入口**（102）、**主入口**（110）、**树荫空间**（171）、**户外设座位置**（241）、**高花台**（245）；根据**空间中心有景物**（126）的规则建立起标志物；根据**小路的形状**（121）确定小路的形状。用**留缝的石铺地**（247）来铺设小路……

模式121 小路的形状*

121 PATH SHAPE*

...paths of various kinds have been defined by larger patterns—PROMENADE(31),SHOPPING STREET(32), NETWORK OF PATHS AND CARS(52), RAISED WALK(55),PEDESTRIAN STREET(100),and PATHS AND GOALS(120).This pattern defines their shape:and it can also help to generate these larger patterns piecemeal,through the very procss of shaping parts of the path.

❧☙

Streets should be for staying in,and not just for moving through,the way they are today.

For centuries,the street provided city dwellers with usable public space right outside their houses.Now,in a number of subtle ways,the modern city has made streets which are for "going through," not for "staying in." This is reinforced by regulations which make it a crime to loiter,by the greater attractions in side the side itself,and by streets which are so unattractive to stay in,that they almost force people into their houses.

From an environmental standpoint,the essence of the problem is this:streets are "centrifugal" not "centripetal" : they drive people out instead of attracting them in.In order to combat this effect,the pedestrian world outside houses must be made into the kind of place where you stay,rather than the kind of place you move through.It must,in short,be made like a kind of outside public room,with a greater sense of enclosure than a street.

This can be accomplished if we make residential pedestrian

……一些较大的模式——**散步场所**（31）、**商业街**（32）、**小路网络和汽车**（52）、**高出路面的便道**（55）、**步行街**（100）及**小路和标志物**（120）已把各种小路确定下来了。本模式规定这些小路的形状：小路各部分形成的过程，也有助于逐个地产生这些较大的模式。

✃✄

街道应该能够让人停留，而不单像现在那样仅供人匆匆而过。

几百年来，街道给城市居民提供其门外的可利用的公共场所。现在，现代化城市已经以许多令人不易察觉的方式把街道用于"过路"而不用于"停留"。这一现象有增无减，因为有徘徊街头作为犯罪的规定，也因为室内的吸引力更大，还因为待在街上毫无兴味，这些情况几乎迫使人们躲进自己的屋子里。

从环境观点看，问题的实质是：街道是"离心的"，而不是"向心的"：它们把人赶跑，而不是吸引人进来。为了消除这一后果，户外环境必须成为行人爱停留的地方，而不只是路过而已。总之，必须使它成为相较街道使人有更大封闭感的一种外部公共空间。

如果我们使住宅区平面图上稍加凸出，在其边缘周围设置些坐椅和回廊，有时甚至还可用梁或棚架结构给这些街道加上顶盖，这样就能实现上述目标。

以下是本模式两个不同规模的实例。第一例，我们出示为秘鲁的 14 幢住宅设计的平面图。图中将这些房子逐步往后退而形成街道形状。结果形成一条可以利用的略呈椭圆形的街道。我们希望，该条街道

由14户住宅住构成的小路形状
The path shape formed by fourteen houses

streets subtly convex in plan with seats and galleries around the edges,and even sometimes roof the streets with beams or trellis-work.

Here are two examples of this pattern,at two different scales.First,we show a plan of ours for fourteen houses in Peru. The street shape is created by gradually stepping back the houses,in plan.The result is a street with a positive,somewhat elliptical shape.We hope it is a place that will encourage people to slow down and spend time there.

The second example is a very small path,cutting through a neighborhood in the hills of Berkeley.Again,the shape swells out subtly,just in those places where it is good to pause and sit.

Therefore:

Make a bulge in the middle of a public path,and make the ends narrower,so that the path forms an enclosure which is a place o stay, not just a place to pass through.

❧❦

Above all,to create the shape of the path,move the building fronts into the right positions,and on no acount allow a set-back between the building and the path— BUILDING FRONTS(122);decide on the appropriate area for the "bulge" by using the arithmetic of PEDESTRIAN DENSITY(123);then form the details of the bulge with ARCADES(119),ACTIVITY POCKETS(124)and STAIR SEATS(125);perhaps even with a PUBLIC OUTDOOR ROOM(69);and give as much life as you can to the path all along its length with windows—STREET WINDOWS(164)...

会成为一个能够让人们在
这里悠闲漫步的地方。

第二例是一条经过伯
克利山岗上一处社区的小
路。它的形状也是有点中
间凸出的，这些凸起恰好
位于那些适合于停留小憩
的地方。

因此：

在伯克利山岗上一条小路中的一
个地方
*A spot along a path in the hills of
Berkeley*

在公共小路的中部，
营造出一个向外凸出的地方，使其两端较窄，这样，这条小
路就形成一种围合，人们可以在此停留，而不仅仅只是经过。

中部鼓胀

两端狭窄

❧❧

最重要的是，为了营造出小路的形状，要把建筑物的
正面移动到合适的地方，而绝不允许在建筑物和小路之间
有退进——**建筑物正面**（122）；用**行人密度**（123）的计算
法选择合适的"凸起区"；然后形成具有**拱廊**（119）、**袋形活
动场地**（124）和**能坐的台阶**（125）的凸起区各点；凸起区
可能还有**户外亭榭**（69）；并利用窗户——**临街窗户**（164）
使整条小路充满生气……

模式122 建筑物正面*

……本模式有助于同时形成小路和建筑物；进而完善**建筑群体**（95）、**有天然采光的翼楼**（107）、**户外正空间**（106）、**拱廊**（119）、**小路的形状**（121）以及**袋形活动场地**（124）。

❧❦

建筑物从街道两边后退，本意是想使每幢建筑物有充沛的阳光和新鲜的空气，以保护公共利益，实际上却是毁坏了作为社会空间的街道。

在**户外正空间**（106）中，我们描述了这样的事实，即建筑物不仅被设置到户外空间中，而且它们实际上形成了户外空间。由于街道和广场对社会是极其重要的，当然必须十分重视它们是如何为建筑物正面所形成的。

20世纪初期提倡不惜代价进行"净化"，进行全民性的动员以清除贫民窟，使热心社会改革的人通过法律确定，必须使建筑物从街道两边后退几英尺，以保证建筑物不拥塞街道，遮挡阳光和妨碍空气流通。

但是，建筑物的后退毁坏了街道。因为建筑物内和街道上充足的空气和阳光有可能利用别的方法来保证——例如，请参阅**不高于四层楼**（21）和**有天然采光的翼楼**（107）——必须使建筑物正面朝向街道，这样才使建筑物形成的街道可以利用。

最后，请注意，仅仅使各建筑物正面相互错开并不能使街道的形状有用。如果要使建筑物正面适应户外空间的形状，它们几乎都会稍稍偏离直角。

122   BUILDING FRONTS*

...this pattern helps to shape the paths and buildings simultaneously;and so completes BUILDING COMPLEX(95), WINGS OF LIGHT(107),POSITIVE OUTDOOR SPACE(106),ARCADES(119),PATH SHAPE(121),and also ACTIVITY POCKETS(124).

&O&

Building set-backs from the street,originally invented to protect the public welfare by giving every building light and air,have actually helped greatly to destroy the street as a social space.

In POSITIVE OUTDOOR SPACE(106)we have described the fact that buildings are not merely placed into the outdoors,but that they actually shape the outdoors.Since streets and squares have such enormous social importance,it is natural to pay close attention to the way that they are shaped by building fronts.

The early twentieth-century urge for "cleanliness" at all costs,and the social efforts to clean up slums,led social reformers to pass laws which make it necessary to place buildings several feet back from the street edge, to make sure that buildings cannot crowd the street and cut off sunshine, light, and air.

But, the set backs have destroyed the streets.Since it is possible to guarantee plenty of air and sun in buildings and streets in other ways—see,for example,FOUR-STORY LIMIT(21)and WINGS OF LIGHT(107)—it is essential to

各建筑物正面角度稍有偏离

Slight angles in the building fronts

因此：

切不可在街道、小路或公共户外空间和面向这些地方的建筑物之间造成后退。后退是毫无价值的，差不多总要毁坏建筑物之间的户外空间。直接把房子盖在路边上；破除许多社区不能照此办理的陈规陋习。让各建筑物正面为适应街道形状可稍稍偏离直角。

没有退进

偏角

❧❧❧

根据**建筑物边缘**（160）这个模式，设计建筑物正面的细部，实际上也是整个建筑物周边。如果在建筑物正面需要有一些户外空间，可以把它做成**私家的沿街露台**（140）或**回廊**（166），使之成为街道的一部分；要使建筑物有许多通向街道的出口——**能坐的台阶**（125）、**室外楼梯**（158）、**临街窗户**（164）、**向街道的开敞**（165）、**大门外的条凳**（242）……

build the front of buildings on the street,so that the streets which they create are usable.

Finally,note that the positive shape of the street cannot be achieved by merely staggering building fronts.If the building fronts are adjusted to the shape of the outdoors,they will almost always take on a variety of slightly uneven angles.

Therefore:

On no account allow set-backs between streets or paths or public open land and the buildings which front on them.The set-backs do nothing valuable and almost always destroy the value of the open areas between the buildings.Build right up to the paths;change the laws in all communities where obsolete by-laws make this impossible.And let the building fronts take on slightly uneven angles as they accommodate to the shape of the street.

Detail the fronts of buildings,indeed the whole building perimeter,according to the pattern BUILDING EDGE(160).If some outdoor space is needed at the front of the building,make it part of the street life by making it a PRIVATE TERRACE ON THE STREET(140) or GALLERY SURROUND(166);and give the building many openings onto the street—STAIR SEATS(125),OPEN STAIRS(158),STREET WINDOWS(164),OPENING TO THE STREET(165),FRONT DOOR BENCH (242)...

模式123　行人密度*

……在各个地方都有步行区，铺成路面供人们集会或散步——**散步场所**（31）、**小广场**（61）、**步行街**（100）、**有顶街道**（101）、**小路的形状**（121）。必须严格限制这些地方的大小，特别是铺设了路面的地段的大小，以便使它们保持生气。

123   PEDESTRIAN DENSITY*

...in various places there are pedestrian areas, paved so that people will congregate there or walk up and down—PROMENADE(31),SMALL PUBLIC SQUARES(61), PEDESTRIAN STREET(100),BUILDING THOROUGHFARE(101), PATH SHAPE(121).It is essential to limit the sizes of these places very strictly,especially the size of areas which are paved,so that they stay alive.

⊰⊱

Many of our modern public squares,though intended as lively plazas,are in fact deserted and dead.

In this pattern,we call attention to the relationship between the number of people in a pedestrian area,the size of the area, and a subjective estimate of the extent to which the area is alive.

We do not say categorically that the number of people per square foot *controls* the apparent liveliness of a pedestrian area.Other factors—the nature of the land around the edge,the grouping of people,what the people are doing—obviously contribute greatly.People who are running,especially if they are making noise,add to the liveliness.A small group attracted to a couple of folk singers in a plaza give much more life to the place than the same number sunning on the grass.

However,the number of square feet per person does give a reasonably crude estimate of the liveliness of a space.Christie Coffin's observations show the following figures for various

我们许多现代的公共广场，虽然原本计划搞得生气蓬勃，但实际上却一片荒寂，死气沉沉。

在本模式中，我们提请读者注意，一个行人区的人数和大小与我们对这个区域生动活泼程度的主观估计之间有多大关系。

我们不是绝对认为，每平方英尺的人数"**决定**"一个行人区其外观生动活泼的程度。其他因素——包括这个区域边缘地方的性质、人们聚集的情况以及人们正在干什么——显然都有很大关系。人们跑步，特别是他们大声喧哗，就会增添生气。一小群人在一个广场上围着两个民歌手，比之同样多的人躺在草地上晒太阳，使这个地方产生的气氛要活跃得多。

可是，每人占有多少平方英尺确能合理地大致估计出一个空间生动活泼的程度。科芬的观察提供了旧金山市内和城郊各种公共场所的下列数字。她对这些地方的生动活泼程度的估计列于下表右边栏内。

	每人平均平方英尺数	
金门广场，中午	1000	死气沉沉
弗雷斯诺林荫散步场	100	活跃
斯普劳尔广场，白天	150	活跃
斯普劳尔广场，傍晚	2000	死气沉沉
联合广场，中央部分	600	半死气沉沉

虽然这些主观估计法显然是有问题的，它们却提供了如下的经验法则：在每人有 150ft^2 时，一个区域是生动活泼的。如每人有 500ft^2 以上时，这区域就开始死气沉沉了。

即使这些数字仅仅在某一定数量级大小范围内是正确的，我们也依然可以利用它们来确定公共行人区——广场、

public places in and around San Francisco.Her estimate of the liveliness of the places is given in the right-hand column.

	Sq.ft.per person	
Golden Gate Plaza,noon:	1000	Dead
Fresno Mall:	100	Alive
Sproul Plaza,daytime:	150	Alive
Sproul Plaza,evening:	2000	Dead
Union Square,central part:	600	Half-dead

Although these subjective estimates are clearly open to question,they suggest the following rule of thumb:At 150 square feet per person,an area is lively.If there are more than 500 square feet per person,the area begins to be dead.

Even if these figures are only correct to within an order of magnitude,we can use them to shape public pedestrian areas—squares,indoor streets,shopping streets,promenades.

To use the pattern it is essential to make a rough estimate of the number of people that are *typically* found in a given space at any moment of its use.In the front area of a market,for example,we might find that typically there are three people lingering and walking.Then we shall want the front of this market to form a little square,no larger than 450 square feet. If we estimate a pedestrian street will typically contain 35 people window shopping and walking,we shall want the street to form an enclosure of roughly 5000 square feet.(For an example of this calculation in a more complicated case—the case of a square in a public building that has yet to be built—see *A Pattern Language Which Generates Multi-Service Centers*, Alexander,Ishikawa,Silver stein,Center for Environmental Structure,1968,p.148.)

室内街道、商业街、散步场所——的形状。

为了利用这一模式，必须粗略估计一下，在任何时刻，一个空间"在典型情况下"使用的人数有多少。例如，在一个市场的前门内，我们可能发现，通常情况下有 3 个人走来走去。于是我们就要在这个市场的门前留出一块小空地，不大于450ft²。如果我们估计一条步行街通常会有 35 人观看商品橱窗和来来往往，我们就要使这条街有一个大约 5000ft² 的有围合的空间。如需这种计算法在更复杂的情况（尚未动工的公共建筑物内的广场就是这种情况）下的例子，请参阅《产生多种服务中心的一种模式语言》。（Alexander，Ishikawa，Silverstein，Center for Environmental Structure，1968，p.148.）

因此：

对于公共广场、庭院、步行街、任何人群聚集的地方，估算在任一时刻该地的平均人数（P），然后使这个地方的面积为人均 150 ～ 300ft²。

平均人数P

人均150～300ft²

❧❦❧❦

利用特别拥挤的边缘地带来增加行人密度并营造出生气蓬勃的氛围——**临街咖啡座**（88）、**袋形活动场地**（124）、**能坐的台阶**（125）、**私家的沿街露台**（140）、**建筑物边缘**（160）、**临街窗户**（164）、**向街道的开敞**（165）、**回廊**（166）……

Therefore:

For public squares,courts,pedestrian streets,any place where crowds are drawn together,estimate the mean number of people in the place at any given moment(P),and make the area of the place between 150P and 300P square feet.

༄༅

Embellish the density and feeling of life with areas at the edge which are especially crowded—STREET CAFE(88), ACTIVITY POCKETS (124), STAIR SEATS(125), PRIVATE TERRACE ON THE STREET(140), BUILDING EDGE(160), STREET WINDOWS(164),OPENING TO THE STREET(165), GALLERY SURROUND(166)...

模式124 袋形活动场地**

124   ACTIVITY POCKETS**

...in many large scale patterns which define public space,the edge is critical:PROMENADE(31),SMALL PUBLIC SQUARES (61), PUDLIC OUTDOOR ROOM(69),PEDESTRIAN STREET(100),BUILDING THOROUGHFARE(101),PATH SHAPE(121).This pattern helps complete the edge of all these larger patterns.

⊗✷⊗

The life of a public square forms naturally around its edge. If the edge fails,then the space never becomes lively.

In more detail:people gravitate naturally toward the edge of public spaces.They do not linger out in the open.If the edge does not provide them with places where it is natural to linger,the space becomes a place to walk through,not a place to stop.It is therefore clear that a public square should be surrounded by pockets of activity:shops,stands,benches, displays, rails,courts,gardens,news racks.In effect,the edge must be scalloped.

Further, the process of lingering is a gradual one;it happens; people do not make up their minds to stay;they stay or go, according to a process of gradual involvement.This means that the various pockets of activity around the edge should all be next to paths and entrances so that people pass right by them as they pass through.The goal-oriented activity of coming and going then has a chance to turn gradually into something more relaxed. And once many small groups form around the edge,it is likely that they will begin to overlap and spill in toward

……在许多规定公共空间的大的模式里，边缘是至关重要的：**散步场所**（31）、**小广场**（61）、**户外亭榭**（69）、**步行街**（100）、**有顶街道**（101）、**小路的形状**（121）。本模式有助于完善所有这些较大模式的边缘。

‽‽‽

广场所具有的生气是在它的边缘周围自然形成的。如果边缘处理得不好，这种空间就绝无生气。

具体点说：人们自然会向着公共空间的边缘走去。他们不会在开阔地停留。如果边缘不能为他们提供自然的逗留场所，这空间就变成路过的地方，而不是可停留的地方。因此显而易见，一个广场其周围应有袋形活动场所，包括商店、小摊、坐椅、展览、栏杆、庭院、花园、阅报栏。实际上，边缘必须呈扇形。

示意图
A conceptual diagram

再则，人们是逐渐停留下来的；只是偶然为之；他们并非刻意停留；他们是停是走，要看他们是否逐渐为环境所吸引。这就是说，边缘周围的各种袋形活动场地全部应靠近路边和入口处，以便人们路过的时候就从它们面前走过。于是，人们有目的的来来往往的活动有可能变得更加轻松愉快。而一旦边缘的人群越来越多，他们有可能挤在一起，于是就涌进广场的中心。因此，我们提出，每隔几个袋形活动场地应设置有可进入广场的入口。

该扇形边缘必须全部包围这个空间。我们可以把这一情况清楚地阐明如下：画一圆圈代表该空间，将周边的某一部分涂黑以表示扇形边缘。接着画出弦来连接被涂黑的周边上各点。随着涂黑边缘的长度变短，被这些弦所覆盖的空间面积急剧变小。这表明，当扇形边缘变短时，空间

the center of the square.We therefore specify that pockets of activity must alternate with access points.

The scalloped edge must surround the space entirely.We may see this clearly as follows:draw a circle to represent the space,and darken some part of its perimeter to stand for the scalloped edge.Now draw chords which join different points along this darkened perimeter.As the length of the darkened edge gets smaller,the area of the space covered by these chords wanes drastically.This shows how quickly the life in the space will drop when the length of the scalloped edge gets shorter. To make the space lively,the scalloped edge must surround the space completely.

When we say that the edge must be scalloped with activity,we mean this conceptually—not literally.In fact,to build this pattern,you must build the activity pockets forward into the square:first rough out the major paths that cross the space and the spaces left over between these paths;then build the activity pockets into these "in-between" spaces,bringing them forward,into the square.

Therefore:

Surround public gathering places with pockets of activity—small,partly enclosed areas at the edges, which jut forward into the open space between the paths, and contain activities which make it natural for people to pause and get involved.

的生气会下降多快。为了使这个空间富有生气，扇形边缘必须完全包围这个空间。

随着空间周围的活动增多，这空间就越富有生气
As the activities grow around the space, it becomes more lively

当我们说边缘的活动角落必须扇形分布时，我们指的是笼统的概念，而不是实际情况。实际上，为了形成本模式，必须把袋形活动场地深入广场里面去：首先，大致确定通过这个空

一个延伸到广场里面的活动角落
A pocket of activity which bulges into the square

间的主要走道和留在这些走道之间的空间；然后，把这些活动角落设置在这些"夹在中间"的空间，使它们往前延伸到广场里面去。

因此：

在公众聚焦的场所周围建立活动角落——小的、边缘地带部分围合的区域，这些区域向前伸入走道之间的空地，使人们自然而然地停留下来参加它们的活动。

Lead paths between the pockets of activity—PATHS AND GOALS (120)—and shape the pockets themselves with arcades and seats,and sitting walls,and columns and trellises— ARCADES(119),OUTDOOR ROOM(163),TRELLISED WALK(174),SEAT SPOTS(241),SITTING WALL(243);above all shape them with the fronts of buildings—BUILDING FRONTS(122);and include,within the pockets,newsstands— BUS STOPS(92),FOOD STANDS(93),gardens,games,small shops,STREET CAFES(88),and A PLACE TO WAIT(150)...

在活动角落之间开辟出小路——**小路和标志物**（120）——利用拱廊和坐椅及能坐的矮墙、柱和棚架——**拱廊**（119）、**有围合的户外小空间**（163）、**棚下小径**（174）、**户外设座位置**（241）、**可坐矮墙**（243）来构成这些角落；最重要的是用建筑物正面来形成它们——**建筑物正面**（122）；在袋形活动场地内还要有阅报栏——**公共汽车站**（92）、**饮食商亭**（93）、花园、娱乐场、小商店、**临街咖啡座**（88）以及一个**等候场所**（150）……

模式125　能坐的台阶*

　　……我们知道，小路和更大的公众集会场所需要有一定的形状和一定程度的围合，使人们往它们里面看，而不是从它们那里往外看——**小广场**（61）、**户外正空间**（106）、**小路的形状**（121）。边缘周围的台阶恰好起到这种作用；它们也有助于美化**各种入口**（102）、**主入口**（110）和**室外楼梯**（158）。

125   STAIR SEATS*

...we know that paths and larger public gathering places
need a definite shape and a degree of enclosure,with people
looking into them,not out of them—SMALL PUBLIC
SQUARES(61), POSITIVE OUTDOOR SPACE(106),PATH
SHAPE(121). Stairs around the edge do it just perfectly;and
they also help embellish FAMILY OF ENTRANCES
(102),MAIN ENTRANCES(110),and OPEN STAIRS(158).

୫୦୯ଷ

**Wherever there is action in a place,the spots which are the
most inviting,are those high enough to give people a vantage
point and low enough to put them in action.**

On the one hand,people seek a vantage point from which
they can take in the action as a whole.On the other hand,they
still want to be part of the action;they do not want to be mere
onlookers.Unless a public space provides for both these
tendencies,a lot of people simply will not stay there.

For a person looking at the horizon,the visual field is far
larger below the horizon than above it.It is therefore dear that
anybody who is "people-watching" will naturally try to take
up a position a few feet above the action.

The trouble is that this position will usually have the effect
of removing a person from the action.Yet most people want to
be able to take the action in and to be part of it at the same time.
This means that any places which are slightly elevated must
also be within easy reach of passers-by,hence on circulation

ᘒᘓ

凡是活动场所，最吸引人的去处是那些高处可让人居高临下、低处使人便于活动的地方。

一方面，人们寻找一个能够纵观活动场景的有利地形。另一方面，他们还想成为这个活动的一部分；他们不愿成为旁观者。一个公共空间，除非这两者兼而有之，不然许多人是不会在此停留的。

一个人水平地看东西，水平线以下的视野比水平线以上的要大得多。因此很清楚：任何想"看看人"的人自然想要占据高于活动地区几英尺的位置。

问题在于这一位置往往使人脱离活动。可是大部分人是希望能够参与这种活动并同时成为其中的一部分。这表明任何略微高出的地方同时必须很容易接近行人，因而必须位于通行的小路旁，而且从下面可以直接走上来。

底层的几级台阶、栏杆柱和台阶两旁的栏杆都恰好是能够满足这些要求的地方。如果台阶足够宽而且吸引人，人们会坐在低一点的台阶边缘上，背靠着栏杆。

有一种很简单的证明，它既证明此处描写的吸引力确有其事，也证明本模式的价值。只要有位置略高且又易接近的公共场所，人们就自然而然会为它们所吸引。台阶上面的露天咖啡座、广场周围的台阶、台阶上的围廊、塑像和座凳都是这样的所在。

因此：

在各个有人漫步的公共场所，从台阶下来或高低变换的边缘地方增设几个踏步。要让人们可以很容易地从下面上到这些高一点的地方，从而可以聚在一起，坐着观看街景。

paths,and directly accessible from below.

The bottom few steps of stairs,and the bahisters and rails along stairs,are precisely the kinds of places which resolve these tendencies.People sit on the edges of the lower steps,if they are wide enough and inviting,and they lean against the rails.

There is a simple kind of evidence,both for the reality of the forces described here and for the value of the pattern.When there are areas in public places which are both slightly raised and very accessible,people naturally gravitate toward them. Stepped cafe terraces,steps surrounding public plazas,stepped porches,stepped statues and seats,are all examples.

Therefore:

In any public place where people loiter,add a few steps at the edge where stairs come down or where there is a change of level.Make these raised areas immediately accessible from below,so that people may congregate and sit to watch the goings-on.

🕸

Give the stair seats the same orientation as SEAT SPOTS(241). Make the steps out of wood or tile or brick so that they wear with time,and show the marks of feet,and are soft to the touch for people sitting on them—SOFT TILE AND BRICK(248);and make the steps connect directly to surrounding buildings—CONNECTION TO THE EARTH(168)...

公共场所

可坐人的台阶

❦

　　将可以坐人的台阶和**户外设座位置**（241）设置在同一方向。用木头或砖来做台阶，以便它们逐渐变旧，显示出人们脚步的印记，而且坐在那里有一种柔软感——**软质面砖和软质砖**（248）；将台阶直接与周围的建筑物连通——**与大地紧密相连**（168）……

模式126　空间中心有景物

……**小广场**（61）、**公共用地**（67）、**有生气的庭院**（115）、**小路的形状**（121）都从它们边缘周围的活动中获得生气——**袋形活动场地**（124）、**能坐的台阶**（125）。但即使那样，中心部分还是空的，它需要充实。

❧❦❧

没有中心的公共场所很可能是空荡荡的。

我们已经讨论过这样的事实，即人们总爱待在自己的后背可以受到局部保护的地方——**外部空间的层次**（114），以及这一事实引起广场边缘周围的活动增多的情况——**袋形活动场地**（124）、**能坐的台阶**（125）。如果这空间很小，在边缘之外不需要任何东西。但如果中心有一个适当的区域打算供公众使用，那就要有树木、纪念碑、坐椅、喷泉——一个人们可以保护他们背后的地方，与他们在广场周围的情况一样，否则这个中心会被白白荒废。为什么要在广场的中心搞点景物，理由是清楚的也是实际的。但也许还有一种更为原始的本能在起作用。

设想你屋子里有一张空桌。请你想象本能的力量，它会告诉你在桌子当中放一支蜡烛或一盆花。再想象一下，你一旦做了这件事情，会产生多大效果。显而易见，这是很有意义的行动；但显然它跟边缘或中心的活动无关。看来效果是纯几何学的。也许就是这么一个简单的事实：先给桌子的空间一个中心，然后利用位于中心的点来组织起其周围的空间，使空间的界面明确，并使它基本上处于稳定状态。对于庭院或公共广场情况也是一样。或许这跟曼荼罗自性有关，这种自性在任何中心对称的图形上都能找到会集点以吸纳梦想、形象和自我的聚合。

126 SOMETHING ROUGHLY IN THE MIDDLE

...SMALL PUBLIC SQUARES(61),COMMON LAND (67),COURT-YARDS WHICHLIVE(115),PATH SHAPE(121) all draw their life from the activities around their edges— ACTIVITY POCKETS(124)and STAIR SEATS (125).But even then,the middle is still empty,and it needs embellishment.

ಸೋಲ

A public space without a middle is quite likely to stay empty.

We have discussed the fact that people tend to take up positions from which they are protected,partly,at their backs—HIERARCHY OF OPEN SPACE(114),and the way this fact tends to make the action grow around the edge of public squares—ACTIVITY POCKETS(124),STAIR SEATS(125).If the space is a tiny one,there is no need for anything beyond an edge.But if there is a reasonable area in the middle,intended for public use,it will be wasted unless there are trees,monuments,seats,fountains—a place where people can protect their backs,as easily as they can around the edge.This reason for setting something roughly in the middle of a square is obvious and practical.But perhaps there is an even more primitive instinct at work.

Imagine a bare table in your house.Think of the power of the instinct which tells you to put a candle or a bowl of flowers in the middle.And think of the power of the effect once you have done it. Obviously,it is an act of great significance;yet

我们认为，这种本能在每一个庭院和每一个广场都在起作用。甚至在少数几个没有明显中心的广场之一的圣马可广场，高耸的钟楼为两个广场创造了一个不规则的共同的中心。

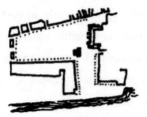

钟楼对两个广场形成一个近似的中心
The campanile forms a rough center to the two piazzas

意大利的大设计师卡米洛·西特在他的《根据艺术原理进行城市规划》（*City Planning According to Artistic Principles*，New York：Random House，1965，pp.20~31）一书中描述了这种中心点的发展及它们的功能意义。但令人感兴趣的是，他声称，把某一景物正好置于广场中心的动机是现时代的"灾难"。

"设想农村的一个小型集市广场，它为冰雪覆盖，车辆和人流在这里开出的大路小径形成天然的交通线路。在这些通道之间留出了人车罕至的几片分布不规则的空地……

恰恰就在这些不受车水马龙干扰的地方，树起了古老社区的喷泉和纪念碑……"

因此：

在跨越一个公共广场、院落或公共土地的自然形成的路径之间，在大致居中处置一景物：一个喷泉、一棵树、一尊塑像、一幢带坐位的钟楼、一台风车、一个音乐台。要使该景物给予广场以强烈和稳定的脉搏，吸引人们走向它的中心。这一景物原来在小路之间的哪个地方，就让它留在那里；不要想方设法把它移到正中。

clearly it has nothing to do with activities at the edge or in the center.Apparently the effect is purely geometrical.Perhaps it is the sheer fact that the space of the table is given a center,and the point at the center then organizes the space around it,and makes it clear,and puts it roughly at rest.The same thing happens in a courtyard or a public square.It is perhaps related to the mandala instinct,which finds in any centrally symmetric figure a powerful receptacle for dreams and images and for conjugations of the self.

We believe that this instinct is at work in every courtyard and every square.Even in the Piazza San Marco,one of the few squares without an obvious center piece,the campanile juts out and creates an off beat center to the two plazas together.

Camillo Sitte,the great Italian planner,describes the evolution of such focal points and their functional significance in his book *City Planning According to Artistic Principles*(New York:Random House,1965,pp.20-31).But interestingly,he claims that the impulse to center something *perfectly* in a square is an "affliction" of modern times.

Imagine the open square of a small market town in the country,covered with deep snow and criss—crossed by several roads and paths that,shaped by the traffic,form the natural lines of communication.Between them are left irregularly distributed patches untouched by traffic...

On exactly such spots,undisturbed by the flow of vehicles,rose the fountains and monuments of old communities...

Therefore:

Between the natural paths which cross a public square or courtyard or a piece of common land choose something to stand roughly in the middle:a fountain,a tree,a

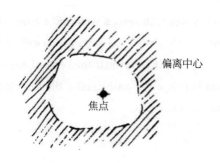

偏离中心

焦点

⁂

　　用小路形成的网络把不同的"景物"相互连接起来——**小路和标志物**（120）。它们可能包括：**眺远高地**（62）、**街头舞会**（63）、**水池和小溪**（64）、**户外亭榭**（69）、**池塘**（71）、**树荫空间**（171）；务必使每一处都用**可坐矮墙**（243）围起来……

　　现在，走道已经布置好了，我们再回到建筑物上来：在任何一个建筑物的各翼楼之内，确定空间的基本层次，并决定，人们的活动如何连接起这些有层次的空间；

　　127. 私密性层次

　　128. 室内阳光

　　129. 中心公用区

　　130. 入口空间

　　131. 穿越空间

　　132. 短过道

　　133. 有舞台感的楼梯

　　134. 禅宗观景

　　135. 明暗交织

statue,a clock-tower with seats,a windmill,a bandstand.Make it something which gives a strong and steady pulse to the square,drawing people in toward the center.Leave it exactly where it falls between the paths;resist the impulse to put it exactly in the middle.

❧✴❧

Connect the different "somethings" to one another with the path system—PATHS AND GOALS(120).They may include HIGH PLACES(62),DANCING INTHESTREETS(63),POOLS AND STREAMS(64),PUBLIC OUTDOOR ROOM(69),STILL WATER(71),TREE PLACES(171);make sure that each one has a SITTING WALL(243)around it...

Now,with the paths fixed,we come back to the building: Within the various wings of any one building,work out the fundamental gradients of space,and decide how the movement will connect the spaces in the gradients;

127. INTIMACY GRADIENT

128. INDOOR SUNLIGHT

129. COMMON AREAS AT THE HEART

130. ENTRANCE ROOM

131. THE FLOW THROUGH ROOMS

132. SHORT PASSAGES

133. STAIRCASE AS A STAGE

134. ZEN VIEW

135. TAPESTRY OF LIGHT AND DARK

模式127　私密性层次**

……如果你已经大致知道自己想在什么地方盖建筑物的翼楼——**有天然采光的翼楼**（107），这些楼有几层——**楼层数**（96）以及**主入口**（110）在哪里，就应该确定每一层楼主要区域的大致配置情况。在每一幢建筑物中确定公共区和私密区之间的关系是很重要的。

❦

除非建筑物的内部空间根据私密性程度按层次排列，不然的话，无论是亲戚朋友或素昧平生的人登门造访都会感到别扭。

在任何一幢建筑物内——住宅、办公楼、公共建筑、避暑别墅，人们都需要有不同私密度的有层次的环境。卧室或闺房是最私密的；起居室或书房次之；公用区或厨房就比较公开了；前门廊或入口则最为公开。当有这种层次存在时，人们可以根据该层次仔细选择地点，赋予每种会面以不同的意义。如果一幢建筑物内各房间交错在一起，以至于没有明确划定的私密性层次，那就不可能为某一特定的会面仔细地选择地点；因此不可能通过选择空间赋予会面以这种重要的附加意义。这种使每个房间的私密性程度都一样的均匀分配的空间，把建筑物中所进行的社会活动的一切可能的差别全都给抹杀了。

我们举一秘鲁的例子来阐明这一带有普遍性的事实，对这个例子我们曾仔细探讨过。在秘鲁，友谊是很受重视的，而且友谊有不同的等级。一般的街坊朋友可能从不登堂入室。正式的朋友，如牧师、女儿的男友和在一起工作的友人可以应邀入内，但通常限于陈设讲究和维护得很好

127 INTIMACY GRADIENT**

...if you know roughly where you intend to place the building wings—WINGS OF LIGHT(107),and how many stories they will have—NUMBER OF STORIES(96),and where the MAIN ENTRANCE(110)is,it is time to work out the rough disposition of the major areas on every floor.In every building the relationship between the public areas and private aeas is most important.

※

Unless the spaces in a building are arranged in a sequence which corresponds to their degrees of privateness,the visits made by strangers,friends,guests,clients,family,will always be a little awkward.

In any building—house,office,public building,summer cottage—people need a gradient of settings,which have different degrees of intimacy.A bedroom or boudoir is most intimate;a back sitting room or study less so;a common area or kitchen more public still;a front porch or entrance room most public of all.When there is a gradient of this kind,people can give each encounter different shades of meaning,by choosing its position on the gradient very carefully.In a building which has its rooms so interlaced that there is no clearly defined gradient of intimacy,it is not possible to choose the spot for any particular encounter so carefully;and it is therefore impossible to give the encounter this dimension of added meaning by the choice of space.This homogeneity of space,where every room

的那部分房间即客厅里。这个房间是同家里杂乱和明显随便得多的其他地方分隔开的。亲朋好友可以到家人常聚的**家庭室**来聊家常。少数亲朋，特别是女客，可以被允许进入厨房、家里的其他工作间，可能还有卧室。这样，家庭既保持私密性，又维护了体面。

私密性层次特别在喜庆的日子表现得尤为明显。即使满屋子是人，有些人却从没有离开过**客厅**；有些人甚至没有迈出过前门门槛一步。而另外一些人则可以长驱直入到厨房，那里正在准备节日的晚餐，他们在那儿可以待上一整个晚上。每个人都心里有数，自己跟这个家庭的亲密程度如何；并且很清楚地知道，根据这种已经确定的亲密程度，自己可以在这个家里走多远。

甚至极为贫穷的人，只要有可能，也总想办法弄出一间客厅：在市郊，我们常常见到这种情况。但秘鲁的现代化住宅和公寓把客厅同家庭室结合在一起以节省空间。差不多每一个和我交谈过的人都对这种情况表示过不满。在我们看来，秘鲁住宅无论如何不该取消私密性层次这一原则。

私密性层次在秘鲁住宅里的重要性是非比寻常的。但这一模式几乎在所有的文化中都以某种形式存在着。我们在各色各样的文化中都能看到它——试比较一个非洲大院、一幢传统的日本住宅以及最早的美国移民时代的住房；它也用于几乎每一建筑类型中——试比较住宅、小商店、大办公楼甚至教堂。私密性层次几乎是所有有人居住的建筑物的典范性的处理原则。所有的建筑物及其各部分，只要住着彼此不同的人群，就需要有一定的层次，它们从"前"到"后"，从前面的最正式的空间到后面的最私密的空间。

在一幢办公楼里，私密性层次可能是：入口门厅、咖啡间和接待室、办公室和工作室、私人休息室。

在一个小商店里，私密性层次可能是：商店入口、顾客进出的地方、顾客观看货物的地方、售货柜台、柜台后方、

has a similar degree of intimacy,rubs out all possible subtlety of social interaction in the building.

We illustrate this general fact by giving an example from Peru—a case which we have studied in detail.In Peru, friendship is taken very seriously and exists at a number of levels.Casual neighborhood friends will probably never enter the house at all.Formal friends,such as the priest,the daughter's boyfriend,and friends from work may be invited in,but tend to be limited to a well-furnished and maintained part of the house,the *sala*.This room is sheltered from the clutter and more obvious informality of the rest of the house.Relatives and intimate friends may be made to feel at home in the family room(*comedor-estar*),where the family is likely to spend much of its time.A few relatives and friends,particularly women,will be allowed into the kitchen,other workspaces,and,perhaps,the bedrooms of the house.In this way,the family maintains both privacy and pride.

The phenomenon of the intimacy gradient is particularly evident at the time of a *fiesta*.Even though the house is full of people,some people never get beyond the *sala*;some do not even get beyond the threshold of the front door.Others go all the way into the kitchen,where the cooking is going on,and stay there throughout the evening.Each person has a very accurate sense of his degree of intimacy with the family and knows exactly how far into the house he may penetrate,according to this established level of intimacy.

Even extremely poor people try to have a *sala* if they can:we saw many in the *barriadas*.Yet modern houses and apartments in Peru combine *sala* and family room in order to save space.Almost everyone we talked to complained about this

工作人员专用的地方。

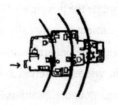

办公楼的私密性层次
Office intimacy gradient

在一幢住宅里，私密性层次依次为：大门、室外门廊、入口、能坐的矮墙、公用空间和厨房、私人花园、床龛。

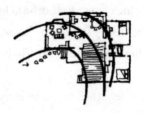

住宅中的私密性层次
Intimacy gradient in a house

在更有气派的住宅里，私密性层次可能开始于类似秘鲁的大厅——客厅或会客室。

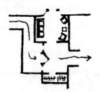

私密性层次中更有气派的前部
Formal version of the front of the gradient

因此：

布置建筑物的空间，使它们具有层次。这个层次首先是建筑物的入口和最公开的部分，其次进入稍为私密一些的区域，最后才到最私密的领域。

situation.As far as we can tell,a Peruvian house must not,under any circumstances,violate the principle of the intimacy gradient.

The intimacy gradient is unusually crucial in a Peruvian house.But in some form the pattern seems to exist in almost all cultures.We see it in widely different cultures—compare the plan of an African compound,a traditional Japanese house,and early American colonial homes—and it also applies to almost every building type—compare a house,a small shop,a large office building,and even a church.It is almost an archetypal ordering principle for all man's buildings.All buildings,and all parts of buildings which house well-defined human groups,need a definite gradient from "front" to "back," from the most formal spaces at the front to the most intimate spaces at the back.

In an office the sequence might be:entry lobby,coffee and reception areas,offices and workspaces,private lounge.

In a small shop the sequence might be:shop entrance, customer milling space,browsing area,sales counter,behind the counter,private place for workers.

In a house:gate,outdoor porch,entrance,sitting wall, common space and kitchen,private garden,bed alcoves.

And in a more formal house,the sequence might begin with something like the Peruvian *sala*—a parlor or sitting room for guests.

Therefore:

Lay out the spaces of a building so that they create a sequence which begins with the entrance and the most public parts of the building,then leads into the slightly more private areas,and finally to the most private domains.

入口　　公共　　半公共　　　私密

❀✿

　　在公用区靠近前门的同时，务必使这些区域也成为各种社会活动的中心和灵魂，并使比较私密的房间之间的所有走道都从该公用区旁边经过——**中心公用区**（129）。在私人住宅里，使入口空间（130）成为最正式和最公开的地方，并安排好最私密的区域，使每个人都有自己的房间，在那里他可以独自安静地休息——**个人居室**（141）。将浴室和厕所设置在公用区和私密区之间，以便大家都能很方便地去使用它们——**浴室**（144）；设置私密程度不同的起居室，根据它们在层次中的地位设计它们的外形——**起居空间的序列**（142）。在办公楼，要把**宾至如归**（149）放在这一层次的前面，把**半私密办公室**（152）放在后面……

At the same time that common areas are to the front,make sure that they are also at the heart and soul of the activity,and that all paths between more private rooms pass tangent to the common ones—COMMON AREASAT THE HEART(129). In private houses make the ENTRANCE ROOM(130)the most formal and public place and arrange the most private areas so that each person has a room of his own,where he can retire to be alone—A ROOM OF ONES OWN(141).Place bathing rooms and toilets half-way between the common areas and the private ones,so that people can reach them comfortably from both—BATHING ROOM(144);and place sitting areas at all the different degrees of intimacy,and shape them according to their position in the gradient—SEQUENCE OF SITTING SPACES(142).In offices put RECEPTION WELCOMES YOU(149)at the front of the gradient and HALF-PRIVATE OFFICE(152)at the back...

模式128　室内阳光*

128   INDOOR SUNLIGHT*

...according to SOUTH FACING OUTDOORS(105),the building is placed in such a way as to allow the sun to shine directly into it,across its gardens.From INTIMACY GRADIENT(127), you have some idea of the overall distribution of public and private rooms within the building. This pattern marks those rooms and areas along the intimacy gradient which need the sunlight most,and helps to place them so that the indoor sunlight can be made to coincide with the rooms in the INTIMCY GRADIENT which are most used.

⊱⊰

If the right rooms are facing south,a house is bright and sunny and cheerful;if the wrong rooms are facing south,the house is dark and gloomy.

Everyone knows this.But people may forget about it,and get confused by other considerations.The fact is that very few things have so much effect on the feeling inside a room as the sun shining into it.If you want to be sure that your house,or building,and the rooms in it are wonderful,comfortable places,give this pattern its due.Treat it seriously;cling to it tenaciously;insist upon it.Think of the rooms you know which do have sunshine in them,and compare them with the many rooms you know that don't.

From the pattern SOUTH FACINH OUTDOORS(105),the building gets an orientation toward the south.Now the issue is the particular arrangement of rooms along this south edge. Here are some examples:(1)a porch that gets the evening sun

……根据**朝南的户外空间**（105），建筑物的朝向应能使阳光直接照进它的内部，遍及它的花园。从**私密性层次**（127）中，相信你对建筑物内部公共空间和私密空间的整体分配已经有了一些概念。本模式规定在私密性层次中最需要阳光的那些房间和区域的位置，使**私密性层次**中使用率最高的那些空间恰好都能够得到阳光。

❀❀❀

如果让该朝南的房间都朝南，住宅就会明亮、阳光充沛和充满欢乐；如果让不该朝南的房间朝南，住宅就暗淡无光。

这是尽人皆知的道理。但人们可能忘记，也可能由于别的考虑而产生混乱。事实是，很少有什么事情像阳光射进房间那样对房间里的人的心情产生那么大的影响。如果你想使你的住宅，即建筑物，以及住宅中的房间同样优美舒适，要给予本模式以应有的地位。认真对待；抓住不放；坚持不懈。想出几个你所知道的阳光充足的空间，把它们同你所知道的许多没有阳光的空间比较一下。

从**朝南的户外空间**（105）这一模式中，我们知道，建筑物的朝向应向着南面。现在的问题是南面空间的具体安排。有以下一些情况：(1) 傍晚能照进夕阳的门廊；(2) 能直接看到溢满晨晖的花园的早餐室；(3) 一间能充分享受朝阳的洗澡间；(4) 白天能从南面照射进光线的工作间；(5) 起居室的一边，阳光能照到它的外墙并洒落在花丛上。

本模式的说明图概括了住宅内各部分与早晨、午后和傍晚的阳光之间的关系。若要在设计中获得合适的阳光，首先要确定你对阳光的要求：你自己制作一个简图，像说明图一样，但要有自己的特殊要求。然后沿着建筑物的南方、东南方和西南方安排空间以获取阳光。特别要注意对

late in the day;(2)a breakfast nook that looks directly into a garden which is sunny in the morning;(3)a bathing room arranged to get full morning sun;(4)a workshop that gets full southern exposure during the middle of the day;(5)an edge of a living room where the sun falls on an outside wall and warms a flowering plant.

The key diagram for this pattern summarizes the relations between parts of the house and the morning,the afternoon and the late afternoon sun.To get the sun right in your design,first decide upon your requirements for sun:make a diagram for yourself,like the key diagram,but with your own special needs. Then arrange spaces along the south,southeast,and southwest of the building to capture the sun.Take special care to detail the south edge properly,so that the sun is working indoors throughout the day.This will most often need a building which is long along the east-west axis.

If we approach the problem of indoor sunlight from the point of view of thermal considerations,we come to a similar conclusion.A long east-west axis sets up a building to keep the heat in during winter,and to keep the heat out during the summer.This makes buildings more pleasant,and cheaper to run.The "optimum shape" of an east-west building is given by the following table,adapted from Victor Olgyay,Design with Climate(New Jersey:Princeton University Press,1963,p.89). Note that it is always best to orient the long axis east-west.

Therefore:

Place the most important rooms along the south edge of the building,and spread the building out along the east-west axis.

Fine tune the arrangement so that the proper rooms

南边作好正确的细部设计，以便阳光能在整个白天照亮室内。这就最需要有沿东西轴走向的长方形建筑物。

如果我们从热原理的观点来对待室内阳光的问题，我们也会得出类似的结论。一个长的东西向的轴能使建筑物在冬季保暖、夏季散热。这会使建筑物更加舒适宜人，而且管理费用低。东西向建筑物的"最佳外形"由下面的图表表示出来，该表依据维克托·奥尔吉埃的《与气候相适应的设计》(*Design with Climate*，New Jersey：Princeton University Press，1963，p.89) 一书改制。注意，向着东西向长轴总是最好的。

适合于不同气候的大致外形

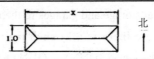

气候	夏季最佳值	冬季最佳值	合成最佳值	合理值
冷（明尼阿波利斯）	1.4	1.1	1.1	1.3
温和（纽约）	1.63	1.56	1.6	2.4
热·干旱（菲尼克斯）	1.26		1.3	1.6
热·潮湿（迈阿密）	1.7	2.69	1.7	3.0

因此：

把最重要的房间设置在建筑物的南边，并使建筑物沿东西轴伸展。

调整好空间的排列，使正式的空间面向东南和西南的阳光。例如：使公用区有充足的南边日照，使卧室朝向东南，门廊朝向西南。对于大多数气候条件来说，这意味着，建筑物的外形呈东西向的长方形。

are exposed to the south-east and the south-west sun.For example:give the common area a full southern exposure, bedrooms south-east,porch south-west.For most climates,this means the shape of the building is elongated east-west.

೮೦Ç೮

When you can,open up these indoor sunny rooms to the outdoors,and build a sunny place and outdoor rooms directly outside—SUNNY PlACE(161),OUTDOOR ROOM(163),WINDOWS WHICH OPEN WIDE(236). Give the bedrooms eastern exposure—SLEEPING TO THE EAST(138),and put storage and garages to the north—NORTH FACE (162).Where there is a kitchen,try to put its work counter toward the sun—SUNNY COUNTER(199);perhaps do the same for any work bench or desk in a HOME WORKSHOP(157),WORKSPACE ENCLOSURE (183)...

卧室　朝阳
门廊　　　早餐室
夕阳　工间　家庭室　厨房
　　花园
朝南的房间

%%%

　　如果有可能，可以将这些室内日照好的房间向外部空间开放，并直接在户外营造出一个有阳光的场地和有围合的户外小空间——**有阳光的地方**（161）、**有围合的户外小空间**（163）、**大敞口窗户**（236）。使卧室朝东——**朝东的卧室**（138），使储藏室和汽车房朝北——**背阴面**（162）。有厨房的地方，设法使它的工作台朝阳——**有阳光的厨房工作台**（199）；在**家庭工作间**（157）、**工作空间的围隔**（183）中，对所有的工作台或书桌大都也可做如此安排……

模式129　中心公共区**

……沿着**私密性层次**（127），在每幢建筑物和建筑物内部的每一社会组织中，都需要设置一些公共区。把它们安排在有阳光的一面以加强**室内阳光**（128）；如果公共区比较大，可以设置**重叠交错的屋顶**（116）中较高的屋顶。

❀❀❀

任何社会组织——无论是家庭、工作小组或是学校——如果在它的成员之间没有经常性的非正式的接触，是不可能存在下去的。

任何社会集团居住的建筑物都是通过提供公共区来维持这种接触的。公共区的形式和地点都有一定的要求。请看一个佳例——对秘鲁工人住宅中的家庭室的描写：

对于一个低收入的秘鲁家庭来说，家庭室是全家生活的中心。一家人在这儿吃饭，看电视，任何到这个家里来的人都要进入这个房间向大家问问好，亲亲他们，跟他们握握手，聊聊家常。人们离开这家时也是这样。

家庭室为这些活动提供场所，它起着家庭生活的中心作用。这房间在住宅中所处的位置使人们进出屋子自然都要从这里经过。他们经过这里可在墙边停留片刻，无须搬张椅子坐下。电视机摆在人们经过地方的对面墙边。看看电视就可以在这儿多停留一会儿。房间里放电视机的地方常常是暗的；家庭室和电视在白天起的作用同晚上一样。

让我们通过这个例子作如下归纳。如果一个公共区设在走廊尽头，人们不得不特意费劲走到那儿，他们不大可能随心所欲地和自发地使用这个公共区。

129   COMMON AREAS AT THE HEART**

...along the INTIMACY GRADIENT(127),in every building and in every social group within the building,it is necessary to place the common areas.Place them on the sunlit side to rein force the pattern of INDOOR SUNLIGHT(128);and,when they are large,give them the higher roofs of the CASCADE OF ROOFS(116).

৪০০৪

No social group—whether a family,a work group,or a school group—can survive without constant informal contact among its members.

Any building which houses a social group supports this kind of contact by providing common areas.Thc form and location of the common areas is critical.Here is a perfect example—a description of the family room in a Peruvian worker's house:

For a low-income Peruvian family,the family room is the heart of family life.The family eat here,they watch TV here,and everyone who comes into the house comes into this room to say hello to the others,kiss them,shake hands with them,exchange news.The same happens when people leave the house.

The family room functions as the heart of the family life by helping to support these processes.The room is so placed in the house,that people naturally pass through it on their way into and out of the house.The end where they pass through it allows them to linger for a few moments,without having to pull out a

在一端
At one end

在另一种情况下，如果过道过多地深入公共区，这块地方则过分暴露，在那里逗留和休息都会令人感到不方便。

通过中心
Through the middle

唯一合适的情况是人们每天行走的公共过道与这些公共区相切，又与它们相通。这样人们就会经常经过这个地方；但由于过道在其一边，人们并不是非在这儿停留不可。如果他们愿意，他们可以径直往前走；如果他们愿意，他们可以在此停留片刻，瞧瞧热闹；如果他们愿意，他们也可以走进来歇歇。

相切
Tangent

有必要指出的是，这个模式已经以某种形式在我们制定的每一个规划中体现出来。在一个具有多种服务项目的中心，我们有一个被称为办公人员休息室的模式，它用的就是这种几何图形（*A pattern language which generates multi-service centers*, C.E.S., 1968, p.241）；我们在规划精神病院时，我们设计了病人可自由选择的公共区，还是以同一种模式用作治病的基本方法。在设计秘鲁住宅时，我们采用家庭室社交——这是我们给家庭提供的一个例子（*Houses generated by patterns* C.E.S., 1969, p.140）；在我们设计大学，即进行俄勒冈实验时，我们为每个系设计"系

chair to sit down.The TV set is at the opposite end of the room from this throughway,and a glance at the screen is often the excuse for a moment's further lingering.The part of the room for the TV set is often darkened;the family room and the TV function just as much during midday as they do at night.

Let us now generalize from this example.If a common area is located at the end of a corridor and people have to make a special,deliberate effort to go there,they are not likely to use it informally and spontaneously.

Alternatively,if the circulation path cuts too deeply through the common area,the space will be too exposed,it will not be comfortable to linger there and settle down.

The only balanced situation is the one where a common path,which people use every day, runs *tangent* to the common areas and is open to them in passing.Then people will be constantly passing the space;but because the path is to one side,they are not forced to stop.If they want to,they can keep going.If they want to,they can stop for a moment,and see what's happening;if they want to,they can come right in and settle down.

It is worth mentioning,that this pattern has occurred,in some form,in every single project we have worked on.In the multi-service center,we had a pattern called *Staff lounge* based on the same geometry(*A pattern language which generates multi-service centers*,C.E.S.,1968,p.241);in our work on mental health centers,we had *Patient's choice of being involved*,the same pattern again,as an essential element in therapy;in our work on Peruvian housing,we *had Family room circulation*— this is the example we have given for a family(*Houses generated by patterns*, C.E.S.,1969,p.140);and in our work on

师生之家"，采用的还是那个模式。这也许是使群体聚集在一起的最基本的模式。

具体地说，一个设计得好的公共区有三个独立的特点：

1. 它必须位于建筑群、建筑物或建筑物翼楼的重心。换言之，它必须处于一个机构的中心位置，以便每一个人都同样容易进去，因而使大家感到它是团体的中心。

2. 最重要的是，它必须位于从入口到私人房间的路旁，这样人们在进出建筑物时总要从它的旁边经过。务必不要使它成为死角上的房间，那种房间即使人们顺路也是到达不了的。由于这个原因，经过它的过道必须同它相切。

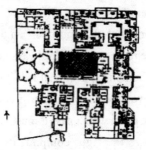

这是我们在加利福尼亚莫德斯托建造的一幢诊所的
公共区，我们使四边的过道都同这个公共区相切
The common area of a clinic we have built in Modesto,
California, where we managed to put tangent paths on all four sides

3. 公共区必须有合适的组成部分——通常是一个厨房和餐厅，因为用餐是最共通的活动之一，以及坐位——至少是一些舒适的椅子，这样人们才会愿意留在那儿。它也应该包括一个室外区——在天气晴朗的日子里大家都想到室外走走——出来抽会儿烟，在草坪上坐坐，讨论点问题。

因此：

为每一社会团体建造一个公共区。把该公共区设置在该社会团体占用的全部空间的重心位置，使进出建筑物的过道都同它相切。

universities,*The Oregon Experiment*,we had a pattern called *Department hearth*,again the same,for each department.It is perhaps the most basic pattern there is in forming group cohesion.

In detail,we have isolated three characteristics for a successful common area:

1.It must be at the center of gravity of the building complex,building,or building wing which the group occupies. In other words,it must be at the physical heart of the organization,so that it is equally accessible to everyone and can be felt as the center of the group.

2.Most important of all,it must be "on the way" from the entrance to private rooms, so people always go by it on the way in and out of the building.It is crucial that it not be a dead-end room which one would have to go out of one's way to get to. For this reason,the paths which pass it must lie tangent to it.

3.It must have the right components in it—usually a kitchen and eating space,since eating is one of the most communal of activities,and a sitting space—at least some comfortable chairs,so people will feel like staying.It should also include an outdoor area—on nice days there is always the longing to be outside—to step out for a smoke,to sit down on the grass, to carry on a discussion.

Therefore:

Create a single common area for every social group. Locate it at the center of gravity of all the spaces the group occupies,and in such a way that the paths which go in and out of the building lie tangent to it.

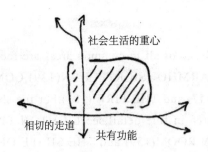

社会生活的重心

相切的走道　共有功能

❀❀

对于公共区来说最重要的是要有食品供应和壁炉。应包括**农家厨房**（139）、**共同进餐**（147）和**炉火熊熊**（181）。关于公共区形状的详细情况，请参阅**两面采光**（159）和**室内空间形状**（191）。一定要设置许多不同的起居空间，它们在不同的场合具有不同的性质——**起居空间的序列**（142）。要包括**有围合的户外小空间**（163）。使过道刚好同公共区相切——**拱廊**（119）、**穿越空间**（131）、**短过道**（132）……

BUILDINGS
建　筑
1195

∞∞

Most basic of all to common areas are food and fire. Include FARMHOUSE KITCHEN(139),COMMUNAL EATING(147),and THE FIRE (181).For the shape of the common area in fine detail,see LIGHTON TWO SIDES OF EVERY ROOM(159) and THE SHAPE OF INDOOR SPACE (191).Make sure that there are plenty of different sitting places,different in character for different kinds of moments—SEQUENCE OF SITTING SPACES(142).Include an OUTDOOR ROOM(163).And make the paths properly tangent to the common areas—ARCADES(119),THE FLOW THROUGH ROOMS(131),SHORT PASSAGES(132)...

模式130　入口空间**

130 ENTRANCE ROOM**

...the position and overall shape of entrances is given by FAMILY OF ENTRANCES(102),MAIN ENTRANCE(110)and ENTRANCE TRANSITION(112).This pattern gives the entrances their detailed shape,their shape and body and three dimensions,and helps complete the form begun by CAR CONNECTION(113),and the PRIVATE TERRACE ON THE STRET(140).

⊰⊱

Arriving in a building,or leaving it,you need a room to pass through,both inside the building and outside it.This is the entrance room.

The most impressionistic and intuitive way to describe the need for the entrance room is to say that the time of arriving,or leaving,seems to swell with respect to the minutes which precede and follow it,and that in order to be congruent with the importance of the moment,the space too must follow suit and swell with respect to the immediate inside and the immediate outside of the building.

We shall see now that there are a tremendous number of miniscule forces which all come together to support this general intuition.All these forces,tendencies,and solutions were originally describe by Alexander and Poyner,in the Atoms of Environmental Structure,Ministry of Public Works,Research and Development,SFB Ba4,London,1966.At that time it seemed important to emphasize the separate and individual patterns defined by these forces. However,at the present writing it seems clear that these original patterns are,in fact,all faces of the one larger and more comprehensive entity,which we call the

……入口的位置和总体形状在**各种入口**（102）、**主入口**（110）和**入口的过渡空间**（112）中已经讲过。本模式要讲各种入口的具体形状，即它们的形状、体量和范围，并有助于完善**与车位的联系**（113）和**私家的沿街露台**（140）的形式。

⁂

走进或离开一幢建筑物，你都需要有一个空间可以通过，无论这空间是在建筑物内部还是外部。这就是入口空间。

为什么需要入口空间？以下的说明最为直观而令人印象深刻：人们进入或离开建筑物的时间都是前后连续的，为了与这一时间的重要性保持一致，空间也应照此处理，使之同建筑物内外相连续。

现在我们会看到，有许多小的因素相加在一起使这个问题更加直观。亚历山大和波纳最初提出这些因素、趋向和解决办法（The Atoms of Environmental Structure，Ministry of Public Works，Research and Development，SFB Ba4，London，1966）。那时，应该是需要强调由这些因素确定的各个模式的独立性。然而，现在看来很明确，原先这些模式事实上是一个更大、更具综合性的整体中的各个方面，这个整体我们称之为**入口空间**（130）。

1. 窗户与入口的关系

（a）在门内应答的人在开门前常常想看看谁在叫门。

（b）人们不想走出来窥视站在门口的人。

（c）如果碰面的是老朋友，可以预料他们在此时会呼喊和招手。

因此入口空间需要有一个窗户，或者几个窗户，窗户开在从家庭室或厨房通向门口的过道上，从一侧面对门外那块地方。

ENTRANCE ROOM(130).

1.*The relationship of windows to the entrance*

(a)A person answering the door often tries to see who is at the door before they open it.

(b)People do not want to go out of their way to peer at people on the doorstep.

(c)If the people meeting are old friends,they seek a chance to shout out and wave in anticipation.

The entrance room therefore needs a window—or windows—on the path from the family room or kitchen to the door,facing the area outside the door from the side.

2.*The need for shelter outside the door*

(a)People try to get shelter from the rain,wind,and cold while they are waiting.

(b)People stand near the door while they are waiting for it to open.

On the outside,therefore,give the entrance room walls enclosing three sides of a covered space.

3.*The subtleties of saying goodbye*

When hosts and guests are saying goodbye,the lack of a clearly marked "goodbye" point can easily lead to endless "Well,we really must be going now," and then further conversations lingering on,over and over again.

(a)Once they have finally decided to go,people try to leave without hesitation.

(b)People try to make their goodbye as nonabrupt as possible and seek a comfortable break.

Give the entrance room,therefore,a clearly defined area,at least 20 square feet,outside the front door,raised with a natural threshold—perhaps a railing,or a low wall,or a step—between

2. 门外需要有遮挡

(a) 人们在等开门的时候要设法躲雨、避风、御寒。

(b) 人们在等待开门的时候总是站在门旁的。

因此在门外，要在入口空间的墙壁三面围合一个有顶的空间。

3. 告别的奥妙

尽管宾主都说了再见，但由于它缺乏明确的"再见"含义，往往还要没完没了地说"好了，我们现在真该走了"，接着又一再地交谈了起来。

(a) 只要人们最终决定告别，就要毫不犹豫地离开。

(b) 人们要设法使告辞不显得唐突，使谈话适可而止。

因此，要把入口空间的区域十分明确地划分出来，至少 $20ft^2$，位于前门外面，地势略高。在入口处和来宾的汽车之间有天然的门槛——或许是栏杆，或者一堵矮墙，或者台阶。

4. 入口旁的搁架

如果有人带一个包进入屋内：

(a) 他得设法攥着这个包：让包直立而又不着地。

(b) 同时他得设法把两只手都空出来在口袋或者手提包里找钥匙。

当他带包离开屋子时：

(c) 人们在离开时往往想到别的事情，这会使他们忘记自己要带上的包。

如果门内外都放有半人高的搁架；一个准备好可以放包的地方；一个开门的时候可以把包放下来的地方，那就可以避免这些麻烦了。

5. 入口空间内部

(a) 当有人进门的时候，门要大开，这是礼貌所要求的。

(b) 人们追求自己住宅内部的私密性。

it and the visitors' cars.

4. *Shelf near the entrance*

When a person is going into the house with a package:

(a)He tries to hold onto the package;he tries to keep it upright,and off the ground.

(b)At the same time he tries to get both hands free to hunt through pockets or handbag for a key.

And leaving the house with a package:

(c)At the moment of leaving people tend to be preoccupied with other things,and this makes them forget the package which they meant to take.

You can avoid these conflicts if there are shelves both inside and outside the door,at about waist height;a place to leave packages in readiness;a place to put them down while opening the door.

5. *Interior of the entrance room*

(a)Politeness demands that when someone comes to the door,the door is opened wide.

(b)People seek privacy for the inside of their houses.

(c)The family,sitting,talking,or at table,do not want to feel disturbed or intruded upon when someone comes to the door.

Make the inside of the entrance room zigzag,or obstructed, so that a person standing on the doorstep of the open door can see no rooms inside,except the entrance room itself,nor through the doors of any rooms.

6. *Coats,shoes,children's bikes...*

(a)Muddy boots have got to come off.

(b)People need a five foot diameter of clear space to take off their coats.

(c)People take prams,bicycles,and so on indoors to protect them from theft and weather;and children will tend to （c）一家人坐着谈天说地，或正在用餐，不希望有不速之客登门打扰。

使入口空间内部蜿蜒曲折，或有隔断，这样，大门敞开时，站在门前台阶上的人看不到里面的房间，看到的只是入口空间，同时也不能通过任何房门往里看。

6. 大衣、鞋、儿童自行车……

（a）沾了泥土的靴子应该拿走。

（b）人们需要有 5ft 直径的空地供脱大衣用。

（c）人们把婴儿车、自行车等杂物拿到屋里去以防失窃和恶劣天气；而儿童们爱把他们常用的零零碎碎的物品——自行车、手推车、旱冰鞋、三轮车、铁铲、球——放在门边。

因此，要留给入口空间一个可存放杂物的死角，挂衣钩的位置能够让人一进前门就能看得见，挂衣钩旁应有 5ft 直径的空间。

因此：

在建筑物的主入口，设置一个明亮的房间，它表明入口的位置，介于建筑物内外之间，既占据室外一些空间，也占据室内一些空间。室外部分可以像老式的门廊；室内部分像前厅或起居室。

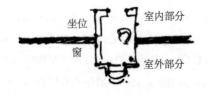

leave all kinds of clutter—bikes,wagons,roller skates, trikes, shovels,balls—around the door they use most often.

Therefore,give the entrance room a dead corner for storage,put coat pegs in a position which can be seen from the front door,and make an area five feet in diameter next to the pegs.

Therefore:

At the main entrance to a building,make a light-filled room which marks the entrance and straddles the boundary between indoors and outdoors,covering some space out doors and some space indoors.The outside part may be like an old-fashioned porch;the inside like a hal or sitting room.

෩෩

Give that part of the entrance which sticks out into the street or garden a physical character which,as far as possible, make it one of the family of entrances along the street— FAMILY OF ENTRANCES (102);where it is appropriate,make it a porch—GALLERY SURROUND(166);and include a bench or seat, where people can watch the world go by or wait for someone—FRONT DOOR BENCH(242).As for the indoor part of the entrance room,above all,make sure that it is filled with light from two or even three sides,so that the first impression of the building is of light—TAPESTRY OF LIGHT AND DARK (135),LIGHT ON TWO SIDES OF EVERY ROOM(159). Put windows in the door itself—SOLID DOORS WITH CLASS(237).Put in BUILTIN SEATS (202)and make the room part of the SEQUENCE OF SITTING SPACES (142);provide a WAISt-HIGH SHELF(201)for packages.And finally,for the overall shape of the entrance room and its construction,begin with THE SHAPE OF INDOOR SPACE(191)...

向街道或花园伸出的那部分入口应能够具有实体性质，并尽可能使入口成为沿街的各种入口之一——**各种入口**（102）；在合适的地方，把入口做成门廊——**回廊**（166）；设置长凳或坐位，在那儿人们可以看见来来往往的人和车辆，或坐着等人——**大门外的条凳**（242）。至于入口空间的室内部分，最重要的，务必使它两面甚至三面采光，这样人们对这个建筑物的第一个印象就是光线充足——**明暗交织**（135）、**两面采光**（159）。在门上装窗——**镶玻璃板门**（237）。建造**嵌墙坐位**（202）并使入口空间成为**起居空间的序列**（142）的一部分；放置一个**半人高的搁架**（201）供人放提包。最后，关于入口空间的总形状和构造施工的问题，请参阅**室内空间形状**（191）及之后几个模式……

模式131　穿越空间

131   THE FLOW THROUGH ROOMS

...next to the gradient of spaces created by INTIMACY GRADIENT (127)and COMMON AREAS AT THE HEART (129), the way that rooms connect to one another will play the largest role in governing the character of indoor space.This pattern describes the most fundamental way of linkig rooms to one another.

೪೦೦೫

The movement between rooms is as important as the rooms themselves;and its arrangement has as much effect on social interaction in the rooms,as the interiors of the rooms.

The movement between rooms,the circulation space,may be generous or mean.In a building where the movement is mean,the passages are dark and narrow—rooms open off them as dead ends;you spend your time entering the building,or moving between rooms,like a crab scuttling in the dark.

Compare this with a building where the movement is generous.The passages are broad,sunlit,with seats in them, views into gardens,and they are more or less continuous with the rooms themselves,so that the smell of woodsmoke and cigars,the sound of glasses,whispers,laughter,all that which enlivens a room,also enlivens the places where you move.

These two approaches to movement have entirely different psychological effects.

In a complex social fabric,human relations are inevitably subtle.It is essential that each person feels free to make connections or not,to move or not,to talk or not,to change

在解决了由**私密性层次**（127）和**中心公用区**（129）所产生的空间层次之后，空间之间彼此连接的方法在决定室内空间的性质方面将起最大的作用。本模式说明空间之间彼此连接的最基本的方法。

৪৩◌ঙ

　　房间之间的流通同房间本身一样重要；如何安排这种流通对在房间内进行活动产生的影响同房间内部布局产生的影响一样大。

　　房间之间的流通，即流通空间，可能舒畅，可能局促。在一个流通空间不大的建筑物里，过道黑暗而狭窄——从房间走到过道就像走进死胡同；你走进建筑物里面或在房间之间穿行的时候，犹如螃蟹在黑暗中奔爬。

　　请将此与一流通空间舒畅的建筑物作一比较。过道宽阔，阳光充足，过道上放着坐椅，可以从这里看到花园，而且过道同房间本身或多或少地连接在一起，这样，木柴燃烧和雪茄的味道，玻璃杯的撞击声和人们的轻声笑语，凡是能使一个房间充满生气的东西，也使你走过的地方充满生气。

　　对待流通空间采取的两种方法具有两种截然不同的心理效果。

　　在一个复杂的社会环境里，人的关系不可避免地是微妙的。每个人是否与他人交往，是否走动，是否谈话，是否改变生活状况，应该让他根据自己的判断自主进行。如果实际的环境抑制他，限制了其行动自由，这将使他不能尽力去医治和改善他所处的社会现状，而那原本是他认为应该做的。

　　流通空间舒畅的建筑物使每个人的本能和直觉充分发挥。而流通空间狭窄的建筑物却压抑了它们。这样的建筑不仅使房间与房间之间彼此分隔，以致让人觉得从一个房

the situation or not,according to his judgment.If the physical environment inhibits him and reduces his freedom of action,it will prevent him from doing the best he can to keep healing and improving the social situations he is in as he sees fit.

The building with generous circulation allows each person's instincts and intuitions full play.The building with ungenerous circulation inhibits them.It not only separates rooms from one another to such an extent that it is an ordeal to move from room to room,but kills the joy of time spent between rooms and may discourage movement altogether.

The following incident shows how important freedom of movement is to the life of a building.An industrial company in Lausanne had the following experience.They installed TV-phone intercoms between all offices to improve communication. A few months later,the firm was going down the drain—and they called in a management consultant.He finally traced their problems back to the TV-phones.People were calling each other on the TV-phone to ask specific questions—but as a result,people never talked in the halls and passages any more— no more "Hey,how are you,say,by the way,what do you think of this idea..." The organization was falling apart,because the informal talk—the glue which held the organization together— had been destroyed.The consultant advised them to junk the TV-phones—and they lived happily ever after.

This incident happened in a large organization.But the principle is just the same in a small work group or a family.The possibility of small momentary conversations,gestures,kindnesses,explanations which clear up misunderstandings,jokes and stories is the lifeblood of a human group.If it gets prevented,the group will fall apart as people's individual relationships go

间走到另一个房间是件苦差事；而且使人在房间之间来往感到兴味索然，还可能使人根本不愿走动。

下面这件事说明，自由流通对于建筑物的生气蓬勃多么重要。瑞士洛桑的一家工业公司有过下面一段经历。他们在所有的办公室之间安装了电视对讲器以改善通信条件。几个月以后，公司却濒临破产——于是他们请来一位经营顾问。这位经营顾问最终把问题归结于电视对讲器。人们遇到大大小小的问题都使用电视对讲器跟对方说话——其结果，人们不再在门厅和过道上交谈了，再听不见人们说："喂，你好啊，你说说，对这个问题你有什么想法……"于是这个机构涣散了，因为轻松随意的聊天——一种使这个机构成为一体的黏合剂——消失了。这位顾问劝告他们撤掉电视对讲器——自那以后他们的日子过得很愉快。

这件事发生在一个大机构里。但其原则对于一个小的工作单位或一个家庭同样适用。小小的简短谈话、手势、体贴、消除误会的解释、玩笑和故事都能加强人们的集体关系。如果这些一概没有，随着人们的个人关系渐渐冷却，集体也将分崩离析。

几乎毫无疑问，流通空间狭小的建筑物较难使人们维持其社会结构。从长远来看，在一个流通空间狭小的建筑物里，社会活动的秩序很可能会被完全毁坏。

流通空间的大小取决于对建筑物内通道的总体布置，而不取决于对各个过道的具体设计。事实上，完全没有过道，只有通过门来互相联系的一连串房间产生的流通空间才是最大的流通空间。

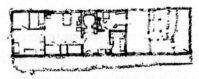

没有过道的空间序列
A sequence of rooms without a passage

gradually downhill.

It is almost certain that the building with ungenerous circulation makes it harder for people to maintain their social fabric.In the long run,there is a good chance that social order in the building with ungenerous circulation will break down altogether.

The generosity of movement depends on the overall arrangement of the movement in the building,not on the detailed design of individual passages.In fact,it is at its most generous,when there are no passages at all and movement is created by a string of interconnecting rooms with doors between them.

Even better, is the case where there is a loop.A loop, which passes through all the major rooms,public and common, establishes an enormous feeling of generosity.With a loop it is always possible to come and go in two different directions.It is possible to walk around and around,and it ties the rooms together.And,when such a loop passes through rooms(at one end so as not to disturb them),it connects rooms far more than a simple passage does.

A building where there is a chain of rooms in sequence also works like this,if there is a passage in parallel with the chain of rooms.

Therefore:

As far as possible,avoid the use of corridors and passages. Instead,use public rooms and common rooms as rooms for movement and for gathering.To do this,place the common rooms to form a chain,or loop,so that it be comes possible to walk from room to room—and so that private rooms open directly off these public rooms.In every case,give this indoor circulation from room to room a feeling of great generosity,passing in a wide and ample loop around the house,with iews of fires and great windows.

有回路的情况就更好了。一条回路，可以经过所有的公共和共用的主要房间，能给人以门户畅通的强烈感觉。有了回路，就有可能从两个不同的方向进出。有可能来来回回地走动，它把各个房间连接在一起了。当这样的回路经过房间时（仅在一个角上，这样才不至于把房间扰乱），它对各个房间所起的连接作用远比一条简单的过道更好。

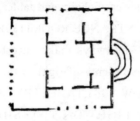

一条宽敞的流通回路
A generous circulation loop

有着按序列编排的一排房间的建筑物，如果有一条过道同这一排房间相平行，也能起到这样的作用。

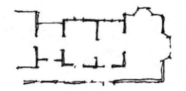

平行过道造成回路
Passage in parallel forms the loop

因此：

只要有可能，应避免使用走廊和过道。代之以公共房间或共用房间作为过往和聚会的空间。其办法是，将共用房间排成一个系列或造成回路，这样便有可能从一个房间走到另一个房间——而这样做的结果可使私人房间直接开向这些公共房间。在所有情况下，都要使这样的房间之间的室内通道给人以畅通的感觉，在一条开阔和宽敞的回路上走过整座房子，看得见熊熊炉火和高大窗户。

‍∞‍

Whenever passages or corridors are unavoidable,make them wide and generous too;and try to place them on one side of the building,so that they can be filled with light—SHORT PASSAGES (132).Furnish them like rooms,with carpets,bookshelves,easy chairs and tables,filtered light,and do the same for ENTRANCE ROOM(130)and STAIRCASE AS ASTAGE(133).Always make sure that these rooms for movement have plenty of light in them and perhaps a view—ZEN VIEW(134),TAPESTRY OF LIGHT AND DARK(135),and LIGHT ON TWO SIDES OF EVERY ROOM(159).Keep doors which open into rooms,or doors between rooms which create the flow through rooms,in the corners of the rooms—CORNER DOORS(196)...

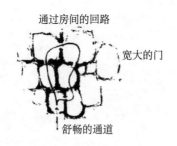

通过房间的回路

宽大的门

舒畅的通道

⚜

如果不可避免地要用到过道和走廊，也应把它们做得开阔宽敞；尽量使它们位于建筑物的一侧，使它们光线充足——**短过道**（132）。把它们装饰得像房间一样，铺上地毯，放置书架、安乐椅和桌子，装上带罩的灯，对**入口空间**（130）和**有舞台感的楼梯**（133）也应照此设置。务必使这些供人们过往用的空间光线充足，多半还要风景优美——**禅宗观景**（134）、**明暗交织**（135）、**两面采光**（159）。使进入房间的门和连通两个房间之间的门开在房间的角落——**墙角的房门**（196）……

模式132　短过道*

　　……穿越空间（131）说明，可用房间之间彼此连通的办法来做到光线充足，走动方便，同时建议不考虑用过道。但当一幢办公楼或住宅必须要有过道而这过道又太小不能充当有顶街道（101）时，则需加以特殊处理，使过道自身可以作为房间使用。本模式指出这些小过道的性质，因而完善了由内部交通领域（98）和有顶街道（101）以及穿越空间（131）奠定的流通系统。

132 SHORT PASSAGES*

...THE FLOW THROUGH ROOMS(131)describes the generosity of light and movement in the way that rooms connect to one another and recommends against the use of passages.But when there has to be a passage in an office or a house and when it is too small to be a BUILDING THOROUGHFARE(101),it must be treated very specially,as if it were itself a room.This pattern gives the character of these smallest passages,and so completes the circulation system laid down by CIRCULATION REALMS(98)and BUILDING THOROUGHFARE101)and THE FLOW THROUGH ROOMS(131).

※

"...long,sterile corridors set the scene for everything bad about modern architecture."

In fact,the ugly long repetitive corridors of the machine age have so far infected the word "corridor" that it is hard to imagine that a corridor could ever be a place of beauty,a moment in your passage from room to room,which means as much as all the moments you spend in the rooms themselves.

We shall now try to pinpoint the difference between the corridors which live,which give pleasure,and make people feel alive,and those which do not.There are four main issues.

The most profound issue,to our minds,is natural light. A hall or passage that is generously lit by the sun is almost always pleasant.The archetype is the one-sided hall,lined with

"……毫无生气的长走廊使现代建筑大煞风景。"

事实上,机器时代难看的千篇一律的长走廊败坏了"走廊"这个字眼,以至于人们很难想象,走廊也可以是一个环境优美的地方,你在走廊上从一个房间走到另一个房间所度过的时光会有可能像和在房间里度过的时光一样美妙。

长走廊
Long corridors

我们现在要找出有生气的、使人感到愉快并给人以活力的走廊跟那些与此相反的走廊的差别。有四个主要问题。

在我们看来,最大的问题是自然光。阳光充沛的过道差不多总是令人愉快的。最好的情况是单边过道,有门窗向外敞开。(注意,从一侧使空间采光是一个好办法,可惜这种地方不多。)

第二个问题是过道与开向过道的房间的关系。由于内窗从这些房间开向过道,使过道增添生气。内窗使房间和过道得到沟通;它们使相互间的交流更加随意;它们使走过过道的人都能感受到房间内部的生活气息。即使在办公楼,只要不过分,只要工作空间各自保持一定距离或者有矮墙隔断,这种接触也是很有意义的——参阅**半私密办公室**(152)、**工作空间的围隔**(183)。

windows and doors on its open side.(Notice that this is one of the few places where it is a good idea to light a space from one side.)

The second issue is the relation of the passage to the rooms which open off it.Interior windows,opening from these rooms into the hall,help animate the hall.They establish a flow between the rooms and the passage;they support a more informal style of communication;they give the person moving through the hall a taste of life inside the rooms.Even in an office,this contact is fine so long as it is not extreme;so long as the workplaces are protected individually by distance or by a partial wall—see HALF-PRIVATE OFFICE(152),WORKSPACE ENCLOSURE(183).

The third issue which makes the difference between a lively passage and a dead one is the presence of furnishings. If the passage is made in a way which invites people to furnish it with book cases,small tables,places to lean,even seats,then it becomes very much a part of the living space of the building,not something entirely separate.

And finally,there is the critical issue of length.We know intuitively that corridors in office buildings, hospitals, hotels,apartment buildings—even sometimes in houses—are far too long.People dislike them:they represent bureaucracy and monotony.And there is even evidence to show that they do actual damage.

Consider a study by Mayer Spivack on the unconscious effects of long hospital corridors on perception, communication, and behavior:

第三个问题：一条过道是生气勃勃还是死气沉沉，要看它有无家具摆设。如果过道的布局使人们乐于用书柜、小桌、供倚靠的地方以及坐椅来充实它，它就会很自然地变成建筑物里起居空间的一部分，而不会显得与建筑物格格不入。

最后，这是一个极为重要的问题，即长度问题。我们根据直觉知道，在办公楼、医院、旅馆、公寓楼——有时候甚至住宅——走廊都实在太长了。人们不喜欢这些走廊：它们呆板单调。而且，甚至有证据表明，在实际生活中它们贻害于人。

迈耶·斯皮瓦克研究了医院的长走廊对人的感觉、相互交流和行为的不知不觉的影响。他的这一研究是值得我们注意的：

"我们研究了四个精神病院的长走廊……可以得出结论，这样的空间由于它们固有的音响性质干扰正常的语言交流。这些过道一般的采光状况使人无法看清人们的体态和面部，并且使人对距离产生错觉。在一条过道中产生的模糊视觉会造成关于空间大小、距离、步行速度和时间的相互关联的、感觉重合的幻觉。对病人行为进行观察表明，狭窄的走廊使人烦躁不安是由于它侵入了人的空间包围层。"(M.Spivack, "Sensory Distortion in Tunnels and Corridors," *Hospital and Community Psychiatry* 18, No.1, January 1967.)

什么情况下一条走廊算是太长了呢？在本模式以前的版本（*Short corridors in A Pattern Language Which Gererates Multiservice Centers*, CES, 1967, pp.179~182）中，我们已经提供的证据表明，在长走廊和短过道之间有一个确定的、可以感觉到的中止点：证据指出，大约50ft这个数字是临界值。超过这个界限，走廊就开始令人感到死气沉沉和单调乏味。

当然，有可能建造甚至很长的宜人的走廊；但如果它们的长度非超过50ft不可，有必要以某种方式限制它们的

Four examples of long mental hospital corridors are examined...it is concluded that such spaces interfere with normal verbal communication due to their characteristic acoustical properties. Optical phenomena common to these passageways obscure the perception of the human figure and face,and distort distance perception.Para-doxical visual cues produced by one tunnel created interrelated,cross-sensory illusions involving room size,distance,walking speed and time. Observations of patient behavior suggest the effect of narrow corridors upon anxiety is via the penetration of the personal space envelope. (M.Spivack, "Sensory Distortion in Tunnels and Corridors," *Hospital and Community Psychiatry*,18,No.1,January 1967.)

When does a corridor become too long?In an earlier version of this pattern(*Short corridors in A Pattern Language Which Generates Multi-Service Centers*,CES,1967,pp.179-182), we have presented evidence which suggests that there is a definite cognitive breakpoint between long corridors and short halls:the evidence points to a figure of some 50 feet as a critical threshold.Beyond that,passages begin to feel dead and monotonous.

Of course it is possible to make even very long corridors in a human way;but if they have to be longer than 50 feet,it is essential to break down their scale in some fashion.For example,a long hall that is lit in patches from one side at short intervals can be very pleasant indeed:the sequence of light and dark and the chance to pause and glance out,breaks down the feeling of the endless dead corridor;or a hall which opens out into wider rooms,every now and then,has the same effect. However,do everything you can to keep the passages really short.

规模。例如，一条长的过道，在一侧每隔一小段距离提供一片一片的照明，的确可能令人感到愉快：一系列明暗相间，可让人停停看看，使人不再觉得这是无穷无尽、死气沉沉的走廊；或者一条过道不时地通往更加宽阔的空间，也能获得同样的效果。不管怎样，你还是要尽一切可能去建造真正的短距过道。

因此：

过道要短。尽可能地使过道像房间一样，地上铺上地毯或地板，摆上家具、书架，装上漂亮的窗户。使它们外形宽大、光线充足；最好的走廊和过道都是沿墙开有窗户的。

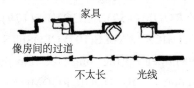

家具

像房间的过道

不太长　　光线

❦❧

在过道装上窗户，放置书架和家具，尽可能使它们像真正的房间，沿着边缘有凹室和坐椅——**两面采光（159）、凹室（179）、窗前空间（180）、厚墙（197）、居室间的壁橱（198）**；使长的一面通向花园或阳台——**有围合的户外小空间（163）、回廊（166）、矮窗台（222）**。在过道和开向过道的房间之间设置内窗——**内窗（194）、镶玻璃板门（237）**。最后，关于过道的形状，详见**室内空间形状（191）**及之后几个模式……

Therefore:

Keep passages short. Make them as much like rooms as possible, with carpets or wood on the floor, furniture, book shelves, beautiful windows. Make them generous in shape, and always give them plenty of light; the best corridors and passages of all are those which have windows along an entire wall.

∞∞∞

Put in windows, bookshelves, and furnishings to make them as much like actual rooms as possible, with alcoves, seats along the edge—LIGHT ON TWO SIDES OF EVERY ROOM(159), ALCOVES(179), WINDOW PLACE(180), THICK WALLS(197), CLOSETS BETWEEN ROOMS (198); open up the long side into the garden or out onto balconies—OUTDOOR ROOM(163), GALLERY SURROUND(166), LOW SILL(222). Make interior windows between the passage and the rooms which open off it—INTERIOR WINDOWS(194), SOLID DOORS WITH GLASS(237). And finally, for the shape of the passages, in detail, start with THE SHAPE OF INDOOR SPACE(191).

模式133　有舞台感的楼梯

……如果入口的位置已经选定——**主入口**（110）；穿过建筑物通道的模式已经确定——**穿越空间**（131）、**短过道**（132），那么，就应该建造主楼梯并赋予它以合适的社会属性。

❧❧❧

楼梯不仅是从一层楼通向另一层楼的过道。楼梯本身占据空间，具有体量，是楼房的一部分；如果不能使这个空间充满生气，则它会变成一个死角，会使楼房上下分隔开来，使它的各种功能遭到破坏。

现在我们对楼梯的一般外形留下的种种印象还是凭借如下的猜想得到的：在社交集会的许多场合高度变化往往举足轻重；高度变化提供各种空间：高出来的地方是人们可以就坐的特殊坐位，人们从高处走进屋来显得风度优雅，富于戏剧性；你可以站在高处对众人讲话，你可以在高处

133 STAIRCASE AS A STAGE

...if the entrances are in position—MAIN ENTRANCE(110); and the pattern of movement through the building is established—THE FLOW THROUGH ROOMS(131),SHORT PASSAGES(132),the main stairs must be put in and given an appropriate social character.

෨෦ඥ

A staircase is not just a way of getting from one floor to another. The stair is itself a space,a volume,a part of the building;and unless this space is made to live,it will be a dead spot,and work to disconnect the building and to tear its processes apart.

Our feelings for the general shape of the stair are based on this conjecture:changes of level play a crucial role at many moments during social gatherings;they provide special places to sit,a place where someone can make a graceful or dramatic entrance,a place from which to speak,a place from which to look at other people while also being seen,a place which increases face to face contact when many people are together.

If this is so,then the stair is one of the few places in a building which is capable of providing for this requirement, since it is almost the only place in a building where a transition between levels occurs naturally.

This suggests that the stair always be made rather open to the room below it,embracing the room,coming down around the outer perimeter of the room,so that the stairs together with the room form a socially connected space.Stairs that are enclosed in stairwells or stairs that are free standing and chop

观看别人活动，同时别人也可一睹你的风采；在众人相聚时，这个地方能增加彼此间面对面接触的机会。

如果情况果真这样，那么在一幢楼房内，楼梯就是仅有的几处能满足这种要求的地方之一。因为它几乎是楼内高度自然发生变化的唯一所在。

这表明，楼梯应该始终向位于它下面的房间完全敞开，环抱着它并绕着它的外周边而下，以便楼梯和房间形成一个连接在一起的社交空间。在楼梯井中封闭的楼梯，或独立而突出于下层空间的楼梯，都根本不具备这一特征。但是直梯、沿下面的墙壁周线建造的楼梯或中间转折的楼梯则都可具有上述功能。

楼梯间数例
Examples of stair rooms

而且，如果楼梯设计得好，其最底部的前四五级台阶会是人们最有可能去坐的地方。为了做到这一点，应将楼梯的底部扩展，使台阶变宽，让人坐在那儿感到舒服。

坐人的台阶
Stair seats

up the space below,do not have this character at all.But straight stairs,stairs that follow the contour of the walls below,or stairs that double back can all be made to work this way.

Furthermore,the first four or five steps are the places where people are most likely to sit if the stair is working well. To support this fact,make the bottom of the staircase flare out,widen the steps,and make them comfortable to sit on.

Finally,we must decide where to place the stair.On the one hand,of course,the stair is the key to movement in a building.It must therefore be visible from the front door;and,in a building with many different rooms upstairs,it must be in a position which commands as many of these rooms as possible,so that it forms a kind of axis people can keep clearly in their minds.

However,if the stair is too near the door,it will be so public that its position will undermine the vital social character we have described.Instead,we suggest that the stair be clear,and central,yes—but in the common area of the building,a little further back from the front door than usual.Not usually in the ENTRANCE ROOM (130),but in the COMMON AREA AT THE HEART(129).Then it will be clear and visible,and also keep its necessary social character.

Therefore:

Place the main stair in a key position,central and visible.Treat the whole staircase as a room(or if it is outside,as a courtyard). Arrange it so that the stair and the room are one,with the stair coming down around one or two walls of the room.Flare out the bottom of the stair with open windows or balustrades and with wide steps so that the people coming down the stair become part of the action in the room while they are on the stair,and so that people below will naturally use the stair for seat.

最后，我们须确定楼梯的位置。楼梯当然是楼房内的交通枢纽。因此，必须一进大门就看得见它；而在一幢拥有许多各式房间的大楼里，楼梯一定要处于能够尽可能多地控制这些房间的位置，这样，它就能在人们的心目中清楚地起着一种轴心作用。

然而，如果楼梯离门太近，那就会处于大庭广众之中，这样的位置势必损害楼梯所具有的上述那种社会属性。我们提出与此不同的方案，楼梯的位置既要醒目，又要居中，但还须位于楼房的公用区，比常见的楼梯距离大门稍后一点。一般不在**入口空间**（130），而是在**中心公用区**（129）。这样，它让人一眼就看得见，并且也保持了所必需的社会属性。

因此：

将主楼梯设置在重要位置，居于中心，又容易看见。把整个楼梯当作一个房间（或者，如果它在室外，则为一个庭院）。设置楼梯时，应使它和房间合为一体，使楼梯绕着房间的一侧或两侧的墙壁下行。加宽楼梯的底部，并设有宽敞的窗户或栏杆，台阶也要相应加宽，这样，从楼梯上下来的人，虽然还在楼梯上，却已经能够参与室内活动；而在楼梯下面的人自然会利用楼梯当坐位。

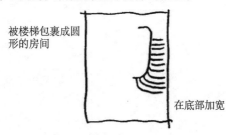

被楼梯包裹成圆形的房间

在底部加宽

❧❀❧

Treat the bottom steps as STAIR SEATS(125);provide a window or a view half-way up the stair,both to light the stair and to create a natural focus of attention—ZEN VIEW(134),TAPESTRY OF LIGHT AND DARK(135); remember to calculate the length and shape of the stair while you are working out its position—STAIRCASE VOLUME(195). Get the final shape of the staircase room and the beginnings of its construction from THE SHAPE OF INDOOR SPACE(191)...

把底部的踏步处理成**能坐的台阶**（125）；在楼梯的中间位置处开一窗户或观景孔，既可供楼梯采光，又可引起人们对自然景物的注意——**禅宗观景**（134）、**明暗交织**（135）；在你精心设计楼梯的位置时，切莫忘记求算它的长度和形状——**楼梯体量**（195）。**从室内空间形状**（191）确定楼梯间的最终形状并开始施工……

模式134　禅宗观景*

134   ZEN VIEW*

...how should we make the most of a view?It turns out that the pattern which answers this question helps to govern not the rooms and windows in a building,but the places of transition.It helps to place and detail ENTRANCE TRANSITION(112),ENTRANCE ROOM(130),SHORT PASSAGES(132),THE STAIRCASE AS A STAGE(133)—and outside,PATHS AND GOALS(20).

❧❧❧

The archetypal zen view occurs in a famous Japanese house,which gives this pattern its name.

A Buddhist monk lived high in the mountains,in a small stone house.Far,far in the distance was the ocean,visible and beautiful from the mountains.But it was not visible from the monk's house itself,nor from the approach road to the house.However,in front of the house there stood a courtyard surrounded by a thick stone wall.As one came to the house,one passed through a gate into this court,and then diagonally across the court to the front door of the house.On the far side of the courtyard there was a slit in the wall,narrow and diagonal,cut through the thickness of the wall.As a person walked across the court,at one spot,where his position lined up with the slit in the wall,for an instant,he could see the ocean.And then he was past it once again,and went into the house.

What is it that happens in this courtyard?The view of the distant sea is so restrained that it stays alive forever.Who,that

……我们应该怎样观景才最为合适呢？看来，回答这一问题的本模式所起的制约作用不在楼房内的房间和窗户，而是在过渡空间。本模式可确定下列模式的位置并对它们作出具体安排：**入口的过渡空间**（112）、**入口空间**（130）、**短过道**（132）、**有舞台感的楼梯**（133）——在外部空间，则有**小路和标志物**（120）。

৪৯৫৪

禅宗观景起源于举世闻名的日本住宅，本模式的名称由此而来。

曾经有一位僧人，隐居在高山之巅的一间小石屋里。从山顶上可以远眺美丽的大海。但从僧人的石屋里和通往石屋的路上都无法看到海景。不过，在石屋前面有一个庭院，院子的四周是一道厚实的石墙。当有人去僧人的石屋时，他穿过石墙的门洞就进入这个庭院，然后斜穿过去，到达石屋的大门。在庭院的较远一侧，石墙上有一条裂缝，它很窄小，斜对着石屋大门。当有人穿过庭院，走到跟院墙上的这条缝隙对准的一个地方，刹那间，他看见了大海。但此景稍纵即逝，接着他走进石屋。

僧人的石屋
The monk's house

这庭院内发生的情况究竟是怎么一回事呢？远处大海的风光如昙花一现，却长久地在人的脑际萦绕。目睹过这一景色的人，又怎能忘怀？它的魅力将永远不会消失。就连小石屋里的这位主人，五十年日复一日地观此景物，仍

has ever seen that view,can ever forget it?Its power will never fade.Even for the man who lives there,coming past that view day after day for fifty years,it will still be alive.

This is the essence of the problem with any view.It is a beautiful thing.One wants to enjoy it and drink it in every day. But the more open it is,the more obvious,the more it shouts,the sooner it will fade.Gradually it will become part of the building,like the wallpaper;and the intensity of its beauty will no longer be accessible to the people who live there.

Therefore:

If there is a beautiful view,don't spoil it by building huge windows that gape incessantly at it.Instead,put the windows which look onto the view at places of transition—along paths,in hallways,in entry ways,on stairs,between rooms.

If the view window is correctly placed,people will see a glimpse of the distant view as they come up to the window or pass it:but the view is never visible from the places where people stay.

৪১৩

Put in the windows to complete the indirectness of the view—NATURAL DOORS AND WINDOWS(221);place them to help the TAPESTRY OF LIGHT AND DARK(135);and build a seat from which a person can enjoy the view— WINDOW PLACE(180).If the view must be visible from inside a room,make a special corner of the room which looks onto the view,so that the enjoyment of the view becomes a definite act in its own right...

觉常见常新。

观景的诀窍就在于此。有一处秀丽景色，它惹人喜爱，吸引人来朝夕观赏，但它越是没遮没拦，越是一览无遗，越是引人入胜，却越易凋零。逐渐地它会变成建筑物的一部分，如同糊墙纸一样；屋子里的人再不会觉得这一景色有什么强烈的感人之美了。

因此：

如果有一处秀丽风光，请不要造个大窗让人在那儿不停地赏景，这样只会破坏景致。要采取别的办法，比如，把观景窗开在过渡空间——沿着走道、在门厅内、在入口处、在楼梯上、在房间与房间之间的地方。

如果观景窗设置得恰到好处，当人们走近窗口或从窗前经过时，他们能够瞥见远处的风景，但在人们停留的地方，远方的景色绝不会映入他们的眼帘。

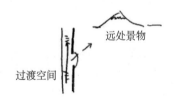

远处景物

过渡空间

∞⋘

筑窗以借景——**借景的门窗**（221）；让窗户造成**明暗交织**（135）；置一让人赏景的座位——**窗前空间**（180）。如果一定要从空间内部看到外界的风景，就必须专辟出面对景物的一角空间，使之成为观景点……

模式135 明暗交织*

135   TAPESTRY OF LIGHT AND DARK*

...passages,entrances,stairs are given their rough
position by THE FLOW THROUGH ROOMS(131),SHORT
PASSAGES(132),STAIRCASE AS A STAGE(133),ZEN
VIEW(134).This pattern helps you fine tune their positions by
placing light corectly.

༺༻

**In a building with uniform light level,there are few
"places"which function as effective settings for human events.
This happens because,to a large extent,the places which make
effective settings are defined by light.**

People are by nature phototropic—they move toward
light,and,when stationary,they orient themselves toward the
light.As a result the much loved and much used places in
buildings,where the most things happen,are places like window
seats,verandas,fireside corners,trellised arbors;all of them defined
by non-uniformities in light,and all of them allowing the people
who are in them to orient themselves toward the light.

We may say that these places become the settings for the
human events that occur in the building.Since there is good
reason to believe that people need a rich variety of settings in their
lives(see for instance,Roger Barker,*The Stream of Behavior:Explo-
rations of its Structure and Content*,New York:Appleton-Century-
Crofts,1963),and since settings are defined by "places," which
in turn seem often to be defined by light,and since light places can
only be defined by contrast with darker ones,this suggests that the
interior parts of buildings where people spend much time should

……过道、入口、楼梯的大致位置已由**穿越空间**（131）、**短过道**（132）、**有舞台感的楼梯**（133）、**禅宗观景**（134）给出，本模式通过合理选择采光，帮你调整它们的位置。

❦

在一幢光线均匀分配的建筑物里，没有多少"地方"可作为人们活动的有效环境。这是因为营造有效环境的空间在很大程度上是由光线来决定的。

人们天生是趋光的——他们总往光亮的地方走，而且，就是不挪动地方，他们也要使自己面向着亮光。结果，在建筑物内大家所喜爱的和经常使用的地方，即人们活动最频繁的空间，是窗前坐位、门廊、炉旁角落、棚架等地方；所有这些地方光线都是不均匀的，所有这些地方都可以使人面向着亮光。

我们可以说，这些地方就是建筑物里人们活动的环境。由于有充分理由认为，人们在生活中需要各种各样的环境（例见 Roger Barker, *The Stream of Behavior*: *Explorations of its Structure and Content*, New York: Appleton-Century-Crofts, 1963），又由于环境是受空间制约的，而空间看来又受光线制约，还由于明亮的空间只有在同较暗的空间对比才能显现出来，这就表明，经常有人居住的建筑物内部空间许多地方都应该明暗交织。建筑物应该有挂毯般明暗交织的鲜明画面。

这种明暗交织的效果还必须同人们流动的情况相适应。我们已经说过，人的本能都是趋光的。因此显而易见，任何入口、流通系统中任何一个枢纽，都必须有计划地做到比它周围的环境明亮——有光线（日光和灯光）照着这些地方，使它们由于明亮而成为目标。理由很简单：如果别

contain a great deal of alternating light and dark.The building needs to be a tapestry of light and dark.

This tapestry of light and dark must then fit together with the flow of movement,too.As we have said,people naturally tend to walk toward the light.It is therefore obvious that any entrance,or any key point in a circulation system,must be systematically lighter than its surroundings—with light(daylight and artificial light)flooded there,so that its intensity becomes a natural target.The reason is simple.If there are places which have more light than the entrances and circulation nodes,people will tend to walk toward *them*(because of their phototropic tendency)and will therefore end up in the wrong place—with frustration and confusion as the only possible result.

*If the places where the light falls are not the places you are meant to go toward,or if the light is uniform,the environment is giving information which contradicts its own meaning.*The environment is only functioning in a single-hearted manner,as information,when the lightest spots coincide with the points of maximum importance.

Therefore:

Create alternating areas of light and dark throughout the building,in such a way that people naturally walk toward the light,whenever they are going to important pla ces:seats,entrances,stairs,passages,places of special beauty, and make other areas darker,to increae the contrast.

ꙮ

Where the light to walk toward is natural light,build seats and alcoves in those windows which attract the movement— WINDOW PLACE(180).If you use skylights,then make

的地方比入口和流通枢纽更明亮，人们（由于人们的趋光性）就会往那些地方走去，结果却走错地方——这样只会使一切都乱了套。

如果光线明亮的地方不是你想去的地方，或者光线是均均分布的，那么环境传递的信息就会同它自身具有的意义相矛盾。只有当最重要的地方恰好最明亮时，环境才真正地起作用。

因此：

在整个建筑物要营造明暗交织的效果，这样一来，每当人们走向重要的地方（坐椅、入口、台阶、过道、布置得特别漂亮的地方）时，他们就会自然而然地向亮处走去，使其他一些地方暗一些，以增强对比度。

强烈的自然光

人们去的地方

❀❀❀

在人们通往自然光照明的地方，在吸引人逗留的窗前设置坐位或筑凹室——**窗前空间**（180）。如果你利用自然光，那么要在自然光周围的表面使用暖色——**暖色**（250）；不然的话，直接来自天空的光线几乎总是冷色的。夜间可使用白炽灯光池照明通道——**投光区域**（252）……

在翼楼及它们内部的空间和交通层次范围内，规定最重要的区域和空间。首先对住宅：

136. 夫妻的领域

137. 儿童的领域

the surfaces around the skylight warm in color—WARM COLORS(250);otherwise the direct light from the sky is almost always cold.At night make pools of incandescent light which guide the movement—POOLS OF LIGHT (252)...

within the framework of the wings and their internal gradients of space and movement,define the most important areas and rooms.First,for a house;

136. COUPLE'S REALM

137. CHILDREN'S REALM

138. SLEEPING TO THE EAST

139. FARMHOUSE KITCHEN

140. PRIVATE TERRACE ON THE STREET

141. A ROOM OF ONE'S OWN

142. SEQUENCE OF SITTING SPACES

143. BED CLUSTER

144. BATHING ROOM

145. BULK STORAGE

模式136　夫妻的领域*

……本模式有助于完善**家庭**（75）、**小家庭住宅**（76）和**夫妻住宅**（77）。它在**私密性层次**（127）上占有一定地位；如果还没有设置该层次的话，可有助于产生这一层次。

⊗⊗⊗

一个家庭有了孩子，常常会破坏夫妻共同需要的那种亲密关系和特殊的私密性。

136   COUPLE'S REALM*

...this pattern helps to complete THE FAMILY(75), HOUSE FOR A SMALL FAMILY(76)and HOUSE FOR A COUPLE(77).It also ties in to a particular position on the INTIMACY GRADIENT(127),and can be used to help generate that gradient,if it doesnt exist already.

ೞೞೞ

The presence of children in a family often destroys the closeness and the special privacy which a man and wife need together.

Every couple start out sharing each other's adult lives. When children come,concern for parenthood often overwhelms the private sharing,and everything becomes exclusively oriented toward the children.

In most houses this is aggravated by the physical design of the environment.Specifically:

1.Children are able to run everywhere in the house,and therefore tend to dominate all of it.No rooms are private.

2.The bathroom is often placed so that adults must walk past children's bedrooms to reach it.

3.The walls of the master bedroom are usually too thin to afford much acoustical privacy.

The result is that the private life of the couple is continually interrupted by the awareness that the children are nearby.Their role as parents rather than as a couple permeates all aspects of their private relations.

每对夫妇都开始享有彼此的成年人生活。但当孩子一出生，父母心往往超过夫妻恩爱，凡事唯宝宝是从。

在大多数住宅中环境的实际设计更助长了这种趋势，具体地说：

1. 孩子可以在屋子里到处乱跑，结果势必主宰一切。哪个房间也无私密性可言。

2. 浴室所在之处，常需大人经过儿童卧室方能到达。

3. 夫妻居室的墙壁通常太薄，容易传声，讲私房话不方便。

结果，夫妻的私房生活因他们觉得有孩子在身边而不断受到打扰。他们作为父母而不是作为夫妻的作用渗透到他们私人生活的全部领域。

当然，另一方面，他们不愿意完全同孩子们的房间分开。他们也愿意亲近孩子们，特别是在孩子还年幼的时候。母亲在幼儿遇到紧急情况时总想立即跑到孩子的床边。

只要住宅里有一个我们称之为夫妻领域的地方，也就是说，有这么一个环境，在那儿夫妻可以亲密相处、悲欢与共，这个问题就能得到解决。这个地方不仅同孩子们的天地分开，而且它本身也是完整的，是一个天地、一个领域。在许多方面，它是**夫妻住宅**（77）的一种，只不过它是一幢有孩子的更大的住宅而已。

夫妻的领域需要成为这么一个地方：人们可以在这儿坐坐，谈谈私事，它可能有通向户外和阳台的出口。这是一间起居室，一个处理私事的地方，一个安排计划的地方；里面有床位，但床放在有窗户的凹室；有一个壁炉则更好；它需要有双层门、一个前厅，以此保护其私密性。

因此：

在住宅中应专设一室，使之同公用区和孩子们的房间分开，供夫妻单独相处。此处应有走道可迅速到达孩子们住的房间，但无论如何应使它成为一个明确分开的独立领域。

On the other hand,of course,they do not want to be completely separated from the children's rooms.They also want to be close to them,especially while the children are young. A mother wants to run quickly to the bed of an infant in an emergency.

These problems can only be solved if there is a part of the house,which we call the couple's realm;that is,a world in which the intimacy of the man and woman,their joys and sorrows,can be shared and lived through.It is a place not only insulated from the children's world,but also complete in itself,a world,a domain.In many respects it is a version of the pattern HOUSE FOR A COUPLE(77),embedded in the larger house with children.

The couple's realm needs to be the kind of place that one might sit in and talk privately,perhaps with its own entrance to the outdoors,to a balcony.It is a sitting room,a place for privacy,a place for projects;the bed is part of it,but tucked away into an alcove with its own window;a fireplace is wonderful;and it needs some kind of a double door,an ante-room,to protect its privacy.

Therefore:

Make a special part of the house distinct from the com mon areas and all the children's rooms,where the man and woman of the house can be together in private.Give this place a quick path to the children's rooms,but,at all costs,make it a distincdy separate realm.

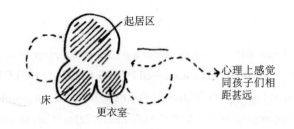

起居区

心理上感觉
同孩子们相
距甚远

床

更衣室

⁂

　　即使夫妻的领域很小，也要给它一个起居空间，以供休息、阅览、谈情说爱、弹琴唱歌之用——**坐位圈**（185）。使它**两面采光**（159）。在夫妻领域的中心地方放置床位——**夫妻用床**（187），使它充满晨曦——**朝东的卧室**（138），隔壁是**更衣室**（189）；如有可能，使浴室开向夫妻卧室——**浴室**（144）。要详细了解这个房间的形状及其构造，请参阅**室内空间形状**（191）。用**低门道**（224）或两扇门使这一领域保持其私密性——**居室间的壁橱**（198）……

∽○∾

Even if it's very tiny,give it a sitting area,a place to relax, read,make love,play music—SITTING CIRCLE(185).Give it LIGHT ONTWO SIDES(159).At the heart of the couple's realm,place the bed—MARRIAGE BED(187)so it has morning light—SLEEPING TO THE EAST(138),and,beside it,the DRESSING ROOM(189);if possible,try to place the bathing room to open off the couple's realm—BATHING ROOM(144). For the shape of this room in fine detail and its construction,see THE SHAPE OF INDOOR SPACE(191).And keep the area private with a LOW DOORWAY(224)or two doors— CLOSETS BETWEEN ROOMS(198)...

模式137 儿童的领域*

……在一幢**小家庭住宅**（76）里，有三个主要区域：**中心公用区**（129）、**夫妻的领域**（136）和儿童的领域，儿童的领域是与公用区交叉的。如果公用区和夫妻领域的位置安排好了，就有可能把这个部分分离、部分重合的儿童领域穿插其间，但我们认为它不是一个分隔的独立区域，而是住宅内专供儿童享用的、实际上仅仅局部隔开的小天地。它是独户住宅内**相互沟通的游戏场所**（68）的一部分。

❧❧❧

如果在儿童们需要的时候，没有空间供他们释放大量能量，他们会疯疯癫癫，把家里人烦个半死。

137 CHILDREN'S REALM*

...in a HOUSE FOR A SMALL FAMILY(76),there are
three main areas:a COMMON AREA AT THE HEART(129),a
COUPLE'S REALM(136),and a CHILDREN'S REALM
which overlaps the common area.If the common area and
couple's realm are in position,it is now possible to weave in
this partly separate,partly overlapping place for children,which
we call a realm,although we recognize that it is not a separate
realm but more an aspect of the house,reserved for children,a
mode of functioning which is physically separate only in
certain parts.It is that component of CONNECTED PLAY(68)
which acts within he individual houses.

❧

**If children do not have space to release a tremendous
amount of energy when they need to,they will drive themselves
and everybody else in the family up the wall.**

For a graphic example,visualize what happens when
children bring in friends after school and have a whole number
of ideas in their heads of what to do or play.They are loud and
boisterous after being pent up in school all day and they need
a lot of indoor and outdoor space to expend all this energy.
Obviously,the mood calls for space which contains long
distances because they suggest the possibility of physical
freedom much more.

And,in general,the child's world is not some single space
or room—it is a continuum of spaces.The sidewalk where he

在餐室中狂乱
A frenzy in the dining room

这是一个生动的例子，孩子们放学时带了朋友回家，脑子里想好了一大堆要做要玩的事情，请看这时他们会干出什么样的事来。在学校关了一整天之后，他们回到家里吵吵闹闹、东奔西跑，需要室内外有大片空间来消耗他们的能量。显然，这种精神状态要求有长距离的大空间，因为长距离使身体活动起来更加方便自由。

一般来说，儿童的天地不限于某一单个的空间或房间——它是空间的连续体。他卖汽水和跟朋友聊天所在的人行道、他邀朋友去玩的自家的室外游乐场、室内玩耍区、他可以单独跟朋友待在一起的屋子里属于他的私密空间、浴室、他母亲在干活的厨房、家里其他人所在的家庭室——对于一个儿童，所有这一切合起来构成他的世界。如果任何其他一个空间打断这个连续体，就会被吞没在儿童世界里，成为他的一部分流通过道。

如果私密性房间、夫妻的领域、安静的起居室都随意分布在儿童的世界，那它们必定会受到侵犯。但如果儿童的世界是一个连续的长条形地带，那么这些安静的、私密的、成年人的地方会得到保护，因为它们不会占据这个地盘。因此我们可以得出结论，凡儿童们需要和使用的空间都应该具有连续的长条形的几何形状，这个空间不包括夫妻的领域、成年人的私密房间或任何正式的、安静的起居空间。这一不间断的游戏场地还需要有某些附加的特征。

sells lemonade and talks with friends,the outdoor play area of his house into which he can invite his friends,the indoor play space,his private space in the house where he can be alone with a friend,the bathroom,the kitchen where his mother is,the family room where the rest of the family is—for the child,all of these together form his world.If any other kind of space interrupts this continuum,it will be swallowed up into the child's world as part of his circulation path.

If the private rooms,the couple's realm,the quiet sitting areas are scattered randomly among the places that form the children's world,then they will certainly be violated.But if the children's world is one continuous swath,then these quiet,private,adult places will be protected by the mere fact that they are not part of the continuum.We therefore conclude that all the places which children need and use should form one continuous geometrical swath,which does not include the couple's realm,the adult private rooms,or any formal,quiet sitting spaces.This continuous playspace needs certain additional properties.

1.Children are apt to be very demanding of everyone's attention when they are in this specially energetic state.The mother is particularly susceptible to being totally swallowed up by them.They will want to show her things,ask her questions,ask her to do things... "Look what I found.Look what I made.Where shall I put this?Where's the clay?Make some paint." The mother must be available for all this,but not forced to be in the thick of it.Her workroom and the kitchen need to be protected,yet tangential to the playspace.

2.The family room is also part of the continuum since it is where children and the rest of the family have contact with

1. 儿童们处于精力特别旺盛的时期，他们希望人人都关心他们。做母亲的特别容易一心扑在孩子们身上。孩子们会拿东西给她看，问她问题，要她干这干那……"瞧我找到什么了？瞧，我做了个什么？这东西往哪儿搁？橡皮泥在哪儿？给我画张画吧。"所有这些要求，母亲应当都能满足他们，但不要陷得太深。她的工作室和厨房需要保护好，但又要紧邻孩子们的游戏场所。

2. 家庭室也是这个连续体的一部分，因为它是孩子们和家庭其他成员彼此接触的地方。因此，游戏场所应当被纳入公用区——最好在其一边——参看**中心公用区**（129）。

3. 孩子们的私密空间（不管它们是凹室还是卧室）可以跟游戏场所分开，但它们必须有门可以关闭。孩子们的天性有时希望不跟家里人在一起——他们常邀请他们的亲密朋友到这样的地方来说说悄悄话，或者拿出一些奖励品来炫耀一番。

4. 通常开辟一个专用的游戏场所太费钱了；但把过道用作室内游戏场所总是可以实现的。它需略宽于边上有凹室和活动场所的一般过道（大约 7ft）。儿童们会利用空间具有的特点——他们看见一个猫儿洞似的空间，就决定玩房子游戏；看见一个高台，他们就打算演戏。因此，无论室内或室外的游戏场所，都需要有不同的高度、小凹角、柜台或桌子等。玩具、服装等物品也应该逐一排开陈列在这些地方。孩子们看见玩具，就可能拿它们玩耍。

5. 邻接室内空间的室外空间有一部分应该有顶，这样便于室内外过渡以强化连续性。

请记住，这类游戏场所既方便孩子，也方便了家里的成年人。在一幢住宅内如果孩子的活动范围逐渐扩大到全家，成年人生活中所需要的平静、珍贵和自由的环境就会遭其扰乱，并受其支配。如果孩子的活动天地具备本模式所提出的那种性质，足够开阔，成年人和孩子们就能在一

each other.The playspace,therefore,should enter the common area—preferably to one side—see COMMON AREA AT THE HEART(129).

3.The children's private spaces(whether they are alcoves or bedrooms)can be off the playspace,but it must be possible to close them off.Children naturally want to be exclusive at times—they often invite their closest friends into such a space for a private chat or to show off some prized possession.

4.It is usually too expensive to create a special playspace; but it is always possible to make a hallway function as the indoor part of the playspace.It needs to be a bit wider than a normal hall(perhaps seven feet)with nooks and stages along the edge.Children take up the suggestive qualities of spaces— on sight of a little cave-like space,they will decide to play house;on sight of a raised platform,they will decide to put on a play.Thus,both indoor and outdoor parts of the playspace need different levels,little nooks,counters,or tables,and so on.A lot of open storage for toys,costumes,and so forth should also be provided in these spaces.When toys are visible,they are more likely to be used.

5.The outdoor space just adjacent to the indoor space should be partially roofed,to provide transition between the two and to reinforce the continuity.

Remember that this kind of playspace is as much in the interest of the adults in the family,as in the interest of the children.If the house is organized so that the children's world gradually spreads throughout the home,it will disrupt and dominate the world of tranquility,preciousness,and freedom that adults need,to live their own lives.If there is an adequate children's world,in the manner described in this pattern,then

起共处，彼此间谁也不会支配谁。

因此：

首先，把完全属于孩子的小区域——他们的多床龛卧室——的位置安排好。把该区域放在住宅后方一个独立的地段，从多床龛卧室到街道之间形成一个连续的游戏场所，很像屋内一条宽阔的长条形地带。这一地带泥泞遍地，玩具成堆，跟孩子们常去的家庭室——主要是浴室和厨房——相通，经过公用区的一侧（但同安静的起居室和夫妻的领域完全分开，丝毫不侵犯它们），通向屋外街道，它或者自立门户，或者由入口空间出去，而止于有围合的户外小空间，然后与街道连通。此处应有顶棚遮盖，顶棚要足够大，使孩子雨天里也能在户外游戏。

在安排孩子床龛和街道之间这个长条形地带的位置时，使**农家厨房**（139）和**家庭工作间**（157）位于该地带的一侧，与它相通，但不受它侵犯。**浴室**（144）的位置也照此安排，并使它同孩子的卧室相通。根据**多床龛卧室**（143）设计多床龛儿童卧室；要让构成这一领域的长过道尽可能明亮温暖——**短过道**（132）；**有围合的户外小空间**（163），面积应足够大，可供孩子们活蹦乱跳地做活动……

both the adults and children can coexist,each without dominating the other.

Therefore:

Start by placing the small area which will belong entirely to the children—the cluster of their beds.Place it in a separate position toward the back of the house,and in such a way that a continuous playspace can be made from this cluster to the street,almost like a wide swath inside the house,muddy,toys strewn along the way,touching those family rooms which children need—the bathroom and the kitchen most of all— passing the common area along one side(but leaving quiet sitting areas and the couple's realm entirely separate and inviolate),reaching out to the street,either through its own door or through the entrance room,and ending in an outdoor room,connected to the street,and sheltered,and large enough so that the children can play in it when it rains,yet still be outdoors.

୫୦୯୬

As you place this swath between the children's beds and the street,place the FARMHOUSE KITCHEN(139) and the HOME WORKSHOP(157)to one side of the path, touching it,yet not violated by it.Do the same for BATHING ROOM(144),and give it some connection to the children's beds.Develop the cluster of children's beds according to BED CLUSTERS(143);make the long passages which form the realm as light and warm as possible—SHORT PASSAGES(132);make the OUTDOOR ROOM(163)large enough for boisterous activity...

模式138　朝东的卧室*

　　……在**私密性层次**（127）的最后部分，**夫妻的领域**（136）和**儿童的领域**（137）的位置可启发我们在何处设置卧室的思路。本模式解决卧室的位置，使它们面朝东，从而体现**室内阳光**（128）的作用,根据室内阳光这一模式，更具公共空间性质的房间都应朝南。

138   SLEEPING TO THE EAST*

...at the back of the INTIMACY GRADIENT(127),the position of the COUPLES REALM(136)and CHILDREN'S REALM(137),give some idea of where bedrooms will be.This pattern settles the position of the bedrooms by placing them to face the east,and thereby complements the effect of INDOOR SUNLIGHT(128),which places the more public rooms toward the south.

∞⁂

This is one of the patterns people most often disagree with.However,we believe they are mistaken.

People's attitude to this pattern often runs along the following lines:"The pattern suggests that I should sleep somewhere where the sun can wake me up;but I don't want the sun to wake me up;I want to be able to sleep late,whenever I can.I guess I have a different style of life;so the pattern doesn't apply to me."

We believe there may be fundamental biological matters at stake here and that no one who once understands them will want to ignore them,even if his present style of life does seem to contradict them.

The facts,as far as we can tell,are these.Our human organism contains a number of very sensitive biological clocks. We are creatures of rhythms and cycles.Whenever we behave in a way which is not in tune with our natural rhythms and cycles, we

　　这是人们常持异议的模式之一。但我们认为，这些人错了。

　　人们对待本模式的态度常常由下面一些话表现出来："这个模式提出，我该睡在阳光能唤醒我的地方，但我不想让太阳把我弄醒；只要可能，我就想睡得晚一点。我认为我的生活方式不同；所以这个模式对我不合适。"

　　我们认为，这里可能存在一些与人体关系重大的生物物质，一个人一旦对这些物质有所了解，就不会对它们置之不理了，即使他目前的生活方式看来与此相悖。

　　我们所能解释的事实如下：人的机体包含许多十分敏感的生物钟。每个人身上都有着自己的节奏和周期规律。每当我们的行为不能与自然界的节奏和周期规律相协调的时候，我们就极有可能打乱自身自然的心理和感情活动。

　　确切点说，这些周期规律跟睡眠有很大关系。太阳的运动周期支配我们的生理机能，致使我们的睡眠摆脱不了这一周期规律。请考虑这一事实，身体内的新陈代谢活动达到最低点大约是在夜里两点。那么看来很可能，有益身心的最好的睡眠其曲线多少与新陈代谢活动的曲线相重合——而新陈代谢活动又是跟太阳运动有关的。

　　伦敦博士最近在旧金山医科学校指出，我们一天的生活在很大程度上取决于我们睡醒时的状态。如果我们在做了一阵子梦（REM 睡眠）之后立即醒来，我们整天都会感到轻松愉快、精力充沛、神志清爽，因为 REM 睡眠刚结束，在我们的血流中会产生某种重要的激素。但是，如果我们在 delta 睡眠（另一类型的睡眠，它发生在两个梦境之间）中醒来，我们整天会感到易怒、昏昏欲睡、没精打采和心不在焉：因为在睡醒那一重要时刻，血流中没有相关的激素。

run a very good chance of disturbing our natural physiological and emotional functioning.

Specifically,these cycles have a great deal to do with sleep. And the cycle of the sun governs our physiology to such an extent,that we cannot afford to sleep out of touch with this cycle.Consider the fact that the body reaches its lowest metabolic activity in the middle of the sun's night,at about 2 A.M.It seems very likely,then,that the most nourishing kind of sleep is a sleep whose curve more or less coincides with the curve of metabolic activity—which is in turn dependent on the sun.

It has recently been shown by Dr.London at the San Francisco Medical School,that our whole day depends critically on the conditions under which we waken.If we wake up immediately after a period of dreaming(REM sleep),we will feel ebullient,energetic,and refreshed for the whole day,because certain critical hormones are injected into the bloodstream immediately after REM sleep.If,however,we wake up during delta sleep(another type of sleep,which happens in between periods of dreaming),we will feel irritable,drowsy,flat,and lethargic all day long:the relevant hormones are not in the bloodstream at the critical moment of awakening.

Now,obviously,anyone who is woken by an alarm clock,will sometimes be woken in the middle of delta sleep and will,on those days,have a lethargic day;and will sometimes wake up just after REM sleep and will,on those days,have an energetic day.Of course this is tremendously oversimplified—many other matters intervene.But if these facts about sleep are correct,they cannot help but have some impact on your waking hours.

Now,the only way to make sure that you wake up at the right time,with the closure of REM sleep,is to wake up

现在，很清楚，被闹钟唤醒的人，有时候是在 delta 睡眠中间醒来的。在这样的一天里，他是没精打采的。有时正好在 REM 睡眠之后醒来，在这样的一天里他就精力充沛。当然，这是把事情过分简单化了——还有其他许多因素在起作用。但是如果关于睡眠的这些事实确系如此，它们不能不对你醒着的时候产生某种影响。

现在，要有把握使你在实现了 REM 睡眠后适时地醒来，唯一的办法就是自然而然地睡醒。但只有当你在日出时醒来，你才是自然醒来，并与新陈代谢活动的另一个、更大的周期规律相一致。太阳使你温暖、使房间明亮，它轻轻地唤醒你——但它的呼唤是如此轻柔，实际上是在最适合于你醒来的时刻，让你在刚刚做完甜梦后醒来。

简而言之，我们认为，若要使自己整天身体健康、精神饱满、精力充沛，采用本模式是十分必要的——任何借口自己不想让太阳弄醒而拒绝这一模式的人，其实对于自己的身体功能的认知是十分错误的。

具体的办法是什么呢？你想看见阳光，但又不愿太阳照到床上，不然的话，你醒来时会感到太热和不舒服。一个最合适的房间应该有晨光弥漫——自然，房间里要有一个从东方射进光线的窗户，在床上能看到阳光，但不直接受到阳光直射。

最后，有必要提一提在床上能看见的景物。人们在黎明时往外张望，想看看这一天天气怎么样。有些景物能很好地反映出天气情况；有些则丝毫不能。一个设计得好的窗户应让人在黎明时分看到某种能反映季节和天气变化的静物或不断生长的东西，从而使人刚一睡醒就知道这一天天气如何。

因此：

使住宅中人们睡觉的地方朝向东方，让人们在太阳升起、光辉满屋时醒来。这清楚地表明，人们的卧室需要在

naturally.But you can only wake up naturally,and in accordance with the other, larger cycle of metabolic activity,if you wake up with the sun.The sun warms you,increases the light,gently nudges you to wake up—but in a way that is so gentle,that you will still actually wake up at the moment which serves you best—that is,just *after* a dream.

We believe,in short,that this pattern is fundamental to the process of having a healthy,active,energetic day—and that anyone who rejects this pattern on the grounds that he does not want to be woken by the sun,is making a serious mistake about the functioning of his or her own body.

What about details?You want to see the sunlight,but you don't want the sun to shine on the bed itself or you'll wake up hot and uncomfortable.The right kind of place is one which provides morning light—consequently a window in the room that lets in the eastern light—and a bed that provides a view of the light without being directly in the light shaft.

And finally,the matter of the view from the bed is worth mentioning.People look out in the morning to see what kind of day it's going to be.Some views give this information very well;others not at all.A good morning window looks out on some kind of constant object or growing thing,which reflects the changes of season and the weather,and allows a person to establish the mood of the day as soon as he wakes up.

Therefore:

Give those parts of the honse where people sleep,an eastern orientation,so that they wake up with the sun and light.This means,typically,that the sleeping area needs to be on the eastern side of the house;but it can also be on the western side provided there is a courtyard or a terrace to the east of it.

住房的东面；但只要有一处庭院或露台朝东，卧室也可以位于西面。

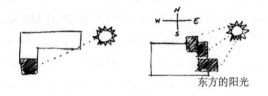

东方的阳光

ଓଡ଼

注意把所有的床位安排好，使它们都能够获得晨光，多床位的卧室如此——**夫妻的领域**（136）、**多床龛卧室**（143），单个的床位也均应如此，务使每张床都能从某一窗户得到从东方进来的阳光——**夫妻用床**（187）、**床龛**（188）。利用**过滤光线**（238）以避免太阳直接照射到床上。如果空间富裕，使窗户有**窗前空间**（180）。仔细选好床前窗的位置，使窗外景物能告诉一觉醒来的人天气变化的信息——**借景的门窗**（221）……

Place all the beds with care,so that they get the morning light,not only as a group—COUPLE'S REALM(136),BED CLUSTER (143),but individually,so that each gets eastern light from some specific window—MARRIAGE BED(187),BED ALCOVE(188).USE FILTERED LIGHT(238)to prevent the sun from shining too directly on the bed.If there is room,make this window function as a WINDOW PLACE(180).Place the window nearest the bed carefully so that it frames a view which tells a person waking what the weather is like—NATURAL DOORS AND WINDOWS(221)...

模式139　农家厨房**

　　……你在建筑物的中心已经布置或者已经设置某种公用区。在许多情况下，特别是在住宅中，这一公用区的核心是厨房或者餐室，因为共同进餐比起几乎其他所有事情都更能产生感情的共鸣——**中心公用区**（129）、**共同进餐**（147）。本模式介绍一种很古老的厨房，人们都在这同一个地方做饭、进餐和起居。

જીભ

　　同家里人分开而另设一处的厨房，只能看作是一个提供饭菜的食品加工厂，但不能给人以乐趣。它是使唤佣人的时代遗留下来的残迹；也是当今妇女们心甘情愿地充当厨娘的产物。

139 FARMHOUSE KITCHEN**

...you have laid out,or already have,some kind of common area at the center of the building.In many cases,especially in houses, the heart of this common area is a kitchen or an eating area since shared food has more capacity than almost anything to be the basis for communal feelings—COMMON AREA AT THE HEART(129),COMMUNAL EATING(147).This pattern defines an ancient kind of kitchen where the cooking and the eating and the living are all in a sigle place.

⋙⋘

The isolated kitchen,separate from the family and considered as an efficient but unpleasant factory for food is a hangover from the days of servants;and from the more recent days when women willingly took over the servants' role.

In traditional societies,where there were no servants and the members of a family took care of their own food,the isolated kitchen was virtually unknown.Even when cooking was entirely in the hands of women,as it very often was,the work of cooking was still thought of as a primal,communal function;and the "hearth," the place where food was made and eaten,was the heart of family life.

As soon as servants took over the function of cooking,in the palaces and manor houses of the rich,the kitchens naturally got separated from the dining halls.Then,in the middle class housing of the nineteenth century,where the use of servants

在传统的社会里没有仆人，家庭成员自己做饭，人们没有听说过隔离的厨房。甚至当炊事工作完全落在妇女手中已是司空见惯的时候，大家还是把厨房活计看做一家人要共同完成的大事；而"炉边"这个制作饭菜和进餐的地方，是家庭生活的中心。

一旦炊事工作由仆人接替时，在宫廷和豪富的庄园主宅第里，厨房自然就同餐厅分开了。后来，在 19 世纪，中产阶级使唤仆役之风甚盛，把厨房隔离开的做法也流行开来，而终于为家家户户所效仿。但是，尽管使唤仆役的时代已经过去，但厨房却依然分离，因为人们认为在看不到食品、闻不到气味的餐室进餐"斯文"而且"舒坦"。富人的住宅还是要用分离的厨房，他们认为餐室理应与厨房分开。

但在家庭中，这种厨房与餐室的分离使妇女们深受其苦。确实，毋庸赘言，由此造成的境遇已经使妇女在 20 世纪中期社会中的地位十分尴尬以致不堪忍受。很简单，负起炊事责任的妇女一旦甘心囹禁在"厨房"——不言而喻，即是情愿使自己成为女佣。

现代美国的住宅，进行所谓的开放性设计，已经朝解决这一冲突的方向前进了。厨房常常跟家庭室半分半合：不隔离，也不完全在家庭室之内。这倒能营造一种环境，使从事烹调的人在制作饭菜的时候仍能同家中其他的人接触。这就避免了由于洗碗处和厨房另设一处所带来的不快。

但这种设计前进的步伐不够大。如果我们透过表面来观察一下，在这类设计中仍然隐约地透露出那种错误的想法，认为厨房劳动是一种家庭杂务，吃饭才是享受。只要这一观念在安排住宅时占据上风，由隔离的厨房产生的矛盾就会始终存在。只有当全家人都能够彻底认识到，自己动手来做饭烧菜如同每日三餐需要吃喝一样，也是生活的

became rather widespread,the pattern of the isolated kitchen also spread,and became an accepted part of any house. But when the servants disappeared,the kitchen was still left separate,because it was thought "genteel" and "nice" to eat in dining rooms away from any sight or smell of food.The isolated kitchen was still associated with those houses of the rich,where dining rooms like this were taken for granted.

But this separation,in a family,has put the woman in a very difficult position.Indeed,it may not be too much to say that it has helped to generate those circumstances which have made the woman's position in mid-twentieth century society unworkable and unacceptable.Very simply,the woman who accepted responsibility for making food agreed to isolate herself in the "kitchen" —and subtly then agreed to become a servant.

Modern American houses,with the so-called open plan,have gone some way toward resolving this conflict.They very often have a kitchen that is half-separated from the family room:not isolated,and not entirely in the family room.This does create a circumstance where the people who are cooking are in touch with the rest of the family,while they are working. And it does not have the obvious stigma and unpleasantness of separated sculleries and kitchens.

But it does not go far enough.If we look beneath the surface,there is in this kind of plan still the hidden supposition that cooking is a chore and that eating is a pleasure.So long as this mentality rules over the arrangement of the house,the conflict which existed in the isolated kitchen is still present. The difficulties which surround the situation will only disappear,finally,when all the members of the family are able to

一部分，这种矛盾所造成的困难才会消除。要做到这一点，只有一家人像原始社会那样重新聚集在厨房的大工作台旁，大家都来处理日常家庭琐事，而不是把家务事都甩给仆人，自己却漠不关心。

我们相信，解决这个问题的办法在于采用老式的农家厨房这个模式。在农家厨房里，炊事工作和家庭活动完全在同一个大空间内进行。家庭活动集中在放在中央的大桌子周围：人们在此吃饭、谈天、打牌以及做各种活计，包括准备饭菜。厨房里的工作大家一起干，有的工作在桌子上干，有的工作在墙边的工作台上进行。还可以在墙角摆一张舒适的老式椅子，有人在干活累了的时候，可以在这儿睡个觉。

因此：

把厨房造得比一般的大一些，大到可以涵盖"家庭室"空间，让它靠近公用区的中心，不要像一般的厨房那样远远地设在住宅的后方。要让厨房里放得下一张可供大家使用的大桌子和一些椅子，椅子有软有硬，在厨房的边角处还要设置工作台、火炉和水池；要把它设计成一间明亮、舒适的用房。

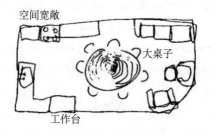

空间宽敞

大桌子

工作台

⊱◌◠⊰

让厨房可以**两面采光**（159）。以后设置厨房工作台时，

accept,fully,the fact that taking care of themselves by cooking is as much a part of life as taking care of themselves by eating. This will only happen when the communal hearth is once more gathered round the big kitchen table,as it is in primitive communities,where the taking care of necessary functions is an everyday part of life,and has not been lost to people's consciousness through the misleading function of the servant.

We are convinced that the solution lies in the pattern of the old farmhouse kitchen.In the farmhouse kitchen,kitchen work and family activity were completely integrated in one big room.The family activity centered around a big table in the middle:here they ate,talked,played cards,and did work of all kinds including some of the food preparation.The kitchen work was done cormnunally both on the table,and on counters round the walls.And there might have been a comfortable old chair in the corner where someone could sleep through the activities.

Therefore:

Make the kitchen bigger than usual,big enough to include the "family room"space,and place it near the center of the commons,not so far back in the house as an ordinary kitchen. Make it large enough to hold a good big table and chairs,some soft and some hard,with counters and stove and sink around the edge of the room;and make it a bright nd comfortable room.

∞∞

Give the kitchen LIGHT ON TWO SIDES(159).When you place the kitchen counters later,make them really long and generous and toward the south to get the light—COOKING LAYOUT(184),SUNNY COUNTER(199);leave room for an alcove or two around the kitchen—ALCOVES(179);make the table in the

要把它们做得确实又长又宽，放在南面以便照到阳光——**厨房布置**（184）、**有阳光的厨房工作台**（199）；在厨房周围留出一两处空间作凹室——**凹室**（179）；在中间摆一张大桌，在中央悬挂一盏漂亮的暖色大吊灯，把全家人都吸引在它周围——**进餐气氛**（182）；在进行细部设计时，在厨房的四壁设置许多敞开的搁架，在搁架上可放锅、杯、瓶和果酱坛——**敞开的搁架**（200）、**半人高的搁架**（201）。放置一把舒适的扶手椅——**起居空间的序列**（142）。关于这个空间的形状和构造，参阅**室内空间形状**（191）及之后几个模式……

middle big,and hang a nice big warm single light right in the middle to draw the family around it—EATING ATMOSPHERE(182);surround the walls,when you detail them,with plenty of open shelves for pots,and mugs,and bottles,and jars of jam—OPEN SHELVES(200),WAISTHIGH SHELF(201).Put in a comfortable chair somewhere—SEQUENCE OF SITTING SPACES(142).And for the room shape and construction,start with THE SHAPE OF INDOOR SPACE(191)...

模式140 私家的沿街露台**

公用区和起居空间——**中心公用区**（129）、**起居空间的序列**（142）——这二者之中至少要有一个能使屋里的人跟屋外街道上的活动相接触。本模式有助于建造**半隐蔽花园**（111），并使街道富有生气——**绿茵街道**（51）或**步行街**（100）。

ଧୋଏଡ଼

房屋与街道的关系往往十分混乱：要么房屋整个向街道敞开而毫无私密性可言，要么房屋背对街道，与街上活动断绝来往交流。

就我们本性而言，我们总是倾向于既有公共性，又有私密性。一幢完美的住宅正是这两者兼而有之：既有私宅

140   PRIVATE TERRACE ON THE STREET**

...among the common areas and sitting spaces—
COMMON AREAS AT THE HEART(129),SEQUENCE
OF SITTING SPACES(142)—there is a need for one,at
least,which puts the people in the house in touch with the
world of the street outside the house.This pattern helps
to create the HALF—HIDDEN GARDEN(111)and gives
life to the street—GREEN STREET(1)or PEDESTRIAN
STREET(100).

୫୦୯୪

**The relationship of a house to a street is often confused:
either the house opens entirely to the street and there is
no privacy;or the house turns its back on the street,and
communion with street life is lost.**

We have within our natures tendencies toward both
communality and individuality.A good house supports both
kinds of experience:the intimacy of a private haven and our
participation with a public world.

But most homes fail to support these complementary
needs.Most often they emphasize one,to the exclusion of
the other:we have,for instance,the fishbowl scheme,where
living areas face the street with picture windows and
the "retreat," where living areas turn away from the street
into private gardens.

The old front porch,in traditional American society,solved

的私密性，又能使我们参与公共活动。

　　但大部分住宅不能满足这种相辅相成的需要。它们往往顾此失彼：例如，我们现在流行"鱼缸"和"退避"两种规划，前者使起居空间朝街，从窗户看到街景；后者使起居空间背街而面向私家花园。

　　在传统的美国社会，古老的大门门廊妥善解决了这个问题。在街道十分安静而住宅临街的情况下，最好的解决办法莫过于此。但如果街道的情况不同，则需稍有不同的解决办法。

　　弗兰克·赖特在他从事建筑工作的早期，曾试验过一种可能的解决办法。当他在繁华的街道旁盖房的时候，他在起居室和街道之间造了一个很宽的露台。

　　据我们所知，格兰特·希尔德布拉德在他的论文中首先提出了赖特创作中的这一模式，这篇论文的题目是：《私密性和参与性：弗兰克·劳埃德·赖特和城市街道》（Privacy and Participation：Frank Lloyd Wright and the City Street，School of Architecture，University of Washington，Seattle，Washington：1970.）希尔德布拉德生动地描述了这一模式在切尼住宅中所起的作用：

私家露台与街道的剖面
Section of private terrace and street

　　"当行人从人行道往住宅张望的时候，露台的砖石墙使观望者视线的最高点落在了露台大门上的精制的铅框玻璃门上部的较低边缘上。因此，从人行道投向起居室的视线是要受到严格控制的。如果住在屋里的人站在门旁，只有他的头部和肩膀通过漫射表面模模糊糊地看得见。如屋子里的人坐着，路上行人则完全看不见他。

　　一方面，行人毫无可能侵犯住户的私密性，另一方面住宅的住户却有许多选择的自由。只要他走到露台上，或站或坐，因露

this problem perfectly.Where the street is quiet enough,and the house near enough to the street,we cannot imagine a much better solution.But if the street is different,a slightly different solution will be necessary.

Early in his career,Frank Wright experimented with one possible solution.When he built beside lively streets he built a wide terrace between the living room and the street.

To our knowledge,Grant Hildebrand first pointed out this pattern in Wright's work,in his paper, "Privacy and Participation:Frank Lloyd Wright and the City Street," School of Architecture,University of Washington,Seattle,Washingt on:1970.Hildebrand gives an interesting account of the way this pattern works in the Cheney house:

As the pedestrian looks toward the house from the sidewalk,the masonry terrace wall is located so that his line of sight over its top falls at the lower edge of the elaborately leaded upper glass zone of the terrace doors.Vision into the living room from the sidewalk thus is carefully controlled.If the occupant within the house is standing near the doors only his head and shoulders are dimly visible through a diffusing surface.If the occupant is sitting he is,of course,completely hidden from the pedestrian's view.

But whereas the pedestrian cannot effectively intrude on the privacy of the house,the inhabitant on the other hand has a number of options available at will.As he stands or sits on the terrace itself,well above the sidewalk,the effect is of easy participation in the full panorama of the street.From the elevated platform vision is unobstructed.Neighbors and friends can be waved at,greeted,invited in for a chat.Thus the terrace,projecting toward the street,linked— and still links—the Cheney house and its inhabitants to the community life of Oak Park.The configuration is so successful that,as in the Robie

台比人行道高了许多，街景就能一览尽收。在高台上视线是通行无阻的。在这里可向街坊和友人招手、问候，或邀其入内小叙。伸向街头的露台就以这种方式将切尼住宅和宅中的人跟橡树园的公共生活联系了起来，这种联系始终在进行着。因为构造合理，如在罗比住宅，从来不需挂窗帘。定位准确的女儿墙和铅框玻璃就起着窗帘的作用。因此，起居室面向街道并不牺牲私密性，而是为宅中人创造了丰富多采的生活环境。"

我们认为，赖特应用这一模式，说明他对人的基本需求有着深切的体会。住宅同外界街道相接触的需要是一种心理上的需求，这一认识确实是以实验为根据的：而它的反面——某些人力图使他们的住宅远离街道，把住宅闭锁起来，把它们同街道分隔开来——却是一种情感的严重畸化的症状——一种自主性消退症。（参阅亚历山大："The City as a Mechanism for Sustaining Human Contact," W. Ewald, ed., *Environment for Man*, Indiana University Press, 1967, pp.60~102.）

下面是一个来自希腊的有关这一模式的实例。很清楚，只要私密性与同街道接触的关系及它们之间的平衡得到维持，这一模式是可以用很多方法表达出来的。

私家的沿街露台
Private terrace on the street

因此：

使公用房间开向朝街的开阔的露台或门廊。使露台略高于街道，并以低墙保护它，如果坐在墙边，你可以从墙的上方望见街景，但墙却阻挡街上行人的视线，使他们看

house,there has never been much need for curtains.The parapets and the leaded glass,carefully placed,do it all.Thus out of the decision to face the living room toward the street has come not a sacrifice of privacy,but a much richer range of alternative experiences for the occupant.

We believe that Wright's use of this pattern was based on accurate intuitions about a fundamental human need. Indeed,there are empirical grounds for believing that the need for a house to be in touch with the street outside is a fundamental psychological necessity:and that its opposite— the tendency some people have to keep their houses away from the street,locked up,barred,and disconnected from the street—is a symptom of a serious emotional disorder— the autonomy-withdrawal syndrome.See Alexander, "The City as a Mechanism for Sustaining Human Contact," W. Ewald,ed.,*Environment for Man*,Indiana University Press, 1967,pp.60-102.

Here is an example of this pattern from Greece.It is clear that the pattern can be expressed in many ways,so long as the relationship,the balance of privacy and street contact,is maintained.

Therefore:

Let the common rooms open onto a wide terrace or a porch which looks into the street.Raise the terrace slightly above street level and protect it with a low wall,which you can see over if you sit near it,but which prevents people on the steet from looking into the common rooms.

不见这边的公用房间。

公共房间　　露台　　　　　　低墙

⊰⊱

　　如有可能，使露台的位置顺应天然的轮廓线——**梯形台地**（169）。如果墙较低，可作**可坐矮墙**（243）；在其他一些对私密性有较多要求的地方，你可以砌上一堵全部围合的花园围墙，墙上开一些像窗户一样的孔，通过它们跟街道联系——**花园围墙**（173）、**半敞开墙**（193）。在任何一种情况下，都要使露台有足够的围合，至少使它具有局部的空间感——**有围合的户外小空间**（163）……

If possible,place the terrace in a position which is also congruent with natural contours—TERRACED SLOPE(169). The wall,if low enough,can be a SITTING WALL(243);in other cases,where you want more privacy,you can build a full garden wall,with openings in it,almost like windows,which make the connection with the street—GARDEN WALL(173),HALF-OPEN WALL(193).In any case,surround the terrace with enough things to give it at least the partial feeling of a room—OUTDOOR ROOM(163)...

模式141　个人居室**

……**私密性层次**（127）清楚地表明，每幢住宅都需要有个人可以独处的房间。在任何一个人以上的家庭，这一需要是基本的和必不可少的——**家庭**（75）、**小家庭住宅**（76）、**夫妻住宅**（77）。

141 A ROOM OF ONE'S OWN**

...the INTIMACY GRADIENT(127)makes it clear that every house needs rooms where individuals can be alone. In any house hold which has more than one person,this need is fundamental and essential—THE FAMILY(75),HOUSE FOR A SMALL FAMILY(76),HOUSE FOR A COUPLE(77). This pattern,which defines the rooms that people can have to themselves,is the natural counterpart and complement to the social activity proided for in COMMON AREAS AT THE HEART(129).

❧

No one can be close to others,without also having frequent opportunities to be alone.

A person in a household without a room of his own will always be confronted with a problem:he wants to participate in family life and to be recognized as an important member of that group;but he cannot individualize himself because no part of the house is totally in his control.It is rather like expecting one drowning man to save another.Only a person who has a well-developed strong personal self,can venture out to participate in communal life.

This notion has been explored by two American sociologists,Foote and Cottrell:

There is a critical point beyond which closer contact with another person will no longer lead to an increase in empathy.(A) Up to a certain point,intimate interaction with others increases the

本模式对人们个人所有的居室作出了规定，它和**中心公用区**（129）所提供的社交活动起到自然的相辅相成的作用。

☢☣

一个人如果没有很多机会独处，是不可能亲近他人的。

一个人在家中没有一间属于自己的居室，总会遇到麻烦：他想参与家庭生活和被承认为家庭的一个重要成员；但他不能有独立的个性，因为家中没有一块地方完全归他所有。这很像希望一个快被淹死的人去援救别人。只有当一个人拥有受过良好培养的坚强的个性时，他才能敢于外出参与公共生活。

两位美国社会学家富特和科特雷尔对这一观点作了研究：

"移情作用存在一个临界点，越过这一点，跟他人接触再密切也不能增加情感的沟通。（A）在一定的点之下，跟其他人的亲密的相互接触会提高同他们情感交融的能力。但当他人出现得过于频繁时，人的机体看来会形成一种保护性的阻力，不再对他们作出反应。移情作用能力的这一局限性，在规划城市人口的最佳容量和密集程度时，以及规划学校和独户家庭住宅时，应当加以考虑。（B）如家庭能为孩子们提供独处的时间和空间，并教育他们从独自遐想中得到好处和满足，那么这样的家庭比起那些没有这样做的家庭表现出更高的平均移情能力。"（Foote，N.and L Cottrell，*Identity and Interpersonal Competence*，Chicago，1955，pp.72~73，79.）

亚历山大·莱顿也发表过类似的观点，着重指出由于经常得不到隐私权而造成的精神损伤（"Psychiatric Disorder and Social Environment，"*Psychiatry*，18（3），p.374，1955）。

就空间而论，解决这个问题需要什么呢？简单地说，每个人都要有个人居室。有一个地方可去，可以把门关起来：有一个安静的地方。有视觉和听觉上的私密性。并要做到使这些房间真正是属于个人的，它们必须位于住宅的最靠

capacity to empathize with them.But when others are too constantly present,the organism appears to develop a protective resistance to responding to them...This limit to the capacity to empathize should be taken into account in planning the optimal size and concentration of urban populations,as well as in plarming the schools and the housing of individual families.(B)Families who provide time and space for privacy,and who teach children the utility and satisfaction of with drawing for private reveries,will show higher average empathic capacity than those who do not.(Foote,N.and L.Cottrell,*Identity and Interpersonal Competence*,Chicago,1955,pp.72-73,79.)

Alexander Leighton has made a similar point,emphasizing the mental damage that results from a *systematic lack of privacy* 〔"Psychiatric Disorder and Social Environment," *Psychiatry*, 18(3),p.374,1955〕.

In terms of space,what is required to solve the problem? Simply,a room of one's own.A place to go and close the door;a retreat.Visual and acoustic privacy.And to make certain that the rooms are truly private,they must be located at the extremities of the house:at the ends of building wings;at the ends of the INTIMACY GRADIENT(127);far from the common areas.

We shall now look at the individual members of the family one at a time,in slightly more detail.

Wife.We put the wife first,because,classically,it is she who has the greatest difficulty with this problem.She belongs everywhere,and every place inside the house is in a vague sense hers—yet it is only very rarely that the woman of the house has a small room which is specifically and exclusively her own. Virginia Woolf's famous essay "A room of one's own" is the strongest and most important statement on this issue—and has given this pattern its name.

两端的地方；在翼楼的各端；在**私密性层次**（127）中的最后面；远离公用区。

现在让我们稍为详细一点来逐一看看家庭各成员的情况。

妻子。我们把妻子放在首位，因为在这个问题上她的困难最多，这情况历来如此。笼统地讲，住宅里面每个地方都是她的——但家庭主妇有一间专门的、完全属于她自己的小房间这种情形是很少见的。弗吉尼亚·伍尔夫的著名随笔《个人居室》是关于这个问题的最有力和最重要的说明——本模式就以此命名。

丈夫。在较老式的住宅里，家里的男人通常有自己的书房或工作间。然而，在现代的住宅和公寓里，这也同主妇的个人居室一样很少见到了，但这却当然是必不可少的。许多男人一提到家就想到小孩的捣乱和他在家庭担负的纷繁的杂务。如果他没有自己的居室，他就得待在自己的办公室里，离开家图个清静。

青少年。对于十几岁的孩子，我们有一个完整的模式来解决这个问题：青少年住所（154）。在那个模式里我们谈到，正是青少年面临着培养坚强个性的问题；但处于成年人中间，年轻人在家中常常得不到一方明确属于他们自己的空间。

孩子。很小的孩子不大觉得有私密性的需要——但他们有时还是有这个需要的。他们需要有地方放他们自己的东西，有时想自个儿待着，有时想要单独跟小伙伴玩耍。参阅多床龛卧室（143）和床龛（188）。约翰·玛奇写了一份关于家庭对私密性空间的需要的精采的调查报告（"Privacyand Social Interaction," Transactions of the Bartlett Society，Vol.3，1964~1965），关于孩子，他说：

"卧室常常是大部分个人财产的储藏所，孩子会因为有了这些财产感到心满意足，并以此显示与自己内部生活圈子里其他人的区别——确实，他宁可常常把这些告诉和他同年龄同性别的孩

Husband.In older houses,the man of the house usually had a study or a workshop of his own.However,in modern houses and apartments,this has become as rare as the woman's own room.And it is certainly just as essential.Many a man associates his house with the mad scene of young children and the enormous demands put on him there.If he has no room of his own,he has to stay at his office,away from home,to get peace and quiet.

Teenagers.For teenage children,we have devoted an entire pattern to this problem:TEENAGER'S COTTAGE(154).We have argued there that it is the teenagers who are faced with the problem of building a firm and strong identity;yet among the adults,it is the young who are most often prevented from having a place in the home that is clearly marked as their own.

Children.Very young children experience the need for privacy less—but they still experience it.They need some place to keep their possessions,to be alone at times,to have a private visit with a playmate.See BED CLUSTER(143)and BED ALCOVE(188).John Madge has written a good survey of a family's need for private space ("Privacy and Social Interaction,"Transactions of the Bartlett Society,Vol.3,1964-1965),and concerning the children he says:

The bedroom is often the repository of most of these items of personal property around which the individual builds his own satisfactions and which help to differentiate him from the other members of the inner circle of his life—indeed he will often reveal them more freely to a peer in age and sex than to a member of his own family.

In summary then,we propose that a room of one's own—an alcove or bed nook for younger children—is essential for each member of the family.It helps develop one's own sense

子，也不愿告诉自己家里的人。"

现在，可以简单地总结一下，我们认为，个人居室——对于年纪尚小的孩子，可以是一个凹室或床龛——对于家庭各个成员都是必不可少的。它有助于培养一个人自己的个性：它加强一个人同家里其他人之间的关系：它营造出了个人的领域，从而建立起跟住宅本身的联系。

因此：

给每个家庭成员（特别是成年人）一间个人居室。最小的个人居室是一个能放书桌、书架和有窗帘的凹室。最大的是一幢住所——青少年住所（154）或老人住所（155）。在所有的情况下，特别是在成年人的情况下，把所有这些居室都安排在私密性层次的最远端——远离公共空间。

私密的空间

死角

⊰⊱

利用本模式来消除由**中心公用区**（129）造成的"大家相聚一起"的极端情况。甚至对很小的儿童，也至少要在共同的卧室里给他们一个凹室——**床龛**（188）：对于夫妻，在他们共有的夫妻领域之外，再各分配一间独立的居室；它可以是一间扩大的更衣室——**更衣室**（189），家庭工作间——**家庭工作间**（157），或者也是同其他某个房间隔开的凹室——**凹室**（179）、**工作空间的围隔**（183）。如果经济上许可，甚至可以为一些家庭成员在主要建筑物之外盖附属的房子——**青少年住所**（154）、**老人住所**（155）。在每种情况下，至少应有空间可放置一张书桌、一张椅子以及**生活中的纪念品**（253）。关于这一空间的具体形状，参阅**两面采光**（159）和**室内空间形状**（191）……

of identity;it strengthens one's relationship to the rest of the family;and it creates personal territory,thereby building ties with the house itself.

Therefore:

Give each member of the family a room of his own, especially adults.A minimum room of one's own is an alcove with desk,shelves,and curtain.The maximum is a cottage— like a TEENAGER'S COTTAGE(154),or an OLD AGE COTTAGE(155).In all cases,especially the adult ones,place these rooms at the far ends of the intimacy gradient—far from the common rooms.

౮)Ⴃ

Use this pattern as an antidote to the extremes of "together ness" created by COMMON AREAS AT THE HEART(129). Even for small children,give them at least an alcove in the communal sleeping area—BED ALCOVE(188);and for the man and woman,give each of them a separate room,beyond the couples realm they share;it may be an expanded dressing room—DRESSING ROOM(189),a home workshop—HOME WORKSHOP(157),or once again,an alcove off some other room—ALCOVES(179),WORK SPACE ENCLOSURE(183). If there is money for it,it may even be possible to give a person a cottage,attached to the main structure—TEENAGER'S COTTAGE(154),OLD AGE COTTAGE(155).In every case there must at least be room for a desk,a chair,and THINGS FROM YOUR LIFE(253).And for the detailed shape of the room,see LIGHT ON TWO SIDES OF EVERY ROOM(159) and THE SHAPE OF INDOOR SPACE(191)...

模式142　起居空间的序列*

……在住宅、办公楼或公共建筑物的**私密性层次**
（127）的各个点上，都需要有起居空间。这些空间中有一
些可能是完全供起居用的房间，如老年人专用的起居室：
其他一些可能只是另外一些空间的某些区域或角落。本模
式讲述这些起居空间的范围和分布，以此来形成私密性
层次。

✺

**建筑物的每个角落都是一个可供利用的起居空间。但
每一个起居空间，依据其在私密性层次中的位置，对舒适
性和围合程度有不同的要求。**

从**私密性层次**（127）中我们知道，一幢建筑物在
其内部有一个自然的空间系列，从入口处外面的公共性
最大的区域到个人居室和夫妻领域这样一些最私密的空
间。以下是起居空间序列，它同**私密性层次**（127）大体
相等：

1. 入口外面——**入口空间**（130）、**大门外的条凳**
（242）；

2. 入口内部——**入口空间**（130）、**宾至如归**（149）；

3. 公共空间——**中心公用区**（129）、**短过道**（132）、
农家厨房（139）、**小会议室**（151）；

4. 半私密性空间——**儿童的领域**（137）、**私家的沿街
露台**（140）、**半私密办公室**（152）、**凹室**（179）；

5. 私密性空间——**夫妻的领域**（136）、**个人居室**（141）、
园中坐椅（176）。

142 SEQUENCE OF SITTING SPACES*

...at various points along the INTIMACY GRADIENT(127) of a house,or office,or a public building,there is a need for sitting space.Some of this space may take the form of rooms devoted entirely to sitting,like the formal sitting rooms of old;others may be simply areas or corners of other rooms. This pattern states the range and distribution of these sitting spaces,and helps create the intimacy gradien by doing so.

৪০৫৪

Every corner of a building is a potential sitting space. But each sitting space has different needs for comfort and enclosure according to its position in the intimacy gradient.

We know from INTIMACY GRADIENT(127)that a building has a natural sequence of spaces in it.ranging from the most public areas,outside the entrance, to the most private, in individual rooms and couples realms.Here is a sequence of sitting spaces that would correspond roughly to the INTIMACY GRADIENT(127) :

1.Outside the entrance—ENTRANCE ROOM(130), FRONT DOOR BENCH(242)

2.Inside the entrance—ENTRANCE ROOM(130), RECEPTION WELCOMES YOU(149)

3.Common rooms—COMMON AREAS AT THE HEART(129), SHORT PASSAGES(132), FARMHOUSE KITCHEN(139), SMALL MEETING ROOMS(151)

4.Half-private rooms—CHILDREN'S REALM(137),

那么，问题在哪里呢？简单点说，它是由下述情况造成的。人们习惯于只考虑起居室，似乎一幢建筑物，特别是一幢住宅，供起居用的只有一个空间。囿于这一想法，人们关心和注意的就只这一个起居室。他们忘记了这样的事实：人们的活动范围自然是遍及整幢房屋的，只不过各处的利用程度和私密程度不同而已——结果，整幢建筑物的起居空间都不能满足真正的起居和休息的需要。

为了解决这个问题，必须认识到，你的房屋应该包括一个有不同程度私密性的起居空间系列，这个系列里的每个空间所需要的围合和舒适程度要与其自身所处的位置相适应。请把注意力放在整个系列上，而不仅在一个房间上。要反躬自问，你所建造和修理的房屋有没有完整的起居空间系列？要创造这一系列，使之丰富多采，需要做些什么？

当然，你可能要建造一个专用的起居室——一个大厅、会客室、图书室或起居室——让它作为你住宅里起居空间中的一个。但别忘记，每个办公室和工作间也都需要有起居空间；厨房、夫妻领域、花园、入口空间，甚至走廊、屋顶、窗前空间也一样需要有起居空间。要精心选择起居空间系列，而在你进一步进行细部设计时，把它标志出来，对该系列中的不同空间，要给予同等的重视。

因此：

在整幢建筑物里设置一系列分级的起居空间，根据它们的围合程度区分它们。将最正式的起居空间完全围合，使之成为房间；把最不正式的设在其他房间的角落，在它们的周围不必加以任何屏挡；把中间的那种周围加以局部围隔，使它们跟某些更大的空间既相互连接，又部分分离。

PRIVATE TER-RACE ON THE STREET(140),HALF-PRIVATE OFFICE(152),ALCOVES(179)

5.Private rooms—COUPLES REALM(136),AROOM OF ONE'S OWN (141),GARDEN SEAT(176).

Now,what is the problem?Simply,it is the following.People have a tendency to think about the sitting room,as though a building,and especially a house,has just one room made for sitting.Within this frame of reference,this one sitting room gets a great deal of care and attention.But the fact that human activity naturally occurs all through the house,at a variety of degrees of intensity and intimacy,is forgotten—and the sitting spaces throughout the building fail to support the real rhythms of sitting and hanging around.

To solve the problem,recognize that your building should contain a sequence of sitting spaces of varying degrees of intimacy,and that each space in this sequence needs the degree of enclosure and comfort appropriate to its position. Pay attention to the full sequence,not just to one room.Ask yourself if the building you are making or repairing has the full sequence of sitting spaces,and what needs to be done to create this sequence,in its full richness and variety.

Of course,you may want to build a special sitting room— a *sala* or a parlor or a library or a living room—as one of the sitting spaces in your house.But remember that each office and workroom needs a sitting space too;so does a kitchen,so does a couple's realm,so does a garden,so does an entrance room,so does a corridor even,so does a roof,so does a window place. Pick the sequence of sitting spaces quite deliberately,mark it,and pay equal attention to the various spaces in the sequence as you go further into the details of the design.

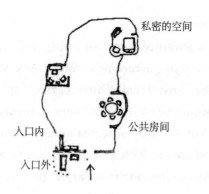

私密的空间

入口内

公共房间

入口外

&∞&

　　把最正式的起居空间设在**中心公用区**（129）和**入口空间**（130）：把半正式的起居空间也设在**中心公用区**（129），以及在**灵活办公空间**（146）、**等候场所**（150）和在**私家的沿街露台**（140）；把最私密的和最不正式的起居空间设在**夫妻的领域**（136）、**农家厨房**（139）、**个人居室**（141）和**半私密办公室**（152）。根据每个空间在起居空间的序列中的地位，在其周围进行围隔——**室内空间形状**（191）；无论在什么地方，都要在壁炉和窗户前合理地放置坐椅，使每个空间都让人感到舒服和闲适——**禅宗观景**（134）、**窗前空间**（180）、**炉火熊熊**（181）、**坐位圈**（185）、**户外设座位置**（241）……

Therefore:

Put in a sequence of graded sitting spaces throughout the building,varying according to their degree of enclosure. Enclose the most formal ones entirely,in rooms by themselves;put the least formal ones in corners of other rooms,without any kind of screen around them;and place the intermediate one with a partial enclosure round them to keep them connected to some larger space,but alsopartly separate.

❧☙

Put the most formal sitting spaces in the COMMON AREAS AT THE HEART(129)and in the ENTRANCE ROOM(130);put the intermediate spaces also in the COMMON AREAS AT THE HEART(129),in FLEXIBLE OFFICE SPACE(146),in a PLACE TO WAIT(150),and on the PRIVATE TERRACE ON THE STREET(140);and put the most intimate and most informal sitting spaces in the COUPLES REALM(136),the FARMHOUSE KITCHEN(139),the ROOMS OF ONE'S OWN(141),and the HALF-PRIVATE OFFICES(152).Build the enclosure round each space,according to its position in the scale of sitting spaces—THE SHAPE OF INDOOR SPACE(191);and make each one,wherever it is,comfortable and lazy by placing chairs correctly with respect to fires and windows—ZEN VIEW(134),WINDOW PLACE(180),THE FIRE(181),SITTING CIRCLE (185),SEAT SPOTS(241)...

模式143 多床龛卧室*

　　……我们已经把有关卧室规定陈述在**夫妻的领域**（136）和**儿童的领域**（137）之内。除此之外，卧室还要朝东以照到早晨的阳光——**朝东的卧室**（138）。本模式规定多床龛卧室，并有助于建造一般卧室。

δ૦∞∞

　　在家庭中每个儿童都需要有一个私密的空间，这个空间通常以床位为核心。但在许多文化中，甚至可能在所有的文化中都如此，如果幼儿睡眠的地方太隐蔽，让他们独自睡觉，他们会感到孤单的。

　　让我们来考虑一下儿童床位的各种可能的结构。一种

143　BED CLUSTER*

...the sleeping areas have.been defined to be inside the
COUPLE'S REALM(136)and CHILDREN'S REALM(137).
Beyond that,they are in places facing east to get the morning
light—SLEEPING TO THE EAST (138).This pattern defines
the grouping of the beds within the sleeping areas,and also
helps to generate the general sleepig areas themselves.

☙❦

**Every child in the family needs a private place,
generally centered around the bed.But in many cultures,
perhaps all cultures,young children feel isolated if they
sleep alone, if their sleeping area is too private.**

Let us consider the various possible configurations of the
children's beds.At one extreme,they can all be in one room—
one shared bedroom.At the other extreme,we can imagine an
arrangement in which each child has a private room.And then,in
between these two extremes,there is a kind of configuration in
which children have their own,small,private spaces,not as large
as rooms,clustered around a common playspace.We shall try to
show that both extremes are bad;and that some version of the
cluster of alcoves is needed to solve the conflict between forces
in a young child's life.

We first discuss the one room version.The problems in this
case are obvious.Children are jealous of one another's toys;they
fight over the light,the radio,the game being played,the door
open or closed.In short,for young children,especially in that age

极端的做法是把它们都放在一个房间里——一间共用的卧室。另一极端则是，我们可以设想，每个儿童都有一间私密的居室。其次，介于这两种极端之间还有一种结构，这就是使儿童有自己的小的私密空间，没有像居室那么大，使其聚集在共同的游戏场所周围。我们会设法说明，上述两种极端做法都不可取；要解决幼儿生活中各种力量之间的冲突，需要采用某种多床龛的卧室。

三种结构：共用卧室、独立卧室、多床龛卧室
Three configurations:the shared bedroom,isolated rooms,a cluster of alcoves

我们首先讨论共用卧室的方案。在这一情况下，问题暴露得很明显。孩子们看见彼此的玩具都想要；他们在做游戏时，房门忽开忽关，你争着要灯，我争着要收音机。总之，对于年龄较小的孩子，特别是对于处在已经懂得占有和控制的年龄的孩子，让他们住在多床位的一间居室里是会产生很多麻烦的。

为了避免这些麻烦，毫不奇怪，许多父母走到另一个极端——只要他们办得到，他们就给每个孩子一间个人居室。但这又产生新的麻烦，不过性质却完全不同：让小孩子独处一隅，使他们感到孤单。

共宿一室的习俗在十分讲究传统文化的国家，如秘鲁和印度，是特别流行的，甚至成年人也喜欢共宿。在这些国家，人们简直不喜爱独处，当他经常处于人群之中时，他们才心里舒坦，有了安全感。但即使像在美国这样一个"个人至上"的国家，虽然独处是司空见惯和认为是理所当然的，可至少孩子的心理还是彼此相同的。他们愿意有人陪着睡觉。例如，我们知道，幼儿喜欢在夜里把门稍稍

when feelings of possession and control are developing,the one room with many beds is just too difficult.

In the effort to avoid these difficulties,it is not surprising that many parents go to the other extreme—if they can afford it—an arrangement in which each child has his own room.But this creates new difficulties,of an entirely different sort:Young children feel isolated when they are forced to be alone.

The need for contact in the sleeping area is particularly true in strongly traditional cultures like Peru and India,where even adults sleep in groups.In these countries,people simply do not like to feel isolated and draw a great deal of comfort and security from the fact that they are constantly surrounded by people.But even in "privacy-oriented" cultures like the United States,where isolation is common and taken for granted,children,at least,feel the same way.They prefer to sleep in the company of others.For instance,we know that little children like to leave their door ajar at night,and to sleep with some light on;they like to go off to sleep hearing the voices of the adults around the house.

This instinct is so strongly developed in children of all cultures,that we believe it may be unhealthy for little children to have whole rooms of their own,regardless of cultural habit. It is very easy for a cultural relativist to argue that it depends on the cultural setting,and that a culture which puts high value on privacy,selfsufficiency,and aloneness,might very well choose to put each child in his own room in order to foster these attitudes.However,in spite of this potentially reasonable cultural relativism,it seems to us that although adults do need their own rooms,the isolation of a private room for a small child may perhaps be fundamentally incompatible with healthy

打开一点，睡觉的时候屋里有点灯光；他们喜欢去睡觉时听到房间周围有大人们说话的声音。

在所有的文化传统中，孩子们身上表现出来的这一特点都是如此强烈，使得我们确信，不管各地的文化习俗如何，让幼儿有一整间只属于个人的居室可能并不明智。文化相对主义者可能会争辩说，这得看文化背景，一个高度珍视隐私权、自主和独处的国家完全可以把每个孩子都安排在他个人的居室里以培养这些品质。然而，尽管文化相对主义者可能有理有据，但在我们看来，虽然成年人确实需要个人居室，但是让幼儿孤零零地独处一室却很可能与健康的社会心理习惯大相径庭；甚至导致身体损害。幸而，这种一个孩子一居室的模式，除了美国和受美国影响的地方外，世界上别的国家都没有广泛实行。而我们的观察确实表明，这一类美国模式造成感情冷漠，夸大了个人自立的概念，其结果，会使一个人形成社交需要与独处需要之间的内在的心理冲突。

因此我们面临两种互相冲突的力量。孩子们需要一些隐私权，需要有某种办法来避免为争夺地盘进行无休止的争吵，需要有一个比成年人的"个人居室"小一点的房间。可是同时，孩子们也需要跟其他人进行广泛的、差不多是身体上直接的接触——同他人谈话，接受他人的关怀和抚摩，聆听他人的声音，闻到他人身上散发的气味。

我们认为，要解决这一矛盾，只能采取二者兼具的办法；给孩子们提供属于他们自己的个人空间，这空间须分布在共同的游戏场所周围，这样，他们彼此可以相见，喧闹之声相闻，绝不会感到寂寞孤单。在那些文化传统中不大讲究隐私权的地方，可设置简单的床龛，给每个床龛装上帘子，这样一种多床龛卧室就可以造成足够的私密性，请参阅**床龛**（188）。在那些受到对隐私权有强烈要求的文化传统所影响的地方，床龛可以安排在各自的小房间里，使它们围绕着一个共同的空间。

psychosocial development;and might even do organic damage. It is significant that there is no culture in the world except the United States,and the offshoots of the United States,where this one-child-one-room pattern is widely practiced.And our observations do certainly suggest that this pattern is correlated with emotional withdrawal,and exaggerated conceptions of the individual's self-sufficiency,which,in the end,bring a person into inner conflicts between the need for contact and the need for withdrawal.

We thus face two conflicting forces.Children need some privacy,some way of retreating from endless squabbles about territory,some way of having a miniature version of the adult's "room of his own." Yet at the same time,they also need extensive,almost animal,contact with others—their talk,their care,their touch,their sound,their smell.

We believe that this conflict can only be resolved in an arrangement which gives them the opportunity for both;an arrangement of individual spaces which they "own," clustered around a common playspace so that they are all in sight and sound of one another,never too alone.In a culture with relatively little need for privacy,the clustered beds can get enough privacy by being set into simple,curtained bed-alcoves,see BED ALCOVE(188).In a culture where people have a strong need for privacy,the clustered beds may be in tiny rooms,surrounding a communal space.

Finally,two examples:One shows the way one lay-designer, working with this pattern language,interpreted this pattern.The other shows a cluster of beds in a Breton farmhouse.

最后，请看两个图例：一个图表明，一位非专业设计师在使用本模式语言时，领会了本模式；另一个图是布雷顿一家农户的一些床龛。

两种家庭自己设计的多床龛卧室
Two homemade bed clusters

因此：

把孩子的床位设在凹室或像凹室的小空间里，让它们围着共同的游戏场所。使每一凹室足够容纳一桌或一椅或一些搁架——至少也得有一些地面面积可供孩子放置他们自己的东西。给每个凹室挂上门帘，可以看到共同房间，但不要设置墙壁和房门，因为这些东西又会使床位跟外部过分隔绝。

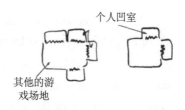

☙❧

共宿（186）给出了本模式的另一种方式，它更加适合于成年人。根据**床龛**（188），在这两种情况下，都要建造单独的凹室；如这些多床龛卧室是供儿童用的，根据**儿童的领域**（137）中的规定，要在中心设置游戏场所，并根据同一模式，还要设计从床龛经过厨房到户外的通道。利用更衣室和壁橱的位置，有助于形成多床龛卧室和单个床龛的形状——**更衣室**（189）、**居室间的壁橱**（198）；包括一些小的角落、旮旯——**儿童猫耳洞**（203）。使整个空间**两面采光**（159）。关于该空间的具体形状和构造，参阅**室内空间形状**（191）及之后几个模式⋯⋯

Therefore:

Place the children's beds in alcoves or small alcove— like rooms,around a common playspace.Make each alcove large enough to contain a table,or chair,or shelves—at least some floor area,where each child has his own things.Give the alcoves curtains looking into the common space,but not walls or doors,which will tend once more toisolate the beds too greatly.

୫୦୯୫

Another version of this pattern,more suitable for adults,is given by COMMUNAL SLEEPING(186).In both cases,build the individual alcoves according to BED ALCOVE(188);if the cluster is for children,shape the playspace in the middle according to the specifications of CHILDREN'S REALM(137),and make the path which leads from the beds,past the kitchen,to the outdoors,according to that pattern too.Use the location of dressing areas and closets to help shape the bed cluster and the individual alcoves—DRESSING ROOM (189),CLOSETS BETWEEN ROOMS(198);include some tiny nooks and crannies—CHILD CAVES(203).Give the entire space LIGHT ON TWO SIDES(159).And for the shape of this space in more detail and its construction,start with THE SHAPE OF INDOOR SPACE(191)...

模式144　浴室*

……本模式规定楼房内的主要浴室并确定它的位置。我们的做法完全改变了浴室的现有性质：它的位置既显眼，又重要，这样它也许有助于形成较大模式规定的睡眠区和共同区：**私密性层次（127）、中心公用区（129）、夫妻的领域（136）、儿童的领域（137）、朝东的卧室（138）、多床龛卧室（143）。**

☙❧

"我们称之为沐浴的活动其实仅仅是洗擦身体，而洗擦身体在过去只是沐浴的前奏。洗擦身体所在的地方虽然可

144   BATHING ROOM*

...this pattern defines and places the main bathroom of a building.It does it by changing the present character of bathing rooms completely:And its position is so dear,and so essential,that it will probably help to form the sleeping areas and public areas given by larger patterns:INTIMACY GRADIENT(127), COMMON AREAS AT THE HEART(129), COUPLES REALM(136),CHILDREN'S REALM(137), SLEEPING TOTHE EAST(138),BED CLUSTER(143).

৪৩েপ্ত

"The motions we call bathing are mere ablutions which formerly preceded the bath.The place where they are performed,though adequate for the routine,does not deserve to be called a bathroom."

Bernard Rudofsky

Rudofsky points out that cleaning up is only a small part of bathing;that bathing as a whole is a far more basic activity,with therapeutic and pleasurable aspects.In bathing we tend to ourselves,our bodies.It is one of the precious times when we are awake and absolutely naked.The relaxation of the bath puts us into sensual contact with water.It is one of the most direct and simple ways of unwinding.And,most astonishing,there is even evidence that we become less warlike when we tend to ourselves and our children in this way.

Cross culturally there is a correlation between the degree to which a society places restrictions on bodily pleasure—particularly

以应付日常生活中的需要，但实在不能叫做浴室。"

伯纳德·鲁道夫斯基

鲁道夫斯基指出，把身体洗干净只不过是沐浴的一个小插曲；整个沐浴过程是一次远比洗净身体重大得多的活动，它有治疗和娱乐的作用。在沐浴时我们照管自己，照管自己的身体。这是一个美妙的时刻，我们头脑清醒而浑身赤裸。洗澡的乐趣是使我们的身体同水接触并因此产生快感。这是使人身心轻松的最直接和最简单的办法之一。而且最令人吃惊的是，甚至有证据表明，当我们以这种方式照管我们自己和我们孩子的时候，我们会变得心平气和。

在各种文化传统中，社会对肉体的愉快施加限制的程度（特别在儿童时代）跟社会对赞美战争和施虐的程度是相互关连的。(Philip Slater，*Pursuit of Loneliness*，Boston：Beacon Press，1970，pp.89~90.)

我们应该记住……古代的大澡堂每天都顾客盈门，人们到那儿去洗澡，就像我们现在去餐馆吃饭一样自然。只不过比餐馆有过之而无不及；这些澡堂被认为是必不可少的。在公元四世纪，光是罗马这一个城市就有澡堂856个；600年以后，科尔多瓦拥有比这更多的澡堂——这使它名噪一时。(Rudofsky，*Behind the Picture Window*，New York：Oxford University Press，1955，p.118.)

芬兰桑拿浴
A Finnish sauna

但是，使人得到乐趣的澡堂沐浴好景不长。随着教会

in childhood—and the degree to which the society engages in the glorification of warfare and sadistic practices.(Philip Slater,*Pursuit of Leneliness*,Boston:Beacon Press,1970,pp.89-90.)

We ought to remember...that the thermae of old,with their routine of daily regeneration,were as much a matter of course to their users as our restaurants are to us.Only more so;they were considered indispensable.In the fourth century,the city of Rome alone counted 856 bathing establishments;six hundred years later,Cordoba boasted an even larger number of public baths—and who ever hears as much as its name?(Rudofsky,*Behind the Picture Window*,New York:Oxford University Press,1955,p.118.)

But bathing for pleasure has had a hard history.It went under ground with the Reformation of the Church,the Elizabethan Era,and Puritanism.It became a "scapegoat" for the evils of society—immorality,ungodliness,and disease.It is strange that we have not yet recovered from such nonsense.Contrast our approach to the bath,tub,and shower with these words, written in 1935 by Nikos Kazantzakis,the Greek novelist and poet,after his first Japanese bath:

I feel unsurpassed happiness.I put on the kimono,wear the wooden sandais,return to my room,drink more tea,and,from the open wail,watch the pilgrims as they go up the road beating drums...I have overcome impatience,nervousness,haste.I enjoy every single second of these simple moments I spend.Happiness,I think,is a simple everyday miracle,like water,and we are not aware of it.

We start,then,with the assumption that there are strong and profound reasons for making something pleasant out of bathing,and that there is something quite wrong with our present way of building several small and separate bathrooms,one for the master bedroom, one for children,perhaps one near

改革、伊丽莎白女王时代的到来和清教主义的出现，它隐入地下。它成了社会罪恶——不道德、否定上帝和疾病——的"替罪羊"。奇怪的是，我们至今尚未从这些荒唐的观念中清醒过来。请将我们对澡堂、浴池、淋浴的态度同希腊小说家和诗人卡赞察基斯于1935年第一次在日本洗完澡之后写下的话对照一下：

我感到无比的幸福。我穿上了和服，登上了木屐，回到自己的房中，再喝点茶，从敞开的墙壁中望着圣徒们敲着鼓在街上走着……我的不耐烦、紧张不安和仓促感都一扫而光。我感到我所度过的那些普通时光中的每一秒都是令人愉快的。我认为，幸福就跟水一样，是普通的日常奇迹，可我们并没有感受到。

那么，我们可以认定，洗澡能给人以愉快是有充分和深刻的理由的，而我们目前建造几个分隔的小浴室的办法是十分错误的，这些浴室一个归主卧室使用，一个给孩子们专用，或许还有一个靠近起居室——每个浴室都是一个紧凑的、效率很高的盒子。这些彼此分隔的、高效率的浴室绝不会给家庭一个机会享受洗澡时大家赤裸着或半裸着身子的亲热和乐趣。不过，显然这种享受是有限度的。家里的客人和偶然来访的人也应该可以使用浴室；如果其中一个人把门插上独自使用的话，一个浴室也不可能让全家人都用。但如果我们设想一个很大的浴室，大到足以使洗澡成为乐趣，那我们就会看到，这样的浴室不是每个家庭都有好几个的。

所有这些问题可以怎样解决呢？为了解决这些问题，我们将列出起作用的各种因素。然后问题就可以迎刃而解。

1. 第一，是一个新近出现的因素，这是我们已经提到过的——人们越来越渴望使洗澡重新成为一项使人身心愉快的有益的活动。

the living room—each one of them a compact efficient box. These separate,efficiency bathrooms never give a family the chance to share the intimacies and pleasures of bathing,of being naked and half-naked together.And yet,of course,this sharing has its limits.House guests and casual visitors must be able to use the bathroom too;and one bathroom will not work for a whole family,if any one person can lock the door and keep it to himself.Yet if we imagine a large bathing room,large enough to make bathing a pleasure,we see that we can certainly not afford more than one of them per family.

How can all these problems be resolved?In order to resolve them,we shall list the various forces which seem to be acting.Then we can untangle them.

1.First,the newly re-emerging force,which we have named already—the growing desire that people have to make their bathin ginto a positive re-generating pleasure.

2.Second,an increasing relaxation about nakedness,which makes it possible to imagine members of a family,and their friends,and even strangers,sharing a bath.

3.Third,the fact that this increasing relaxation has its limits;and that the limits are different for every person.Some people still want to be able to keep their nakedness private:they must be able to have a shower,or use the toilet,unseen,when they want to.

4.The fact that the habit of putting toilets in bathrooms(not next to them as they used to be),springs from the convenience of passing to and fro between the toilet and the bath—or shower—without dressing and undressing to go out into a passage.People want to be comfortably naked while they are in the bathroom going into the bathroom,going from the toilet

2. 第二，人们对待裸体的态度在日益放松，这就有可能设想一家人、他们的朋友甚至包括陌生人，一起来入浴。

3. 第三，对待裸体的态度的放松有其限度：而限度因人而异。有些人还是情愿把赤身裸体作为私密：他们愿意淋浴，或在厕所洗澡，不希望让别人看见。

4. 把浴室兼作厕所使用的习惯（在过去是厕所靠近浴室），是因为考虑在厕所和浴室（或淋浴间）之间来回走动的方便，这样不必为了走到过道上而穿衣服和脱衣服。人们在浴室里面喜欢舒舒服服地光着身子——走进浴室，从恭桶走到澡盆，刮脸等。仅仅因为做这样一类的事情就得穿衣服就太麻烦了。

5. 而且，家庭成员必须能够脱了衣服从卧室走到浴室而不经过公用区，对成年人尤其如此。

6. 客人能够使用浴室，因而应该能够不经过私密性空间或卧室就到达浴室。

这些因素中的基本矛盾看来是在公开性和私密性之间产生的。有理由把浴室的功能综合起来，也有理由使它们各自独立。这说明，应将浴室的所有功能进行综合以形成一个套间，这一套间或洗澡间应被视作住宅中的唯一浴室。但这个套间应开辟出来一些私密性领域，在那里人们可以关起门来或拉下帘子享受他们的隐私权。

我们设想单独盖一个浴室，它跟住宅其他部分分开，也跟公共户外建筑物分开。在这个空间内，有可能做到使浴缸和浴室其他部分恰当地联系起来，还要做到浴室向只使用水盆、喷头和马桶的人开放。我们建议，浴室应紧邻夫妻的领域——他们使用得最多——也要位于住宅的公共部分和私密部分之*间*，这样从家庭餐桌到浴室的走道不通过卧室或个人工作间。还要做到使从卧室到浴室的走道不经过从公用空间能看得见的区域。

to the bath,shaving,and so on.It is a nuisance to have to dress simply to negotiate any one of these connections.

5.And yet,the members of the family must be able to pass between bedrooms and bathroom,in various stages of undress,without passing through public areas.This is especially true of the adults.

6.And visitors must be able to use the bathing room,and must therefore be able to reach it without passing through the private rooms or bedrooms.

The fundamental conflict in these forces seems to be between openness and privacy.There are reasons to draw the functions of the bathroom together,and reasons to keep them separate.This suggests that all the functions of the bathroom be drawn together to form a suite,that this suite or bathing room be conceived of as the only bathroom in the house,but that private realms be created within this suite,where people can shut a door or pull a curtain and be private.

We imagine the entire bathing room tiled and protected from other parts of the house,and the public outdoors.Within this space it is possible to achieve the right connections between the bath itself and the other parts of the bathing room,and yet keep the bathing room proper open to people who want to use only the sink,the shower,or the toilet.We suggest that the room be placed next to the couple's realm—they will use it most— but also *between* the public part of the house and the private part of the house,so that the path from the family commons to the bathing room does not pass through the bedrooms or private workspace.And make sure paths from bedrooms to bathroom do not pass through any area which is visible from the common rooms.

解决裸体和浴衣这样微妙问题的一个简单方法是在浴室中几个地方都装一些显眼的毛巾架子，每个架子上挂几条大浴巾，这样浴巾就可以把人整个裹住。在这种情况下，洗澡的人如果对赤裸着身体感到不自在，可以很简单地把浴巾往身上一围，然后用它把身体包裹起来，不然，就甩掉它。它比正式的浴衣好得多，浴衣总没有一个合适的地方可放，而且太像衣服了。

浴缸的大小应该能够使两三个人舒舒服服地入浴——这样你才感到你是泡在水里，而不是从中进出。光线很起作用。如果私密性是一个问题，可用毛玻璃将自然光过滤；不然，用透明玻璃的窗户可以面对私人花园。

最后，说一下门的问题：使门装在合适的位置很重要，因为门对建立公开性和私密性之间的微妙平衡方面起的作用最大。我们设想，浴室的门总体上是不能关死的实心门；也许可用旋转门以满足那里的流动需要；然后在淋浴间有一道半透明的玻璃门或一个帘子；马桶间——这是最私密的处所——有一道小门；放浴缸的隔室是敞开的。洗手盆和毛巾、搁架及其他杂物都放在瓷砖铺砌的外间。

因此：

把住宅内的厕所、淋浴间和水盆都集中在一个瓷砖铺砌的区域。让浴室位于夫妻的领域旁边——有私密的通道——住宅中的私密区和公用区之间的中心地段；如有可能，浴室可以通向户外；也可通向小阳台和有围墙的花园。

放置一个大浴缸——大到至少可以使两个人完全没于水中；一个实用的喷头和水盆用来洗净身体；两三个挂大浴巾的架子——一个在门边，一个在喷头旁，一个靠近水池。

A simple way to cope with the subtleties of nakedness and gowns is to give the bathing room prominent towel racks in several places,each with a few giant towels,towels that people can wrap up in.Under these circumstances a person can simply throw a towel around himself and twist it together when he is uneasy about his nakedness,and otherwise let it drop.This is far better than the formal robes,which are always in the wrong place,and are too much like dressing.

The bath itself should be large enough so that two or three people can get themselves comfortably in the water so that you feel like staying,not rushing in and out.Light helps a lot. If privacy is an issue,natural light can filter through translucent glass;or a window with clear glass can overlook a private garden.

Finally,a word about the doors:It is important to place them correctly,as they do the most to establish the subtle balance between openness and privacy.We imagine solid unlockable doors to the bathing room as a whole;perhaps swinging doors to establish the fluidity of the area;and then opaque glass doors or curtains on the shower stall;a simple door for the toilet stalls—this is the most private spot;and an open doorway to the alcove which contains the bath.The sinks and the towels,the shelves,and all the other odds and ends are in the tiled outer zone.

Therefore:

Concentrate the bathing roon,toilets,showers,and basins of the house in a single tiled area.Locate this bathing room beside the couple's realm—with private access—in a position half-way between the private secluded parts of the house and the common areas;if possible,give it access to the outdoors;perhaps a tiny

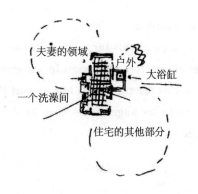

夫妻的领域

户外

大浴缸

一个洗澡间

住宅的其他部分

∞∞

　　最重要的是，务必有充足的光线——**两面采光**（159）和**过滤光线**（238）；设法把浴室的位置定好，让它开向花园的私密部分——**花园围墙**（173），或者甚至直接通向某一个当地的游泳池——**池塘**（71）。将厕所安排在堆肥间旁边——**堆肥**（178）；关于浴室的具体形状和构造，参阅**室内空间形状**（191）及之后几个模式……

balcony or walled garden.

Put in a large bath—large enough for at least two people to get completely immersed in water;an efficiency shower and basins for the actual business of cleaning;and two or three racks for huge towels—one by the door,one by the shower,one by the sink.

∞⊙∞

Above all,make sure that there is light,plenty of light—LIGHT ON TWO SIDES OF EVERY ROOM(159)and FILTERED LIGHT(238);try to place the bathing room so that it opens out into a private part of the garden—GARDEN WALL(173),and perhaps even gives direct access to some local swimming pool—STILLWATER(71).Line up the toilet with the compost chamber—COMPOST(178);and for the detailed shape of the room and its construction,start with THE SHAPE OF INDOOR SPACE(191)...

模式145 大储藏室

……本模式有助于对任何**小家庭住宅**（76）、**自治工作间和办公室**（80）以及**个体商店**（87）予以完善。推而广之，它也为充实每一**建筑群体**（95）所需要。

⋘⋙

在住宅和工作场所总需要有大的储藏空间；这里可存放手提箱、旧家具、老的案卷、木箱子等——所有那些你不打算扔掉而又不天天使用的杂物。

某些古老的建筑物拥有阁楼、地窖和库房等现成的储藏室。但这种储藏空间经常不受重视。例如，我们发现，在一些精心设计的建筑物里这个问题却被忽视，因设计师十分注意单方造价，他们不赞成建造属于非"起居空间"的多余的空间。

然而，根据我们的经验，大物件储藏空间是十分重要的。如果没有这样的空间，那就得腾出其他空间以供人们储藏所有那些大件的杂物。

大储藏室应占多大面积？当然，它不应占地过多。它只是供我们用来保存我们早已不用的旧东西。但它又必不可少。每个家庭、工作间或住宅团组都会有一些旧家具要暂时存放起来等待修理；都会有旧汽车轮胎、书籍、柜子和只是偶然动用一下的工具；家庭的杂物越多，需要的储藏空间越大。在极端的情况下，甚至需要有储藏建筑材料的空间！储藏室所需的面积不能少于建筑面积的 10%——有时高达 50%——通常为 15%～20%。

145 BULK STORAGE

...this pattern helps to complete any HOUSE FOR A SMALL FAMILY (76)， SELF-GOVERNING WORKSHOPS AND OFFICES(80),and INDIVIDUALLY OWNED SHOPS (87).More generally,it is needed to fill out every BUILDING COMPLEX(5).

❧❧❧

In houses and workplaces there is always some need for bulk storage space;a place for things like suitcases,old furniture,old files,boxes—all those things which you are not ready to throw away,and yet not using everyday.

Some old buildings provide for this kind of storage automatically,with their attics,cellars,and sheds.But very often this kind of storage space is overlooked.We find it neglected,for example,in carefully designed buildings,where the designer is watching the square foot costs closely and cannot justify an extra room that is not "living space."

In our experience,however,bulk storage space is terribly important;and when it is not provided,it usually means that some other space becomes the receptacle for all the bulky,marginal things that people need to store.

How much bulk storage should be provided?Certainly there should not be too much of it.That only invites us to keep old things that we have long since finished with.But some bulk storage is essential.Any household or workshop or cluster will have old furniture to store until it can be fixed,old tires,books,chests,tools that are only occasionally used;and the

因此：

不要到最后才想到大储藏室或干脆把它忘掉。在建筑物中应包括大储藏室的体量——它的地面面积至少应占总建筑面积的 15% ~ 20%——不能少于这个数量。将储藏室安排在建筑物中造价比别的空间低的地方——当然是因为它毋需装修。

＊＊＊

如果建筑物的屋顶是坡屋顶——**带阁楼的坡屋顶**（117）；可以把储藏室设置在屋顶的尖端，如果有坡形地基，则将储藏室设在地下室——**梯形台地**（169）、**底层地面**（215）：不然的话，将其设在以后也许要改建为住房的棚屋里——**出租房间**（153）。不管是阁楼、地下室还是棚屋，通常应遵照**背阴面**（162）中的原则，将大储藏间设置在建筑物的北边，而把向阳的空间留给房间和花园……

对于办公间、车间和公共建筑物，也要采取同样的办法。

146. 灵活办公空间

147. 共同进餐

148. 工作小组

149. 宾至如归

150. 等候场所

151. 小会议室

152. 半私密办公室

more self sufficient the household is,the more space it needs.In the extreme case,it is even necessary to have space for storing building materials!The amount needed is never less than 10 per cent of the built area—sometimes as high as 50 per cent—and normally 15 to 20 per cent.

Therefore:

Do not leave bulk storage till last or forget it.Include a volume for bulk storage in the building—its floor area at least 15 to 20 per cent of the whole building area—not less.Place this storage somewhere in the building where it costs less than other rooms—because,of course,it doesn't nee a finish.

ॐ∕ষ

Put the storage in the apex of the roof if the roof has a steep pitch—SHELTERING ROOF(117) ; if there is a sloping site, put it in a basement—TERRACED SLOPE(169), GROUND FLOOR SLAB (215) ; otherwise, put it in a shed which can perhaps be made into a cottage later—ROOMS TO RENT(153).No matter whether it is an attic, cellar, or shed, it is usually good advice to follow NORTH FACE(162)and situate bulk storage to the north of the building, leaving the sunny spaces for rooms and gardens...

then the same for offices,workshops,and public buildings.

146. FLEXIBLE OFFICE SPACE

147. COMMUNAL EATING

148. SMALL WORK GROUPS

149. RECEPTION WELCOMES YOU

150. A PLACE TO WAIT

151. SMALL MEETING ROOMS

152. HALF-PRIVATE OFFICE

模式146 灵活办公空间

……假定你已经安排好车间和办公间的基本面积——**自治工作间和办公室**（80）、**办公室之间的联系**（82）。同在住宅中一样，最基本的布局还是通过**私密性层次**（127）和**中心公用区**（129）进行的。在它们的总框架内，本模式有助于详细地规定工作空间、从而完善这些更大的模式。

❦❧

是否有可能创造这样一种空间，它能满足人们工作的特定要求，在其内部还能进行各式各样的多种安排和组合呢？

每一个人员组织都要经历一系列的变化。在办公间、各工作团组，它们的规模和功能都在发生变化——这些变化又常常是不可预料的。该如何设计办公空间以应付这一情况呢？

使办公空间灵活的标准做法是：（1）用组件隔板（全高或半高）造成连续空间；（2）用低天花但不用隔板形成整个连续空间的地面（称为"办公室景观"）。

但这两种解决办法都不现实。它们并不真正灵活。让我们逐一加以剖析。

首先我们讨论用隔板的解决办法。根据比较单纯的想法，看来这个问题显然可以通过活动隔板来解决。然而，实际上有着一系列严重的困难。

1. 如果隔板很容易移动，它们一定是用轻质材料做成的，因此隔音效果不好。

2. 如果隔板既易移动，隔音效果又好，它们通常价格昂贵。

146   FLEXIBLE OFFICE SPACE

...imagine that you have laid out the basic areas of a work shop or office—SELF GOVERNING WORKSHOPS AND OFFICES(80),OFFICE CONNECTIONS(82).Once again,as in a house,the most basic layout of all is given by INTIMACY GRADIENT(127)and COMMON AREAS AT THE HEART (129). Within their general framework,this pattern helps to define the working space in more detail,and so completes these largr patterns.

❧❧

Is it possible to create a kind of space which is specifically tuned to the needs of people working,and yet capable of an infinite number of various arrangements and combinations within it?

Every human organization goes through a series of changes.In offices,the clusters of work groups,their size and functions,are all subject to change—often unpredictably.How must office space be designed to cope with this situation?

The standard approaches to the problem of flexibility in office spaces are:(1)uninterrupted modular space with modular partitions(full height or half-height)and(2)entire floors of uninterrupted space with low ceilings and no partitions(known as "office landscape").

But neither of these solutions really work.They are not genuinely flexible.Let us analyze them in turn.

We discuss the partition solution first.In a naive sense,it seems obvious that the problem can be solved by movable partitions. However,in practice there are a number of serious difficulties.

3. 移动一个隔板的实际费用通常很高，甚至在高度"灵活"及"模块化"的系统里，隔板实际上也是很少移动的。

4. 最严重的是，在隔板系统中做些细小的变动往往也不可能。当一个工作小组在扩大而需要更多的空间的同时，邻室的工作小组正在缩小，这种情况是极为少见的。为了给正在扩大的小组腾出更多的空间，这个办公间有一大部分得重新布置，但这会引起许多麻烦，所以许多办公室管理部门采取比较简单的解决办法——不挪动隔板而调动人员。

5. 最后，由于办公空间的性质，某些非正式的、半永久性的安排，随着时间的推移，变得更具永久性（例如，家具布置、档案归置方法、占有某些空间或窗边的坐位）。这些都使得占用者不愿变动。虽然，当他们自己的工作小组扩大已迫在眉睫时，他们可能愿意搬迁，但是如果因为其他某个工作小组扩大或缩小而引起的办公间大调整影响到他们，他们就会强烈地抵制搬迁。

组件隔板法不可取，因为隔板实际上成了一般的墙壁；但它们不如真正的墙壁在隔断空间和隔声方面的用处大；况且，隔板不一定能满足半围隔工作空间的需要，在**工作空间的围隔**（183）中将对此予以讨论。这就很清楚了，安置活动隔板的办法并不能真正解决问题。

因为办公室景观没有隔板，所以其解决办法确实比较灵活。然而，这种方法仅适合于这样一些工作类型：它们既不需要高度私密性，也不需要几个工作小组之间有许多内部联系。再则，布赖恩·韦尔斯所作的研究表明，办公人员十分喜欢小工作空间，而不喜欢大的——参阅**工作小组**（148）。韦尔斯指出，如果有不同大小的办公间供大家选择，人们会选择小办公间的办公位置，而不会选择大的。他还指出，小办公室的工作小组比较团结（这是通过内部社会群体心理测

1.If partitions are made easy to move,they become lightweight and provide inadequate acoustic insulation.

2.If the partitions are both easy to move and acoustically insulated,they are usually very expensive.

3.The actual cost of moving a partition is usually so high that even in highly "flexible" and "modular" systems,the partitions are in fact very rarely moved.

4.Most serious of all:it is usually not possible to make minor changes in a partition system.At the moment when one working group expands and needs more space,it is only by rare accident that the working group next door happens at this same moment to be contracting.In order to make room for the expanding group,a large part of the office must be reshuffled,but this causes so much disruption that many office managements adopt the simpler solutions—they leave the partitions as they are and move the people.

5.Finally,it is in the nature of office space that certain informal,semi-permanent arrangements *grow more permanent over time*(for example,furnishing,filing systems, "ownership"of special spaces or windows).This makes the occupants resistant to change.Though they may be willing to move when the growth of their own working group is at stake,they will resist moving strongly,as part of any general office reshuffle,caused by the expansion or contraction of some other working group.

The modular partition system fails because the partitions become,in effect,ordinary walls;yet they are less useful than real walls for defining territory and for sound insulation;and what is more,the partitions do not necessarily satisfy the need for a semienclosed workspace,discussed in WORKSPACE ENCLOSURE(183).It is clear,then,that systems of movable

定法选择所占的较大百分比确定的)，大办公室的工作小组不如他们。(Pilkington Research Unit, *Office Design：A Study of Environment*, Department of Building Science, University of Liverpool, 1965, pp.113~121.)

那么，看来，灵活隔板或办公室景观两种方法无一可取。它们无一能够创造既完全适应某种特定的工作安排，又确实灵活的空间。某些机构利用改建过的住房作办公间，在这个问题上却从未遇到困难，这个事实无疑给我们以启示，使我们可以用完全不同的方法来解决灵活性问题。确实，看来这些老房子相较于表面上很灵活的组件隔板办公间，实际上能够提供更多的真正的灵活性。这个道理很简单。在这些老房子中，有许多小房间、少数大房间，还有许多局部隔断的空间，通常它们以各种方式相互连接着。

大小空间相结合
Mixture of room sizes

虽然这些空间是为维持家庭生活而设计的，但看来它们也可以维持工作小组的自然结构：有供私密或半私密办公室用的小空间，供 2 ～ 6 人的工作小组用的稍大空间，通常有一个可容纳 12 个人在一起的空间，还有一个周围有厨房和餐室的公共食堂。而且，在每一空间内部通常有各种墙、半截墙、窗前坐位，这些使空间内部富于变化。

partitions do not really solve the problem.

The office landscape solution,since it has no partitions,is more genuinely flexible.However,this system is only suitable for types of work which require neither a high degree of privacy nor much internal cohesion within individual working groups. Moreover,studies by Brian Wells have made it clear that office workers strongly prefer small work spaces to larger ones—see SMALL WORK GROUPS(148).Wells shows that,when given a choice among different sized offices,people choose desks in small offices rather than large ones.And he shows that working groups in small offices are much more cohesive(defined by a larger per centage of internal sociometric choices),than the working groups in large offices.(Pilkington Research Unit,*Office Design:A Study of Environment*,Department of Building Science,University of Liverpool,1965,pp.113-121.)

It seems then,that neither flexible partitions nor office landscape,really works.Neither creates space that is both well-adapted to specific work arrangements and truly flexible.A clue to an altogether different approach to flexibility comes from the fact that organizations which use converted houses as office space have no difficulty with this problem at all.Indeed,it appears that these old buildings actually provide more real flexibility than the apparent flexibility of modular partitioned offices.The reason is simple.In these old houses,there are many small rooms,a few large rooms,and many partially defined spaces,usually interconnected in a variety of ways.

Though these spaces were designed to support family life,they turn out also to support the natural structure of work groups:there are small spaces for private and half-private offices,slightly larger spaces for work groups of two to six,usually one space where up to

虽然墙壁不能说移动就移动——房子却是灵活的。工作小组可以在几分钟内就进行变动，不需付出代价。只要把一些门打开，一些门关上就行。而且隔音效果很好——因为大部分墙体是实心的，且多半是承重墙。

偶尔也可以把办公间或工作空间建成住宅模样——如果你预先对工作小组有足够的了解，可以根据房间和更大的空间的特殊性能使它们结合在一起。但是更可能出现的情况是，你对要占用这个空间的工作小组在建造该空间时其实并不了解。在这种情况下，不可能进行特定的"住宅模样"的设计。而只能设计和建造一种空间，它能逐渐地和系统地改为使用时所需要的住宅式的空间。

有这种可能性的空间不是"仓库"空间或"办公室景观"空间，而是人们以柱子和天花板高度变化的形式使他们在使用时可以改装的空间。如果柱子的位置这样安排，使得一些连接在柱上的隔板开始构成空间变化和室中之室，那么我们可以有把握地说，一旦人们开始在那里工作，他们能够在实际生活中改造这一空间以适应他们的需要。

关于柱子的几何布局，我们发现，如要让它起到最好作用，基本上要有一个中央空间（边上有过道）及一种把过道的开间形成工作空间的可能性。下图表示这个设想的概况以及这一模式在几年后可能被改造的方式。

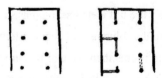

增设隔板
Adding partitions

当然，你可以用几乎无穷多的方法来增加不同大小的房间，并把各个空间结合起来，而总的轮廓不变。这些方

12 people can gather,and a commons centered around the kitchen and dining room.Furthermore,within each space there are usually a variety of walls,half-walls,window seats,which allow for changes within the rooms.

Although the walls cannot be moved at a moment's notice——the house is genuinely adaptable.Changes in work groups can be made in a few minutes,at no cost,just by opening and closing doors.And the acoustic characteristics are excellent——since most of the walls are solid,often load-bearing walls.

It is occasionally possible to build an office or a workspace like a house——when you know enough about the working group ahead of time to base the mix of rooms and larger spaces on their specific nature.But,*far more often*,the work groups which will occupy the space are unknown at the time the space is built.In this case,no specific "house-like" design is possible. Instead,it is necessary to design and build a type of space which can gradually,and systematically,be turned into this needed house-like kind of space once it is occupied.

The kind of space which will create this possibility is not "warehouse" space or "office landscape" space but instead,a kind of space which contains the possibility that people need,in the form of columns and ceiling height variety,to encourage them to modify it as they use it.If there are columns,so placed,that a few partitions nailed to the columns will begin to form differentiations and rooms within rooms,then we can be sure that people will actually transform it to meet their needs once they begin to work there.

As far as the geometrical layout of the columns is concerned, we have found that it works best when there is essentially a central space——with aisles down the sides——and the

法有时可能很简单，只需把开间布置成一排。在另一种情况下，开间可能弯弯曲曲，它们之间的房间和空间形状怪异。但这些细枝末节的问题是无关紧要的。重要的是柱子的总体位置，当然，在安排柱子的时候，要保证内部空间有充沛的阳光——**两面采光**（159）。

因此：

把办公间布置成开敞空间的一翼，其边沿上有独立柱，以便界定互相连通的半私密空间和公用空间。要设置足够多的柱子，以便人们能够以许多不同的方式（但始终是半永久性的）在多年内填满它们。

如果你在建造这个空间之前，恰好已经对将使用此空间的工作小组有所了解，那么可以把它造得更像住宅一点，使之更适合于他们的需要。不管是哪一种情况，都要在整个办公楼营造出一系列的空间——其样式之多足可与一幢规模宏大的老式住宅中大小不同、种类各异的空间相媲美。

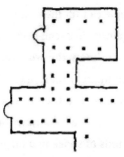

产生多种大小不同的空间的可能性
Possibility of many different sized rooms

❀❀❀

光线是至关重要的。这类工作空间一定要独立（以便在凹室后面有光线），或者整个开间的进深必须很浅，以便从两端引进足够的光线——**两面采光**（159）。利用**天花板**

possibility of forming the bays of the aisles into workspaces. The illustration below shows the general idea,together with the ways this pattern may be transformed after a few years.

Of course,you can add rooms of different sizes and combine spaces to follow this general outline in an almost endless variety of ways.In one case they may be rather simple,with bays laid out in rows.In another case,the bays may twist and turn,with odd sized rooms and spaces in between.The details are irrelevant.What matters is the general position of the columns and,*of course*,the guarantee that they are placed in such a way that there is plenty of natural light inside—LIGHT ON TWO SIDES OF EVERY ROOM(159).

Therefore:

Lay out the office space as wings of open space,with free standing columns around their edges,so they define half-private and common spaces opening into one another.Set down enough columns so that people can fill them in over the years,in many different ways—but always in a semi-permanent fashion.

If you happen to know the working group before you build the space,then make it more like a house,more closely tailored to their needs.In either case,create a variety of space throughout the office—comparable in variety to the different sizes and inds of space in a large old house.

❧❧

Light is critical.The bays of this kind of workspace must either be free-standing(so that there is light behind the alcoves),or the entire bay must be short enough to bring enough light in from the two ends—LIGHT ON TWO SIDES OF

高度变化（190）和**柱旁空间**（226）使可能形成的空间适当地混合在一起。最重要的是，要把工作空间布置成能够容纳两三个人工作，使他们得以部分接触，部分私密——**工作小组**（148）和**半私密办公室**（152）。把接待室安排在前方——**宾至如归**（149）；在中心公用区安排一个地方供大家每日共同进餐——**共同进餐**（147）……

EVERY ROOM(159).Use CEILING HEIGHT VARIETY(190) and COLUMN PLACES(226)to define the proper mix of possible spaces.Above all,lay the workspace out in such a way to make it possible for people to work in twos and threes,always with partial contact and partial privacy—SMALL WORK GROUPS(148)and HALF-PRIVATE OFFICE(152). Place a welcoming reception area at the front—RECEPTION WELCOMES YOU(149);and in the common areas at the heart arrange a place where people can eat together,everyday—COM MUNAL EATING(147)...

模式147　共同进餐*

……本模式有助于完善所有在内部拥有**中心公用区**（129）的那些人群和机构，它尤其有助于完善工作间、办公室和大家庭——**家庭**（75）、**自治工作间和办公室**（80）。其中，公用区因供人们同食共饮而显示活力。本模式将对此予以详述，并且指出它如何有助于产生一种更大的社会秩序。

∞∞

如果不共同进餐，人群则无法聚集在一起。

共同进餐的重要性在所有的人类社会都是显而易见的。

147 COMMUNAL EATING*

...this pattern helps complete all those human groups and institutions which have COMMON AREAS AT THE HEART(129)in them,and most of all it helps to complete workshops and offices and extended families—THE FAMILY(75), SELF-GOVERNING WORKSHOPS AND OFFICES (80).In all of them,the common area will draw its strength from the sharing of food and drink.This pattern defines it in detail,and shows also how ithelps to generate a larger social order.

⋇⋇⋇

Without communal eating, no human group can hold together.

The importance of communal eating is clear in all human societies.Holy communion,wedding feasts,birthday parties, Christmas dinner,an Irish wake,the family evening meal are Western and Christian examples,but every society has its equivalents.There are almost no important human events or institutions which are not given their power to bind,their sacral character,by food and drink.The anthropological literature is full of references.For example: "Food and Its Vicissitudes:A Cross-Cultural Study of Sharing and Nonsharing," in Yehudi A.Cohen,*Social Structure and Personality:A Casebook*,New York:Holt,1961.Audrey I.Richards,*Hunger and Work in a Savage Tribe:A Functional Study of Nutrition Among the Southern Bantu*.Glencoe,Ill.:Free Press,1932.

Thomas Merton summarizes the meaning of communal eating beautifully:

圣餐、喜酒、生日聚餐、圣诞晚宴、爱尔兰守夜、家庭晚餐，这些都是西方和基督教的习俗，但每个社会都有与此相当的活动。人们的重大事件和风俗习惯所具有的团聚人的力量和神圣的性质无一不是借酒宴表现出来的。此类例子在人类学文献中比比皆是。参看 "Food and Its Vicissitudes : A Cross-Cultural Study of Sharing and Nonsharing," in Yehudi A.Cohen, *Social Structure and Personality : A Casebook*, New York : Holt, 1961.Audrey I.Richards, *Hunger and Work in a Savage Tribe : A Functional Study of Nutrition Among the Southern Bantu*.Glencoe, Ⅲ . : Free Press, 1932.

托马斯·默顿出色地总结了共同进餐的意义：

"会餐具有这样的性质，它把人们吸引到餐桌上来，使他们把其他事情丢在一边来参与聚餐的快乐。同桌共聚会使人留下跟友人一起共享欢乐的记忆。共同进餐这个行动，与正式的宴会和其他一些节日请客很不一样，就其性质而论，是友谊和'共享'的象征。

现在，我们看不到这样的事实，即甚至我们日常生活中最普通的行动在本质上也带有深刻的精神意义。餐桌在一定意义上是家庭生活的中心，是家庭生活的表现。在这里，孩子们和父母在一起吃饭，饭菜体现着父母的爱。

因此，宴会也是一样。拉丁字 *convivium* 比我们的"宴会"或"筵席"更富神秘感。把筵席叫做 convivium，就是把它叫做"生活中共享的神秘"——客人们正是在这种神秘感中分享他们的主人以爱为他们准备的和赠给他们的美好东西，在这种神秘感中友谊和感激的气氛扩大为思想和感情的共鸣，并以共同的欢乐结束。"（Thomas Merton, *The Living Bread*, New York, 1956, pp.126~127.）

那么，很明显，共同进餐在差不多所有的人类社会中都起着极重要的作用，它把人们结合在一起，使他们更加感受到自己是集体的"一分子"。

A feast is of such a nature that it draws people to itself,and makes them leave everything else in order to participate in its joys.To feast together is to bear witness to the joy one has at being with his friends. The mere act of eating together,quite apart from a banquet or some other festival occasion,is by its very nature a sign of friendship and of "communion."

In modern times we have lost sight of the fact that even the most ordinary actions of our everyday life are invested,by their very nature,with a deep spiritual meaning.The table is in a certain sense the center of family life,the expression of family life.Here the children gather with their parents to eat the food which the love of their parents has provided....

So,too,with a banquet.The Latin word *convivium* contains more of this mystery than our words "banquet" or "feast." To call a feast a "convivium" is to call it a "mystery of the sharing of life" — a mystery in which guests partake of the good things prepared and given to them by the love of their host,and in which the atmosphere of friendship and gratitude expands into a sharing of thoughts and sentiments,and ends in common rejoicing.(Thomas Merton,*The Living Bread*,New York,1956,pp.126-127.)

It is clear,then,that communal eating plays a vital role in almost all human societies as a way of binding people together and increasing the extent to which they feel like "members" of a group.

But beyond this intrinsic importance of communal eating, as a way of binding the members of a group together,there is another important reason for maintaining the pattern,which applies especially to modern metropolitan society.

Metropolitan society creates the possibility of meeting a wonderful variety of people,a possibility almost entirely new in human history.In a traditional society,one learns to live with the people he knows,but the people he knows form a relatively

但是，除了共同进餐作为把一个集体的各个成员团结在一起的方式这一固有的重要性而外，维持这一模式还有另一重要理由——它特别适用于现代的大都市社会。

大都市社会创造同各种各样的人相遇的可能性，一种在人类历史上几乎全新的可能性。在一个传统的社会里，一个人学会跟他熟悉的人生活在一起，但他熟悉的只是有限的一群人，不大可能把数量扩充得很大。在现代都市社会中，每个人都有可能在城市中找到他很想与其在一起的其他少数几个人。从理论上讲，在一个五百万人口的城市中，一个人有可能正好遇到所有这五百万人中他愿意与之相处的那五六个人。

但这仅仅是理论而已。实际上很难做到。很少能有人信心十足地说，在他们居住的城市中，他们已经遇到他们可能最亲近的伙伴或找到他们愿意归属的非正式团体。事实正好相反，人们常常抱怨说，他们遇到的人不够多，与人交往的机会太少。人们远不能自由自在地了解社会上所有的人的本性，也远不能自由自在地跟其他那些有着最大的天然的相互吸引力的人在一起，相反，他们还要勉强和他们碰巧遇上的少数几个人在一起相处。

如何使大都市社会的巨大潜力得以实现呢？一个人如何才能找到对之有巨大吸引力的其他的人呢？

为了回答这个问题，我们必须弄清楚人们在社会上遇到陌生人的过程是如何发生的。对这个问题的回答取决于以下三个重要的假设：

1. 这个过程完全取决于社会上人群的相交，以及一个人能够通过这些人群扩大其联系面的方法。

2. 只要社会上不同的团体都拥有"团体领地"，在那里可以相互见面，那么这个过程就能发生。

3. 相遇过程看来尤其要取决于共同进餐饮酒，因而特别容易发生在那些至少已使共同进餐部分制度化了的团体。

closed group;there is little possibility of expanding it greatly.In a modern metropolitan society,each person has the possibility of finding those few other people in the city he really wants to be with.In theory,a man in a city of five million people has the possibility of meeting just those half dozen people who are the people he most wants to be with,in all of these five million.

But this is only theory.In practice it is very hard.Few people can feel confident that they have met their closest possible companions or found the informal groups they want to belong to in the cities they inhabit.In fact,on the contrary,people complain constantly that they cannot meet enough people,that there are too few opportunities for meeting people.Far from being free to explore the natures of all the people in society,and free to be together with those others who have the greatest natural and mutual affinities,instead people feel constrained to be with the few people they happen to have run into.

How can the great potential of metropolitan society be realized?How can a person find the other people for whom he has the greatest possible affinity?

To answer this question,we must define the workings of the process by which people meet new people in society.The answer to this question hinges on the following three critical hypotheses:

1.The process hinges entirely on the *overlap* of the human groups in society,and the way a person can pass through these human groups,expanding his associations.

2.The process can only take place if the various human groups in society possess "group territories" where meeting can take place.

3.The process of meeting seems to depend especially on communal eating and drinking and therefore takes place

如果这三个假设都如我们认为的那样是正确的话，那么很清楚，人们互相认识的过程很大部分要看他们能够在多大程度上从一个团体走到另一个团体，互相访问和串门，共同进餐。要实现这一点，每个机关和社会团体都得有自己定时开放的公共食堂，每个工作人员都可邀请客人来用膳，反过来也可随意应他曾相遇的客人的邀请，在其他的社交场合与人共餐。

因此：

使每个机关和社会团体都有一个可供人们在一起进餐的地方。使共同进餐制度化。特别是，要在每个工作地点都有公共午餐，要使大家坐在公共餐桌旁（不是在箱子、机器或提包上）吃顿像样的午饭成为重要的、令人心情舒畅的惯例。如有客人来，也可邀其入席。在我们自己位于中心的工作小组内，我们轮流做午饭，这件工作可以搞得很出色。午饭成了一件大事，它是一次聚会：轮到我们做饭的那一天，我们每个人都把爱和能力统统贡献了出来。

饭桌　　　　　　　　　　定时开饭

人们自己轮流做饭

ഔന

如果机构很大，想些办法把它分为若干小组，这些小组可以共同进餐，务使共同进餐的每个小组的人数不多于12个人——**工作小组**（148）、**小会议室**（151）。在食堂周围建造厨房，使之像**农家厨房**（139）；使餐桌成为重要的中心——**就餐气氛**（182）。

especially well in those groups which have at least partly institutionalized common food and drink.

If these three hypotheses are correct,as we believe,then it is plain that the process by which people meet one another depends very largely on the extent to which people are able to pass from group to group,as visitors and guests,at communal meals.And this of course can happen only if each institution and each social group has its own common meals,regularly,and if its members are free to invite guests to their meals and in turn are free to be invited by the guests they meet to other meals at other gatherings.

Therefore:

Give every institution and social group a place where people can eat together.Make the common meal a regular event.In particular,start a common lunch in every work place,so that a genuine meal around a common table(not out of boxes,machines,or bags)becomes an important, comfortable,and daily event with room for invited guests. In our own work group at the Center,we found this worked most beantifully when we took it in turns to cook the lunch. The lunch became an event:a gathering:something that each of ns put our love and energy into,on our day to cook.

ꕥ

If the institution is large,find some way of breaking it down into smaller groups which eat together,so that no one group which eats together has more than about a dozen people in it—SMALL WORK GROUPS(148),SMALL MEETING ROOMS(151).Build the kitchen all around the eating place like a FARMHOUSE KITCHEN(139);make the table itself a focus of great importance—EATING ATMOSPHERE (182)...

模式148 工作小组**

······在一个机构的工作空间——**自治工作间和办公室**
（80）、**灵活办公空间**（146）内部，还需要进一步细分。
最重要的是，正如本模式所要说明的，最小工作小组也都
必须有自己的有形空间。

❧❧❧

当至少有6个人在同一个地方工作时，一定不要让他
们勉强在一个一统的大空间内工作，而应使他们有可能把
工作空间加以分割，从而形成较小的小组。

148   SMALL WORK GROUPS**

...within the workspace of an institution—SELF-
GOVERNING WORKSHOPS AND OFFICES(80),
FLEXIBLE OFFICE SPACE(146),there need to be still further
subdivisions.Above all,as this pattern shows,it is essential
that the smallest human working groups each have their own
physical space.

∞⊗

**When more than half a dozen people work in the same
place,it is essential that they not be forced to work in one huge
undifferentiated space,but that instead,they can divide their
workspace up,and so form smaller groups.**

In fact,people will feel oppressed,both when they are
either working in an undifferentiated mass of workers and when
they are forced to work in isolation.The small group achieves
a nice balance between the one extreme in which there are so
many people,that there is no opportunity for an intimate social
structure to develop,and the other extreme in which there are so
few,that the possibility of social groups does not occur at all.

This attitude toward the size of work groups is supported by
the findings of the Pilkington Research Unit,in their investigations
of office life(*Office Design:A Study of Environment*, ed.Peter
Manning,Department of Building Science, University of
Liverpool,1965,pp.104-128).In a very large study indeed,office
workers were asked their opinions of large offices and small
offices. The statements they chose most often to describe their

事实上，许多工作人员彼此间不加区别地在一起工作，或者他们不得不分开独自工作，这两种情况都会使他们感到压抑。组织工作小组会使两种极端得到很好的平衡，一种极端是，很多人在一起没有可能形成一种亲密的社会结构；另一种极端是，人数太少，根本构不成社会团体。

皮尔金汤研究小组在调查办公室生活中的发现使这一对待工作小组规模的态度得到肯定。(*Office Design：A Study of Environment*，ed.Peter Manning，Department of Building Science，University of Liverpool，1965，pp.104~128) 在一项规模十分巨大的调研中，办公人员被询问关于他们对大办公室和小办公室的意见。他们在陈述他们的意见时最常说的话是："大办公室使人感到自己不怎么重要"和"在大办公室里时时让别人瞧见，感到很不自在。"当请他们把办公室的五种不同的可能布置加以比较时，工作人员坚定地选择工作小组最小的那几种。

按照选择顺序的五种布局
The five layouts in order of preference

对结果的分析也表明，"在小办公室工作的人反对大办公室比在大办公室里工作的人要多。"显然，只要人们有了在小办公室里工作的体验，他们会觉得去想象重新回到大办公室的环境中是很不愉快的。

我们调查对待工作空间的态度——在伯克利市政厅工作人员中听取意见——发现人们喜欢在 2～8 人的工作小组里工作。当人数超过 8 人时，小组作为人们的集合体会失去维系人们关系的作用；同时，几乎谁也不想一个人单独工作。

日本建筑师 T·高野在研究日本的工作小组时也报告了类似的发现。在他研究的办公室里，他发现，工作效率最高

opinions were: "The larger offices make one feel relatively unimportant" and "There is an uncomfortable feeling of being watched all the time in a large office." And when asked to compare five different possible layouts for offices,workers consistently chose those layouts in which workgroups were smallest.

Analysis of the results also showed that "the people who work in small office areas are more opposed to large office areas than those who actually work in them." Apparently,once people have had the experience of working in small groups,they find it very uncomfortable to imagine going back to the larger office settings.

In our own survey of attitudes toward workspace— taken among workers at the Berkeley City Hall—we found that people prefer to be part of a group that ranges from two to eight.When there are more than eight,people lose touch with the group as a human gathering;and almost no one likes working alone.

A similar finding is reported by the Japanese architect,T. Takano,in his study of work groups in Japan.In the offices he studied,he found that five persons formed the most useful functional group.(Building Section,Building and Repairs Bureau,Ministry of Construction:The Design of Akita prefectural government office,Public Buildings,1961.)

How should these small groups be related to each other? Brian Wells points out that while small offices support an intimate atmosphere,they do not support communications between groups. "The Psycho-Social Influence of Building Environment" (*Building Science,*Vol.1,Pergamon Press, 1965,p.153). It would seem that this problem can be solved

的小组是由 5 个人组成的。(Building Section, Building and Repairs Bureau, Ministry of Construction: The Design of Akita prefectural government office, Public Buildings, 1961.)

这些工作小组彼此间如何联系呢？韦尔斯指出，虽然小办公室形成了亲密气氛，但它们并不能使小组之间彼此沟通。"The Psycho-Social Influence of Building Environment"(*Building Science*, Vol.1, Pergamon Press, 1965, p.153)。看来，解决这个问题要靠对工作小组进行安排，使其中几个小组共用一些公共设施：饮水器、厕所、办公室设备，或许还有共同的前厅和花园。

因此：

把机构区划成小规模的、所占的空间面积一样的工作小组，每个小组的人数都在 6 人以下。把这些工作小组加以安排，让每个人至少部分

二至六人

相互可见

公共入口

地处于本小组中其他成员的视线之内；对几个小组做出统一安排，让它们共用一个公共入口、餐室、办公设备、饮水器、卫生间。

⁊❣⁋

对工作小组彼此之间的位置进行布置，使小组间的距离处于**办公室之间的联系**（82）的限度之内，并给每一小组的办公空间留有余地，以便扩大和缩小——**灵活办公空间**（146）；提供一个公用区，或者供本小组内部使用，或者供几个小组共用，或二者兼而有之——**中心公用区**（129）。把每个工厂或办公楼中的工作小组视作学习场所——**师徒情谊**（83）。使它自己有楼梯可直通街道——**室外楼梯**（158）。根据**半私密办公室**（152）和**工作空间的围隔**（183）来安排工作小组内部的单独的工作空间。

by arranging the small work groups so that several of them share common facilities:drinking fountains,toilets,office equipment,perhaps a common anteroom and garden.

Therefore:

Break institutions into small,spatially identifiable work groups, with less than half a dozen people in each. Arrange these work groups so that each person is in at least partial view of the other members of his own group; and arrange several groups in such a way that they share a common entrance, food,office equipment,drinking fountains, bathrooms.

৪০০৩

Lay the workgroups out with respect to each other so that the distances between groups is within the constraints of OFFICE CONNECTIONS(82),and give each group office space which leaves room to expand and to contract—FLEXIBLE OFFICE SPACE(146);provide a common area,either for the group itself or for several groups together or both— COMMON AREA AT THE HEART(129).Treat each small work group,in every kind of industry and office,as a place of learning—MASTER AND APPRENTICES(83).Give it its own stair,directly to the street—OPEN STAIRS(158).Arrange the individual workspaces within the small work group according to HALF-PRIVATE OFFICE(152)and WORKSPACE ENCLOSURE(183)...

模式149　宾至如归

……在一幢有许多人进出的公共建筑物或办公楼里，**自治工作间和办公室**（80）、**小服务行业**（81）、**旅游小客栈**（91）、**灵活办公空间**（146）——**入口空间**（130）内部的空间起着重要的作用；应在建造它的时候从一开始就使它具备合适的气氛。本模式最初是国立精神病研究所的克莱德·多塞特在一个为社区精神病诊所所制订的计划里提出来的。

♠♥♣♦

你是否曾经走进一幢公共建筑物被那里的招待员像处理一个包裹那样对待过？

要让一个人感到心情舒畅，你必须像欢迎他到你家做客那样对待他；走到他跟前，向他打招呼，请他坐在椅子上，请他吃些点心或喝点饮料，接过他的大衣。

在大多数机关内，来宾得走到招待员的面前；而招待员却一点也不主动，态度冷淡。招待员如想表现出热情好客，就应主动采取行动——迎上前去招呼客人，端出食品，让客人坐在炉旁，喝杯咖啡。因为这是最初的重要印象，应该首先让人感受到这样一种气氛。

我们知道的一个佳例是伦敦布朗旅馆的服务台。你进入这个旅馆时要走过一个不惹人注目的小入口，它很像一家住家的门口。你经过两三间房间；然后来到中心房间，那儿有两张老式的写字台。服务员从里面的办公间走出来，邀你坐在靠近登记台的一张舒适的椅子上，在你填写旅馆登记本的时候，他就坐在你的旁边。

大多数接待区完全不具备这种格调，其原因是招待员

149   RECEPTION WELCOMES YOU

...in a public building,or an office where there are many
people coming in,SELF-GOVERNING WORKSHOPS
AND OFFICES(80),SMALL SERVICES WITHOUT RED
TAPE(81),TRAVELERS INN(91),FLEXIBLE OFFICE
SPACE(146)—the place inside the ENTRANCE ROOM(130)
plays an essential role;it must be built from the very start with
the right atmosphere.This pattern was originally proposed by
Clyde Dorsett of the National Institute of Mental Health,in a
program for community mental health clinics.

ಬಿೂೞ

**Have you ever walked into a public building and been
processed by the receptionist as if you were a package?**

To make a person feel at ease,you must do the same for
him as you would do to welcome him to your home;go toward
him,greet him,offer him a chair,offer him some food and
drink,and take his coat.

In most institutions the person arriving has to go toward
the receptionist;the receptionist remains passive and offers
nothing.To be welcoming the receptionist must initiate the
action—come forward and greet the person,offer a chair,food,a
seat by the fire,coffee.Since it is first impressions which
count,this whole atmosphere should be the first thing a person
encounters.

A beautiful example we know is the reception desk at
Browns Hotel in London.You pass into the hotel through a
small,unassuming entrance,not unlike the entrance to a house.

的服务台形成一道障碍，服务台和设备助长了气氛的衙门化，令人完全没有宾至如归的感觉。

因此：

在刚走进入口的地方，准备好迎宾用的整套物品——沙发椅、壁炉、食品、咖啡。不要把服务台设置在服务员和来宾之间，而要把它放在角落的一边——这样，男女服务员能够站起来迎接来宾，向他们致意，请他们坐下来。

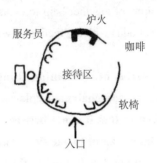

* * *

要精心安排壁炉的位置，使它成为中心——**炉火熊熊**（181），使服务员有一个舒适的工作空间，而且还营造出宾至如归的气氛——**工作空间的围隔**（183）；使这个空间**两面采光**（159）；多半可以设一凹室或窗前坐位供等候的人使用——**等候场所**（150）、**凹室**（179）、**窗前空间**（180）。务必使接待点比周围地方明亮——**明暗交织**（135）。关于接待空间的形状，参阅**室内空间形状**（191）及之后几个模式……

You pass through two or three rooms;then come to the central room in which there are two old writing desks.The receptionist comes forward from an inner office,invites you to sit down in a comfortable chair at one of these writing desks,and sits down with you while you fill out the hotel register.

The reasons most reception areas fail completely to have this quality,is that the receptionist's desk forms a barrier,so that the desk and equipment together help to create an institutional atmosphere,quite at odds with the feeling of welcome.

Therefore:

Arrange a series of welcoming things immediately inside the entrance—soft chairs,a fireplace,food,coffee.Place the reception desk so that it is not between the receptionist and the welcoming area,but to one side at an angle—so that she, or he,can get up and walk toward the people who come in, greet them,and then invite tem to sit down.

୫୬୯୪

Place the fireplace most carefully,to be a focus—THE FIRE(181)give the receptionist a workspace where she can be comfortable in her own work,and still make visitors feel welcome—WORKSPACE ENCLOSURE(183);give the space LIGHT ON TWO SIDES(159);perhaps put in an alcove or a window seat for people who are waiting—A PLACE TO WAIT(150),ALCOVES(179),WINDOW PLACE(180).Make sure that the reception point itself is lighter than surrounding areas—TAPESTRY OF LIGHT AND DARK(135).And for the shape of the reception space start with THE SHAPE OF INDOOR SPACE(191)...

模式150 等候场所*

……任何一个办公楼、工作间、公共服务部门、车站或诊所，凡是人们需要等候的地方——**换乘站**（34）、**保健中心**（47）、**小服务行业**（81）、**办公室之间的联系**（82），都要提供一个专用的等候场所，尤其应该让这一场所不会带有通常候人场所那种肮脏的、闭塞的、令人感到苦熬时间的性质。

150 A PLACE TO WAIT*

...in any office,or workshop,or public service,or station,
or clinic,where people have to wait—INTERCHANGE(34),
HEALTH CENTER(47),SMALL SERVICES WITHOUT
RED TAPE(81),OFFICE CONNECTIONS(82),it is essential
to provide a special place for waiting,and doubly essential that
this place not have the sordid,enclosed,timeslowed character of
orinary waiting rooms.

❧❧❧

The process of waiting has inherent conflicts in it.

On the one hand,whatever people are waiting for—
the doctor,an airplane,a business appointment—has built in
uncertainties,which make it inevitable that they must spend a
long time hanging around,waiting,doing nothing.

On the other hand,they cannot usually afford to enjoy
this time.Because it is unpredictable,they must hang at the
very door.Since they never know exactly when their turn
will come,they cannot even take a stroll or sit outside.They
must stay in the narrow confine of the waiting room,waiting
their turn.But this,of course,is an extremely demoralizing
situation:nobody wants to wait at somebody else's beck and
call.Kafka's greatest works,*The Castle and The Trial*,both deal
almost entirely with the way this kind of atmosphere destroys a
man.

等候过程天生令人心烦意乱。

一方面，不管人们在等候什么——医生、飞机、商务约见——总不是有十分把握，这就不可避免地导致他们长时间地徘徊、等待、无所事事。

另一方面，这段时间常常令人感觉百无聊赖。因为它不可预测，人们只好在门边待着。由于他们心里不确定，不知什么时候轮到他们，他们甚至不能去遛个弯儿，或在外面坐一会儿。他们只能待在窄小的等候室的角落里，等着轮到他们。但这种环境当然是极为烦人的：没有人愿意听任别人摆布。卡夫卡的两本巨著《城堡》和《审判》差不多全是描写这种气氛是如何摧残人的。

传统的"等候室"全然无济于事。一间拥挤、沉闷的小屋，人们彼此你瞧我望，焦急不安，翻弄着一两本杂志——正是这种环境使人心烦意乱。斯科特·布赖厄对这种环境的窒息人的作用提供了证据。（"Welfare From Below : Recipients' Views of the Public Weffare System," in Jacobus Tenbroek, ed., *The Law and the Poor*, San Francisco : Chandler Publishing Company, 1966, p.52）。大家都知道，当我们感到百无聊赖或心情焦躁不安时，时间看起来总是过得很慢。布赖厄发现，在福利机关等候的人总认为他们已经等了很长时间，其实没有那么长。有些人认为他们等候时间有真正时间的四倍那么长。

儿科诊所的候诊区
Wating room at the pediatrics clinic

可见，根本问题就在于此。如何使等待的人专心致志地度过他们的时间呢？——使他们

The classic "waiting room" does nothing to resolve this problem.A tight dreary little room,with people staring at each other, fidgeting,a magazine or two to flip—this is the very situation which creates the conflict.Evidence for the deadening effect of this situation comes from Scott Briar ("Welfare From Below:Recipients 'Views of the Public Welfare System," in Jacobus Tenbroek,ed.,*The Law and the Poor*,San Francisco:Chandler Publishing Company,1966,p.52).We all know that time seems to pass more slowly when we are bored or anxious or restless.Briar found that people waiting in welfare agencies consistently thought they had been waiting for longer than they really had.Some thought they had been wating four times as long.

The fundamental problem then,is this.How can the people who are waiting,spend their time wholeheartedly—live the hours or minutes while they wait,as fully as the other hours of their day—and yet still be on hand,whenever the event or the person they are waiting for is ready?

It can be done best when the waiting is fused with some other activity:an activity that draws in other people who are not there essentially to wait—a cafe,pool tables,tables,a reading room,where the activities and the seats around them are within earshot of the signal that the interviewer(or the plane,or whatever)is ready.For example,the Pediatrics Clinic at San Francisco General Hospital built a small playground beside the entrance,to serve as a waiting area for children and a play area for the neighborhood.

能像利用一天内其他的时间一样利用这段等候的时光——而且还要做到，当他们所等候的人或事一出现，他们能立即迎上前去。

只要等候期间融合了某些其他活动，这个问题是可以得到圆满解决的：安排一种活动，它能把并非在那里等候的其他人也吸引进来——如咖啡座、台球桌、餐桌、阅览室，人们在那些地方参加活动或在周围坐着同时还能听得见接见人（或飞机，或任何别的东西）所发出的信号。例如，旧金山总医院的儿科诊所在离入口不远的地方盖了一个小游乐场，给儿童们作候诊区，也给住在附近的居民提供了一处游乐场地。

我们知道的另一个例子是，在人们来等候约会的平台旁造一个蹄铁游戏场。等候的人必然开始掷蹄铁游戏，其他的人也参加进来，约会时间一到他们就离开了——在蹄铁游戏场、平台和办公楼之间人来人往。

也可以使等候的人觉得悠闲，在周围环境的保护下能够自得其乐，可以安静地沉思——这种境况与上述那种活动恰好相反。

只要等候场所能提供一些安静的、有保护的且又不必为等候焦虑的地方，自然就能出现良好的气氛。请看这样一些例子：在公共汽车站旁树荫底下设立一些坐位，跟街道分隔开；窗前放些椅子，让人们看到下面的街景；在花园里安排一些隐蔽的坐椅，一个秋千或吊床；坐在一个光线暗淡的地方，喝杯啤酒，离过道尽量远些，使人不必老得看着人们来来往往；还可以在金鱼缸旁安排一个私密的坐位。

总之，必须使等候的人能自由地做他们要做的事。如果他们愿意

安静的等候
Quiet waiting

In another example we know, a horseshoe pit was built alongside a terrace where people came to wait for appointments. The people waiting inevitably started pitching horseshoes, others joined in, people left as their appointments came up—there was an easy flow between the horseshoe pit, the terrace, and the offices.

Waiting can also be a situation where the person waiting finds himself with free time, and, with the support of the surroundings, is able to draw into himself, become still, meditative—quite the opposite of the activity described above.

The right atmosphere will come naturally if the waiting area provides some places that are quiet, protected, and do not draw out the anxiety of the wait. Some examples: a seat near a bus stop, under a tree, protected from the street; a window seat that looks down upon a street scene below; a protected seat in a garden, a swing or a hammock; a dark place and a glass of beer, far enough away from passages so that a person is not always looking up when someone comes or goes; a private seat by a fish tank.

In summary, then, people who are waiting must be free to do what they want. If they want to sit outside the interviewer's door, they can. If they want to get up and take a stroll, or play a game of pool, or have a cup of coffee, or watch other people, they can. If they want to sit privately and fall into a daydream, they can. And all this without having to fear that they are losing their place in line.

坐在准备接见他们的人的门外，他们可以这么做。如果他们想站起来溜达一下，或玩一下弹子游戏，或喝一杯咖啡，或观察一下他人，他们可以这么做。如果他们想安安静静地坐着神游遐想，他们也可以这么做。而所有这些都不会使他们因为错过在排队序列中的位置而担惊受怕。

因此：

在人们等候（公共汽车、约会、飞机）的地方，营造一种不会使等候乏味的环境。把等候跟某些其他的活动结合在一起——阅报、喝咖啡、玩弹子戏、掷蹄铁；也跟一些能够吸引不专门等候的人的活动结合起来。也要有与此相反的环境：创造一种场所使等候的人得以遐想；安静；愉快地沉默。

人们聚集在一起的活动　　　　　　　如有信号也听得见

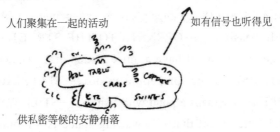

供私密等候的安静角落

⊰⊱

使环境活跃的因素可能有临街的窗户——**临街窗户** (164)、**窗前空间**（180），咖啡座——**临街咖啡座**（88），游戏，与过路行人一起活动——**向街道的开敞**（165）。使环境安静的因素可能有安静的园中坐椅——**园中坐椅** （176），供人打盹的地方——**在公共场所打盹**（94），或许还有养鱼池——**池塘**（71）。如这一等候空间是一个房间或一组房间，它的具体形状参阅**两面采光**（159）和**室内空间的形状**（191）……

Therefore:

In places where people end up waiting(for a bus,for an appointment,for a plane),create a situation which makes the waiting positive.Fuse the waiting with some other activity—newspaper,coffee,pool tables,horseshoes;something which draws people in who are not simply waiting.And also the opposite:make a place which can draw a person waiting int a reverie;quiet;a positive silence.

୫୦୧୫

The active part might have a window on the street—STREET WINDOWS(164),WINDOW PLACE(180),a cafe—STREET CAFE(88),games,positive engagements with the people passing by—OPENING TO THE STREET(165). The quiet part might have a quiet garden seat—GARDEN SEAT(176),a place for people to doze—SLEEPING IN PUBLIC(94),perhaps a pond with fish in it—STILL WATER (71).To the extent that this waiting space is a room,or a group of rooms,it gets its detailed shape from LIGHT ON TWO SIDES OF EVERY ROOM(159)and THE SHAPE OF INDOOR SPACE(191)...

模式151　小会议室*

　　……在一些机关和工作单位——**像市场一样开放的大学**（43）、**地方市政厅**（44）、**师徒情谊**（83）、**灵活办公空间**（146）、**工作小组**（148）内部，不可避免地要有各种会议室、小组活动室和教室。对会议室所作的调查表明，很难预先确定既根据大小，又根据位置配置会议室的最佳方案。

✴✴✴

　　会议规模越大，获益者越少。但许多机关常常热衷于花钱盖大会议室和演讲厅。

151   SMALL MEETING ROOMS*

...within organizations and workplaces—UNIVERSITY AS A MARKETPLACE(43),LOCAL TOWN HALL(44), MASTER AND APPRENTICES(83),FLEXIBLE OFFICE SPACE(146), SMALL WORK GROUPS(148),there will, inevitably, be meeting rooms,group rooms,classrooms,of one kind or another.Investigation of meeting rooms shows that the best distribution—both bysize and by position—is rather unexpected.

სოცა

The larger meetings are,the less people get out of them. But institutions often put their money and attention into large meeting rooms and lecture halls.

We first discuss the sheer size of meetings.It has been shown that the number of people in a group influences both the number who never talk,and the number who feel they have ideas which they have not been able to express.For example,Bernard Bass (*Organizational Psychology*,Boston:Allyn,1965,p.200)has conducted an experiment relating group size to participation.The results of this experiment are shown in the following graph.

There is no particularly natural threshold for group size;but it is clear that the number who never talk climbs very rapidly.In a group of 12,one person never talks.In a group of 24,there are six people who never talk.

首先，我们单就会议的规模进行讨论。事实表明，一个小组的人数多寡既会影响从不发言的人的数目，也会影响到那些有话要说但未能说出来的人的数目。例如：伯纳德·巴斯（Bernard Bass）（*Organizational Psychology*，Boston：Allyn，1965，p.200）做了一个有关参加会议的小组规模的实验。这一实验结果如下图所示：

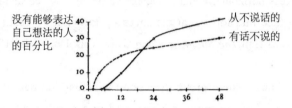

当小组的规模增大时，越来越多的人不能说话
A size of group grows, more and more people hold back

小组规模没有一个特定的一成不变的限度；但从图中看出，一言不发的人的数目迅速上升。在一个 12 个人的小组里，有一个人不发言。在一个 24 个人的小组里，一言不发的人就有 6 个。

当我们考虑谈话的合适的距离时，我们也会发现同样的限度。霍尔确定，平常的高声谈话最远可传到大约 8ft；一个具有 20/20 视力的人能在 12ft 内看到别人细致的面部表情；两个头部距离为 8 ~ 9ft 的人，如果他们都伸出手来，可以彼此传递物体；清晰的视觉（即黄斑区视觉）包括水平 12° 和垂直 3°——这时在达到大约 10ft 的距离时能看到一个人的面孔，而不是两个人。（参阅 Edward Hall，*The Silent Language*，New York：Doubleday，1966，pp.118~119.）

因此，小组讨论效果最好，如果小组成员大致分布成圆圈形，最大直径大约 8ft。具有这个直径的圆周的长度是 25ft。由于每人的坐位大约要占 27in，围坐成这样一个圆

We get similar thresholds when we consider comfortable distances for talking.Edward Hall has established the upper range for full casual voice at about 8 feet;a person with 20/20 vision can see details of facial expression up to 12 feet;two people whose heads are 8 to 9 feet apart,can pass an object if they both stretch;clear vision(that is,macular vision)includes 12 degrees horizontally and 3 degrees vertically—which includes one face but not two,at distances up to about 10 feet.(See Edward Hall,*The Silent Language*,New York:Doubleday,1966, pp.118-119.)

Thus a small group discussion will function best if the members of the group are arranged in a rough circle,with a maximum diameter of about 8 feet.At this diameter,the circumference of the circle will be 25 feet.Since people require about 27 inches each for their seats,there can be no more than about 12 people round the circle.

Next we shall present evidence to show that in institutions and workgroups,the natural history of meetings tends also to converge on this size.

The following histograms show the relative numbers of different sized classes held at the University of Oregon in the Fall of 1970 and the relative numbers of available classrooms in the different size ranges.We believe these figures are typical for many universities.But it is obvious at a glance that there are too many large classrooms and too few small classrooms. Most of the classes actually held are relatively small seminars and "section" meetings,while most of the classrooms are in the 30 to 150 size range.These large classrooms may have reflected the teaching methods of an earlier period,but apparently they do not conform to the actual practice of teaching in the 1970's.

圈不可能多于 12 个人。

接着我们要提出证据来证明，在一个机关和工作小组中，会议的现成经历也趋向于拥有这个规模。

下面的矩形图表示在 1970 年秋季俄勒冈大学开办的大小不等的班级的有关人数以及可供利用的大小不同的教室的有关数字。我们相信，这些数字也是许多大学里经常见到的。但一眼就能看出，大教室太多而小教室太少。实际上的课堂大多数是小的讨论会和"分组会议"，但大部分教室却可容纳 30 ～ 150 人。这些大教室可能反映早期的教学方法，但它们显然不符合 20 世纪 70 年代教学的实际情况。

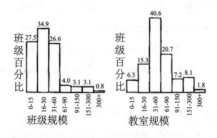

矩形图：班级和教室不相适应
Histogram:classes don't fit the classrooms

我们发现，伯克利行政人员委员会、商会和市政委员会的会议有着类似的分配情况。在该市各商会、委员会中间，有 73 ％的会议其平均参加人数只有 15 人或不到 15 人。可是，这些会议毫无疑问大部分都在设计标准远远超过 15 人的房间里举行。还是同样的问题，因为会议在过大的房间里举行，房间都是半空着的，人们自然往后排坐，发言人面对着的是一排排空坐位。在这种情况下营造不出一个开得好的会议所特有的那种亲切和紧凑的气氛。

We found that the meetings of official committees, boards,and commissions in the City of Berkeley have a similar distribution. Among the various city boards,commissions,and committees,73 per cent have an average attendance of 15 or less.Yet of course,most of these meetings are held in rooms designed for far more than 15 people.Here again,most of the meetings are held in rooms that are too large;the rooms are half-empty;people tend to sit at the back;speakers face rows of empty seats.The intimate and intense atmosphere typical of a good small meeting cannot be achieved under these circumstances.

Finally,the spatial distribution of meeting rooms is often as poorly adapted to the actual meetings as the size distribution. The following histograms compare the distribution of classrooms in different sectors of the University of Oregon with the distribution of faculty and student offices.

Once again,this discrepancy has a bad effect on the social life of small meetings.The meetings work best when the meeting rooms are fairly near the participants'offices. Then discussions which begin in the meeting rooms are able to continue in the office or the laboratory.When the meeting rooms are a long walk from offices,the chances of this kind of informal business are drastically reduced.

Therefore:

Make at least 70 per cent of all meeting rooms really small— for 12 people or less.Locate them in the most public parts of the uilding,evenly scattered among the workplaces.

最后，会议室的空间分配也像面积分配那样并不匹配会议的实际情况。以下的矩形图将俄勒冈大学各区的教室分布同教员办公室和学生工作室的分布作一比较。

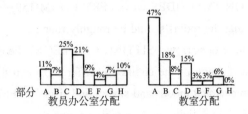

部分　　教员办公室分配　　　　　教室分配

会议室的地点不在人们工作的地方
The meeting rooms are not located where people work

还是这个问题，这种不一致对小会议的社会生活产生不良影响。当会议室同与会者的办公地点很接近时，会议效果最好。这样，在会议室内开始的讨论可以在办公室或实验室继续展开。如果会议室离办公室很远，进行这类非正式活动的机会就大为减少了。

因此：

要使全部会议室中至少有70%是名副其实的小会议室——供至多12个人使用。把这些会议室设在大楼的公共区域，均匀分布在各工作地点之间。

70%的小会议室

均匀分布于工作区

Shape meeting rooms like any other rooms,perhaps with special emphasis on the fact that there must be no glare— LIGHT ON TWO SIDES OF EVERY ROOM(159)—and on the fact that the rooms should be roughly round or square,and not too long or narrow—SITTING CIRCLE(185).People will feel best if many of the chairs are different,to suit different temperaments and moods and shapes and sizes—DIFFERENT CHAIRS(251).A light over the table or over the center of the group will help tie people together—POOLS OF LIGHT (252). For the shape of the room in detail,start with THE SHAPE OF INDOOR SPACE(191)...

使会议室的形状同任何其他房间一样，也许要强调一下，一定要让会议室没有眩光——**两面采光**（159）——还要强调另一点：会议室应基本上是圆形或正方形的，不要太长或太窄——**坐位圈**（185）。如果有各式各样的坐椅以适合于不同的气质和心情，不同的体形和大小，人们将会感到心情舒畅——**各式坐椅**（251）。悬挂于桌子上方或小组会中心上方的吊灯会有助于把人们聚集在一起——**投光区域**（252）。关于会议室形状的详细情况，参阅**室内空间形状**（191）及之后几个模式……

模式152 半私密办公室

……**私密性层次**（127）、**灵活办公空间**（146）、**工作小组**（148）对小组活动空间和个人工作空间作了全面安排，在此范围内，本模式规定单个房间和办公室的形状。本模式也有助于构成这些较大的模式。

⊰⊱

怎样才能使办公室工作既具有私密性，又能够彼此间互相联系，使二者得到恰当的平衡？

完全私密的办公室对一个工作小组内部人们相互交流具有破坏性的影响，并使讨厌的办公室等级制度固定化。同时，有些时候需要有私密性；而且几乎每种工作都在一定程度上要求不能被随意打断。

凡是体验过办公室工作的人都能对这个问题发表一些看法。作为建筑师工作小组的成员，以我们自己的经历来说，我们在许多方面遇到这个问题。我们所能提供的最好证据就是自己的工作小组的亲身经历。

在过去 7 年内，我们的办公室曾几度搬迁。有一次我们搬到一座很大的老房子里：因为房子很大，我们当中有的一个人独用一间，有的几个人合用一间。在几个月后，我们这个小组的社会联系濒于中断。小组的工作徒具形式；轻松随意的交往不复存在；整个小组的气氛发生了变化，原来那种使我们的小组欣欣向荣的环境变成了办公衙门，在这里人们彼此预约见面，在特制的信箱里留条，很不自在地敲彼此的房门。

过了一段时间，我们的工作变得兴味索然。

后来我们逐渐意识到，这幢房屋的环境对小组内部的

152 HALF-PRIVATE OFFICE

...within the overall arrangement of group space and individual working space provided by INTIMACY GRADIENT (127),FLEXIBLE OFFICE SPACE(146),and SMALL WORK GROUPS(148),this pattern shapes the individual rooms and offices.The pattern also helps togenerate the organization of these larger patterns.

❧❦

What is the right balance between privacy and connection in office work?

The totally private office has a devastating effect on the flow of human relationships within a work group,and entrenches the ugly quality of office hierarchies.At the same time,there are moments when privacy is essential;and to some extent nearly every job of work needs to be free from random interruption.

Everyone who has experienced office work reports some version of this problem.In our own experience—as members of a working team of architects—we have faced the problem in hundreds of ways.The best evidence we have to report is our own experience as a work group.

Over the last seven years we moved our offices on several occasions.At one point we moved to a large old house:large enough for some of us to have private rooms and others to share rooms.In a matter of months our social coherence as a group was on the point of breakdown.The workings of the group became formalized;easy going communication vanished;the entire atmosphere changed from a setting which sustained our

联系有很大破坏性。当我们开始予以重视时，我们注意到，那些仍然发挥着良好作用的房间——我们大家集合在一起谈论工作的地方——有着一种特殊的性质：它们只是**半私密**的，即使它们内部的工作空间边界分明。

我们因此发现，凡是我们觉得在工作中合作得很好的地方，看来几乎都具有这样的一些性质：没有一间办公室是完全私密的；大部分办公室都有一个人以上在工作；但即使一间办公室只供一个人使用，它的前部也有一个简单的公共区，每个人都可以随便进来待一会儿。用办公桌布置成一个内部的私密领域，它们设在办公室的边角上，这样办公室的门可随时敞开。最后，我们对自己的工作状况重新加以调整，使每个人都能以某种方式符合本模式的要求。

本模式用处很大，我们谨将它推荐给所有处于类似境况下的人们。

因此：

避免使办公室封闭、隔离和私密。每个工作室，不管里面的办公人员是两个人一组或三个人一组，抑或独自一人，都要对其他的工作小组和临近这个办公室的周围环境开放。在办公室前部刚进门的地方，放置舒适的坐椅，实际办公的地方应远离房门，在房间的最后方。

坐位区

宽大的门道　　　　　　　　　　工作空间

growth as a group to an office bureaucracy,where people made appointments with each other,left notes in special boxes,and nervously knocked on each other's doors.

For a while we were virtually unable to produce any interesting work.

It gradually dawned on us that the environment of the house was playing a powerful role in the breakdown.As we started to pay attention to it,we noticed that those rooms which were still functioning—the places where we would all gather to talk over the work—had a special characteristic:they were only *half*-private,even though the workspaces within them were strongly marked.

As we thought it out,it seemed that almost every place where we had found ourselves working well together had these characteristics:no office was entirely private;most offices were for more than one person;but even when an office was only for one,it had a kind of simple common area at its front and everyone felt free to drop in and stay for a moment.And the desks themselves were always built up as private domains within and toward the edges of these offices,so that doors could always be left wide open.Eventually we rearranged ourselves until each person had some version of this pattern.

The pattern works so well,that we recommend it to everyone in similar circumstances.

Therefore:

Avoid closed off,separate,or private offices.Make every workroom,whether it is for a group of two or three people or for one person,half-open to the other workgroups and the world immediately beyond it.At the front,just inside the door,make comfortable sitting space,with the actual workspace(s)away from the door,and further back.

根据**室内空间形状**（191）具体规划办公室的形状；使它至少两面有窗——**两面采光**（159）；把个人的工作空间安排在房间的角落——**工作空间的围隔**（183），在这里能看到窗外——**俯视外界生活之窗**（192）；使靠近房门的坐位尽可能舒适——**坐位圈**（185）。

加盖一些小的外屋，它们必须稍微独立于主要建筑物，并设置从楼房上层通往街道和花园的引道。

153. 出租房间

154. 青少年住所

155. 老人住所

156. 固定工作点

157. 家庭工作间

158. 室外楼梯

Shape each office in detail,according to THE SHAPE OF INDOOR SPACE(191);give it windows on at least two sides— LIGHT ON TWO SIDES OF EVERY ROOM(159);make individual workspaces in the corners—WORKSPACE ENCLOSURE(183),looking out of windows—WINDOWS OVERLOOKING LIFE(192);make the sitting area toward the door as comfortable as possible—SITTING CIRCLE(185)...

add those small outbuildings which must be slightly independent from the main structure,and put in the access from the upper stories to the street and gardens;

153. ROOMS TO RENT

154. TEENAGER'S COTTAGE

155. OLD AGE COTTAGE

156. SETTLED WORK

157. HOME WORKSHOP

158. OPEN STAIRS

模式153　出租房间

……本模式是确定外屋结构的第一个模式。如果使用得当,它将有助于创造**项链状的社区行业**（45）、**家庭**（75）、**自治工作间和办公室**（80）、**小服务行业**（81）、**灵活办公空间**（146）、**青少年住所**（154）、**老人住所**（155）、**家庭工作间**（157）：总之,它使所有建筑物在许多情况下更加灵活而适用。

❧❧❧

随着房屋内的生活发生变化,对于空间的需要周而复始地时而减少、时而增加。建筑物必须能够适应对空间需要的这种无规则的增加和减少的现象。

很简单,当一个家庭或工作小组因为有一两个人离开而缩小时,空出来的空间应能得到利用。否则,留下来的人就得生活在这个对他们说来太大的空壳里。他们甚至可能被迫卖掉他们的房子而搬家,因为他们不可能保养、维护这么大的地方。

此外,由于扩大和缩小几乎总是不可预测的,空间的这种分离应该是可逆的。一些房间在无人使用时让给外人使用或出租,但可能有一天情况变了,工作小组和家庭的规模又增大了,它们又会重新派上用场。

要使房屋具有这样一种灵活性,就有必要使它们的一部分相对独立。事实上,应预先考虑把某些房间作为备用房间,在小组人数发生变化时可供出租。这些房间同房子的其他部分具备某种联系,但又能与后者分离而自成一统,然后,又能轻易地重新连接在一起。在通常情况下,这意味着要有一个与外界相通的专用入口,一个专用的卫生间

153   ROOMS TO RENT

...this pattern is the first which sets the framework for the outbuildings.Used properly,it can help to create NECKLACE OF COMMUNITY PROJECTS(45),THE FAMILY(75),SELF-GOVERNING WORKSHOPS AND OFFICES(80),SMALL SERVICES WITHOUT RED TAPE(81),FLEXIBLE OFFICE SPACE(146),TEENAGER'S COTTAGE(154),OLD AGE COTTAGE(155),HOME WORKSHOP(157):in general it makes any building flexible,useful in a greater variety of circumstances.

❀❀❀

As the life in a building changes,the need for space shrinks and swells cyclically.The building must be able to adapt to this irregular increase and decrease in the need for space.

Very simply,when a family or a workgroup shrinks because one or two people leave,the space which becomes empty should be able to find a use.Otherwise,the people who stay behind will rattle around in a hollow shell which is too big for them.They may even be forced to sell their property and move because they cannot afford the upkeep of so big a place.

And by the same token,since swelling and shrinking is almost always unpredictable,this splitting off of space should be reversible.The rooms which are given to outside use or let out when they are not used,may one day be needed again when circumstances change and the workgroup or family swells in size again.

或一条直接通往卫生间的过道，可能还要有通往厨房的过道。

在丹麦，迪布勒曾经制定了一个采用本模式作为主要住宅形式的建筑规划。他发表在 "*Enfamiliehust 1970*"（Landsbankerns Reallanefond，Stiftedes den 9.maj 1959）上的住宅在慢慢地造起来，住宅的每一部分既可以同大家庭连成一体，也可以作为独立的单元供人居住。以下是迪布勒的"四部分"住宅的平面图。

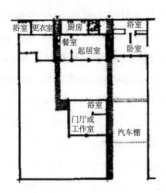

迪布勒的四部分住宅
Dybbroe's four-part house

虽然出租房子通常对环境有着破坏性的影响——参阅**自己的家**（79），但根据我们的经验，在主人占用主要房间的情况下，当面出租仍不失为一种合理的健康的租赁关系。房东事实上住在那儿，因此他直接关心自己周围的生活福利和环境，这跟不居住在产权所在地的房主不同，后者拥有房产只是为了赚钱。而房客通常是短期房客，他们只是想租一间房子而并不想拥有它。即使这样，更加理想的情况还是房东将其房屋某一部分的所有权出让给别人，而又保留收回的权利。然而，由于所有权的这种微妙形式目前尚无法律根据，我们认为，当面出租是不会引起社会和环境遭到破坏的唯一租赁形式。

To give buildings this flexibility,it is essential that parts of them be relatively independent.In effect,some rooms should be conceived in advance as potential rooms to let if the size of the group should change.These rooms need a kind of connection to the rest of the house,which allows them to be closed off and separated,and then,just as easily,joined up again.Generally,this means a private entrance from the outside,either a private hath or direct access to a bathroom,and perhaps access to the kitchen.

In Denmark,Ole Dybbroe has developed a scheme for housing that takes this pattern as a crucial generator of the form of the house.The houses he shows in *Enfamiliehuset 1970* (Lands bankernes Reallanefond,stiftedes den 9.mai 1959) grow slowly,and each part of them can either be united with the larger household or inhabited as an independent unit.Here is his plan for a "four part" house.

Though renting in general has a devastating impact on the environment—see YOUR OWN NOME(79)our experience has been that face-to-face rental,with the owners occupying the main structure,is the one kind of rental relationship that is reasonably healthy.The landlord is actually there,so he is directly concerned with the wellbeing of the life around him and with the environment,unlike the absentee landlords,who own property only for the money which it makes.And the tenants are usually short-term tenants,who prefer to rent a room rather than take on burdens of ownership.Even here a more ideal situation would be for the owner to share out ownership over some part of the building,with certain options for taking back the space.However,in the absence of such subtle forms of legal ownership,face to face renting is,we believe,the only form of renting that is not socially and physically destructive.

因此:

至少使房屋的某一部分可供出租:除了使之在同房屋的其余部分保持正常的联系之外,还有一个专用的入口。务必使这个通常的入口可轻易关闭,且又不会损害房屋中的畅通,还要使这个出租房间可以直接通往卫生间而不需要经过主要住房。

❧❧❧

安排出租房间时,要使它们能成倍扩大以用作**青少年住所**(154)、**老人住所**(155)或**家庭工作间**(157);使专用的入口有一个**入口的过渡空间**(112),如果此空间在楼房的上层,给它一个直通廊道,使它可以从**室外楼梯**(158)通往街道。使这些房间都能**两面采光**(159)并具有**室内空间形状**(191)……

Therefore:

Make at least some part of the building rentable:give it a private entrance over and above its regular connection to the rest of the house.Make sure that the regular entrance can be easily closed off without destroying the circulation in the house,and make sure that a bathroom can be directly reached from this room without having to go through the main hous.

৪০৫৪

Place the rooms to rent in such a way that they can double as a TEENAGER'S COTTAGE(154),or an OLD AGE COTTAGE(155), or a HOME WORKSHOP(157);give the private entrance an ENTRANCE TRANSITION (112),and if the space is on an upper floor,give it direct access to the street by means of OPEN STAIRS(158).And give the rooms themselves LIGHT ON TWO SIDES(159)and THE SHAPE OF INDOOR SPACE (191)...

模式154　青少年住所*

……在任何一幢有青少年的住宅里——**家庭**（75）、**小家庭住宅**（76）——需要对青少年的居室予以特别的考虑——**个人居室**（141）。如有可能，这些居室既应附属于主建筑，又应与之分开，以便日后有可能用作**出租房间**（153）。

⋙⋘

如果在家庭中一个青少年的住所不能反映出他对某种程度的独立生活的需要，他会同自己的家庭闹出许多矛盾。

在大多数的家庭住房里儿童和青少年的用房基本上是一样的。但当儿童成长为青少年时，他们跟家庭的关系发

154   TEENAGER'S COTTAGE*

...in any house which has teenagers in it—THE FAMILY(75), HOUSE FOR A SMALL FAMILY(76)—it is necessary to give special consideration to their rooms—A ROOM OF ONE'S OWN(141).If possible,these rooms should be attached but separate,and made to help create the possibility of later being ROOMS TO RENT(53).

⍿⍿⍿

If a teenagers place in the home does not reflect his need for a measure of independence,he will be locked in conflict with his family.

In most family homes the rooms for children and adolescents are essentially the same.But when children become adolescents,their relationship to the family changes considerably.They become less and less dependent on the family;they take on greater responsibilities;their life outside the home becomes richer,more absorbing.Most of the time they want more independence;occasionally they really need the family to fall back on;sometimes they are terrified by the confusion within and around them.All of this places new demands on the organization of the family and,accordingly,on the organization of the house.

To really help a young person go through this time,home life must strike a subtle balance.It must offer tremendous opportunities for initiative and independence,as well as a constant sense of support,no matter what happens.But

生很大的变化。他们对家庭的依赖越来越少；他们担负更大的责任；他们在家庭之外的生活变得更加丰富、更有吸引力。在大部分时间内青少年想要有更多的独立性；偶而他们也确实需要家庭的帮助；有时他们自身或周围的混乱使他们感到恐惧。所有这些都对家庭的结构提出新的要求，因而相应地对住房布局提出新的要求。

为了能够切实帮助年轻人度过这一时光，家庭生活必须起到微妙的平衡作用。家庭应该为创新精神和独立精神提供大量的机会，同时，不管发生什么情况，都能给青年人以不懈的支持。但美国人的家庭生活似乎从来起不到这样的平衡作用。对青少年的家庭生活的研究表明，这个时期有着无数的小冲突、专制、犯罪和默许。青春期看来更多的是改变青年男女的心理状态，而不能帮他们找到他们在世界上的位置。（例见 Jules Henry，*Culture Against Man*，New York：Random House，1963.）

就物质条件来讲，这些问题可归纳如下：一个青少年需要住宅里面有一个能体现较多的自主和个性的地方，这个地方不是儿童的卧室或床龛，而是能够进行独立活动的地盘。他需要有一个任他自由来去的处所，一个他的隐私权得到尊重的处所。同时他需要有机会跟他的家庭建立起亲密关系，这种关系比过去任何时候都更加体现相互关怀，而更少依赖性。看来所需要的是一处住所，通过它的布局和位置使新的独立要求和同家庭的新的联系得到调和。

青少年住所可以由旧有的儿童卧室改建而成，男孩和他的父亲在墙上开出一道门来，于是就扩大了这个房间。它也可以是从无到有的新建房子，目的是使它今后可用作工作间，或可供祖父安度晚年，或可作出租房间。这处住所甚至可以是一个坐落在花园中完全分隔开的建筑物，但在这种情况下跟主房间需要有一种密切的联系：大多是有一条短的有顶甬道从住所通往主厨房。甚至在联排式住宅

American family life never seems to strike this balance.The studies of adolescent family life depict a time of endless petty conflict,tyranny,delinquency,and acquiescence.As a social process,adolescence,it seems,is geared more to breaking the spirit of young boys and girls,than to helping them find themselves in the world.(See,for example,Jules Henry,*Culture Against Man*,New York:Random House,1963.)

In physical terms these problems boil down to this.A teenager needs a place in the house that has more autonomy and character and is more a base for independent action than a child's bed room or bed alcove.He needs a place from which he can come and go as he pleases,a place within which his privacy is respected.At the same time he needs the chance to establish a closeness with his family that is more mutual and less strictly dependent than ever before.What seems to be required is a cottage which,in its organization and location,strikes the balance between a new independence and new ties to the family.

The teenager's cottage might be made from the child's old bedroom,the boy and his father knocking a door through the wall and enlarging the room.It might be built from scratch,with the intention that it later serve as a workshop,or a place for grandfather to live out his life,or a room to rent.The cottage might even be an entirely detached structure in the garden,but in this case,a very strong connection to the main house is essential:perhaps a short covered path from the cottage into the main kitchen.Even in row housing,or apartments,it is possible to give teenagers rooms with private entry.

Is the idea of the teenage cottage acceptable to parents? Silverstein interviewed 12 mothers living in Foster City,a

或公寓中，也有可能使青少年住所有专用的入口。

关于建立青少年住所的想法能为父母所接受吗？西尔弗斯坦向住在旧金山郊区福斯特城的 12 位母亲作了了解，问她们是否愿意在她们的家中建立青少年住所。她们对这个想法的抵触情绪主要由三方面的反对意见引起的。

1. 这种住所只能用几年，随后就会闲置起来。

2. 这种住所会把家庭弄得支离破碎；把青少年隔离了起来。

3. 它使青少年出入过分自由。

针对这些反对意见，西尔弗斯坦于是提出三条修改建议。

针对第 1 条反对意见，使该空间扩大一倍作为工作间、客房、画室、祖母居室；这个空间用木材作建筑材料，这样用手工工具就能很容易将其改建。

针对第 2 条反对意见，将该住所同住宅相连，但有自己的出入口；通过短廊或门道将该住所同住宅相连，或使该住所建于屋后空地后方。

针对第 3 条反对意见，安排好该住所的位置，使它通往街上的走道经过住宅的一个重要的公共区——厨房或庭院。

他把这些修改意见整理后再跟那 12 位母亲讨论。其中 11 位觉得修改过的方案有某些可取之处，值得去试试。这一材料见于西尔弗斯坦的报告《男孩的房间：12 位母亲对一个建筑模式的反应》，加利福尼亚大学，建筑系，1967 年 12 月。

以下是包含这些修改意见的几种可能的方案：

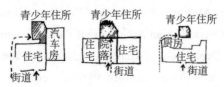

青少年住所的不同设计方案
Variations of teenager's cottage

suburb of San Francisco,and asked them whether they would like a teenage cottage in their family.Their resistance to the idea revolved around three objections:

1.The cottage would be useful for only a few years,and would then stand empty.

2.The cottage would break up the family;it isolates the teenager.

3.It gives the teenager too much freedom in his comings and goings.

Silverstein then suggested three modifications,to meet these objections:

To meet the first objection,make the space double as a workshop,guest room,studio,place for grandmother;and build it with wood,so it can be modified easily with hand tools.

To meet the second objection,attach the cottage to the house,but with its own entrance;attach the cottage to the house via a short hall or vestibule or keep the cottage to the back of the lot, behind the house.

To meet the third objection,place the cottage so that the path from the room to the street passes through an important communal part of the house—the kitchen,a courtyard.

He discussed these modifications with the same twelve mothers.Eleven of the twelve now felt that the modified version had some merit,and was worth trying.This material is reported by Murray Silverstein,in "The Boy's Room:Twelve Mothers Respond to an Architectural Pattern," University of California,Department of Architecture,December 1967.

Here are some possible variants containing these modifications.

Among the Comanches, "...the boy after puberty was given

在科曼契人（美国印第安人）中间，"……男孩子一到青春期就给他一个独立的圆锥形帐篷。他在帐篷里面睡觉，招待他的朋友和度过他的大部分时光。"（Abram Kardiner, *Psychological Frontiiers of Society*, New York : Columbia University Press，1945，p.75.）

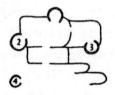

非洲扬格人大住宅的平面图；
2是家长卧室；3是女儿住所；4是儿子住所
Plan of a Yungur Compound,Africa;
2 is the master bedroom;3 is the daughter's hut;4 is the son's
hut

最后，引述博瓦著作中的一段话如下：

"当我12岁的时候，我为自己在家中没有一个私密安静的处所而苦恼。在翻阅《我的日记》时，我发现一个关于英国女学生的故事，羡慕地凝视着那张描画她房间的彩色插图。房间里有一张书桌、长沙发和摆满书的书架。在这儿，在这四墙色彩绚丽的房间里，她读书、干活、饮茶，而谁也不会看着她——多么令我羡慕啊！我平生第一次看到了比我自己的生活要美满得多的生活。而现在，时隔许久，我终于也有了我自己的房间了。我的祖母把她客厅里的安乐椅、不常用的桌子以及一些小摆设统统拿走。我搬来一些未上油漆的家具，我的姐姐帮我把它们漆成棕色。我有了一张桌子、两张椅子、一个大箱子（它又可以当凳子坐，又可以搁下所有的杂物），还有放书的书架。我把墙裱糊成桔黄色，还弄了一张靠墙的长沙发，颜色与墙壁很相称。从我的五楼阳台上我可以看到白尔浮耳的狮子和唐菲尔·罗斯荷街上的悬铃木。用一个气味难闻的煤油炉来取暖。但是它的臭味毕竟使我得以独处，这使我喜爱它。我能够关上门，使自己的日常生活

a separate tepee in which he slept,entertained his friends,and spent most of his time." (Abram Kardiner, *Psychological Frontiers of Society*,New York: Columbia University Press,1945,P.75.)

And finally,from Simone De Beauvoir:

When I was twelve I had suffered through not having a private retreat of my own at home.Leafing through *Mon Journal* I had found a story about an English schoolgirl,and gazed enviously at the colored illustration portraying her room.There was a desk,and a divan,and shelves filled with books.Here,within these gaily painted walls,she read and worked and drank tea,with no one watching her—how envious I felt!For the first time ever I had glimpsed a more fortunate way of life than my own.And now,at long last,I too had a room to myself.My grandmother had stripped her drawing room of all its armchairs,occasional tables,and knick knacks.I had bought some unpainted furniture,and my sister had helped me to give it a coat of brown varnish.I had a table,two chairs,a large chest which served both as a seat and as a hold-all,shelves for my books.I papered the walls orange,and got a divan to match.From my fifth-floor balcony I looked out over the Lion of Belfort and the plane trees on the Rue Denfert-Rochereau.I kept myself warm with an evil-smelling kerosene stove. Somehow its stink seemed to protect my solitude,and I loved it.It was wonderful to be able to shut my door and keep my daily life free of other people's inquisitiveness.For a long time I remained indifferent to the decor of my surroundings.Possibly because of that picture in *Mon Journal* I preferred rooms that offered me a divan and bookshelves,but I was prepared to put up with any sort of retreat in a pinch.To have a door that I could shut was still the height of bliss for me...I was free to come and go as I pleased.I could get home with the milk,read in bed all night,sleep till midday,shut myself up for forty-eight hours at

不受别人好奇心的干扰，这就是至上至美了。在一个长时间内我不热衷于装饰自己的生活环境。可能由于《我的日记》里的那张画的影响，我偏爱能够让我拥有长沙发和书架的房间，但如果必要，我也准备接受任何一个僻静的处所。对我来说，至高无上的幸福莫过于有一扇门可以让我关闭……我可以凭自己的喜欢自由自在地进出。我可以带牛奶回家，彻夜躺在床上读书，一觉睡到正午时分，一连48个小时把自己关在屋里，或者心血来潮想出去就出去……我的最大愉快在于我爱干什么就干什么"。(Simone De Beauvoir, *The Prime of Life*, New York : Lancer Books, 1966, pp.9~10.)

因此：

为了表明一个儿童已经长大，把他在家中原来的生活环境改变为一个住所，使之在物质上表示独立生活的开始。使该住所同家宅相连，但要明显地看出它是独立的，远离父母卧室，有自己专用的出入口，最好有自己单独的房顶。

能过公共区的甬道　　　青少年住所

分开的出入口

⁂

使青少年住所有**坐位圈**（185）和**床龛**（188），但没有专用的卫生间和厨房——这些东西必须要跟家里人共用：这样会使男女青少年跟他们的家庭保持充分的联系。使这个地方最终能够用作客房、出租房间、工作间等——**出租房间**（153）、**家庭工作间**（157）。若它在楼上，则使它有一个单独的专用的**室外楼梯**（158）。关于青少年住所的形状和构造，参阅**室内空间形状**（191）和**结构服从社会空间的需要**（205）……

a stretch,or go out on the spur of the moment..,my chief delight was in doing as I pleased.(Simone De Beauvoir,*The Prime of Life*,New York:Lancer Books,1966,pp.9-10.)

Therefore:

To mark a child's coming of age,transform his place in the home into a kind of cottage that expresses in a physical way the beginnings of independence. Keep the cottage attached to the home,but make it a distinctly visible bulge, far away from the master bedroom,with its own private entrance, perhaps it own roof.

❧❧

Arrange the cottage to contain a SITTING CIRCLE(185) and a BED ALCOVE(188)but not a private bath and kitchen— sharing these is essential:it allows the boy or girl to keep enough connection with the family.Make it a place that can eventually become a guest room,room to rent,workshop,and so on— ROOMS TO RENT(153),HOME WORKSHOP(157).If it is on an upper story,give it a separate private OPEN STAIR(158). And for the shape of the cottage and its construction, start with THE SHAPE OF INDOOR SPACE(191)and STRUCTURE FOLLOWS SOCIAL SPACES(205)...

模式155　老人住所**

　　……在**老人天地**（40）里，我们解释了，有必要使每一社区的老年人人数取得平衡，一部分集中在公共场所的周围，但大部分分散在社区的其他住宅中。本模式更具体地规定老年人住房的性质：它们既是住宅团组的一部分，又独立、隐蔽地间杂在一些较大的房屋之间。我们将看到，似有必要使每个家庭都有一个这样的附属于它的住所——**家庭**（75）。像**出租房间**（153）和**青少年住所**（154）一样，一旦遇到困难，这样的住所也可以出租或另作别用。

155 OLD AGE COTTAGE**

...we have explained,in OLD PEOPLE EVERYWHERE(40), that it is essential to have a balanced number of old people in every neighborhood,partly centered around a communal place,but largely strung out among the other houses of the neighborhood.This pattern now defines the nature of the houses for old people in more detail:both those which are a part of clusters and those which are tucked,autonomously, between the larger houses.As we shall see,it seems desirable that every family should have a cottage like this, attached to it—THE FAMILY(75).Like ROOMS TO RENT(153)and TEENAGER'S COTTAGE(154),this cottage can be rented out or used for other purposes in time of trouble.

৪০৩

Old people,especially when they are alone,face a terrible dilemma.On the one hand,there are inescapable forces pushing them toward independence:their children move away;the neighborhood changes;their friends and wives and husbands die.On the other hand,by the very nature of aging,old people become dependent on simple conveniences,simple connections to the society about them.

This conflict is reflected often in their children's conflict. On the one hand,children feel responsible for their parents, because,of course,they sense their growing need for care and comfort.On the other hand,as families are whittled down,parent-child conflicts become more acute,and few people can imagine

🙐🙑

老年人，特别是当他们独自生活的时候，处境往往进退两难。一方面，某些不可避免的天灾人祸迫使他们独居一隅：他们的子女搬走了；左邻右舍发生了变化；他们的友人、丈夫（或妻子）去世了。另一方面，由于年老力衰，老人需要依靠容易获得的方便设施和跟他们周围社会的简便联系来过日子。

这种矛盾常常反映在他们子女的矛盾上。一方面，子女感到对他们的父母有责任，这当然是因为他们意识到老年人越来越需要照顾和舒适的生活条件。另一方面，随着家庭人口的逐渐减少，父母和子女的矛盾日益尖锐化，很少有人真正有能力或心甘情愿去照顾他们衰老的父母。

这种矛盾可以得到部分解决，如果一个小家庭住宅附近有一个小住所，使祖父或祖母可以住在那儿，有一定的距离使它独立，但又离得不远而便于联系，在老人有困难或临终之前可以有所照顾。

但矛盾是更普遍地存在着的。即使我们完全忽略父母和子女之间错综复杂的关系，事实上，大部分老人还面临由日益衰老带来的巨大困难。高福利国家企图以付款的方式（社会保障或养老金）来取代大家庭可给老人提供的方便。但这项收入往往微乎其微；而通货膨胀更使其变得微不足道。在美国，四分之一的 65 岁以上的老人依靠不足 4000 美元的年收入过活。在我们的社会里，许多老人被迫住在窄小的破旧房屋里，躲避在某些年久失修的老人公寓的背后。他们没有像样的住房，因为收入低微又深居简出的人住不起像样的小屋。

这第二个矛盾，就是需要一个简朴的小处所和需要社会接触、看见过往行人跟人点个头以及有个晒太阳的地方之间的矛盾。也可以像第一个矛盾那样，通过老人住所来

actually being able or willing to take care of their parents in their dotage.

The conflict can be partly resolved,if each house which houses a nuclear family has,somewhere near it,a small cottage where a grandparent can live,far enough away to be independent,and yet close enough to feel some tie and to be cared for in a time of trouble or approaching death.

But the conflict is more general.Even if we ignore, altogether, the complexities of parent-child relationships, the fact is that most old people face enormous difficulties as they grow older.The welfare state tries to replace the comfort of the extended family with payments—social security or pensions. This income is always tiny;and inflation makes it worse.In the United States,one-quarter of the population over 65 lives on less than $ 4000 a year.Many of the old people in our society are forced to live in miserable tiny rooms,way in the back of some run-down old folks hotel.They cannnot have a decent house,because there are no decent tiny houses compatible with a small income and reduced activity.

This second conflict,between the need for someplace really small and modest and the need for social contact,a view of passing people,someone to nod to,a place in the sun,can also be resolved,like the first conflict,by cottages.It can be resolved,if there are many tiny cottages,dotted among the houses of communities and always strung along pedestrian paths—tiny enough to be really cheap.

解决。如果有许多小住所，分布在社区的住宅之间，而且都排列在有行人经过的小路旁——这些住所很小，价格确实便宜，那么问题就可以得到解决了。

因此：

专门给老人盖些小住所。将其中一些盖在较大住宅的空地上，供祖父或祖母居住；将另一些盖在分散的地段上，这些地段比通常的小得多。在任何情况下，都要把这些住所盖成平房，靠近人们来来往往的马路，并接近社区内的服务性行业和公共场所。

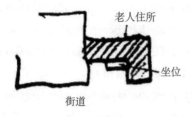

老人住所

坐位

街道

&❀❀&

老人住所最重要的部分也许是大门门廊和靠近街边的大门外的板凳——**私家的沿街露台**（140）、**大门外的条凳**（242）；其他方面，老人住所多半要根据**单人住宅**（78）的布局进行安排；为**固定工作点**（156）作好准备；使老人住所有**临街窗户**（164）。关于老人住所的形状，参阅**室内空间形状**（191）和**结构服从社会空间的需要**（205）……

Therefore:

Build small cottages specifically for old people.Build some of them on the land of larger houses,for a grandparent;build others on individual lots,much smaller than ordinary lots. In all cases,place these cottages at ground level,right on the street,where people are walking by,and close to neighborhood services and common land.

৪০৫৪

Perhaps the most important part of an old age cottage is the front porch and front door bench outside the door, right on the street—PRIVATE TERRACE ON THE STREET(140),FRONT DOOR BENCH (242);for the rest, arrange the cottage pretty much according to the layout of any HOUSE FOR ONE PERSON(78);make provisions for SETTLED WORK(156);and give the cottage a STREET WINDOW(164).And for the shape of the cottage start with THE SHAPE OF INDOOR SPACE(191)and STRUCTURE FOLLOWS SOCIAL SPACES(205)...

模式156　固定工作点*

　　……随着人们年龄的增大，做一些有益身心的使人感到愉快的简单工作变得越来越重要。本模式表明，家家户户都会有这种需要。它有助于形成**家庭**（75），它有助于形成**老人住所**（155），它自然地美化了**个人居室**（141）。

156   SETTLED WORK*

...as people grow older,simple satisfying work which nour-
ishes,becomes more and more important.This pattern specifies
the need for this development to be a part of every family.
It helps to form THE FAMILY(75),it helps form OLD AGE
COTTAGE(155),and it is a natural embellishment of A ROOM
O ONE'S OWN(141).

⊱⊰

**The experience of settled work is a prerequisite for peace
of mind in old age.Yet our society undermines this experience
by making a rift between working life and retirement,and
between workplace and home.**

First of all,what do we mean by "settled work"?It is the
work which unites all the threads of a person's life into one
activity:the activity becomes a complete and wholehearted
extension of the person behind it.It is a kind of work that one
can not come to overnight;but only by gradual development.
And it is a kind of work that is so thoroughly a part of one's
way of life that it most naturally occurs within or very near the
home:when it is free to develop,the workplace and the home
gradually fuse and become one thing.

It may be the same kind of work that a man has been doing
all his life—but as settled work it becomes more profound,more
concrete,and more unique.For example,there is the bureaucrat
who finally breaks through all the paper work and finds the
underlying organic function in his work.Then he begins to let

有固定工作可以成为使老年人心情安定的前提。但我们的社会由于无法处理好工作与退休之间、工作空间与家宅之间的关系，致使老年人得不到固定工作的机会。

首先，"固定工作"指的是什么呢？它是一个人集中全部精力进行的一项活动，从事这项活动的人一心一意地把它持续地干下去。这种工作不是一下子开始的；而是逐步建立起来的。这种工作完全成为人的一种生活方式，因而工作地点自然就在家里或在离家很近的地方；只要任其自由发展，工作地点和住家会渐渐融为一体。

它可能就是一个人一辈子从事的那种工作——但作为固定工作，它的意义更为深远，更为具体，也更为独特。例如，一位机关工作人员，他最终在文书工作上有所突破，发现他工作中的潜在的固有功能。然后他将这种功能公诸于世。这是黑泽明导演的最美好的影片《活下去》的主题。或者它可能是一个人在其业余时间开始的工作，跟他从事的职业无关，但这种工作范围逐渐扩大，变得越来越吸引人，以至完全取代了他原来的职业。

问题在于，许多人从未体验过固定工作的乐趣。这主要由于一个人在其工作中，既无时间也无空间来完善它。在当今的商业领域，大多数人被迫让自己的工作去适应办公室、工厂和机关的规程。通常这个工作弄得人筋疲力尽——当周末来到时，人们已无力去开始做一种新的、要求很高的工作。即使在自治工作间和办公室里，尽管工作程序是工作人员在工作中自己制定的，但工作本身通常也是为了商业的需要。它不允许有时间逐步开展"固定工作"——这工作是内在的，对商业不见得有什么重要意义。

为了解决这个问题，首先我们得创造一个环境，一个人在这样的环境中，比方从中年开始，有机会慢慢从事一

this function into the world.This is the theme of Kurosawa's most beautiful film,*Ikiru:To Live*.Or it may be work that a person begins in his spare time,away from his occupation,and it gradually expands and becomes more involving,until it replaces his old occupation altogether.

The problem is that very many people never achieve the experience of settled work.This is essentially because a person,during his working life,has neither the time nor the space to develop it.In today's marketplace most people are forced to adapt their work to the rules of the office,the factory,or the institution.And generally this work is all-consuming— when the weekends come people do not have the energy to start a new,demanding kind of work.Even in the self-governing workshops and offices,where working procedures are created ad hoc by the workers as they go,the work itself is generally geared to the demands of the marketplace.It does not allow time for the slow growth of "settled work" —which comes from within and may not always carry its weight in the marketplace.

To solve the problem,we must first of all create a working environment,where a person,from say middle age,has the opportunity of slowly developing a kind of settled work that is right for him.For instance,if people were able to take off one day a week,with half-time pay,beginning at the age of 40,they could gradually set up for themselves a workshop in their home or in their neighborhood.If the time is increased gradually over the years,a person can explore various kinds of work;and,then,gradually let the settled work replace his working life.

We make special mention of settled work as the work of old age,because,even though it must begin early on in a

种适合于他的固定工作。例如，如果人们一星期能够有一天不工作，这一天只领一半工资，从 40 岁开始，他们逐渐在家中或街坊上为自己建立起一个工作间。如果在数年内这样的时间逐渐增多，一个人就能够钻研各种工作；然后，他目前的工作逐渐会被固定工作所代替。

我们特别要指出，固定工作是老年人的工作，因为，即使它必须在人生的早期就开始，也只是到了晚年才使从事这种工作成为必需。老年的危机，生活的完美同失望和愤世嫉俗之间的矛盾，只有通过一个人从事某种形式的固定工作来解决——参阅**人生的周期**（26）。有机会掌握这种工作并以某种合适的方式将其同他们周围世界发生联系的人，当他们年老的时候，会找到可以成功解决这一危机的方法；其他的人则会陷入失望。

因此：

每一个人，特别是当他年老的时候，都应会有机会在自己的家里或在家附近建立起属于自己的工作间。使它逐步完善，开始的时候也许只是周末的业余活动，而后来则渐渐发展成为一个完整的、富有成果的、舒适的工作间。

固定工作点

&OCG

按**家庭工作间**（157）确定的线路具体安排这种工作间，使工作间向街道开敞，参与当地的街道生活——**私家的沿街露台**（140）、**向街道的开敞**（165）……

person's life,it is in old age that having such work becomes a necessity.The crisis of old age,life integrity versus despair and cynicism,can only be solved by a person engaged in some form of settled work—see LIFE CYCLE(26).People who have the opportunity to develop such work and to relate it in some appropriate way to the world about them,will find their way to a successful resolution of this crisis as they grow old;others will sink into despair.

Therefore:

Give each person,especially as he grows old,the chance to set up a workplace of his own,within or very near his home. Make it a place that can grow slowly,perhaps in the beginning sustaining a weekend hobby and gradually becoming a complete,productive,and confortable work shop.

☙❧

Arrange the workshop,physically,along the lines defined by HOME WORKSHOP(157),and make the workshop open to the street,a part of local street life—PRIVATE TERRACE ON THE STREET(140),OPENING TO THE STREET(165)...

模式157　家庭工作间

　　……在每一**住宅团组**（37）的中心和在**自己的家**（79）需要有一个房间或外屋，它既跟住宅相连又可随意由外面出入。这是家庭工作间。下面的模式告诉我们：工作间有多重要，它们应该分布多广，如何使它们随处都有，当这些工作间盖起来的时候，如何使之便于出入，如何使之始终开敞。它有助于加强**分散的工作点**（9）、**学习网**（18）及**男人和女人**（27）这些模式的应用。

157 HOME WORKSHOP

...at the center of each HOUSE CLUSER(37)and in YOUR OWN HOME (79)there needs to be one room or outbuilding,which is freely attached and accessible from the outside.This is the workshop.The following pattern tells us how important work shops are,how widely they ought to be scattered,how omnipresent,and when they are built,how easy to reach,and how public they should always be.It helps to reinforce the patterns of SCATTERED WORK(9),NETWORK OF LEARNING(18),and MEN AND WOMEN(27).

❧❧❧

As the decentralization of work becomes more and more effective,the workshop in the home grows and grows in importance.

We have explained in SCATTERED WORK(9), NETWORK OF LEARNING(18),MEN AND WOMEN(27), SELF-GOVERNING WORKSHOPS AND OFFICES(80),and other patterns that we imagine a society in which work and family are far more intermingled than today;a society in which people—businessmen,artists,craftsmen,shopkeepers,profession als—work for themselves,alone and in small groups,with much more relation to their immediate surroundings than they have today.

由于分散的工作越来越有成效，家庭工作间变得越来越重要。

在**分散的工作点**（9）、**学习网**（18）、**男人和女人**（27）、**自治工作间和办公室**（80）以及其他的模式中，我们解释过，在我们所设想的社会里，工作和家庭相互交融的程度远非今日可比；在那个社会里，人们——实业家、艺术家、手工业者、店主、脑力劳动者——都为自己工作，他们独自或在一个小团体内工作，但跟他们周围环境的联系比现在密切得多。

在这样一个社会里，家庭工作间远不是一个设在地下室或汽车房里的业余修理间。它成了每个住家的不可分割的一部分；在住宅中所起的作用不亚于厨房和卧室。我们认为它最重要的特征是同公共街道的联系。因为我们大多数人所从事的工作，相对来说都具有公共性质。无疑，同家庭生活的私密性相比较，工作是一种公共事务。即使在公共联系很少的地方，只要扩大工作者同社区的联系，就能使两者都有所获益。

在家庭工作间的情况下，工作的公共性质尤为重要，它使工作间脱离了后院的业余爱好的小天地而面向大庭广众。在工作间工作的人对街景一览尽收；他们自己也处于行人的视线之内。行人可以了解关于该社区的一些特点。特别是儿童们通过这种接触变得更加活跃。根据工作的性质，公共联系的形式可以是工作间铺面、装卸货物的车道、露天工作台、小会客室……

因此我们主张，使一个事实上的工作间具有真正的工作场所的全部特征，并使它跟公共街道发生一定程度的联系：至少是一种间接的联系，使工作间里外的人能彼此目视；或者是全面的联系，像一个敞开的商店店面。

In such a society, the home workshop becomes far more than a basement or a garage hobby shop. It becomes an integral part of every house; as central to the house's function as the kitchen or the bedrooms. And we believe its most important characteristic is its relationship to the public street. For most of us, work life is relatively public. Certainly, compared to the privacy of the hearth, it is a public affair. Even where the public relationship is slight, there is something to be gained, both for the worker and the community, by enlarging the connection between the two.

In the case of the home workshop, the public nature of the work is especially valuable. It brings the workshop out of the realm of backyard hobbies and into the public domain. The people working there have a view of the street; they are exposed to the people passing by. And the people passing learn something about the nature of the community. The children especially are enlivened by this contact. And according to the nature of the work, the public connection takes the form of a shopfront, a driveway for loading and unloading materials, a work bench in the open, a small meeting room...

We therefore advocate provision for a substantial workshop with all the character of a real workplace and some degree of connection to the public street: at least a glancing connection so that people can see in and out; and perhaps a full connection, like an open shop front.

因此：

　　在家里开辟出一个地方，在那里可以做些实实在在的事情；不仅是消遣，更是工作。改变地区规章以鼓励在邻里兴办小型的、安静的工作间。留给工作间大约几百平方英尺的面积；把它置于从街上能看得见的地方，工作间主人也可挂出一个招牌来。

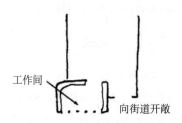

工作间　　　　　　　　向街道开敞

&ROCB&

　　使工作间有一个工作起来特别舒适的角落——**两面采光**（159）、**工作空间的围隔**（183）；使它同街道紧密联系——**向街道的开敞**（165）、**俯视外界生活之窗**（192）；也许还可提供一个地方以便在天气暖和的时候可以在太阳底下工作——**有阳光的地方**（161）。关于工作间的形状和构造，参阅**室内空间形状**（191）……

Therefore:

Make a place in the home,where substantial work can be done;not just a hobby,but a job.Change the zoning laws to encourage modest,quiet work operations to locate in neighborhoods.Give the workshop perhaps a few hundred square feet;and locate it so it can be seen from the street and the owner can hang out a shingle.

๛๛

Give the workshop a corner where it is especially nice to work—LIGHT ON TWO SIDES(159),WORKSPACE ENCLOSURE(183);a strong connection to the street—OPENING TO THE STREET(165),WINDOWS OVERLOOKING LIFE(192);perhaps a place to work in the sun on warm days—SUNNY PLACE(161).For the shape of the workshop and its construction,start with THE SHAPE OF INDOOR SPACE(191)...

模式158　室外楼梯*

　　……前几个模式——**出租房间**（153）、**青少年住所**（154）、**固定工作点**（156）、**家庭工作间**（157）——大多数可以安置在楼上，只要它们同街道有直接的通路可以连接即可。以前的模式中出现的家庭、公共服务行业和工作小组中的大多数，如果在楼上，也同样可行，只要给它们以直通街道的通路。例如，在一个工作社区内，**自治工作间和办公室**（80）、**小服务行业**（81）、**工作小组**（148），当它们在楼房的上层时，都要求直接通往公共街道。独立

158   OPEN STAIRS*

...most of the last patterns—ROOMS TO RENT(153), TEENAGER'S COTTAGE(154),SETTLED WORK(156), HOME WORKSHOP(157)—can be upstairs,provided that they have direct connections to the street.Far more generally,it is true that many of the households,public services,and workgroups given by earlier patterns can be successful when they lie upstairs,only if they are given direct connections to the street.For instance,in a work community SELFGOVERNING WORKSHOPS AND OFFICES(80),SMALL SERVICES WITHOUT RED TAPE(81),SMALL WORK GROUPS(148) all require direct access to the public street when they are on the upper storys of a building.And in the individual house holds—HOUSE FOR A SMALL FAMILY(76),HOUSE FOR A COUPLE(77),HOUSE FOR ONE PERSON(78)also need direct connections to the street,so people do not need to go through lower floors to get to them.This pattern describes the open stairs which may be used to form these many individual connections to the street.They play a major role in helping to create PEDESTRIAN STREETS(10).

❧☙

Internal staircases reduce the connection between upper stories and the life of the street to such an extent that they can do enormous social damage.

The simple fact of the matter is that an apartment on the second floor of a building is wonderful when it has a direct

的住宅——**小家庭住宅**（76）、**夫妻住宅**（77）、**单人住宅**（78）也需要同街道有直接的联系，这样人们出入不需要经过下面的楼层。本模式所描述的室外楼梯，可以用来营造许多这种独自向街道开敞的通道。它们对**步行街**（100）的建设起关键作用。

❧

楼房内部的楼梯减少了上面楼层和街上生活的联系，它们产生的社会效果可能具有巨大的破坏性。

如果二楼的一个单元房有一条直接通往街道的楼梯，那是十分理想的，但如果几个单元一起共用一条室内楼梯，那就不怎么理想了。这个事实是显而易见的。我们通过下面这些颇费周折的讨论，力图解释这一极端重要但又十分平常的直觉认识。

在传统文化中，楼房一旦建成，通往楼房上层的室外楼梯也随之出现，这是很普遍的现象。同样普遍的还有半室外楼梯——受到墙和屋顶保护，但仍然通向街道的楼梯。

美观的室外楼梯
The beauty of open stairs

相反，在工业化的、权力主义的社会里，大部分楼梯是室内楼梯。到达这些楼梯要经过楼房内的门厅和走廊；楼房上层同街道生活的直接联系被切断了。

stair to the street,and much less wonderful when it is merely one of several apartments served by an internal stair.The following,perhaps rather laborious discussion,is our effort to explain this vital and commonplace intuition.

In a traditional culture where buildings are built incrementally,outdoor stairs leading to upper stories are common. And half "outdoor" stairs—protected by walls and roofs,but nonetheless open to the street—are also common.

By contrast,in industrialized,authoritarian societies most stairs are indoor stairs.The access to these stairs is from internal lobbies and corridors;the upper stories are cut off from direct access to the life of the street.

This difference is not an incidental by-product of fire laws or construction techniques.It is fundamental to the difference between a free anarchical society,in which there is a voluntary exchange of ideas between equals,and a highly centralized authoritarian society,in which most individuals are subservient to large government and business organizations.

In effect we are saying that a centralized entrance,which funnels everyone in a building through it,has in its nature the trappings of control;while the pattern of many open stairs,leading off the public streets,direct to private doors,has in its nature the fact of independence,free comings and goings.

We can see this most easily in the cases where the centralized door is,without question,a source of social control. In work places with a central entrance and a time-clock,workers punch in and out,and they have to make excuses when they are

这不是室外楼梯——不要上当
This is not an open stair—don't be fooled

这一差别不是消防规范或建造技术的偶然性的产物。它是自由的无政府主义社会（在这个社会中平等的人之间可以自由交换意见）和高度集中的权力社会（在这个社会里大部分人都从属于大的政府和企业组织）之间差别的主要表现。

事实上，我们认为，一个集中出入口迫使楼房里的每个人都得经过这里，使人的行动受到约束；而许多室外楼梯从街道直达私家门口，具有使人自由出入的独立性。

在以集中出入口作为社会控制手段的情况下，我们可以很容易了解这一点。在一个有集中出入口和装有打卡钟的工作单位，工作人员上下班用打卡钟打卡，一旦他们在不到下班的时间离开工作岗位，他们就得找些借口。在某些学生宿舍，人们出入都需登记，如果他们在大门上锁前不回来，他们就会遇到麻烦。

在另一些情况下，控制的方法更为巧妙。在一幢公寓楼或办公楼里，每个人都可随心所欲地自由进出，但主要的一道门却时常上锁。当然，住户进入大楼可用钥匙开门；可是他们的朋友没有钥匙。一旦大门上锁（比如说在规定时间之后），他们就被阻挡在外面不能随意进来了。只有直通私宅门口的所有走道都是公共的，客人的自由访问才不至成为问题。

更加微妙的事实是，甚至在集中出入口并不对社会实行明显控制的地方——比方说大门始终开着——它也仍然

leaving at a time that is not normal.In some kinds of student housing,people are asked to sign in and out;and if they are not back by "lock-out" time,they are in trouble.

Then there are cases where the control is more subtle. In an apartment house or a workplace where everyone is free to come or go as he pleases it is not uncommon for the main door to be kept locked.Of course the residents have a key to the building;but their friends do not.When the front door is locked—after normal hours,say—they are effectively cut off from the spontaneous "dropping in" that can occur freely only where all paths are public right up to the thresholds of private territory.

Then there is the still more subtle fact that,even where the centralized entrance carries with it no explicit policy of social control—let us say that it is a door that is always open—it still has an uneasy feeling about it for people who cherish basic liberties.The single,centralized entrance is the precise pattern that a tyrant *would* propose who wanted to control people's comings and goings.It makes one uneasy to live with such a form,even where the social policy is relatively free.

This may very easily sound paranoid.But the point is this:socially,a libertarian society tries to build for itself structures which cannot easily be controlled by one person or one group "at the helm." It tries to decentralize social structures so that there are *many* centers,and no one group can come to have excessive control.

A physical environment which supports the same libertarian ideal will certainly put a premium on structures that allow people freedom to come and go as they please.And it will

给要求基本自由的人以不自在的感觉。这种单一的、集中的出入口毕竟是企图控制人们自由进出的专制主义者所能想出来的招数。即使这个地方对公众采取比较自由的政策，这种形式的居住方式是不会令人感到自在的。

上面这些话听起来很像是胡言乱语。但问题在于：从社会的角度看，一个自由的社会总是力图给自己建造一些建筑物，它们不易受控于一个人或一个"领导"集团。它力图使社会建筑物分散，使之有许多中心，从而没有一个集团能够具有过多的控制权。

一个能够支持这一自由主义理想的物质环境，一定会对建造人们能随意自由进出的建筑物起促进作用。它会力图保护这一权利，使之体现在建筑物和城市建设的基本规划之中。我们在一个空间上过分集中和受权力控制的建筑物里感到不自在，这是因为在这种情况下我们感到得不到保护；我们感到，我们的一种基本权利可能会受到侵犯，而不能得到环境的物质结构的充分保障。

室外楼梯解决了这个问题，它作为公共环境的延伸物，直达各家的家门和各个工作小组自己的空间。于是这些空间都直接跟外部环境充分地沟通了起来。平民百姓把每个这样的入口看作是真正的人们的活动领域，而不是企业和机关的势力范围，后者是有着实际的和潜在的权力实行专权的。

因此：

只要有可能，在社会事业机构里不要采用室内楼梯。使所有建筑物上层独立的住户、公共服务行业和工作团体都直接同地面建立联系。这可通过建造直接从街道走上来的室外楼梯来实现。根据气候条件，楼梯可以有顶，也可以没有，但无论如何使楼梯在地面开敞，不设门户，这样楼梯实际上就可以成为街道的延续。也不必在楼上建造走廊。可代之以室外的楼梯休息平台或室外拱廊，使楼上几个单元共用一个楼梯。

try to protect this right by building it into the very ground plan of buildings and cities.When we feel uneasy in a building that is spatially overcentralized and authoritarian,it is because we feel unprotected in this way;we feel that one of our basic rights is potentially vulnerable and is not being fully affirmed by the physical structure of the environment.

Open stairs which act as extensions of the public world and which reach up to the very threshold of each household's and each workgroup's own space solve this problem.These spaces are then connected directly to the world at large.People on the street recognize each entry as the domain of real people—not the domain of corporations and institutions,which have the actual or potential power to tyrannize.

Therefore:

Do away,as far as possible,with internal staircases in institutions.Connect all autonomous households,public services,and workgroups on the upper floors of buildings directly to the ground.Do this by creating open stairs which are approached directly from the street.Keep the stair roofed or unroofed,according to climate,but at all events leave the stair open at ground level,without a door,so that the stair is functionally a continuation of the street.And build no upstairs corridors.Instead,make open landings or an open arcade where upstairs units shre a single stair.

公共的室外楼梯

8003

在楼梯通往地面的地方，搭建一个出入口使之有助于补足街上已有的各种入口——**各种入口**（102）；建造楼梯休息平台和通往屋顶的楼梯顶端，从这里可以进入屋顶花园，在那里可种植花木，也可让人坐着晒太阳——**屋顶花园**（118）、**有阳光的地方**（161）。别忘了**能坐的台阶**（125），并根据**楼梯体量**（195）建造楼梯……

准备把建筑物内部同外部连接起来，把内外之间的边缘作为具有特殊用途的空间来处理，使人们可以在这里进行各种活动。

159. 两面采光

160. 建筑物边缘

161. 有阳光的地方

162. 背阴面

163. 有围合的户外小空间

164. 临街窗户

165. 向街道的开敞

166. 回廊

167. 六英尺深的阳台

168. 与大地紧密相连

Where the stair comes down to the ground, make an entrance which helps to repair the family of entrances that exist already on the street—FAMILY OF ENTRANCES(102);make the landings and the top of the stair, where it reaches the roof, into gardens where things can grow and where people can sit in the sun—ROOF GARDEN (118),SUNNY PLACE(161). Remember STAIR SEATS(125),and build the stair according to STAIRCASE VOLUME(195)...

prepare to knit the inside of the building to the outside. by treating the edge between the two as a place in its own right, and making human details there;

159. LIGHT ON TWO SIDES OF EVERY ROOM

160. BUILDING EDGE

161. SUNNY PLACE

162. NORTH FACE

163. OUTDOOR ROOM

164. STREET WINDOWS

165. OPENING TO THE STREET

166. GALLERY SURROUND

167. SIX-FOOT BALCONY

168. CONNECTION TO THE EARTH

模式159 两面采光**

......建筑物的主要房间一旦安排就绪，我们就该把建筑物的实际形状确定下来：这样必定涉及边缘的位置。从建筑物的总体形状中边缘的位置已经大体明确——**有天然采光的翼楼**（107）、**户外正空间**（106）、**狭长形住宅**（109）、**重叠交错的屋顶**（116）。本模式完善**有天然采光的翼楼**（107）的工作，将每一个单独的房间都恰好安排在能获取阳光的地方。根据这些单个的房间的位置，本模式形成了建筑物边缘的准确线条。下一个模式开始讨论边缘的形状。

159   LIGHT ON TWO SIDES OF EVERY ROOM**

...once the building's major rooms are in position,we have to fix its actual shape:and this we do essentially with the position of the edge.The edge has got its rough position already from the overall form of the building—WINGS OF LIGHT(107),POSITIVE OUTDOOR SPACE(106),LONG THIN HOUSE(109),CASCADE OF ROOFS(116).This pattern now completes the work of WINGS OF LIGHT(107),by placing each individual room exactly where it needs to be to get the light.It forms the exact line of the building edge,according to the position of these individual rooms.The next pattern stats to shape the edge.

ஐௌௗ

When they have a choice,people will always gravitate to those rooms which have light on two sides,and leave the rooms which are lit only from one side unused and empty.

This pattern,perhaps more than any other single pattern, determines the success or failure of a room.The arrangement of daylight in a room,and the presence of windows on two sides,is fundamental.If you build a room with light on one side only,you can be almost certain that you are wasting your money.People will stay out of that room if they can possibly avoid it.Of course,if all the rooms are lit from one side only,people will have to use them.But we can be fairly sure that they are subtly

ဆၢ

　　当人们有机会选择的时候，他们总会为有两面采光的房间所吸引，而将那些只能单面采光的房间闲置不用，让它们空着。

　　本模式或许比任何其他一个模式都更能决定一个房间的成败得失。一个房间是否光线充足，是否两面有窗，这是一个根本的问题。如果你盖的房间只能单面采光，可以肯定地说，你是在浪费自己的钱财。如果人们可以避免住这种房间，那么他们宁可不搬进去。当然，如果所有的房间都是单面采光，人们就不得不使用它们。但我们可以相当有把握地说，他们住在那里是会感到有些不舒服的，总希望自己最好不待在那儿，等待着离开那儿——正是因为我们确信，当人们有选择余地的时候，他们会作出怎样的选择。

　　我们对本模式所做的实验一直是非正规的，前后经历了若干年。同许多建筑工作者一样，我们产生这种想法也颇有时日了。(我们曾听说过"两面采光"是老的巴黎美术学院设计传统的原则)。在任何情况下，我们的实验都是简单的：一次一次地做，一幢建筑物一幢建筑物地做，凡是我们待过的地方，我们总要核实一下，这个模式是否起作用。如果人们事实上在避开单面采光的房间，而其实是要两面采光的房间——他们会如何看待本模式呢？

　　我们跟朋友一起在办公室和许多家庭对此做过试验——在绝大多数情况下，两面采光的模式是有效的。人们对这一模式是了解的或有所了解的——他们完全明白我们的用意。

uncomfortable there,always wishing they weren't there,wanting to leave—just because we are so sure of what people do when they do have the choice.

Our experiments on this matter have been rather informal and drawn out over several years.We have been aware of the idea for some time—as have many builders.(We have even heard that "light on two sides" was a tenet of the old Beaux Arts design tradition.)In any case,our experiments were simple:over and over again,in one building after another,wherever we happened to find ourselves,we would check to see if the pattern held.Were people in fact avoiding rooms lit only on one side,preferring the two-sided rooms—what did they think about it?

We have gone through this with our friends,in offices,in many homes—and overwhelmingly the two-sided pattern seems significant.People are aware,or half-aware of the pattern—they understand exactly what we mean.

If this evidence seems too haphazard,please try these observations yourself.Bear the pattern in mind,and examine all the buildings you come across in your daily life.We believe that you will find,as we have done,that those rooms you intuitively recognize as pleasant,friendly rooms have the pattern;and those you intuitively reject as unfriendly,unpleasant,are the ones which do not have the pattern.In short,this one pattern alone,is able to distinguish good rooms from unpleasant ones.

The importance of this pattern lies partly in the social atmosphere it creates in the room.Rooms lit on two sides,with natural light,create less glare around people and objects;this lets us see things more intricately;and most important,it allows us to

有两面采光   没有两面采光
With light on two sides...and without

　　如果说这一证据看起来具有过多的偶然性，那就请你自己作这些观察吧。心里想着这个模式，考察你在日常生活中看到的所有建筑物。相信你会发现，正如我们已经发现的那样，那些你凭直觉感到令人愉快的、使你感到可亲的房间都具有这一模式；而那些你凭直觉感到不可亲、使你感觉不愉快的房间则不具有这一模式。总之，仅根据这一模式，你就能区分出房间的优劣，判断出它们是否宜人。

　　本模式的重要意义部分在于它在房间里所创造的社会气氛。有两面采光的房间，光线自然，在人和物的周围较少造成眩光；这使我们有可能把事物看得更加仔细；而最重要的，它使我们能仔细地看到人们面部一闪而过的细微表情，他们的手部动作……因而更清楚地了解到这些表情和动作所包含的意思。**两面采光使人们能够更好地互相了解。**

　　在单面采光的房间里，光线在室内墙壁和地板上的倾斜度很大，因此离窗户最远的部分同靠近窗户的地方比较起来，显得黑乎乎的，令人不舒服。甚至更糟的是，由于在房间的内表面很少有反射光，紧靠窗户的内墙通常是黑暗的，使人感到不舒服，并产生与此光线对比度很强的眩光。**在一个单面采光的房间里，在人们脸部周围的眩光妨碍人们相互了解。**

　　虽然这种眩光通过补充人造光及精心设计窗洞可以有所减弱，但克服眩光的最简单、最基本的办法是使每个房间能够有两个窗户，从每一窗户射进来的光线恰好照在

read in detail the minute expressions that flash across people's faces,the motion of their hands...and thereby understand,more clearly,the meaning they are after.*The light on two sides allows people to understand each other.*

In a room lit on only one side,the light gradient on the walls and floors inside the room is very steep,so that the part furthest from the window is uncomfortably dark,compared with the part near the window.Even worse,since there is little reflected light on the room's inner surfaces,the interior wall immediately next to the window is usually dark,creating discomfort and glare against this light.*In rooms lit on one side,the glare which surrounds people's faces prevents people from understanding one another.*

Although this glare may be somewhat reduced by supplementary artificial lighting,and by well-designed window reveals,the most simple and most basic way of overcoming glare,is to give every room two windows.The light from each window illuminates the wall surfaces just inside the other window,thus reducing the contrast between those walls and the sky outside.For details and illustrations, see R.G.Hopkinson,*Architectural Physics:Lighting*,London:Building Research Station,1963,pp.29,103.

A supreme example of the complete neglect of this pattern is Le Corbusier's Marseilles Block apartments.Each apartment unit is very long and relatively narrow,and gets all its light from one end,the narrow end.The rooms are very bright just at the windows and dark everywhere else.And,as a result,the glare created by the light-dark contrast around the windows is very disturbing.

另一窗户里边的墙面上，从而减弱了内墙和室外天空之间的对比。细节和实例请参看 R.G.Hopkinson, *Architectural Physics : Lighting*, London : Building Research Station, 1963, pp.29, 103.

完全忽视本模式的最极端的例子是勒·柯布西耶的马赛公寓。它的每一住宅单元都很长，又比较窄，仅能从一端即窄端摄取阳光。房间仅在窗户旁的地方很明亮，其他地方都是暗的。结果，窗边明暗对比造成的眩光叫人心烦。

在一幢小建筑物里，很容易让每个房间两面采光：建筑物中处于四个角上的每个房间自然都可以有两个窗户。

在稍大一点的建筑物里，就得把房屋边缘折起来，把屋角转个方向，以获得相同效果。大房间和小房间并排也能做到这一点。

在更大的建筑物里，可能需要有系统地加宽平面或进一步折叠边缘，以便使每个房间两面采光。

但是，毫无疑问，不管我们设计得多么巧妙，不管我们多么精心地折叠边缘，有时我们仍然会束手无策。在这样的情况下，有两个手段可使房间获得两面采光的效果。如果房间很浅——不深于大约8ft——至少并排用两个窗户可使它获得两面采光的效果。光线从后墙反射过来，向两窗之间的侧面反射，

把边缘折起来
Wrinkle the edge

因此这种光线仍具有不会产生眩光的两面采光的性质。

最后，如果房间的深度不得不超过 8ft，而又不能两面采光——这时可以通过下面的一些办法来解决这个问题：把天花板高度加得很高，把墙粉刷得很白，在墙上筑起很高的窗户，把窗洞造得很深以抵消眩光。乔治亚式宅邸中的伊丽莎白式餐厅和起居室就常按此格局建造。但请记住，

In a small building,it is easy to give every room light on two sides:one room in each of the four corners of a house does it automatically.

In a slightly larger building,it is necessary to wrinkle the edge,turn corners,to get the same effect.Juxtaposition of large rooms and small,helps also.

In an even larger building,it may be necessary to build in some sort of systematic widening in the plan or to convolute the edge still further,to get light on two sides for every room.

But of course,no matter how clever we are with the plan, no matter how carefully we convolute the building edge, sometimes it is just impossible.In these cases,the rooms can get the effect of light on two sides under two conditions.They can get it,if the room is very shallow—not more than about eight feet deep—with at least two windows side by side.The light bounces off the back wall,and bounces sideways between the two windows,so that the light still has the glare—free character of light on two sides.

And finally,if a room simply has to be more than eight feet deep,but cannot have light from two sides—then the problem can be solved by making the ceiling very high,by painting the walls very white,and by putting great high windows in the wall,set into very deep reveals,deep enough to offset the glare.Elizabethan dining halls and living rooms in Georgian mansions were often built like this.Remember,though,that it is very hard to make it work.

要做到使它产生两面采光的效果是很困难的。

因此：

安排好每个房间的位置，使其外部至少两面有室外空间，然后在这些室外墙上开出两个窗户，使自然光从不同的方向照射进每个房间。

每个房间有两面采光

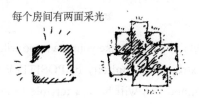

⚜

别让本模式把你的设计弄得太狂野——不然你会毁了单纯的**户外正空间**（106），而且还会为盖这种房屋的屋顶伤透脑筋——**屋顶布置**（209）。请记住，可以在一面多开窗来维持本模式的实质，如果房间特别高，如果同窗墙的长度比较起来房间很浅，如果窗户很大，房间的墙壁很白；还有，如果窗上有大的深窗洞以保证虽然有明亮的大窗对着天空，因此并不会产生眩光。

从每个窗户都能看到某种美丽的景色——**俯视外界生活之窗**（192）、**借景的门窗**（221）；将房间里的一个窗户作特殊处理，使其周围有一片空间——**窗前空间**（180）。**使用深窗洞**（223）和**过滤光线**（238）……

Therefore:

Locate each room so that it has outdoor space outside it on at least two sides,and then place windows in these out door walls so that natural light falls into every rom from more than one direction.

৪০৫৪

Don't let this pattern make your plans too wild—otherwise you will destroy the simplicity of POSITIVE OUTDOOR SPACE(106),and you will have a terrible time roofing the building—ROOF LAYOUT(209).Remember that it is possible to keep the essence of the pattern with windows on one side,if the room is unusually high,if it is shallow compared with the length of the window wall,the windows large,the walls of the room white,and massive deep reveals on the windows to make quite certain that the big windows,bright against the sky,do not create glare.

Place the individual windows to look onto something beautiful—WINDOWS OVERLOOKING LIFE(192), NATURAL DOORS AND WINDOWS(221);and make one of the windows in the room a special one,so that a place gathers itself around it—WINDOW PLACE(180).Use DEEP REVEALS(223)and FILTERED LIGHT(238)...

模式160 建筑物边缘**

……假定建筑物边缘的位置已经确定——最近的模式是为**两面采光**（159）所确定——在这之前为建筑物翼楼及其内部空间的位置以及建筑物之间的庭院、花园和街道所确定——**有天然采光的翼楼**（107）、**户外正空间**（106）。本模式现在准备利用室内外之间的地带。这一"地带"常常被看作是边缘，是没有厚度的画在纸上的线条，是墙壁。但这完全错了……

⊷⊶

人们常常认为，建筑物的目的是利用内部空间——即它的房间。人们通常并不认为建筑物也必须面向外部环境。

但如果不能认真而有效地使建筑物像面向内部空间那样面向它周围的外部环境，建筑物周围的空间便毫无用处，

160   BUILDING EDGE**

...assume that the position of the building edge is fixed—most recently by LIGHT ON TWO SIDES OF EVERY ROOM(159)—and before that by the position of the building wings and their interior spaces and by the courts and gardens and streets between the buildings—WINGS OF LIGHT(107),POSITIVE OUTDOOR SPACE(106).This pattern now sets the stage for the development of the zone between the indoors and the outdoors.Often this "zone" is thought of as an edge,a line on paper without thicknss,a wall.But this is altogether wrong...

❀❀❀

A building is most often thought of as something which turns inward—toward its rooms.People do not often think of a building as something which must also be oriented toward the outside.

But unless the building is oriented toward the outside, which surrounds it,as carefully and positively as toward its inside,the space around the building will be useless and blank—with the direct effect,in the long run,that the building will be socially isolated,because you have to cross a no-man's land to get to it.

Look,for example,at this machine age slab of steel and glass.You cannot approach it anywhere except at its entrance—because the space around it is not made for people.

And compare it with this older,warmer building,which has a continuous surrounding of benches,galleries,balconies,flow

荒芜一片——从长远来看，直接
产生的后果是使建筑物同社
会隔绝，因为要去那里你就必须
要经过一个无人的地段。

它的边缘荒无人迹
The edge cannot support any life

例如，请看这一机器时代
用钢板和玻璃筑成的大楼。除
了在它的入口处，你在哪儿也
接近不了它——因为它周围的空
间不是供人使用的。

一个可利用的建筑物
边缘
An edge that can be used

再来将它跟下面这幢老式的、
但亲切得多的房屋比较一下，它有
一个连续的周围环境，这里有长凳、
走廊、露台、花丛、供休憩的角落、
可停留的空间。这个建筑物边缘是
有生气的。它形成使人们能享受到
乐趣的有用空间，从而使自己同外
部环境连成一片。

请考虑这一微小差别的效果。像
机器一样的建筑物同周围环境是割
裂的，无依无靠，形同孤岛。相反，
边缘充满生机的建筑物则同周围连
成一片，它是社会结构的一个组成部分，是城市的组成部分，
也是在其周围居住和活动的所有的人生活的组成部分。

从下面的例子我们可以看出这一对比是有事实根据
的：很明显，人们喜欢停留在露天空间的边缘——如果这
些边缘生机盎然，人们就会对之留连忘返。例如，当观察
人们在室外空间的行为时，简·吉尔发现，"无论是站着的
或是坐着的人，都有一个显著的倾向，即总要靠近某一样
东西——房屋外墙、柱子、家具等"［"Mennesker til Fods
(Pedestrians)," *Arkitekten*, No.20, 1968.］人们这种愿

ers,corners to sit,places to stop.This building edge is alive.It is connected to the world around it by the simple fact that it is made into a positive place where people can enjoy themselves.

Think of the effect of this small difference.The machine—like building is cut off from its surroundings,isolated,an island. The building with a lively building edge,is connected,part of the social fabric,part of the town,part of the lives of all the people who live and move around it.

We get empirical support for this contrast from the following:apparently people prefer being at the edges of open spaces—and when these edges are made human,people cling to them tenaciously.In observing people's behavior in outdoor spaces, for example,Jan Gehl discovered that "there is a marked tendency for both standing and sitting persons to place themselves near something—a facade, pillar, furniture, etc." ["Mennesker til Fods(Pedestrians)," *Arkitekten*, No.20,1968.] This tendency for people to stay at the edges of spaces,is also discussed in the pattern ACTIVITY POCKETS(124).

If this propensity were taken as seriously outdoors as it is indoors,then the exterior walls of buildings would look very different indeed from the way they look today.They would be more like places—walls would weave in and out,and the roof would extend over them to create little places for benches,posters,and notices for people to look at.For the niches to have the right depth,they would have to be occasionally as much as 6 feet deep—see the arguments for SIX-FOOT BALCONY(167).

When it is properly made,such an edge is a realm between realms:it increases the connection between inside and outside, encourages the formation of groups which cross the

意停留在空间边缘的倾向也在袋形**活动场地**（124）中讨论过。

假如人们的这一倾向在室外也像在室内一样得以被认真看待，建筑物外墙的面貌与其今日的模样就会大不相同了。它们会更像场所——墙会在有的地方凹进、有的地方凸出，屋顶会在它们的上方向外伸出而形成一些小空间以放置长凳，供人招贴和让人观看布告。壁龛的合适深度有时达 6ft——参阅在**六英尺深的阳台**（167）中的论述。

如果安排得当，这种边缘就成为区域之间的区域：它加强内外空间之间的联系，促使形成跨越边界的许多小空间，使人们可以从这一边走到那一边，并使人有可能既在边界之上，也在边界之内进行活动。这是一个很重要的观念。

因此：

务必把建筑边缘看作一件"实物"，一个场所，一个有体量的区域，而不是没有厚度的一条线或边界。使建筑物边缘的某些部分凸出，以营造出引人逗留的场所。要形成一些有深度和有遮挡的场所，可以使人坐着、倚靠和散步的场所，这些场所最好位于周边那些能看到丰富多采的室外生活的地方。

突出的地方

沿边缘的深度

有遮蔽的处所

* * *

建造拱廊、回廊、门廊和棚架来进行这项工作——**拱廊**（119）、**有围合的户外小空间**（163）、**回廊**（166）、**六英尺深的阳台**（167）、**与大地紧密相连**（168）；特别要考虑阳光——**有阳光的地方**（161）、**背阴面**（162）；设置坐位和窗户使之加强彼此间的连接感——**能坐的台阶**（125）、**临街窗户**（164）、**户外设座位置**（241）、**大门外的条凳**（242）……

boundary,encourages movement which starts on one side and ends on the other,and allows activity to be either on,or in the boundary itself.A very fundamental notion.

Therefore:

Make sure that you treat the edge of the building as a"thing," a"place,"a zone with volume to it,not a line or interface which has no thickness.Crenelate the edge of buildings with places that invite people to stop.Make places that have depth and a covering,places to sit,lean,and walk, especially at those points along the perimeter which look onto interesting outdoor life.

☙❧

Do it with arcades,galleries,porches,and terraces— ARCADES(119),OUTDOOR ROOM(163),GALLERY SURROUND(166), SIX-FOOT BALCONY(167), CONNECTION TO THE EARTH(168);take special account of the sun—SUNNY PLACE(161),NORTH FACE(162);and put in seats and windows which complete the feeling of connection— STAIR SEATS (125),STREET WINDOWS(164),SEAT SPOTS(241),FRONT DOOR BENCH (242)...

模式161　有阳光的地方**

161   SUNNY PLACE**

...this pattern helps to embellish and give life to any SOUTH FACING OUTDOORS(105);and,in a situation where the outdoors is not to the south,but east or west,it can help to modify the building so that the effective part of the outdoors moves towards the south.It also helps to complete BUILDING EDGE(160),and to place OUTDOOR ROOM(163).

୫୬୯ଓଔ

The area immediately outside the building,to the south— that angle between its walls and the earth where the sun falls— must be developed and made into a place which lets people bask in it.

We have already made the point that important outdoor areas should be to the south of buildings which they serve,and we presented the empirical evidence for this idea in SOUTH FACING OUTDOORS(105).But even if the outdoor areas around a building are toward the south,this still won't guarantee that people actually will use them.

In this pattern,we shall now discuss the subtler fact that a south-facing court or garden will still not work,unless there is a functionally important sunny place within it,intently and specifically placed for sun,at a central juncture between indoors and out-doors and immediately next to the indoor rooms which it serves.

We have some evidence—presented in SOUTH-FACING OUT-DOORS (105)—that a deep band of shade between a

……本模式有助于美化所有**朝南的户外空间**（105），使之富有生气；在户外空间不朝南，而是朝东或朝西的情况下，它对建筑物能起调节作用，将户外空间的有效部分往南转移。它也有助于完善**建筑物边缘**（160）和安置**有围合的户外小空间**（163）。

✂︎✂

房屋外边的朝南地段——阳光照到的墙和地面之间的角落——必须加以利用，使它成为人们能够晒太阳的地方。

我们曾经指出，重要的户外地区应该在它们所属房屋的南面，而且我们还在**朝南的户外空间**（105）中为这一想法提供了实际的证据。但即使房屋周围的户外地区朝南，也并不能保证人们实际上会利用它们。

在本模式中，我们将讨论这一相当微妙的事实，即朝南的庭院或花园仍旧可能无用，除非其内部有一个实际上很重要的有阳光照到的地方，这个地方是专门为得到阳光而特意设计的，它位于室内外之间的中心交界处，并紧邻它为之服务的房间。

我们有一些证据——**朝南的户外空间**（105）提供的——说明，在房屋和有阳光的区域之间一条很深的阴影可能是一个障碍，使这一区域不能得到充分利用。正是这一证据使我们认为，最重要的有阳光的地方总是紧邻房屋的外墙，人们从室内可以看到这个地方，并可直接出来走到阳光底下，倚靠在房屋的门口。再则，我们观察到，这些地方会更加吸引人，如果它们位于房屋或墙的弯曲处，那里的篱笆、矮墙、柱子可以形成足够的围合，从而提供一个背景，一处可以靠着它坐着和晒太阳的地方。

当然，如果最终要使这个地方确实发挥作用，需有充

building and a sunny area can act as a barrier and keep the area from being well used.It is this evidence which makes us believe that the most important sunny places occur up against the exterior walls of buildings,where people can see into them from inside and step directly out into the light,leaning in the doorway of the building.Furthermore,we have observed that these places are more inviting if they are placed in the crook of a building or wall,where there is just enough enclosure from a hedge,a low wall,a column,to provide a backdrop,a place to sit up against and take in the sun.

And finally,of course,if the place is really to work,there must be a good reason for going there:something special which draws a person there—a swing,a potting table for plants,a special view,a brick step to sit upon and look into a pool— whatever,so long as it has the power to bring a person there almost without thinking about it.

Here is an example—a sunny place at the edge of a building, directly related to the inside,and set in a nook of the building. Someone comes there every day to sit for a moment,water the hanging plants,see how they are doing,and take in some sun.

A particularly beautiful version of this pattern can be made when several sunny places are placed together—perhaps for a HOUSE CLUSTER(37)or a WORK COMMUNITY(41).If the places can be set down so that they form a south-facing half-necklace of sunny spots,each within hailing distance of all the others,it makes the act of coming out into the sun a communal affair.

Therefore:

Inside a south-facing court,or garden,or yard,find the spot between the building and the outdoors which gets the

分的理由使人愿意到那里去：需有某些能吸引人去的特别的东西——一架秋千、一张摆盆花的台子、一处特别的景致、一个可让人坐着休息和观看池子的砖砌台阶——什么东西都行，只要它具备一种吸引力，能够使人毫不迟疑地到那儿去。

例如，建筑物边缘一个有阳光的地方直接连接室内空间，坐落在房屋的凹角处。有人每天到这里来坐一会儿，给垂挂的植物浇点水，观察它们的生长情况,然后晒晒太阳。

有阳光的地方
Sunny place...

当几个有阳光的地方连在一起的时候——多半在**住宅团组**（37）或**工作社区**（41），本模式就可以变得美不胜收。如果这些地方能确定下来，使有阳光的区域形成朝南的半条项链状，每个区域都同其他各区域相距不远，彼此呼应，这样就使走出户外晒太阳成为居者们的一项共同性的活动。

因此：

在一个朝南的庭院、花园或场院里，要找到在房屋和户外空间之间的能获得最佳阳光的区域。把这个区域用作专门的有阳光的地方——使它成为一个重要的有围合的户外小空间，一个可以在阳光下工作的地方，或可以荡秋千

best sun. Develop this spot as a special sunny place—make it the important outdoor room,a place to work in the sun,or a place for a swing and some special plants,a place to sunbathe.Be very careful indeed to place the sunny place in a position where it is sheltered from the wind.A steady wind will prevent you from using the most beautiful place.

ଚ୍ଚ

Make the place itself as much as possible like a room— PRIVATE TERRACE ON THE STREET(140),OUTDOOR ROOM(163);always at least six feet deep,no less—six-FOOT BALCONY(167);perhaps with foliage or a canvas to filter the light on hot days—FILTERED LIGHT (238),TRELLISED WALK(174),CANVAS ROOF(244).Put in seats according to SEAT SPOTS(241)...

和种植某些特殊植物的地方、沐浴阳光的地方。要精心地将此有阳光的地方设置在背风处。经常受风会使你无法利用这个最美的地方。

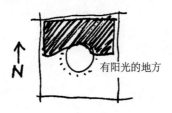

有阳光的地方

⊱⊰

　　使这个地方本身能够尽可能像一间房间——私家的**沿街露台**（140）、**有围合的户外小空间**（163）；至少总要有6ft 深，不能少于此数——**六英尺深的阳台**（167）；在热天，也许要用叶饰或帆布来过滤光线——**过滤光线**（238）、**棚下小径**（174）、**帆布顶篷**（244）。根据**户外设座位置**（241）设置坐位……

模式162 背阴面

……根据朝南的**户外空间**（105），即使房屋的坐向得到了恰当的安排，使朝北的空间很小，但一般总还有某种区域或空间处在房屋的背阴面。因此有必要处理好这一背阴空间，作为**室内阳光**（128）和**有阳光的地方**（161）的补充。

❧❦❧

请观察一下你所知道的建筑物的背阴面。无论在哪里你都会发现，这些背阴的地方总是死寂、潮湿、阴冷和无用。但在一个城市中有成千上万英亩的土地在房屋的背阴面；而只要有房屋存在，就不可避免地会有背阴的地方。

如果一幢房屋有一面完全背阴，一年中有许多个月份该房屋背后会留下一条很长的阴影。

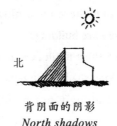

背阴面的阴影
North shadows

这些死寂、阴暗的背阴面不仅浪费大量的土地；它们还会对更大的环境造成损害，因为没有人愿意走过阴影区，阴影区使环境各部分彼此分割，从而切断环境。需要找到一种办法，至少根据背阴地方自身的条件，赋予它们以生气，这样使它们对周围的环境有所补益，而不至于分解这些背阴面土地。

背阴面投下的阴影基本上是三角形的，要使这三角形阴影不至成为一个荒凉的地方，必须用不需阳光的实物和

162   NORTH FACE

...even if the building has been placed correctly according to SOUTH-FACING OUTDOORS(105) and there is little outdoor space toward the north,there is usually still some kind of area or volume on the north face of the building.It is necessary to take care of this north-facing place to supplement the work of INDOOR SUNLIGHT(128) and SUNNY PACE(161).

⊱⊰

Look at the north sides of the buildings which you know. Almost everywhere you will find that these are the spots which are dead and dank,gloomy and useless.Yet there are hundreds of acres in a town on the north sides of buildings; and it is inevitable that there must always be land in this position,wherever there are buildings.

If a building has a sheer north face,during many months of the year it will cast a long shadow out behind it.

These dead and gloomy north sides not only waste enormous areas of land;they also help to kill the larger environment,by cutting it up with shadow areas which no one wants to cross,and which therefore break up the various areas of the environment from one another.It is essential to find a way of making these north-facing areas alive,at least in their own terms,so that they help the land around them instead of breaking it apart.

The shadow cast by the north face is essentially triangular. To keep this triangle of shade from becoming a forlorn place,it is necessary to fill it up with things and places which do not

空间来填充它。例如，背阴区域可以建造一个平缓的阶梯式建筑，里面可停放小汽车，也许还可容纳浴室、储藏室、画室和放垃圾箱。如果这一阶梯式建筑造得恰当，一年内大部分时间它外面的朝北户外

北

背阴面的阶梯式建筑
North cascade

空间将会使花园、温室、私密的园中坐椅、工作间、步行道得到充足的阳光。

再则，如果不可避免出现背阴的朝北房间，置一反射墙就能使情况大为改善：这堵墙位于北面，与房屋有一定距离，刷成白色或黄色，处于能获得阳光的位置，并把阳光反射到房屋里面。这种墙可以是相邻房屋的墙、花园墙等。

因此：

在房屋的背阴面造一阶梯式建筑，使之向地面倾斜，这样可使通常向北投射很长阴影的阳光能够照射到紧靠房屋的地面。

背阴面
North face

※○※

将这背阴面的三角形阶梯式建筑用来作汽车房、垃圾箱、储藏间、库房、需要北面光线的画室，以及没有室内阳光也无所谓的盥洗室和厕所——**与车位的联系**（113）、**大储藏室**（145）、**堆肥**（178）、**居室间的壁橱**（198）。如果条件允许，可以利用房屋北面的一堵白色或黄色的墙把阳光反射到背阴的房间里——**室内阳光**（128）、**两面采光**（159）、**花园墙**（173）……

need the sun.For example,the area to the north may form a gentle cascade which contains the car shelter,perhaps a bath suite,storage,garbage cans,a studio.If this cascade is properly made,then for most of the year the outdoors beyond it to the north will have enough sun for a garden,a greenhouse,a private garden seat,a workshop,paths.

Furthermore,if there are north rooms that are inevitably gloomy,it helps enormously to make a reflecting wall:a wall standing some ways to the north of the building,painted white or yellow,and set in a position which gets the sun and reflects it back into the building.This wall might be the wall of a nearby building,a garden wall,etc.

Therefore:

Make the north face of the building a cascade which slopes down to the ground,so that the sun which normally casts a long shadow to the north strikes the ground immediately beside thebuilding.

❧❦

Use the triangle inside this north cascade for car, garbage,storage,shed,a studio which requires north light, closets—those parts of the building which can do very well without interior sunlight—CAR CONNECTION(113),BULK STORAGE(145),COMPOST(178),CLOSETS BETWEEN ROOMS(198).If it is at all practical,use a white or yellow wall to the north of the building to reflect sunlight into the north-facing rooms—INDOOR SUNLIGHT(128),LIGHT ON TWO SIDES OF EVERY ROOM(159),GARDEN WALL(173)...

模式163　有围合的户外小空间**

　　……每幢建筑物都有供人们逗留、居住和交谈的空间——**中心公用区**（129）、**农家厨房**（139）、**起居空间的序列**（142）。无论何时，只要有可能，这些空间还需要进一步用一种户外的"空间"来美化。这类户外空间也有助于形成部分**户外亭榭**（69）、**半隐蔽花园**（111）、**私家的沿街露台**（140）或**有阳光的地方**（161）。

 howcow

　　在花园里人们可以躺在草坪上、荡秋千、玩槌球、种花、和狗扔球玩。但人们还有另外一种户外活动：这种户外活动的需要是花园绝对满足不了的。

163　OUTDOOR ROOM**

...every building has rooms where people stay and live and talk together—COMMON AREAS AT THE HEART(129), FARMHOUSE KITCHEN (139),SEQUENCE OF SITTING SPACES(142).Whenever possible,these rooms need to be embellished by a further "room" outdoors. This kind of outdoor room also helps to form a part of any PUBLIC OUTDOORROOM(69),HALF-HIDDEN GARDEN (111),PRIVATE TERRACE ON THE STEET(140),or SNNY PLACE(161).

❧⬧❧

A garden is the place for lying in the grass,swinging, croquet,growing flowers,throwing a ball for the dog.But there is another way of being outdoors:and its needs are not met by the garden at all.

For some moods,some times of day,some kinds of friendship,people need a place to eat,to sit in formal clothes,to drink,to talk together,to be still,and yet outdoors.

They need an outdoor room,a literal outdoor room—a partly enclosed space,outdoors,but enough like a room so that people behave there as they do in rooms,but with the added beauties of the sun,and wind,and smells,and rustling leaves,and crickets.

This need occurs everywhere.It is hardly too much to say that every building needs an outdoor room attached to it,between it and the garden;and more,that many of the special

当人们或在一天的某个时间段产生某种情绪时，或为了某种友谊，他们需要找个地方用餐，穿着整齐地坐着，喝酒，交谈，或安静地待着，但这一切都要在户外进行。

他们需要一个户外小空间，一个真正的户外空间——这个空间部分围合，它虽然在户外，但酷似一个房间，人们在那里活动，就像在房间里一样，只是多了阳光、风、户外的气息、树叶的沙沙响声和蟋蟀的鸣叫声，因而增添了美趣。

这样的需求无处不有。应该说，每幢建筑物都需要有一个附属的户外有围合的小空间，它位于房屋和花园之间；而且，花园中许多特殊空间——阳光照到的地方、露台、亭台——也需要做成户外有围合的小空间。

产生这一模式的灵感来自鲁道夫斯基的作品《在风景窗后面》（*Behind the Picture Window*，New York：Oxford Press，1955）中的一章"适合于户外的空间"。

"一个布局精妙的私家花园，理应使人们能够在里面工作和睡觉，烹调和用餐，玩耍和游逛。毫无疑问，偏爱室内生活的人对此是不以为然的，因此需要加以详细说明。

在我们这样的气候条件下，居民照例是不会迈出家门一步的。他们最远不过走到有围屏的门廊。花园（如果有的话）在不举行花园晚宴时是无人问津的。他们在谈到户外空间时，很少是指花园。他们没有把花园看作可利用的起居空间。如同祖母们的接待室一样，花园受到过多的关照。花园也像接待室那样，不是为了住人用的。在一个注重实用的时代这是极不正常的。虽然这可能让人觉得反常，但近年来使用玻璃墙确实使花园变得不那么可亲了。甚至'风景窗'（对家用橱窗的一种称呼）也造成了室内外空间的分离；花园成了观赏用的园子。

历史上私家花园的概念全然不同。千百年来我们所知道的私家花园其最为可贵之处在于可以住人和它的私密性，但这两种性质在现代花园里均已荡然无存。私密性，在现代人们对之已无所

places in a garden—sunny places,terraces,gazebos—need to be made as outdoor rooms,as well.

The inspiration for this pattern comes from Bernard Rudofsky's chapter, "The Conditioned Outdoor Room," in *Behind the Picture Window*(New York:Oxford Press,1955).

In a superbly layed out house-garden,one ought to be able to work and sleep,cook and eat,play and loaf.No doubt,this sounds specious to the confirmed indoor dweller and needs elaboration.

As a rule,the inhabitant of our clirnate makes no sallies into his immediate surroundings.His farthest outpost is the screened porch.The garden—if there is one—remains unoccupied between garden parties.Indeed,when he talks about the outdoors,he seldom means his garden.He does not think of gardens as potential living space....Like the parlor of our grandmothers,the garden is an object of excessive care.Like the parlor,it is not meant to be lived in.In an age that puts a premium on usefulness this is most irregular. Paradoxical though it may sound,the use of glass walls in recent years alienated the garden.Even the "picture window," as the domestic version of the show-window is called,has contributed to the estrangement between indoors and outdoors;the garden has become a spectator garden.

The historical concept of the house-garden is entirely different. Domestic gardens as we have known them through the centuries were valued mostly for their habitableness and privacy,two qualities that are conspicuously absent in contemporary gardens.Privacy,so little in demand these days,was indispensable to people with a taste for dignified living.The house-gardens of antiquity furnish us,even in their fragmentary and dilapidated state,perfect examples of how a diminutive and apparently negligible quantity of land can,with some ingenuity,be transformed into an oasis of delight.Miniature

求，在过去却是人们为维护生活的尊严所不可或缺的。古老的私家花园是完美的范例，它向我们昭示，一小块微不足道的土地，通过巧夺天工之手，能够变为欢乐的绿洲，即使花园已经残破，其作用也复如此。花园虽小，却给人无穷乐趣。

这些花园是住宅的必不可少的一部分；请注意，它们包含在住宅之内。我们可以把它们说成是没有顶棚的房间。它们是真正的户外起居室，居住在那里的人就是这样看待它们的。例如，罗马花园的园墙和地面建筑材料比之住宅内部的用料毫不逊色。综合利用石质玛赛克、大理石石板、拉毛粉饰浮雕、壁饰（从最简单的几何图案到最精致的壁画）造成一种特别有利于心灵安宁的气氛。至于天花，那是千姿百态的天空。"（pp.157～159）

当户外空间以房屋的墙、树叶的墙、柱子、棚架以及天空作为屏障时，它就形成了一个特别的户外空间；该户外空间同室内空间连在一起，形成了真正连续的起居空间。

这里有几个关于户外空间的实例。每一个例子都是把不同的材料结合在一起造成围合的；每一个例子中的户外空间都是以稍有不同的方式同其所属房屋相连接的。鲁道夫斯基在我们援引过的那本书中还举了许多其他的例子。例如，他描述一个大门前的草坪可以如何改建为有围合的户外小空间。

两个有围合的户外小空间
Two outdoor rooms

最后，有一则注意事项。由于另一模式有一个很相似的名称——**户外亭榭**（69）——我们想提请你注意以下的区别：在一定的意义上说，这两者是相反的。有围合的户

gardens though they were,they had all the ingredients of a happy environment.

These gardens were an essential part of the house;they were,mind you,contained *within* the house.One can best describe them as rooms without ceilings.They were true outdoor living rooms,and invariably regarded as such by their inhabitants.The wall-and floor-materials of Roman gardens,for example, were no less lavish than those used in the interior part of the house.The combined use of stone mosaic,marble slabs,stucco reliefs,mural decorations from the simplest geometric patterns to the most elaborate murals established a mood particularly favorable to spiritual composure.As for the ceiling,there was always the sky in its hundred moods.(pp.157-159)

An outdoor space becomes a special outdoor room when it is well enclosed with walls of the building,walls of foliage,columns,trellis,and sky;and when the outdoor room, together with an indoor space,forms a virtually continuous living area.

Here are several examples of outdoor rooms.Each one uses a different combination of elements to establish its enclosure;each one is related to its building in a slightly different way.Rudofsky gives many other examples in the book we have cited.For instance,he describes how a front lawn can be rebuilt to become an outdoor room.

Finally,a note.Since there is another pattern with a rather similar name—PUBLIC OUTDOOR ROOM(69)—we want to remind you of the following distinction:in a certain sense,the two are opposites.An OUTDOOR ROOM has walls around it and is only partially roofed;while a PUBLIC OUTDOOR ROOM has a roof,but essentially no walls.

外小空间周围有墙，它只是部分有顶；而户外亭榭有顶，但基本上没有墙。

因此：

在户外营造一个空间，即使该空间是露天的，也要在其周围许多地方形成围合，让人感到它像一个房间。即使它是露天的。要做到这一点，需用柱子把它围合在屋角，多半可用棚架或活动帆布将其部分遮盖，用篱笆、能坐人的矮墙、屏风、树篱或房屋本身的外墙在其周围造成"围墙"。

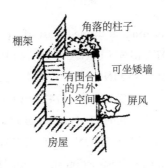

这一有围合的户外小空间的形成常常有赖于独立式立柱——**柱旁空间**（226）、墙——**花园围墙**（173）、**可坐矮墙**（243），棚架——**棚下小径**（174）或半透明的帆布篷——**帆布顶篷**（244），及建造与**大地紧密相连**（168）的地面。和其他所有房间一样，关于它的构造，请参阅**室内空间形状**（191）和**结构服从社会空间的需要**（205）……

Therefore:

Build a place outdoors which has so much enclosure round it,that it takes on the feeling of a room,even though it is open to the sky.To do this,define it at the corners with colunms,perhaps roof it partially with a trellis or a sliding canvas roof,and create "walls"around it,with fences,sitting walls,screens,hedges, or the exterior walls of the building itself.

෫෨෬෪

This outdoor room is formed,most often,by free standing columns—COLUMN PLACE(226),walls—GARDEN WALL(173), low SITTING WALLS(243),perhaps a trellis overhead—TRELLISED WALK (174),or a translucent canvas awning—CANVAS ROOFS(244),and a ground surface which helps to provide CONNECTION TO THE EARTH (168).Like any other room,for its construction start with THE SHAPE OF INDOOR SPACE(191)and STRUCTURE FOLLOW SOCIAL SPACES (205)...

模式164　临街窗户*

164 STREET WINDOWS*

...wherever there are GREEN STREETS(51),SMALL PUBLIC SQUARES (61),PEDESTRIAN STREETS(100), BUILDING THOROUGHFARES(101)—in short,any streets with people in them,these streets will only come to life if they are helped to do so by the people looking out on them,hanging out of widows,laughing,shouting,whistling.

❧❧

A street without windows is blind and frightening.And it is equally uncomfortable to be in a house which bounds a public street with no window at all on the street.

The street window provides a unique kind of connection between the life inside buildings and the street.Franz Kafka wrote a short commentary entitled "The Street Window," which expresses beautifully the power of this relationship.

Whoever leads a solitary life and yet now and then wants to attach himself somewhere,whoever,according to changes in the time of day,the weather,the state of his business,and the like,suddenly wishes to see any arm at all to which he might cling—he will not be able to manage for long without a window looking onto the street. And if he is in the mood of not desiring anything and only goes to his window sill a tired man,with eyes turning from his public to heaven and back again,not wanting to look out and having thrown his head up a little,even then the horses below will draw him down into their train of wagons and tumult,and so at last into the human harmony.(Franz Kafka,*The Complete Stories*,ed.Nahum N.Glatzer,New York:Schocken Books,1972,p.384).

……凡是有**绿茵街道**（51）、**小广场**（61）、**步行街**（100）、**有顶街道**（101）的地方——简言之，在任何有人行走的街道，只有当人们向街道探头张望，把上身伸出窗外，欢笑、喊叫、吹口哨时，这些街道才能变得有生气。

※

一条街道，如果看不见窗户，是一条盲街，令人感到害怕。住在临街的房屋里，如果完全没有临街窗户，也会同样令人感到不舒服。

临街窗户是使楼房内的生活和街上生活发生联系的唯一途径。弗朗兹·卡夫卡在一则题为《临街窗户》的简短评论中完美地表达了这种联系的力量，他写道：

有人过着独居生活，但又不时地使自己跟某些地方建立起一些联系；有人因为时间、天气、工作状况等发生变化就会突然想看见同他关系密切的人——如果没有面向街道的窗户，这对他说来谈何容易。如果他百无聊赖，没精打采地走到窗台，眼睛从他周围的人转向天空，然后又转回来，但不想往外看，而是把脑袋微微扬起，即使这时，马车的喧闹声也会吸引他往街上瞧，最终使他适应人间的生活。(Franz Kafka, *The Complete Stories*, ed.Nahum N.Glatzer, New York : schocken Books, 1972, p.384)

在传统的秘鲁文化中，有一种"凸肚窗"使得居者可以从楼房上层眺望街景，它是从利马的许多殖民时代的楼房里向街道上方伸出的装饰得很漂亮的阳台。秘鲁的少女特别喜爱看大街，但她们自己却不愿意抛头露面。她们从阳台上看大街是再合适不过了，这是她们在大门口不容易做到的。假如有人一直看她们，她们可以从窗口抽身回去。

The process of watching the street from upper story windows is strongly embedded in traditional Peruvian culture in the form of the *mirador*,the beautiful ornamented gallery which sticks out over the street from many of the colonial buildings in Lima.Peruvian girls especially love to watch the street,but only if they are not too visible.They can watch the street from the mirador without any impropriety,something they cannot do so easily from the front door.If anyone looks at them too hard,they can pull back into the window.

Street windows are most successful on the second and third floors.Anything higher,and the street becomes a "view" —the vitality of the connection is destroyed.From the second and third floors people can shout down to the street,throw down a jacket or a ball;people in the street can whistle for a person to come to the window,and even glimpse the expressions on a person's face inside.

At ground level,street windows are less likely to work. If they are too far back from the street,they don't really give a view onto the street—though of course they still give light. If they are too close to the street,they don't work at all,because they get boarded up or curtained to protect the privacy of the rooms inside—see the empirical findings presented in *Houses Generated by Patterns*,Center for Environmental Structure,1969,pp.179-180.

One possible way of making a street window at ground level might be to build an alcove,two or three steps up,with a window on the street,its window-sill five feet above the street. People in the alcove,can lean on the window sill,and watch the street;people in the street can see them,without being able to see into the room behind them.It is even easier,of course,if the

供眺望的凸肚窗
The mirador—the lookout

　　临街窗户在二层楼和三层楼最为有用。只要再高一点，街道便变成了"景物"——起不到联系作用了。从二层楼和三层楼上，人们可以对着街道喊叫，往下面扔件外套或一个球；人们在街上也可以吹声口哨，招呼一个人走到窗口来，甚至可以瞥见楼上的人脸部的表情。

　　在底层，临街窗户起不了太大作用。如果它们离街太远，它们实在看不到街上的东西——虽然它们无疑还是有助于采光。如果这些临街窗户离街太近，则完全不起作用，因为为了保护楼房里面房间的私密性，它们得钉上木板或挂上窗帘——参阅 *Houses Generated by Patterns*，Center for Environmental Structure，1969，pp.179~180 中的实验发现。

　　建造底层临街窗户的一种可能的方法可以是建一凹室，两三级台阶高度，在临街处筑窗，其窗台高出街面 5ft。凹室中的人可倚靠窗台观看街道；街上行人能看到他们，但看不到他们背后的房间。如果楼房的底层本身高出街面 2 或 3ft（许多底层都是这样的），做到这一点当然就更容易了。

底层的一个凹室的临街窗户
An alcove street window at ground level

ground floor of the house is two or three feet above the street,as many ground floors are.

Finally,on whatever floor it is,a street window must be placed in a position which the people inside pass often,a place where they are likely to pause and stand beside the window:the head of a stair,the bay window of a favorite room,a kitchen,bedroom,or window in a passage.

Therefore:

Where buildings run alongside busy streets,build windows with window seats,looking out onto the street. Place them in bedrooms or at some point on a passage or stair,where people keep passing by.On the first floor,keep these windows high enough to be private.

≈≈≈

On the inside,give each of these windows a substantial place,so that a person feels encouraged to sit there or stand and watch the street—WINDOW PLACE(180);make the windows open outward—WINDOWS WHICH OPEN WIDE(236);enrich the outside of the window with flower boxes and climbing plants—then people,in the course of caring for the flowers,will have the opportunity for hanging out—FILTERED LIGHT(238),CLIMBING PLANTS(246)...

最后，不管在哪层楼，临街窗户必须位于楼内的人经常走过的地方，他们有可能停留的地方和在窗边站立的地方，如楼梯顶口、人们爱去的房间、厨房、卧室里的凸窗，或是过道上的窗户。

因此：

在楼房位于繁华街道两旁时，建造面前有坐位的窗户，面对街道。把窗户开在卧室或人们经常走过的过道或楼梯上某些地方。在第一层楼，则务必使这些窗户保持一定的高度，使之具有私密性。

楼上的活动
室内活动
下面的街道

&oc8

在楼房内，给每一个这样的窗户以宽敞的空间，这样人们会兴致勃勃地坐在或站在那里观看街景——**窗前空间**（180）；使窗户向外开——**大敞口窗户**（236）；用花盒子和攀援植物来美化窗外——于是，人们在欣赏花卉的时候，就会把脑袋探出窗外——**过滤光线**（238）、**攀援植物**（246）……

模式165 向街道的开敞*

　　……城市中许多地方成功的秘诀在于它们完全向行人敞开——比**临街窗户**（164）要敞开得多。**像市场一样开放的大学**（43）、**地方市政厅**（44）、**项链状的社区行业**（45）、**综合商场**（46）、**保健中心**（47）、**临街咖啡座**（88）、**有顶街道**（101）等都是范例。本模式规定敞开的形式。

৪৩৫৪

　　看到别人怎么做，自己也会照着去做。当人们从街上能看到空间的内部时，他们的天地开阔了，知识丰富了，理解的事情多了；因而就有可能进行交流、学习。

165 OPENING TO THE STREET*

...many places in a town depend for their success on complete exposure to the people passing by—far more exposure than a STREET WINDOW(164)can provide.UNIVERSITY AS A MARKET-PLACE(43),LOCAL TOWN HALL(44), NECKLACE OF COMMUNITY PROJECTS(45), MARKET OF MANY SHOPS(46),HEALTH CENTER(47), STREET CAFE(88),BUILDING THOROUGHFARE(101)are all examples.this pattern defines the form of the exposure.

❧❧❧

The sight of action is an incentive for action.When people can see into spaces from the street their world is enlarged and made richer,there is more understanding;and there is the possibility for communication,learning.

The service center is a storefront,with the windows all along the front.A man walks past the door.As he does so,he looks into the center,but only turns his head for a second—apparently unwilling to show too great a curiosity.He then sees a notice on the window,stops to read the notice.As he stands reading it,he looks past the notice to see what is going on inside.After a few seconds he retraces his steps,and comes into the center.(*A Pattern Language Which Generates Multi-Service Centers*,C.E.S.,1968,p.251.)

There are many ways of establishing connection with the street.

这个服务中心是沿街的商店铺面，大门面全是窗户。有人打门口经过。他经过时往中心瞧了瞧，但转过头去的时间只有一秒钟——显然是不愿意表现出来过于好奇。这时他看见橱窗上的一张招贴，他停下来看招贴。在他站着看招贴的时候，他透过招贴看看里面在干什么。几秒钟之后，他又掉转脚跟回来，进入中心。

(*A Pattern Language Which Generates Multi-Service Centers*, C.E.S., 1968, p.251.)

有许多方法可以建立同街道的联系。

1. 首先，是显眼的环境：沿街的墙基本上是玻璃的，玻璃里面的景象是诱人的。伯克利一社区中心从一间背街的翻修过的屋子搬迁到沿街一间完全透明的翻修过的家具陈列室。搬迁后来客猛增。部分原因是新的店址坐落在繁华得多的步行街上。但透明的玻璃窗在吸引来客方面也起了作用：经过中心的人中约有66%回头往里张望，约有7%的人停了下来——或者看看招贴，或者更仔细地瞧瞧里面的东西。

2. 然而，靠玻璃窗建立联系相对地讲还是消极的办法。比较起来，一道真正敞开的墙——带有推拉墙或百叶窗——会产生有价值得多和紧密得多的联系。若墙是敞开的，就有可能听到里面的声音，闻到里面的气味，可以跟里面的人说话，乃至从敞开的门口长驱直入。临街的咖啡座、露天食品摊、车间的大门敞开的汽车修理厂都属此例。

每天我们从学校回家的路上都要经过这个工厂。这是一家家具厂。我们往往站在门口看工人做椅子和桌子，锯末飞溅，在车床上做出桌椅的架子来。那里有堵矮墙，工头告诉我们待在墙外；但他让我们坐在那里，我们也就坐着，有时一坐几个钟头。

1.First,the obvious case:the wall along the street is made essentially of glass,and the view in is of some inviting activity. A community center in Berkeley moved from a rennovated house which was away from the street,to a rennovated furniture showroom which was completely transparent to the street.The number of people dropping in soared after the move.It was partly because the new location was on a much busier pedestrian street.But transparency also played a role in bringing people in:of the people walking past the center about 66 per cent turned and looked in,and about 7 per cent stopped—either to read a notice or to look into the interior more carefully.

2.However,a glass connection creates relatively passive involvement.By comparison,a wall which is actually open— with a sliding wall or shutter—creates a far more valuable and involving connection.When the wall is open it is possible to hear what is going on inside,to smell the inside,to exchange words,and even to step in all along the opening.Street cafes,open food stalls,workshops with garage door openings are examples.

We passed the workshop every day on our way home from school.It was a furniture shop,and we would stand at the opening and watch men building chairs and tables,sawdust flying,forming legs on the lathe.There was a low wall,and the foreman told us to stay outside it;but he let us sit there,and we did,sometimes for hours.

3.The most involving case of all:activity is not only open to sight and sound on one side of the path,but some part of the activity actually crosses the path,so that people who walk down the sidewalk find themselves walking *through* the activity.The extreme version is the one where a shop is set up to straddle the path,with goods displayed on either side.A more modest version is the one

3. 与街道联系最多的情况是：活动不仅在道旁可以为行人看得见和听得见，而且它的某些部分实际上跨越走道，使在人行道上行走的人可以直接从活动中间穿过。最突出的情况是，一间商店横跨走道，两边陈列商品。稍为节制点的情况是，商店的屋顶遮住道路，墙却是完全敞开的，使走道跟店铺的"内部"连成一片。

不管开敞的方式如何，最重要的是使内部平常的活动开放，设法吸引行人参与这种活动，跟它发生某种哪怕是不大的联系。派克黑姆的先锋保健中心的医生们认为这一原则是很重要的，因而他们特意建造了该中心的体育馆、游泳池、舞池、自助餐厅以及剧场，以这种方式使路过的行人往往看见他们的熟人在里面活动：

……在夜晚，整幢楼房灯火通明，吸引着行人的注意，舞会正在进行着，在主楼的地面上可以看到翩翩起舞的身影……

……必须记住，在这个中心，人们看到的不仅是能人的行动，而且是熟练程度不一的各类人的行动。这一点是至关重要的，它有助于如何使观看能够促成聚集的人群产生行动。在平常的生活中，任何一种活动，都要有专门人才表演才能吸引观众；这一趋势日益令人不安。成千的观众聚集起来观看专家表演，但由于"明星"们富有才华，他人跃跃欲试则会被斥为不当，这使众人越发不能参与活动。因此，并非所有的活动形式都邀请人们去参加：观众所能做到的只是观看活动，而这最终对他产生不可抗拒的诱惑力。我们试验的时间虽短，但这个事实却是有充分根据的，这也被中心所进行的活动所证实。(*The Peckham Experiment*, I.Pearse and L.Crocker, New Haven : Yale University Press, 1947, pp.67~72.)

where the roof of the space covers the path,the wall is entirely open,and the paving of the path is continuous with the "interior" of the space.

No matter how the opening is formed,it is essential that it expose the ordinary activity inside in a way that invites people passing to take it in and have some relationship,however modest,to it.The doctors of the Pioneer Health Center in Peckham believed this principle to be so essential that they deliberately built the center's gymnasium,swimming pool,dance floor,cafeteria,and theater in such a way that people passing could not help but see others,often people they knew,inside:

...dancing goes on there and moving figures can be seen on the floor of the main building at night when the whole building is lit up attracting the attention of the passers by...

...it must be remembered that it is not the action of the skilled alone that is to be seen in the Centre,but *every degree* of proficiency in all that is going on.This point is crucial to an understanding of how vision can work as a stimulus engendering action in the company gathering there.In ordinary life the spectator of any activity is apt to be presented *only* with the exhibition of the specialist;and this trend has been gathering impetus year by year with alarming progression. Audiences swell in their thousands to watch the expert game,but as the "stars" grow in brilliance,the conviction of an ineptitude that makes trying not worth while,increasingly confirms the inactivity of the crowd.It is not then all forms of action that invite the attempt to action:it is the sight of action that is within the possible scope of the spectator that affords a temptation eventually irresistible to him.Short though the time of our experiment has been,this fact has been amply substantiated,as the growth of activities in the Centre demonstrates. (*The Peckham Experiment*,I.Pearse and L.Crocker,New Haven:Yale

因此：

在任何一个其成功的关键在于向街道开敞的空间，都应使之敞开，使之有一道可以开得很大的充分敞开的墙，如有可能，包括步行街上较远一侧的某部分的活动，使之实际上横跨走道，人们沿街走的时候会从它那儿经过。

建造这样的开放空间有许多方法。例如，可以利用简单的胶合板制做的悬挂百叶窗在头顶上方的轨道上滑动，一旦将其卸掉就完全敞开了，夜间则装回原处锁好，用这种办法可以制造价格便宜的墙。

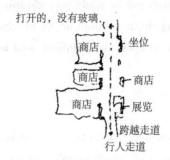

打开的，没有玻璃

商店 — 坐位

商店 — 商店

商店 — 展览

— 跨越走道

行人走道

&CB

当空间完全敞开时，利用可以坐人的实心矮墙来界定敞开的空间——**可坐矮墙**（243）；在越过这空间的小路的外边筑一有围合的户外小空间——**小路的形状**（121）、**有围合的户外小空间**（163）……

University Press,1947,pp.67-72.)

Therefore:

In any public space which depends for its success on its exposure to the street,open it up,with a fully opening wall which can be thrown wide open,and if it is possible,include some part of the activity on the far side of the pedestrian path, so that it actually straddles the path,and people walk through it as they walk along the path.

There are dozens of ways to build such an opening. For example,a wall can be made very cheaply with a simple plywood hanging shutter sliding on an overhead rail,which can be removed to open up completely,and locked in place at night.

❧❧❧

Give the opening a boundary,when it is entirely open, with a low solid wall which people can sit on—SITTING WALL(243); and make an outdoor room out of the part of the path which runs past it—PATH SHAPE(121),OUTDOOR ROOM(163)...

模式166　回廊*

　　……我们继续充实**建筑物边缘**（160）。假定在需要
建造拱廊的地方已经建造了拱廊——**拱廊**（119）；在建筑
物边缘内仍有大片地方，这些地方根据建筑物边缘可以
成为"正空间"——但迄今未有一个模式来讲解具体的做
法。本模式告诉你可以怎样使边缘完善。它是对**屋顶花园**
（118）及**拱廊**（119）的补充，并有助于使**步行街**（100）
充满生气。

166   GALLERY SURROUND*

...we continue to fill out the BUILDING EDGE(160).
Assume that arcades have been built wherever they make sense—
ARCADES(119);there are still large areas within the building edge
where BUILDING EDGE tells you to make something positive—
but so far no patterns have explained how this can be done
physically.This pattern shows you how you can complete the edge.
It complements ROOF GARDEN (118)and ARCADES(119)and
helps to enliven the PEDESTRIAN STREET (10).

ഔന

**If people cannot walk out from the building onto
balconies and terraces which look toward the outdoor space
around the building,then neither they themselves nor the
people outside have any medium which helps them feel the
building and the larger public world are intertwined.**

We have discussed the importance of the building edge in two
other patterns:BUILDING EDGE(160)itself,and ARCADES(119).
In both cases,we explained how the arcades and the edge help to
create space which people who are *outside* the building can use to
help them feel more intimately connected with the building.These
patterns,in short,look at the problem of connection from the point
of view of the people *outside* the building.

In this pattern we discuss the same problem—but from
the point of view of the people *inside* the building.We believe,
simply,that every building needs at least one place,and
preferably a whole range of places,where people can be still
within the building,but in touch with the people and the scene

如果人们不能从楼房中走到面向楼房周围户外空间的阳台和露台上来，那么无论他们自己或户外的人都没有任何办法使他们感到，这楼房和更大的公共环境是相互沟通的。

我们已在另外两个模式中讨论过建筑物边缘的重要性：**建筑物边缘**（160）和**拱廊**（119）。在这两种情况下，我们讲解了拱廊和边缘如何能用来创造一种空间，它能够为建筑物外面的人所利用，使他们同建筑物的联系更加密切。简言之，这些模式是从建筑物外面的人的视角来看待联系问题的。

在本模式中我们讨论同一个问题——但是从建筑物里面的人的视角来看待这个问题的。简单地说，我们认为，每一幢楼房至少需要一个空间，最好是整个连续的空间，人们在这里仍然是在楼房内，但却接触到了户外的人群和景物。这个问题也在**私家的沿街露台**（140）中讨论过。但那个模式只涉及这一需求的一个十分重要又很特殊的情况。本模式提出把这种需要普遍化：显然，这是可以反复应用于所有建筑物的基本的、无所不包的需要。

这种需要有丰富的文献资料可供参考，例见《住宅和社会结构》《一千个家庭的适居性问题》。

临街窗户，虽然有它们自己的用处，但根本不能满足这一需要。它们通常只占有墙的很小一部分，而且只有当一个人站在房间边缘上的时候才能得到利用。人们所需要的环境远比这丰富、有趣。我们需要有沿着楼房上层边缘的一些地方让我们可以舒舒服服地待上几个钟头，接触到街道——玩牌，天热时把活拿到露台上去干，吃饭，同孩子们嬉戏打闹，装一辆电气火车，晾晒和叠衣服，用黏土进行雕刻，整理单据。

outside.This problem has also been discussed in PRIVATE TERRACE ON THE STREET(140).But that pattern deals only with one very important and highly specific occurrence of this need.The present pattern suggests that the need is completely general:very plainly,it is fundamental,an allembracing necessity which applies to all buildings over and again.

The need has been documented extensively.(See, for example, Anthony Wallace,*Housing and Social Structure*,Philadelphia Housing Authority,1952;Federal Housing Authority,*The Livability Problem of 1,000 Families*, Washington,D.C.,1945.)

Windows on the street,while they have their own virtues,are simply not enough to satisfy this need.They usually occupy a very small part of the wall,and can only be used if a person stands at the edge of the room.The kinds of situations that are needed are far more rich and engrossing.We need places along the upper stories of the building's edge where we can live comfortably,for hours,in touch with the street-playing cards,bringing work out on the terrace on a hot day,eating,scrambling with children or setting up an electric train,drying and folding the wash,sculpting with clay,paying the bills.

In short,almost all the basic human situations can be enriched by the qualities of the gallery surround.This is why we specify that each building should have as many versions of it as possible along its edge—porches,arcades,balconies,awnings,ter races,and galleries.

Therefore:

Whenever possible,and at every story,build porches, galleries,arcades,balconies,niches,outdoor seats, awnings, trellised rooms,and the like at the edges of buildings—

总之，几乎所有的人们的基本生活环境通过建立回廊都能变得丰富多采。因此我们认为，每幢房屋沿其边缘应有各式各样的回廊——门廊、拱廊、阳台、凉篷、露台和长廊。

本模式的四种例子
Four examples of this pattern

因此：

只要有可能，在每一层楼的建筑物边缘——特别在有门户通往公共空间和街道的地方——建造游廊、长廊、拱廊、阳台、壁龛、大门外坐位、凉篷、棚架空间等，并用门户把它们直接同内部空间连接在一起。

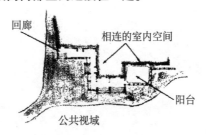

especially where they open off public spaces and streets,and connect them by doors,directly to the rooms iside.

৪০০৪

A warning:take care that such places are not stuck artificially onto the building.Keep them real;find the places along the building edge that offer a direct and useful connection with the life indoors—the space outside the stair landing,the space to one side of the bedroom alcove,and so on.

These places should be an integral part of the building territory,and contain seats,tables,furniture,places to stand and talk,places to work outside—all in the public view— PRIVATE TERRACE ON THE STREET(140),OUTDOOR ROOM(163);make the spaces deep enough to be really useful—SIX-FOOT BALCONY(167)—with columns heavy enough to provide at least partial enclosure—HALF-OPEN WALL (193),COLUMN PLACES(226)...

一则注意事项：请注意不要将这些地方人工地黏合到建筑物上去。使它们保持原本状态；要在建筑物的边缘处找出这样的空间，使它们同室内生活能够发生直接的和有效的联系——楼梯平台外面的空间、床龛另一边的空间等。

这些空间应当是楼房整体的一部分，在它们里面设有坐位、桌子、家具，有供站立和说话的地方，有供在户外工作的地方——一切都在公共视域之内——**私家的沿街露台**（140）、有围合的**户外小空间**（163）；使这些空间有足够的进深以便真正可以利用——**六英尺深的阳台**（167）——有较粗的柱子以便至少形成局部围合——**半敞开墙**（193）、**柱旁空间**（226）……

模式167　六英尺深的阳台＊＊

　　……在不同的地方，**拱廊**（119）和**回廊**（166）有助于构想在沿建筑物边缘或进入建筑物的中途建造某种阳台、游廊、露台、门廊、拱廊。本模式只是规定这种拱廊、门廊或阳台的深度，以便保证它们真正具有使用价值。

167   SIX-FOOT BALCONY**

...in various places ARCADES(119) and GALLERY SURROUND(166) have helped you to imagine some kind of a balcony,veranda,terrace,porch,arcade along the building edge or halfway into it.This pattern simply specifies the depth of this arcade or porch or balcony,to make sure that it relly works.

❧✿❦

Balconies and porches which are less than six feet deep are hardly ever used.

Balconies and porches are often made very small to save money;but when they are too small,they might just as well not be there.

A balcony is first used properly when there is enough room for two or three people to sit in a small group with room to stretch their legs,and room for a small table where they can set down glasses,cups,and the newspaper.No balcony works if it is so narrow that people have to sit in a row facing outward.The critical size is hard to determine,but it is at least six feet.The following drawing and photograph show roughly why:

Our observations make it clear that the difference between deep balconies and those which are not deep enough is simply astonishing.In our experience,almost no balconies at all which are 3 or 4 feet deep manage to gather life to them or to get used. And almost no balconies which are more than six feet deep are

◎◎◎

小于 6ft 深的阳台和门廊几乎是没有用处的。

为了省钱，阳台和门廊常常做得很小；但当它们太小时，有没有它们是一样的。

如一个阳台有足够的空间供两三个人一块儿坐下，使他们能够伸开腿脚，还可摆上一张小桌，桌上可放玻璃杯、茶杯和报纸，这样的阳台首先是适用的。如果一个阳台很窄，人们只能朝外坐成一排，这种阳台用处不大。很难对阳台规定一个临界尺寸，但它至少应有 6ft。下面的插图和照片大致可以说明这个原因：

6ft进深
Six feet deep

我们的观察表明，深的阳台和进深不够的阳台之间简直有天壤之别。根据我们的经验，差不多没有一个 3ft 或 4ft 深的阳台是有生气的、可供人们使用的。而差不多所有超过 6ft 深的阳台则都是的。

窄的阳台没有用处
Narrow balconies are useless

not used.

Two other features of the balcony make a difference in the degree to which people will use it:its enclosure and its recession int othe building.

As far as enclosure goes,we have noticed that among the deeper balconies,it is those with half-open enclosures around them—columns,wooden slats,rose-covered trellises—which are used most.Apparently,the partial privacy given by a half-open screen makes people more comfortable—see HALF-OPEN WALL(193).

And recesses seem to have a similar effect.On a cantilevered balcony people must sit outside the mass of the building;the balcony lacks privacy and tends to feel unsafe.In an English study("Private Balconies in Flats and Maisonettes," *Architect's Journal*,March 1957,pp.372-376), two-thirds of the people that never used their balconies gave lack of privacy as their reason,and said that they preferred recessed balconies,because,in contrast to cantilevered balconies,the recesses seemed more secure.

Therefore:

Whenever you build a balcony,a porch,a gallery,or a terrace always make it at least six feet deep.If possible, recess at least a part of it into the building so that it is not cantilevered ont and separated from the building by a simple line,and enclos it partially.

阳台的另外两个特点使人们在利用它的程度上产生分歧：阳台的围合程度；阳台向建筑物凹进的程度。

就围合程度而言，我们注意到，在较深的阳台之中，使用得最多的是半开半合的那些阳台——其周围有柱子、木条、蔷薇棚架。显然，由半开放的围屏造成的部分私密性使人们感到比较舒适——参阅**半敞开墙**（193）。

凹进看来具有相似的效果。在悬臂式阳台上人们必须坐在建筑物实体的外部；阳台缺少私密性，而且令人感到不安全。在一份英国的研究报告（"Private Balconies in Flats and Maisonettes," *Architect's Journal*，March 1957, pp.372~376）中，从来不使用他们阳台的人中有 2/3 把缺乏私密性作为理由，他们说他们喜欢凹进的阳台，因为跟悬臂式阳台比较，凹进的阳台看起来更为安全。

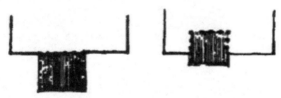

不是这个…………是这个
Not this……this

因此：

在建造阳台、门廊、长廊或露台时，要把它做成至少 6ft 深。如果可能，至少使它的一部分向建筑物内凹进，这样它不会悬挂在外面，而以一条简单的线同建筑物相隔，还要将它加以部分围合。

6ft深

Enclose the balcony with a low wall—SITTING WALL(243),heavy columns—COLUMN PLACES(226),and half-open walls or screens—HALF-OPEN WALL(193).Keep it open toward the south-SUNNY PLACE(161).Treat it as an OUTDOOR ROOM(163),and get the details of its shape and its construction from THE SHAPE OF INDOOR SPACE(191)...

　　用矮墙——**可坐矮墙**（243），粗大的柱——**柱旁空间**（226），半敞开的墙或围屏——**半敞开墙**（193）使阳台围合。使它开向南面——**有阳光的地方**（161）。把它处理成**有围合的户外小空间**（163），关于它的形状和构造，详见**室内空间形状**（191）……

模式168　与大地紧密相连**

　　……本模式详述房屋的地面如何延伸至它周围的大地和花园，借以创造**建筑物边缘**（160）和它的**拱廊**（119）、私家的**沿街露台**（140）、**回廊**（166），以及**六英尺深的阳台**（167）。

168   CONNECTION TO THE EARTH**

...this pattern helps to create the BUILDING EDGE(160) and its ARCADES(119),PRIVATE TERRACE ON THE STREET(140),the GALLERY SURROUND(166),and SIX-FOOT BALCONY(167),by specifying the way the floor of the building reaches out into the land and gardens ound about it.

❧❧❧

A house feels isolated from the nature around it,unless its floors are interleaved directly with the earth that is around the house.

We shall understand this best by contrasting those houses which are sharply separated from the earth with those in which there is a continuity between the two.

Look first at this house where there is no continuity.

The inside and the outside are abruptly separate.There is no way of being partly inside,yet still connected to the outside;there is no way in which the inside of the house allows you,in your bare feet,to step out and feel the dew collecting or pick blossoms off a climbing plant because there is no surface near the house on which you can go out and yet still be the person that you are inside.

Compare it with the house in our main picture,where there is continuity.Here,there is an intermediate area,whose surface is connected to the inside of the house—and yet it is in plain outdoors.

This surface is part of the earth—and yet a little smoother,a little more beaten,more swept—stepping out on it is not like stepping out into a field in your bare feet—it is as if the earth itself

一幢房屋，如果其地面不能同宅旁的大地直接相连，就会令人感到它同其周围的自然界是隔绝的。

　　如果将同大地完全隔绝的那些房屋与两者彼此相连的房屋相比较，此道理就不讲自明了。

　　先请看这幢房屋，在这里没有连续性。

一幢普通的住宅——可是请仔
细瞧瞧，它与本模式毫不相干
*An average house but look at it closely.It lacks this pattern
utterly*

　　这幢房屋内外截然分离。你无法做到部分在里面，却仍然跟外面相连；你也无法从屋内光着脚走出来去领略浓重的露珠和采摘爬墙植物上的花朵，因为房屋周围没有一个表面使你可以走出来但仍感到你还在屋里。

　　试将此屋同本模式主要插图作一比较，在那张图中有一种连续性。那里有一个中间地带，其表面跟房屋内部相通——但却明明白白是在户外。这个表面是大地的一部分——但却更为光滑、平整，扫得更干净——你光着脚从屋里走到它上面不像走到野地上——仿佛在那一小块地方，大地本身变成你的室内地面的一部分。

　　当我们比较这些例子时，看来毫无疑问，这里面包含着一种深切的感受，因而我们相信可将本模式作为重要模式来介绍。但它的起因或者说为什么这么重要，我们还只

becomes in that small area a part of your indoor terrain.

When we compare the examples,there seems little doubt that some deep feeling is involved,and we are confident in presenting this pattern as a fundamental one.But we can only speculate about its origins or why it is important.

Perhaps the likeliest of all the explanations we are able to imagine is one which connects the earth boundness and rootedness of a man or a woman to their physical connection to the earth. It is very plain,and we all discover for ourselves,that our lives become satisfactory to the extent that we are rooted, "down to earth," in touch with common sense about everyday things—not flying high in the sky of concepts and fantasies.The path toward this rootedness is personal and slow—but it may just be true that it is helped or hindered by the extent to which our physical world is itself rooted and connected to the earth.

In physical terms,the rootedness occurs in buildings when the building is surrounded,along at least a part of its perimeter,by terraces,paths,steps,gravel,and earthen surfaces, which bring the floors outside,into the land.These surfaces are made of intermediate materials more natural than the floors inside the house—and more man-made than earth and clay and grass.Brick terraces,tiles,and beaten earth tied into the foundations of the house all help make this connection;and,if possible,each house should have a reasonable amount of them,pushing out into the land around the house and opening up the outdoors to the inside.

Therefore:

Connect the building to the earth around it by building a series of paths and terraces and steps around the edge.Place them deliberately to make the boundary ambiguous—so that it is impossible to say exactly where the

能进行猜测。

也许我们能够想象的所有解释中最有可能的一种是，人们受制于大地，人们的根也来自大地，因而跟大地有千丝万缕的联系。十分明显，我们大家自己都能发现，我们生活得满意是因为我们落地生根、实实在在、明察世情——而不是想法怪异，在半空飘浮。根植于大地由个人亲历，其过程是缓慢的——但我们的生活环境自身根植于大地和同大地连接的程度却能对此起到促进或阻碍的作用，这一点可能是千真万确的。

当建筑物周围，至少沿其周边的一部分，有露台、小道、台阶、砂砾和土质表面而使地面往外延伸，连接大地时，这种建筑物就是根植的物质体现。这些表面由中间材料所构成，它们比屋内地面要自然一些，比大地、黏土和草地却要更加人工化。与房屋地基连成一片的铺砖露台、瓮砖和夯压过的土地都有助于形成这种连接；如果可能，每幢房屋都应有一些这样的东西，伸展到房屋周围的大地，开辟出通向内部的户外空间。

因此：

在建筑物边缘建造一些小路、露台和台阶，使建筑物与其周围的大地相连。在进行布局时要有意识地使边界模糊不清——这样就不能准确说清楚建筑物和大地的边界在哪里。

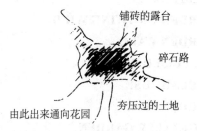

铺砖的露台

碎石路

由此出来通向花园　　夯压过的土地

building stops nd earth begins.

୫୨୦ଓ

Use the connection to the earth to form the ground for outdoor rooms,and entrances,and terraces—ENTRANCE ROOM(130),PRIVATE TERRACE ON THE STREET(140), OUTDOOR ROOM(163),TERRACED SLOPE(169);prepare to tie the terraces continuously into the wall which forms the edge of the ground floor slab,to make the very structure of the building feel connected to the earth—GROUND FLOOR SLAB(215);and where you come to form the terrace surfaces,use things like hand-made bricks and softbaked crumbling biscuit-fired tile—SOFT TILE AND BRICK(248);and further out,along the paths a little distance from the house,leave cracks between the tiles to let the grass and flowers grow between them—PAVING WITH CRACKS BETWEEN THE STONES(247)...

decide on the arrangement of the gardens,and the places in the gardens;

169.TERRACED SLOPE

170.FRUIT TREES

171.TREE PLACES

172.GARDEN GROWING WILD

173.GARDEN WALL

174.TRELLISED WALK

175.GREENHOUSE

176.GARDEN SEAT

177.VEGETABLE GARDEN

178.COMPOST

利用与大地的连接来形成有围合的户外小空间、入口和露台的地面——**入口空间**（130）、**私家的沿街露台**（140）、**有围合的户外小空间**（163）、**梯形台地**（169）；准备把露台跟形成底层地面边缘的墙垣相连接，使它们成为连续的整体，以便使人感到房屋自身结构与大地连接——**底层地面**（215）：在你打算形成露台表面的地方，请使用手工制作的砖和软质的素坯烧制的砖——**软质面砖和软质砖**（248）；稍远一点，沿着离房屋不远的小径，在铺砌的砖块之间留出空隙让花草在其间生长——**留缝的石铺地**（247）……

决定花园的布置和花园内各种处所；

169. 梯形台地

170. 果树林

171. 树荫空间

172. 花园野趣

173. 花园围墙

174. 棚下小径

175. 温室

176. 园中坐椅

177. 菜园

178. 堆肥

模式169　梯形台地*

　　……本模式有助于完善**基地修整**（104）。在有建筑物的地方，它连接**建筑物边缘**（160）并能有助于它的形成；它有助于形成**与大地紧密相连**（168）。如果地面是完全倾斜的，本模式告诉你如何处理好梯形地面使之有利于房屋中的人和地面上的树木花草。

୧୦୧

　　在梯形地面上，地表径流所产生的侵蚀会破坏土壤。它也造成地面上雨水的不均匀分布，这比起均匀分布当然使植物更难受益。

169   TERRACED SLOPE*

...this pattern helps to complete SITE REPAIR(104).Where there are buildings,it ties into the BUILDING EDGE(160)and can help form it;and it helps create the CONNECTION TO THE EARTH(168).If the ground is sloping at all,this pattern tells you how to handle the slope of the ground in a way that makes sense for the people in the building,and for the plants and grasss on the ground.

⁸ᴑᏏ

On sloping land,erosion caused by run off can kill the soil.It also creates uneven distribution of rainwater over the land,which naturally does less for plant life than it could if it were evenly distributed.

Terraces and bunds,built along contour lines,have been used for thousands of years to solve this problem.Erosion starts when the water runs down certain lines,erodes the earth along these lines,makes it hard for plants to grow there,then forms rills in the mud and dust,which are then still more vulnerable to more runoff,and get progressively worse and worse.The terraces control erosion by slowing down the water,and preventing the formation of these rills in the first place.

Even more important,the terraces spread the water evenly over the entire landscape.In a given area,each square meter of earth gets the same amount of water since the water stays where it falls.Under these conditions,plants can grow everywhere—on the steepest parts of hillsides as easily as in the most luscious valleys.

沿等高线建造的台地和堤岸千百年来都是用来解决这个问题的。水沿着一定的路线往下奔流,侵蚀沿线的土壤,使植物难以在那里生长,然后在烂泥和干土中形成沟纹,有了这些沟纹更易遭受径流,使情况日益恶化,于是就开始发生侵蚀。台地通过使流水减慢而控制侵蚀,首先防止这些沟纹的形成。

更为重要的是,台地将水均匀地分布在整个地面。由于水停留在落下的地方,在一定的面积内,每平方米的土地获得的水量是相等的。在这样的条件下,植物到处可以生长——在陡峭的山坡如同在芬芳的山谷一样容易。

梯形台地这个模式对于小块房屋基地也如对于峡谷周围的山坡一样具有同等意义。在一小块土地上做成合适的台地可以创造出稳定的小型排水系统,并为当地园林保护了顶部土壤。本模式的主要照片拍的就是建在台地上的一幢小房屋。一旦台地筑成,房屋即可在此兴建,它随台地的路线而延伸。

在房屋基地上和在山坡上用这种办法保养土地并使之健康,这是自古就有的。"直到最近,现代化的防侵蚀措施,例如通过等高线耕种,其效果才可同如日本和秘鲁一般相距甚远的国家实行了很久的传统的梯田法相媲美。"(M.Nicholson, *The Environmental Revolution*, New York: McGraw Hill, 1970, p.192.)

在丘陵和峡谷地带,中国以这种方法正在做出令人印象深刻的努力来开垦其受到侵蚀的土地。例见约瑟夫·阿尔索普所著《中国的梯田》[1]一书:

在中国农村,人们一直在不遗余力地利用现有资源获得最大限度的农作物。即使这样,我对他们带我去看的在重庆附近的农

---

[1] 译注:本书成文较早,凡书中提及中国的事情,有些已发生变化,请读者体察。

The pattern of terracing makes as much sense on a small house lot as it does on the hills around a valley.Proper terracing on a small lot creates a stable micro-system of drainage,and protects the top soil for the local gardens.Our main photograph shows a small building that is built on a terraced site.Once the terracing has been accomplished,the building can fit to it,and stretch across the lines of the terrace.

At both scales—the house lot and the hills—this method of conserving the land and making it healthy is ancient. "Only very lately has modern anti-erosion practice,for example, through contour ploughing,managed to match the effectiveness of traditional methods of terracing long practiced in countries as far apart as Japan and Peru." (M.Nicholson,*The Environmental Revolution*,New York:McGraw Hill,1970,p.192.)

At the scale of hillsides and valleys,China is making an impressive attempt to reclaim her eroded land in this way.For instance,Joseph Alsop, "Terraced Fields in China" :

In the Chinese countryside,no effort has ever been spared to get a maximum crop with the resources available.Even so,I was hardly prepared for the "terrace fields" that they took me to see in the farming communes around Chungking.

The countryside hereabouts is both rocky and largely composed of such steep hills that even Chinese would not think of trying to grow rice on them.The old way,ruinously eroding,was to grow as much rice as possible in the valleys:and then plant the hillsides,too,where soil remained.

The new way is to make "terrace fields." The rocks are dynamited to get the needed building materials.Heavy dry-stone walls are then built to heights of six or seven feet,following the contours of the land.And earth is finally brought to fill in

业公社的"梯田"还是几乎毫无思想准备。

这个农村到处是石头，并且大部分是陡峭的山坡，甚至中国人也不会想在那里种稻子。具有破坏性侵蚀作用的老方法是尽可能多地把稻子种在峡谷，然后种在还有土壤的山坡。

新方法是种"梯田"。把岩石炸碎以获取所需的建筑材料。大块的干石砌成的墙高达 6 ～ 7ft，达到土地的等高线。把泥土搬运过来填在石墙里面从而形成梯田。

因此：

在所有梯形的土地上——在田野、在公园、在公共花园，甚至在住宅旁的私家花园——建造达到等高线的台地和堤岸系统。沿着等高线建造低墙，然后用土填塞它们以形成台地。

没有理由让房屋适应台地——它可以顺利地跨越等高线。

在梯形台地上种植蔬菜和果树——**菜园**（177）、**果树林**（170）；在形成梯形台地的墙边种植花木，使其有足够的高度以便触及花枝闻到花香，当然也要造墙使人能坐在上面——**可坐矮墙**（243）……

behind the stone walls,thereby producing a terrace field.

Therefore:

**On all land which slopes—in fields,in parks,in public
gardens,even in the private gardens around a house—make
a system of terraces and bunds which follow the contour
lines.Make them by building low walls along the contour
lines,and then backfilling them with earth to form the
terraces.**

**There is no reason why the building itself should fit
into the terraces—it can comfortbly cross terrace lines.**

᠍ৡৡৡ

Plant vegetables and orchards on the terraces—
VEGETABLE GARDEN(177),FRUIT TREES(170);along the
walls which form the terraces,plant flowers high enough to
touch and smell—RAISED FLOWERS(245).And it is also very
natural to make the walls so people can sit on them—SITTING
WALL(243)...

模式170　果树林*

170   FRUIT TREES*

...both the COMMON LAND(67)outside the workshops, offices and houses,and the private gardens which belong to individual buildings—HALF-HIDDEN GARDEN (111),can be helped by planting fruit trees.After all,a garden,whether it is public or private,is a thing of use.Yet it is not a farm.That half way kind of garden which is useful,but also beautiful in spring and autumn,and a marvelous place to walk because it smells so onderful,is the orchard.

❧☙

In the climates where fruit trees grow,the orchards give the land an almost magical identity:think of the orange groves of Southern California,the cherry trees of Japan,the olive trees of Greece.But the growth of cities seems always to destroy these trees and the quality they possess.

The fact that the trees are seasonal and bear fruit has special consequences.The presence of orchards adds an experience that has all but vanished from cities—the experience of growth,harvest,local sources of fresh food;walking down a city street,pulling an apple out of a tree,and biting into it.

Fruit trees on common land add much more to the neighborhood and the community than the same trees in private backyards:privately grown,the trees tend to produce more fruit than one household can consume.On public land,the trees concentrate the feeling of mutual benefit and responsibility.And because they require yearly care,pruning,and harvesting,the

……工作间、办公楼和住宅外面的**公共用地**（67）及属于独立住宅的私家花园——**半隐蔽花园**（111）此两者都可用种植果树来改善环境。归根结蒂，一个花园，不管是公共的还是私人的，目的是供人使用。但它不是农场。这类不完整的花园就是果园，它具有使用价值，但春秋两季又很美丽，又因芳香四溢而成为散步的好地方。

৪৩৫১

在气候适宜于果树生长的地方，果园赋予大地近乎神奇的品性：请想一想南部加利福尼亚州的橘子林、日本的樱桃树、希腊的橄榄树。但城市的发展看来总要破坏这些果树和果树所具有的品性。

果树是有季节性的，它们生产水果，这一事实具有特别的意义。有了果园，生活就变得丰富多彩，我们可以看到果实生长、收获，供给当地以新鲜果品；可以漫步在城市街道，从果树上摘下一个苹果，咬一口品尝品尝——可惜城市中已经看不到果园。

在公共用地上的果树，比起私家后院种植的同类果树能为邻里和社区增添更多的东西，虽然私家种的果树生产的果子往往一家人吃不完。但在公共用地上，果树却体现了互利精神和责任感。由于果树每年都需要管理、修剪和收获，它们自然使居民们同他们的公共用地发生联系。这是一个众目睽睽的地方，在这里人们可以为其本区的公共用地尽一份责任，为它的成果感到骄傲，把他们自己及其孩子们的部分时间花在这上面。

请设想一下，一个社区逐步能够生产它所需要的一部分水果、果汁或果酱。开始时，可能真正只有小部分，但作为开端这就不错了。活是大家一起干的，个人要做的事不是很多，但却得到无穷的乐趣。

fruit trees naturally involve people in their common land.It is an obvious place where people can take responsibility for their local common land,have pride in the results,employ themselves and their children part time.

Imagine a community gradually being able to produce a portion of its own need for fruit,or cider,or preserves.In the beginning it would be a small portion indeed,but it would serve as a beginning.There is not much work involved if it is tackled communally,and the satisfaction is great.

Therefore:

Plant small orchards of fruit trees in gardens and on common land along paths and streets,in parks,in neighborhoods:wherever there are well-established groups that can themselves care for the trees and harvest the fruit.

❧❧❧

If you have an especially nice fruit tree,make a TREE PLACE (171)under it,with a GARDEN SEAT(176),or arrange a path so the tree can provide a natural goal along the path— PATHS AND GOALS (120)...

因此：

在花园，在小路和街道旁的公共用地上，在公园，在社区：凡是群众很好组织起来能自己管理果树、自己收获水果的地方，都可以种植小片的果树林。

果树

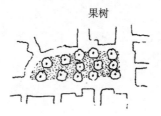

※◎◎◎

如果你有一棵特别好的果树，要让树下成为**树荫空间** (171)，放一张**园中坐椅** (176)，或铺设一条小路使果树提供一个沿着小路的天然标志物——**小路和标志物** (120) ……

模式171 树荫空间**

171   TREE PLACES**

...trees are precious.Keep them.Leave them intact.If you have followed SITE REPAIR(104),you have already taken care to leave the trees intact and undisturbed by new construction;you may have planted FRUIT TREES(170);and you may perhaps also have other additional trees in mind.This pattern reemphasizes the importance of leaving trees intact,and shows you how to plant them,and care for them,and use them,in such a way that the spaces which they form are useful as extensions of the building.

❧

When trees are planted or pruned without regard for the special places they can create,they are as good as dead for the people who need them.

Trees have a very deep and crucial meaning to hmnan beings.The significance of old trees is archetypal;in our dreams very often they stand for the wholeness of personality: "Since...psychic growth cannot be brought about by a conscious effort of will power,but happens involuntarily and naturally,it is in dreams frequently symbolized by the tree,whose slow,powerful involuntary growth fulfills a definite pattern." (M.L.yon Franz, "The process of individuation," in C.G.Jung,*Man and his Symbols*,New York:Doubleday,1964, pp.161,163-164.)

There is even indication that trees,along with houses and other people,constitute one of the three most basic

……树木是宝贵的。要保护树木使它们免遭损害。如果你秉持了**基地修整**（104）中的原则，一定已经注意使树木不受建筑施工的损害和破坏；你可能已经种植了**果树林**（170）；而且也许你还可能打算种些其他树木。本模式再次强调保护树木的重要性，并告诉你怎样种植、养护和利用它们，把树木形成的空间作为房屋的延伸来使用。

୨୦ଓଷ

如果光种植树木和修剪树木，但对树木可能产生的特殊空间却不予重视，那么对需要它们的人来说，树木如同废物。

树木对人类有着十分深远而重大的意义。古树是有典范意义的；在我们的梦境中它们常常是完美品格的化身："……由于……心理成长不能靠意志力有意识的努力来产生，而是无意识地自然而然地形成，在梦境中它常常以树木为标志，是树木的缓慢的、有力的自然成长塑造出一个特定的模型。"（M.L.von Franz,"The process of individuation,"in C.G.Jung, *Man and his Symbols*, New York: Doubleday, 1964, pp.161, 163~164.）

甚至有迹象表明，树木，与房屋以及人，是人类环境的三大基本要素。由心理学家约翰·巴克所创立的房屋—树木—人这种方法把一个人对这三个"整体"的描画作为投射测验的基础。仅就树木被认为具有跟人和房屋同等意义这一事实本身就足以说明它的重要性。（V.J.Bieliauskas, *The H-T-P Research Review*, 1965 Edition, Western Psychological Services, Los Angeles, California, 1965; and Isaac Jolles, *Catalog for the Qualitative Interpretation of the House-Tree-Person*, Los Angeles, California: Western Psychological Services, 1964, pp.75~97.）

parts of the human environment.The House-Tree-Person Technique,developed by Psychologist John Buck,takes the drawings a person makes of each of these three "wholes" as a basis for projective tests.The mere fact that trees are considered as full of meaning,as houses and people,is,alone,a very powerful indication of their importance(V.J.Bieliauskas,*The H-T-P Research Review*,1965 Edition,Western Psychological Services,Los Angeles,California,1965;and Isaac Jolles,*Catalog for the Qualitative Interpretation of the House-Tree-Person*,Los Angeles,California:Western Psychological Services, 1964,pp.75-97.).

But for the most part,the trees that are being planted and transplanted in cities and suburbs today do not satisfy people's craving for trees.They will never come to provide a sense of beauty and peace,because they are being set down and built around *without regard for the places they create.*

The trees that people love create special social places: places to be in,and pass through,places you can dream about,and places you can draw.Trees have the potential to create various kinds of social places:an *umbrella*—where a single,low-sprawling tree like an oak defines an outdoor room;a *pair*—where two trees form a gateway;a *grove*—where several trees cluster together;a *square*—where they enclose an open space;and an *avenue*—where a double row of trees,their crowns touching,line a path or street.It is only when a tree's potential to form places is realized that the real presence and meaning of the tree is felt.

The trees that are being set down nowadays have nothing of this character—they are in tubs on parking lots and along streets,in specially "landscaped areas" that you can see but

但是，目前在城市和郊区种植和移栽的树木大多不能满足人们对树木的要求。它们绝不会给人以美感和使人得到安宁，因为人们种下树木，在它们周围盖起建筑物，*而不考虑它们会形成什么样的空间*。

人们所喜爱的树木能够形成特殊的社会空间：人们愿意停留在这些地方，从它们那里经过，它们会进入你的梦境，供你入画。树木有可能创造各种各样的社会空间：伞形——这儿一棵枝叶低垂四处延伸的橡树构成有围合的户外小空间；双树对峙——这两棵树形成一个大门；一个小树林——这里的几棵树簇拥在一起；一个广场——这里的树木把一块空地围合起来；一条林荫道——这里两行树木树冠交错，排列在小路和街道两旁。只有当人们认识到树木创造空间的能力时，他们才会体会到树木的真正存在和意义。

现在种植下来的一些树木丝毫不具备这些性质——它们种在木桶里，放置在停车场和街道旁，放在一些你可望而不可及的特殊"风景区"里。它们并不构成我们所说的那种空间——因而对人们毫无意义。

现在，存在一种很大的危险性，一个读了我们论点的人，可能会错误地把它理解为树木应该作为使人们得到好处的手段来"利用"。不幸得很，今天在城市有一种很强烈的倾向正是要这么做——即把树木作为手段来对待，把它们当作我们得到乐趣的工具。

可是我们的论点却刚好相反。在城市，在房屋四周，在公园或花园中的树木不是森林中的树木。它们需要养护。一旦我们决定在城市里植树，我们必须明白，这里的树木有着不同的生态环境。譬如，在森林中，树木是在对它们有利的地形上生长的：它们的密度、阳光、风、湿度都由淘汰法来决定。但在城市，一棵树种在哪里便生长在那里，而除非十分精心地养护它们——修剪、观察，在蜕皮的时候要看管好，否则它们是存活不了的。

cannot get to.They do not form places in any sense of the word—and so they mean nothing to people.

Now,there is a great danger that a person who has read this argument so far,may misinterpret it to mean that trees should be "used" instrumentally for the good of people.And there is,unfortunately,a strong tendency in cities today to do just that—to treat trees instrumentally,as means to our own pleasure.

But our argument says just the opposite.Trees in a city,round a building,in a park,or in a garden are not in the forest.They need attention.As soon as we decide to have trees in a city,we must recognize that the tree becomes a different sort of ecological being.For instance,in a forest,trees grow in positions favorable to them:their density,sunlight,wind,moisture are all chosen by the process of selection.But in a city,a tree grows where it is planted,and it will not survive unless it is most carefully tended—pruned,watched,cared for when its bark gets pierced...

But now we come to a very subtle interaction.The trees will not get tended unless the places where they grow are liked and used by people.If they are randomly planted in some garden or in the shrubbery of some park,they are not near enough to people to make people aware of them;and this in turn makes it unlikely that they will get the care they need.

So,finally,we see the nature of the complex interactive symbiosis between trees and people.

1.First,people need trees—for the reasons given.

2.But when people plant trees,the trees need care(unlike the forest trees).

这样就有一个微妙的相互影响的关系。除非树木生长的空间为人们所喜爱和利用，否则它们得不到管理。如果把树木随意种植在某个花园里或某个公园的灌木丛生的地方，它们远离人们，不被人们想起；这就会使它们得不到所需要的管理。

因此，最后，我们看到在树木和人之间存在着复杂的相互影响的共生关系。

1. 首先，人需要树木——理由已如上述。

2. 但人们种植树木时，树木需要养护（这跟森林中的树木不同）。

3. 除非树木生长在人们喜爱的地方，否则它们得不到所需要的管理。

4. 而这又反过来要求树木形成社会空间。

5. 一旦树木形成社会空间，它们才有可能自然地生长。

这样我们看到，经过奇特的曲折关系，城市中的树木只有在人类的合作下，有助于形成人们所需要的空间时，才能按照它们自己的天性健康地生长。

因此：

如果你打算植树，要根据树木的特性来种植它们，使它们形成围合、林荫道、广场、小树林以及向着空地的中心伸展的一些单棵的树木。使附近建筑物的造型跟树木相协调，这样树木本身以及树木和建筑物一起，形成人们可以利用的空间。

伞形
umbrella

小树林
grove

林荫道
avenue

3.The trees won't get the care they need unless they are in places people like.

4.And this in turn requires that the trees form social spaces.

5.Once the trees form social spaces,they are able to grow naturally.

So we see,by a curious twist of circumstances,trees in cities can only grow well,and in a fashion true to their own nature,when they cooperate with people and help to form spaces which the people need.

Therefore:

If you are planting trees,plant them according to their nature,to form enclosures,avenues,squares,groves,and single spreading trees toward the middle of open spaces.And shape the nearby buildings in response to trees,so that the trees themselves,and the trees and buildings together,form places which people ca use.

⁎⁎⁎

Make the trees form "rooms" and spaces,avenues,and squares,and groves,by placing trellises between the trees,and walks,and seats under the trees themselves—OUTDOOR ROOM(163),TRELLISED WALK(174),GARDEN SEAT (176),SEAT SPOTS(241).One of the nicest ways to make a place beside a tree is to build a low wall,which protects the roots and makes a seat-SITTING WALL(243)...

　　通过设置树木间的棚架、散步场地、树荫下的坐椅使树木形成"房间"和空间、林荫道、广场和小树林——**有围合的户外小空间**（163）、**棚下小径**（174）、**园中坐椅**（176）、**户外设座位置**（241）。在树旁创造一种空间的最佳方法之一是造一道矮墙，它能保护树根并可让人坐在上面——**可坐矮墙**（243）……

模式172 花园野趣**

172 GARDEN GROWING WILD**

...with terracing in place and trees taken care of—
TERRACED SLOPE (169),FRUIT TREES(170),TREE
PLACES(171),we come to the garden itself—to the ground
and plants.In short,we must decide what kind of garden to
have,what kind of plants to grow,what style of gardening is
compatible with both artifice andnature.

શ્

**A garden which grows true to its own laws is not a
wilderness, yet not entirely artificial either.**

Many gardens are formal and artificial.The flower beds are
trimmed like table cloths or painted designs.The lawns are clipped like
perfect plastic fur.The paths are clean,like new polished asphalt.The
furniture is new and clean,fresh from the department store.

These gardens have none of the quality which brings a garden
to life—the quality of a wilderness,tamed,still wild,but cultivated
enough to be in harmony with the buildings which surround it
and the people who move in it.This balance of wilderness and
cultivation reached a high point in the oldest English gardens.

In these gardens things are arranged so that the natural
processes which come into being will maintain the condition of
the garden and not degrade it.For example,mosses and grasses will
grow between paving stones.In a sensible and natural garden,the
garden is arranged so that this process enhances the garden and
does not threaten it.In an unnatural garden these kinds of small
events have constantly to be "looked after" —the gardener
must constantly try to control and eradicate the processes of

……筑好了台地、管好了树木——**梯形台地**（169）、**果树林**（170）、**树荫空间**（171），我们开始研究花园本身——研究地面和植物。简言之，我们必须确定，我们应该建造何种花园，种植何种植物，哪种风格的园林既独具匠心又合乎自然。

✽✽✽

一个符合园林规律的花园既非只有野趣，也非完全人工雕饰。

许多花园一本正经，处处都是人工痕迹。花坛修整得如同台布或画出的图案。草坪推剪得犹如完美的塑料绒毛。小径一尘不染，好比新铺的柏油路；设备崭新、光洁，是刚从百货商店买来的。

这些花园一点不具有使花园充满生机的那种气质——这种气质就是野生野长，虽经修整，仍不失野趣，但又进行过充分加工使之同周围的建筑物及走进园中的人相协调。在古老的英国花园里，这种野趣和人工培养的平衡达到很高的程度。

在这样的花园里事物经过人们安排会使自然产生的东西为园林增辉，而无损于它。比方说，在铺路石子之间会长出青苔和小草。在一个实用而又天然的花园里，花园的布局要促使草木的自然生长以美化花园而不败坏它。在一个没有天然情趣的花园里总是对这类小事"关怀备至"——园丁一定要经常想方设法不让杂草萌芽，铲除莠草，刨掉草根，修剪草坪。

在具有野趣的花园里选栽植物和划定边界都要考虑使植物生长可进行自我调节。植物不需要在人工的控制下进行调节。但它并不乱长而破坏种植计划。例如，把天然的野生植物种在花草之间，这样所谓的莠草其实并不能占据

seeding,weeds,the spread of roots,the growth of grass.

In the garden growing wild the plants are chosen,and the boundaries placed,in such a way that the growth of things regulates itself.It does not need to be regulated by control.But it does not grow fiercely and undermine the ways in which it is planted.Natural wild plants,for example,are planted among flowers and grass,so that there is no room for so—called weeds to fill the empty spaces and then need weeding.Natural stone edges form the boundaries of grass so that there is no need to chop the turf and clip the edge every few weeks.Rocks and stones are placed where there are changes of level.And there are small rock plants placed between the stones,so that once again there is no room for weeds to grow.

A garden growing wild is healthier,more capable of stable growth,than the more clipped and artificial garden.The garden can be left alone,it will not go to ruin in one or two seasons.

And for the people too,the garden growing wild creates a more profound experience.The gardener is in the position of a good doctor,watching nature take its course,occasionally taking action,pruning,pulling out some species,only to give the garden more room to grow and become itself.By contrast,the gardens that have to be tended obsessively,enslave a person to them;you cannot learn from them in quite the same way.

Therefore:

Grow grasses,mosses,bushes,flowers,and trees in a way which comes close to the way that they occur in nature:intermingled,without barriers between them, without bare earth,without formal flower beds,and with all the boundaries and edges made in rough stone and brick and wood which become a part of the naturl growth.

这些空间，因而不需要锄掉莠草。天然的石头边缘形成草坪的边界，这样就不需要每隔几周铲除草皮和修剪边缘。岩块和石头垒放在高度变换的地方。在石块之间有苔藓植物，这样莠草又无处生长了。

野趣花园比起修剪整齐的人工化的花园来更为健康，能够更加稳定地生长。这样的花园可以任植物自然生长，一两个季度也不会衰败。

对于人来说也一样，野趣花园使人有机会增长见识。园丁有可能成为好医生，他可以观察自然界的时序变化，有时可以动动手，修枝剪叶，拔掉某些品种，结果使花园有更多的发展余地，使它更具花园特色。相反，需要着意加以管理的花园使人变成花园的奴隶；你不能以同样的方法从它们那里学到东西。

因此：

种植青草、苔藓、花卉和树木，让它们如同在大自然中一样生长：把它们混种在一起，不使它们之间界线分明，不使泥土裸露，不做正式的花坛，使所有的边界和边缘都由粗糙的石块、砖头和树木做成，它们同天然环境浑然一体。

培植的花木野生野长

粗糙的天然边缘

Include no formal elements,except where something is specifically called for by function—like a greenhouse—GREENHOUSE(175),a quiet seat—GARDEN SEAT(176), some water—STILL WATER(71),or flowers placed just where people can touch them and smell them—RAISED FLOWERS(245)...

ᘍᏮᏳ

不必使用讲究的材料，除了有时有功能上的特别要求——如温室——**温室**（175）、一个安静的坐位——**园中坐椅**（176）、一些水——**池塘**（71）或种在人们够得着和闻得到的地方的鲜花——**高花台**（245）……

模式173　花园围墙*

　　……在私人住宅里，**半隐蔽花园**（111）和**私家的沿街露台**（140）都需要围墙。推而广之，不仅私人花园，包括公共花园，甚至小公园和草地——**僻静区**（59）、**近宅绿地**（60），也需要在它们周围有一种围合，使它们尽可能美观和安静。

⊰⊱

　　花园和小公园如果不加以很好地保护，则难以避开市尘的喧嚣。

173   GARDEN WALL*

...in private houses,both the HALF-HIDDEN GARDEN(111) and the PRIVATE TERRACE ON THE STREET(140) require walls.More generally,not only private gardens,but public gardens too,and even small parks and greens—QUIET BACKS(59),ACCESSIBLE GREEN(60),need some kind of enclosure round them,to make them as beautiful and quiet aspossible.

⋘⋙

Gardens and small public parks don't give enough relief from noise unless they are well protected.

People need contact with trees and plants and water.In some way,which is hard to express,people are able to be more whole in the presence of nature,are able to go deeper into themselves,and are somehow able to draw sustaining energy from the life of plants and trees and water.

In a city,gardens and small parks try to solve this problem;but they are usually so close to traffic,noise,and buildings that the impact of nature is entirely lost.To be truly useful,in the deepest psychological sense,they must allow the people in them to be in touch with nature— and must be shielded from the sight and sound of passing traffic,city noises,and buildings.This requires walls,substantial high walls,and dense planting all around the garden.

人们需要接触树木、花草和水。从某种意义上讲（虽然这很难讲清楚），人们在大自然中会更加完美，能够更加深入地考察他们自己，并能从花草、树木和水的生命中汲取鼓舞的力量。

在城市中，花园和小公园力图解决这个问题；但它们通常离交通、闹市和房屋太近，致使大自然的影响完全消失。它们若要真正有用处，从最深刻的心理学意义上讲，必须使园中的人跟大自然接触——因而必须屏挡起来看不到和听不见过往的行人车辆、建筑物及城市的噪声。这就需要花园四周有围墙、牢固的高墙和茂密的植物。

城市中向公众开放的有围墙的小花园为数不多，在它们当中，阿尔汉伯拉的宫殿、哥本哈根皇家图书馆花园几乎始终闻名遐迩。人们理解并珍视它们创造的宁静环境。

莫卧儿人的有围墙的花园
Walled gardens—Mughal

……你的砖砌花园墙或公园墙……在外面看起来让人感到不友好，但与其说它不友好不如说它是自谦。这就是说，并非主人不想让你看见他的花园，而是不想让你看到他本人：这坦率地表明，他需要一些属于他自己的时间，因而他需要一些属于他自己的地面。当他脱掉外衣在园中挖地，或跟放学回家的儿子做跳背游戏，或跟他的妻子在晚霞中漫步，共忆往昔岁月时，一定不可

In those few cases where there are small walled gardens in a city,open to the public—Alhambra,Copenhagen Royal Library Garden—these gardens almost always become famous.People understand and value the peace which they create.

...your garden or park wall of brick...has indeed often an unkind look on the outside,but there is more modesty in it than unkindness. It generally means,not that the builder of it wants to shut you out from the view of his garden,but from the view of himself:it is a frank statement that as he needs a certain portion of time to himself,so he needs a certain portion of ground to himself,and must not be stared at when he digs there in his shirtsleeves,or plays at leapfrog with his boys from school,or talks over old times with his wife,walking up and down in the evening sunshine.Besides,the brick wall has good practical service in it,and shelters you from the east wind,and ripens your peaches and nectarines,and glows in autumn like a sunny bank. And,moreover,your brick wall,if you build it properly,so that it shall stand long enough,is a beautiful thing when it is old,and has assumed its grave purple red,touched withmossy green...(John Ruskin,*The Two Paths*,New York:Dutton,1907,pp.202-205.)

This pattern applies to all private gardens and to small parks in cities.We are not convinced that it applies to all small parks—but it is hard to differentiate precisely between the places where a walled garden is desirable and the places where it is not.There are definitely situations where a small park,and perhaps even a small garden that is open to the rush of life around it,is just right.However,there are far more parks and gardens left open,that need to be walled,than vice versa,so we emphasize the walled condition.

以被人窥探。此外，砖墙对花园还有实用价值，它能挡住东风，催熟桃子和曲桃，在秋天像洒满阳光的堤岸那样鲜艳夺目。再则，你的砖墙如果建造得好，坚固耐久，当它变旧了的时候，本身还是美景，它会呈现暗紫色，点缀着青苔的绿色……（John Ruskin，*The Two Paths*.New York：Dutton，1907，pp.202~205.）

本模式可应用于城市中的所有私家花园和小公园。我们并不认为它适合于所有的小公园——但很难精确地区分哪里需要有围墙的小公园，哪里不需要。肯定会有这样一些情况，某个小公园，也许甚至是小花园，应该向其周围的繁华地区敞开。然而，至今仍然开敞的公园和花园需要围墙，这比相反的情况要多得多，所以我们强调建造围墙。

因此：

建造某种形式的围墙以保护花园内部的安宁，使花园能够避开过路的行人和车辆的视线和声音。如果这是一个大花园或公园，围墙可以采用灵活的方式，可以用灌木丛、小树林、斜坡等。但是，花园越小，围墙越要坚固和定型。在非常小的花园里，用房屋或墙来做围墙；甚至篱笆和树篱也难以阻挡外来的喧闹声。

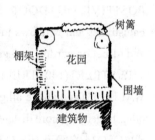

※※※

利用花园围墙使之有助于形成**户外正空间**（106）；但要在墙上打开出口，设置栏杆和窗户来进行花园和街道或花园和花园之间的联系——**私家的沿街露台**（140）、**棚下小径**（174）、**半敞开墙**（193），而最重要的，要使它有出口以看到其他的更大和更远的空间——**外部空间的层次**（114）、**禅宗观景**（134）……

Therefore:

Form some kind of enclosure to protect the interior of a quiet garden from the sights and sounds of passing traffic.If it is a large garden or a park,the enclosure can be soft,can include bushes,trees,slopes,and so on.The smaller the garden,however,the harder and more definite the enclosure must become.In a very small garden,form the enclosure with buildings or walls;even hedges and fences will not be enough to keep ut sound.

༄༅༆

Use the garden wall to help form positive outdoor space—POSITIVE OUTDOOR SPACE(106);but pierce it with balustrades and windows to make connections between garden and street,or garden and garden—PRIVATE TERRACE ON THE STREET(140),TRELLISED WALK (174),HALF-OPEN WALL(193),and above all,give it openings to make views into other larger and more distant spaces—HIERARCHY OF OPEN SPACE(114),ZEN VIEW(134)...

模式174 棚下小径**

……假设花园内各主要空间已经确定——**有围合的户外小空间**（163）、**树荫空间**（171）、**温室**（175）、**果树林**（170）。现在，在特别需要强调路径的地方——**小路和标志物**（120）——或者，甚至更加重要一些，在一个花园的

174   TRELLISED WALK**

...suppose the main spots of the garden have been defined—OUTDOOR ROOM(163),TREE PLACES(171), GREENHOUSE(175),FRUIT TREES(170).Now,where there is a special need to emphasize a path—PATHS AND GOALS(120)—or,even more important,where the edges between two parts of a garden need to be marked without making a wall,an open trellised walk which can enclose space,is required.Above all,these trellised walks help to form the POSITIVE OUTDOOR SPACES(106)in a garden or a park;and may perhaps help to form an ENTRANCE TRANSIION(112).

❧❧

Trellised walks have their own special beauty.They are so unique,so different from other ways of shaping a path, that they are almost archetypal.

In PATH SHAPE(121),we have described the need for outdoor paths to have a shape,like rooms.In POSITIVE OUTDOOR SPACE (106),we have explained the need for larger outdoor areas to have positive shape.A trellised walk does both.It makes it possible to implement both these patterns at the same time—simply and elegantly.But it does it in such a fundamental way that we have decided to treat it as a separate pattern;and we shall try to define the places where a trellised structure over a path is appropriate.

1.Use it to emphasize the path it covers,and to set off one part of the path as a special section of a longer path in order to

两部分之间的边界不需要通过筑墙来做标记的地方，要求有一条能围合空间的棚下小径。最重要的是，这些棚下小径有助于形成花园或公园中的**户外正空间**（106）；还可有助于形成入口的**过渡空间**（112）。

❧❦

棚下小径具有自身独特的美。它们独具一格，有别于形成路径的其他方法，堪称完美的典型。

在**小路的形状**（121）中，我们描述过了，户外小路需要有如房间那样的形状。在**户外正空间**（106）中，我们解释过了，更大的户外空间需要有"正"的形状。棚下小径两者都需要。它有可能同时体现这两种模式——朴素而又优雅。但它表现的方式很重要，因此我们决定把它作为一个独立的模式来研究；我们将力图确定路径上适合于构筑棚架的地方。

1. 利用它以突出它所覆盖的那段小路，把小路的一部分分离出来作为较长的小路上的一个特殊路段，使它成为异常优美的地方，吸引人们来这里散步。

棚架使户外空间具备形状
A trellis gives shape to an outdoor area

2. 由于棚下小径使它邻近的空间造成围合，可利用它作户外空间的隔墙。例如，利用棚下小径围绕或局部围绕花园，能够形成一个巨大的有围合空间。

make it an especially nice and inviting place to walk.

2.Since the trellised path creates enclosure around the spaces which it bounds,use it to create a virtual wall to define an outdoor space.For example,a trellised walk can form an enormous outdoor room by surrounding,or partially surrounding,a garden.

Therefore:

Where paths need special protection or where they need some intimacy,build a trellis over the path and plant it with climbing flowers.Use the trellis to help shape the outdoor spaces on either sde of it.

∞∞

Think about the columns that support the trellis as themselves capable of creating places—seats,bird feeders—COLUMN PLACES (226).Pave the path with loosely set stones—PAVING WITH CRACKS BETWEEN THE STONES(247).Use climbing plants and a fine trellis work to create the special quality of soft,filtered light underneath the trellis—FILTERED LIGHT(238),CLIMBING PLANTS(246)...

因此：

在小路需要特殊保护或需要一些私密性的地方，在它上面构筑棚架，植以攀援花木。利用棚架促使其两边形成户外空间。

棚架

⊰⊱

应该考虑到支撑棚架的柱子，它们自身也能构成一些空间——坐位、喂鸟亭——**柱旁空间**（226）。用石块稀疏地铺砌路面——**留缝的石铺地**（247）。利用攀援植物和精工细作的棚架使棚下有一种特别柔和的过滤光线——**过滤光线**（238）、**攀援植物**（246）……

模式175 温室

　　……为使花园保持盎然生机，几乎必备一个"工作间"——一间位于花园和住宅之间的屋子，小苗能够在这里发芽，在温暖的气候条件下即使在冷天花木也照样在这里生长。在**住宅团组**（37）或**工作社区**（41）中，这一工作间对**公共用地**（67）能作出很大贡献。

175 GREENHOUSE

...to keep a garden alive,it is almost essential that there be a
"workshop" —a kind of halfway house between the garden and
the house itself,where seedlings grow,and where,in temperate
climates,plants can grow in spite of cold.In a HOUSE
CLUSTER(37)or a WORK COMMUNITY(41),this workshop
makes an essential contribution to the COMMN LAND(67).

❧❧❦

Many efforts are being made to harness solar energy
by converting it into hot water or electric power.And yet the
easiest way to harness solar energy is the most obvious and
the oldest:namely,to trap the heat inside a greenhouse and
use it for growing flowers and vegetables.

Imagine a simple greenhouse,attached to a living
room,turned to the winter sun,and filled with shelves for
flowers and vegetables.It has an entrance from the house—
so you can go into it and use it in the winter without going
outdoors.And it has an entrance from the garden—so you can
use it as a workshop while you are out in the garden and not
have to walk through the house.

This greenhouse then becomes a wonderful place:a source
of life,a place where flowers can be grown as part of the life
of the house.The classic conservatory was a natural part of
countless houses in the temperate climates.

For someone who has not experienced a greenhouse as
an extension of the house,it may be hard to recognize how
fundamental it becomes.It is a world unto itsdf,as definite and

为将太阳能转化为热水和电力，人们在进行多方面的努力。但利用太阳能的最简易的方法却是人人看得见的老方法：那就是把能量保持在温室之内，用它促进鲜花和蔬菜生长。

请设想这样一个简易的温室：它连接起居室，朝向冬天里的太阳，温室里面是一架架鲜花和蔬菜。从屋里可以走进它的入口——因此在冬天你可以到它里面去使用它而无须经过户外。它还有一个由花园进来的入口——因此当你在花园里的时候可以利用它作工具间而无须经过室内。

于是这个温室就成为一个绝妙的去处：它充满生机，屋子里的人可以在这里养花休闲。在气候温和的地方，这种传统的温室家家户户无有不备。

对于那些不曾把温室作为住宅的附属建筑的人来说，可能不容易看到它的重要意义。它自身构成一个世界，犹如火或水一样不容置疑和异常美妙，它使人感受到从其他任何模式都得不到的东西。我们曾借助这一模式语言在莫德斯托为精神病学家休伊特·赖恩建造过诊所，他认为温室极为重要，因此盖了一个温室作为诊所的基本部分：它靠近公共区，病人在这里种植树苗，这些树苗经过移栽逐渐为诊所形成一座花园，通过这些活动使他们自己得到康复。

由于受到这一生态活动的启示，近来有好几个"能源组织"打算把温室当作新住宅区的基本部分来建造。例如，格雷厄姆·盖因斯的独户的生态住宅里有一个庞大的温室作为热量和食物的来源。（参阅 *London Observer*，October 1972.）查劳迪栽培洞——一个供冬天种蔬菜用的装有玻璃的凹进去的地下温室——是另一类型的温室。（*Progressive Architecture*，July 1970，p.85.）

wonderful as fire or water,and it provides an experience which can hardly be matched by any other pattern.Hewitt Ryan,the psychiatrist for whom we built the clinic in Modesto with the help of this pattern language,thought greenhouses so essential that he included one as a basic part of the clinic:a place beside the common area,where people could reintegrate themselves by growing seedlings that would be gradually transplanted to form gardens for the clinic.

Several recent "energy-systems" inspired by the ecology movement have sought to make greenhouses a fundamental part of human settlements.For example,Grahame Gaines' self-contained ecohouse includes a large greenhouse as a source of heat and food.(See *London Observer*,October 1972.)And Chahroudi's Grow Hole—a glazed sunken pit for growing vegetables in winter—is another kind of greenhouse(*Progressive Architecture*,July 1970,p.85.)

Therefore:

In temperate climates,build a greenhouse as part of your house or office,so that it is both a"room"of the house which can be reached directly without going outdoors and a part of the garden which can be reached directly rom the garden.

❧❦

Place the greenhouse so that it has easy access to the VEGETABLE GARDEN(177)and the COMPOST(178). Arrange its interior so that it is surrounded with WAIST-HIGH SHELVES(201)and plenty of storage space—BULK STORAGE(145);perhaps give it a special seat,where it is possible to sit comfortably—GARDEN SEAT(176),WINDOW PLACE (180)...

因此：

在气候温和的地方，建一温室作为住宅或办公间的一部分，这样它既可用作住宅的一个"房间"，你可以不经过户外就能直接到那儿去，也可作为花园的一部分，你可以从花园直接进去。

和房屋连接一起的温室

🙟🙠

将温室建在很容易通往**菜园**（177）和**堆肥**（178）的位置。布置温室内部使其四周摆满**半人高的搁架**（201）和使它拥有大量的储藏空间——**大储藏室**（145）；在有可能坐得舒服的地方多半可以放置一张专用的坐椅——**园中坐椅**（176）、**窗前空间**（180）……

模式176　园中坐椅

　　……随着花园性质的确定——**花园野趣**（172），我们开始考虑一些特殊的角落，它们会使花园更加宝贵并带点神秘感。其中最重要的是**有阳光的地方**（161），前面已有模式予以描述，因为这对建筑物极为重要。在此之外，现在我们再设一个坐位，它比较私密，可供一个人来此歇息、沉思和遐想。

176   GARDEN SEAT

...with the character of the garden fixed—GARDEN GROWING WILD (172),we consider the special corners which make the garden valuable and somewhat secret.Of these,the most important is the SUNNY PLACE(161),which has already been described,because it is so fundamental to the building. Now we add to this another seat,more private,where a person can go to sit and hink and dream.

❧❧❧

Somewhere in every garden,there must be at least one spot,a quiet garden seat,in which a person—or two people—can reach into themselves and be in touch with nothing else but nature.

Throughout the patterns in this pattern language we have said,over and again,how very essential it is to give ourselves environments in which we can be in touch with the nature we have sprung from—see especially CITY COUNTRY FINGERS(3)and QUIET BACKS (59).But among all the various statements of this fact there is not one so far which puts this need right in our own houses,as close to us as fire and food.

Wordsworth built his entire politics,as a poet,around the fact that tranquility in nature was a basic right to which everyone was entitled.He wanted to integrate the need for solitude-in-nature with city living.He imagined people literally stepping off busy streets and renewing themselves in private gardens—every day. And now many of us have come to

ℰℬℭℰ

每一座花园至少应该有一个地方有一张安静的园中坐椅，在这里一个人——或两个人——能够安心养神，完全沉浸在大自然中。

在本模式语言的所有模式中我们都一再说过，最重要的是要为我们自己创造环境，使我们能够接触养育我们的大自然——尤其请参阅**指状城乡交错**（3）和**僻静区**（59）。但在对这一事实的各种陈述中，迄今没有一种把这一需要在我们自己的住宅中正确地体现出来，把它看作火和食物一样不可缺少。

沃兹沃斯作为一名诗人，其全部政见都在于说明，在大自然中得到安宁是人人都应享有的基本权利。他希望把在大自然中独处的需要同城市生活融成一体。他设想，人们一离开繁华的大街就能到幽静的花园中呼吸新鲜的空气——每天都如此。现在我们许多人都开始认识到，若无这样一块地方简直无法在城市中生活。活动如此频繁，一天到晚忙于工作、照顾家庭、访亲会友、处理杂务——能独自待一会儿的时间太少。但我们的生活越是没有养成安静的习惯，我们就越是忙忙碌碌，一安静下来独自待着就会感到不习惯和心烦：谁都知道城里人一个劲儿地忙个不停，不能无忧无虑地安静一会儿。

正是在这种背景下我们提出设置僻静的园中坐椅：它在花园隐蔽处，周围花木繁荫，一两个人可以在这里独自坐着，不受打扰。这个地方可以在房顶上，在平地上，甚至也可以半凹在堡坎中。

可为本模式作佐证的论述园林的古籍多至千百种。其中一本是希尔德加德·霍索恩的《花园的魅力》（*The Lure of the Garden*, New York : The Century Co., 1911）。我们援引其中一段，它描写在安静的园中坐椅上进行一种特别的谈话。

learn that without such a place life in a city is impossible. There is so much activity,days are so easily filled with jobs,family,friends,things to do—that time alone is rare.And the more we live without the habit of stillness,the more we tie ourselves to this active life,the stranger and more disquieting the experience of stillness and solitude becomes:city people are notoriously busy-busy,and cannot be alone,without "input," for a moment.

It is in this context that we propose the isolated garden seat:a place hidden in the garden where one or two people can sit alone,undisturbed,near growing things.It may be on a roof top,on the ground,perhaps even half-sunken in an embankment.

There are literally hundreds of old books about gardens which testify to this pattern.One is Hildegarde Hawthorne's *The Lure of the Garden*,New York:The Century Co.,1911.We quote from a passage describing the special kind of small talk that is drawn out of people by quiet garden seats:

Perhaps,of all the various forms of gossip overheard by the garden,the loveliest is that between a young and an old person who are friends.Real friendship between the generations is rare,but when it exists it is of the finest.That youth is fortunate who can pour his perplexities into the ear of an older man or woman,and who knows a comradeship and an understanding exceeding in beauty the facile friendships created by like interests and common pursuits;and fortunate too the girl who is able to impart the emotions and ideas aroused in her by her early meetings with the world and life to some one old in experience but cornprehendingly young in heart.Both of them will remember those hours long after the garden gate has closed behind their friend forever;as long,indeed,as they remember anything that went to the making of the best in them.

也许，在所有从花园旁听到的各式各样的谈天说地中，最可爱的是一老一少两个友人之间的谈话。两代人之间的真诚友谊本属罕见，但如确有此种友谊，则它最为难能可贵。这样一个小伙子是幸运的，如果他能将其心事向年长的男朋女友倾诉，如果他懂得友谊和同情，而这是比那种由相似利益和共同追求建立起来的一拍即合的交情更为美好的；这样的一位少女也是幸运的，如果她能把最初接触社会和生活在其内心激起的感情和思绪向一位阅历深而心灵年轻的长者倾诉。这两位年轻人在园门已经于身后关闭了很久之后仍将长久地记住那些时刻；这种记忆如同他们记得生活中最美好的东西那样天长地久。

因此：

在花园中开辟一个幽静的处所——一个私密的有围合的小空间，这里有舒适的坐椅、茂盛的花木和充沛的阳光。精心选择放置坐椅的地方；使这个地方能让你安静地独处。

幽静的处所

∞ C3

置一园中坐椅，如同其他户外坐椅一样，在这里可观赏风景，可晒太阳，又可避风——**户外设座位置**（241）；也许可在光线柔和及斑斑点点的灌木林和树丛下面——**过滤光线**（238）……

Therefore:

Make a quiet place in the garden-a private enclosure with a comfortable seat,thick planting,sun.Pick the place for the seat carefully;pick the place that will give you the most intense kind of solitude.

❀❀❀

Place the garden seat,like other outdoor seats,where it commands a view,is in the sun,is sheltered from the wind-SEAT SPOTS(241);perhaps under bushes and trees where light is soft and dappled-FILTERED LIGHT(238)...

模式177　菜园*

　　……我们已经有一个模式说明公共和私家花园的实用性——**果树林**（170）；现在我们补充一个较小的模式，它也是花园的一个重要部分，是每一个公共和私家花园都应具有的：利用公共土地——**公共用地**（67）和私家花园——**半隐蔽花园**（111）开辟出一块人们可种植蔬菜的地方。

෪෬

　　在一个社会功能健全的市镇里，每个家庭都能够种植自家食用的蔬菜。把种菜当作热心人的业余爱好的时代已经过去；它已是人们生活中不可缺少的一部分。

177   VEGETABLE GARDEN*

...we have one pattern,already,which brings out the useful character of gardens—both public and private ones—FRUIT TREES (170);we supplement this with a smaller,but as important aspect of the garden-one which every public and private garden should contain:enhance common land-COMMON LAND(67)and private gardens -HALF-HIDDEN GARDEN(111)with a patch where people can grow vegetables.

৪৩৫৪

In a healthy town every family can grow vegetables for itself.The time is past to think of this as a hobby for enthusiasts;it is a fundamental part of human life.

Vegetables are the most basic foods.If we compare dairy products,vegetables and fruits,meats,and synthetic foods,the vegetables play the most essential role.As a class,they are the only ones which are by themselves wholly able to support human life.And,in an ecologically balanced world,it seems almost certain that man will have to work out some balanced relationship with vegetables for his daily food. (See,for example,F.Lappe,*Diet for a Small Planet*,New York:Ballantine,1971.)

Since the industrial revolution,there has been a growing tendency for people to rely on impersonal producers for their vegetables;however,in a world where vegetables are central and where selfsutfficiency increases,it becomes as natural for families to have their own vegetables as their own air.

蔬菜是人类最基本的食物。如果我们比较一下乳制品、蔬菜、水果、肉及合成食品，蔬菜最为必需。这是一类唯一能以自身的力量整体地维持人们的生活的食物。在一个生态平衡的世界，看来几乎确定无疑，人们须设法用蔬菜对自己每天的食品建立一种平衡关系。（例见 F.Lappe，*Diet for a Small Planet*，New York：Ballantine，1971.）

自从工业革命以来，人们需要的蔬菜要依靠别人来生产的这种趋势与日俱增。然而，在把蔬菜作为最重要的食物和自给自足的风气兴盛的社会，对于家庭来说，自己生产蔬菜如同自己呼吸空气一样理所当然。

一个家庭种菜所需要的土地小得令人难以置信。一个四口之家大约有十分之一英亩的土地就可种出足够一年需要的蔬菜。显然，利用同等数量的能量——阳光和劳动，蔬菜比任何其他食物能产生更多的养分。这就是说，每个住宅或住宅团组都可以实现蔬菜自给自足，而每一个没有私人土地的住户应该在近处的公共菜园占有一块土地。

对城市菜园来说，除了这一根本需求之外，还有一种更加微妙的需求。公园、街旁树木、修剪平整的草地对建立我们和土地之间的这种联系贡献甚少。它们没有告诉我们土地能为我们生产出东西，土地能发挥它的潜力。许多人在城市出生、长大，生活一辈子，却根本不知道他们吃的食物来自何处，也不知道真正的菜园是什么样的。他们唯一接触到的大地的产物是来自超级市场货架上的包装好的西红柿。但接触大地及其滋生万物的过程绝非仅仅是我们心血来潮的雅兴。很可能，这是进行机体保护的一项基本工作。更进一步看，在完全依靠从超市购买食品的城市居民一定有一种不安全感。

建立社区菜园也不需要花很多钱。当圣巴巴拉的居民在1970年5月决定办起一个城市中心菜园时，他们运用了他们的智慧。他们租用了闹市的一块空地（租金为六个月一美元），市政当局

The amount of land it takes to grow the vegetables for a household is surprisingly small.It takes about one-tenth of an acre to grow an adequate year round supply of vegetables for a family of four.And apparently vegetables give a higher "nutrient return" for fixed quantities of energy-sun,labor-than any other food.This means that every house or house cluster can create its own supply of vegetables,and that every household which does not have its own private land attached to it should have a portion of a *common vegetable garden close* at hand.

Beside this fundamental need for vegetable gardens in cities,there is a subtler need.Parks,street trees,and manicured lawns do very little to establish the connection between us and the land.They teach us nothing of its productivity,nothing of its capacities.Many people who are born,raised,and live out their lives in cities simply do not know where the food they eat comes from or what a living garden is like.Their only connection with the productivity of the land comes from packaged tomatoes on the supermarket shelf.But contact with the land and its growing process is not simply a quaint nicety from the past that we can let go of casually.More likely,it is a basic part of the process of organic security.Deep down,there must be some sense of insecurity in city dwellers who depend entirely upon the supermarkets for their produce.

Commtmity gardens needn't be expensive propositions either. When Santa Barbara residents decided to start a downtown garden back in May,1970,they used their ingenuity.A vacant downtown lot was acquired(at a cost of one dollar for 6 months),and the city provided free water and a tractor with operator for two days.Compost was no problem.The group got leaves from the park department,hard

免费供水，并开来了一辆拖拉机为他们耕了两天地。肥料不成问题。居民们从公园管理处弄来树叶，从当地的下水道挖来干硬的淤泥，从附近的马术俱乐部弄来马粪。工具和种子则是大家捐赠的。("Community Gardens," Bob Rodale, *San Francisco Chronicle*, May 31, 1927, p.16.)

由儿童们种植的阿姆斯特丹学校菜园
School garden in Amsterdam, worked by the children

因此：

在私人花园或公共用地上辟出一块土地作为菜园。大约十分之一英亩即可供四口之家的需要。一定要使菜园见到阳光，并处于中心地带以便可以供应各家。用篱笆把它围起来，在园旁修建一个储藏工具的小棚屋。

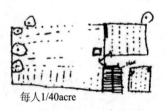

每人1/40acre

⊰⊱

给蔬菜施肥时，要使用家庭和邻里产生的天然堆肥——**堆肥**（178）；如果可能，尽量用井水和下水道的水来浇地——**浴室**（144）……

sludge from the local sanitation district,and horse manure from a nearby riding club.Tools and seeds were donated.("Community Gardens," Bob Rodale,*San Francisco Chronicle*,May 31,1972,p.16.)

Therefore:

Set aside one piece of land either in the private garden or on common land as a vegetable garden.About one-tenth of an acre is needed for each family of four.Make sure the vegetable garden is in a sunny place and central to all the households it serves.Fence it in and build a small storage shed for gardening tools beside it.

⋘⋙

To fertilize the vegetables,use the natural compost which is generated by the house and the neighborhood... COMPOST(178);and if possible,try to use water from the sinks and drains to irrigate the soil-BATHING ROOM(144)...

模式178　堆肥*

在浙江省，同在中国其他省份一样，有很多路边厕所。
它们是农民们用以吸引行人为他们贡献宝贵的肥料而盖的。
In Zhejiang Province,as in many other parts of China,roadside
toilets abound.They are built by farmers to entice passersby
into favoring them with a gift of valued fertilizer.

……花园是住宅中有实用价值的一个部分，因为它能
供你种植水果和蔬菜——**果树林**（170）、**菜园**（177）。但
它必须获取养料才能欣欣向荣，而这种以堆肥形式出现的
养料只有将各家庭和**住宅团组**（37）的垃圾和**动物**（74）
的粪便适当地混合才能制造出来。

178   COMPOST*

...the garden is a valuable part of the house,because it can help you grow fruit and vegetables-FRUIT TREES(170), VEGETABLE GARDEN(177).But it can only flourish if it gets nourishment;and this nourishment,in the form of compost,can only be created when the garbage and the wastes from the individual houses and HOUSE CLUSTERS(37)and from the ANIMALS(74)are properly organize.

❧☙

Our current ways of getting rid of sewage poison the great bodies of natural water,and rob the land around our buildings of the nutrients they need.

To the average individual in the city,it probably appears that the sewage system works beautifully-no muss,no fuss.Just pull the toilet chain,and everything is fine.In fact,city dwellers who have had the experience of using a smelly outhouse would probably argue that our modem system of sewage disposal is a tremendous advance over earlier practices.Unfortunately,this is simply not the case.Almost every step in modern sewage disposal is either wasteful,expensive,or dangerous.

We can start by remembering that every single time a toilet is flushed,seven gallons of drinking water go down the drain. In fact,around half of our domestic water consumption goes to flushing out the toilet.

Beyond the cost of the water,there is an enormous cost in the hardware of the sewer system.The average new homeowner today,living on a 50×150 foot lot in the city,has paid $1500 as

我们现行的排污方法污染了大量的天然水源，而且夺走了房屋周围的土地所需的养分。

对于城市的普通居民来说，也许看起来下水道是再方便不过了——它简直完美无缺。只要拉一下抽水马桶的链条就万事大吉。事实上，凡是用过臭气熏天的屋外厕所的城市居民都会认为我们现代的污水处理系统比过去的做法不知要好多少倍。很不幸，情况并非如此。现代化的污水处理几乎每一步骤都是既浪费又价格昂贵，还充满危险。

我们可以想一想，抽水马桶冲一次水，七加仑的自来水跑进了下水道。事实上，大约有一半的家庭用水是用来冲刷马桶的。

除了水要花费本钱，污水系统的设备花费的本钱也很大。今天，平均每个新住宅的主人，在城市中的住房面积为50ft×150ft，要付1500美元作为污水从其住宅被引向污水处理厂的排污费用。在人口密度较低的居民区，这一费用可能是2000美元或6000美元。每幢住宅还得另付500美元作为污水处理工厂的费用。这就很清楚了，现今污水系统的原始价格至少是每幢住宅2000美元，通常还要更多。而且这个价格还不包括自来水和污水设备的每月维修费：每户每年大约50美元。

此外，我们还得加上另外一些费用，这些费用不大容易用美元和美分来计算，但从长远讲，它们甚至比上述费用还要高。这些费用包括：(1) 流到江河和海洋里的养分的价值，这些养分从土壤中产生出来，本来是可以用来加强土壤的肥力的；(2) 花于处理污染的费用——污水造成"海藻污染"——下水道消耗水中的氧气使水中长满藻类。

应该怎么办呢？污水的某些成分可以经过再循环成为泥浆送回到地面。但民用污水常常跟含有剧毒化学元素的

his share of the collection system which takes sewage from his house to the sewage treatment plant.In lower density residential areas,this cost may be $2000 or even $6000.Each house pays an additional $500 toward the cost of the sewage treatment plant.We see then that the initial cost of today's sewage system is at least $2000 per house,often more.And these prices do not include monthly service charges for water and sewer facilities:around $50 per year for a single family household.

In addition we must add those costs which are less easily measurable in dollars and cents,but which may,in the long run,prove even higher than those already discussed.These include:(1)the value of lost nutrients which are allowed to flow away into the rivers and oceans-nutrients which could have been used to build up the soil they came from;and(2)the cost of the pollution:effluents cause "eutrophication" —the sewage depletes the oxygen in water and causes it to become clogged with algae.

What can be done?Some of this effluent might be recycled back to the land in the form of sludge.But residential sewage is usually mixed with industrial waste which often contains extremely noxious elements.And even if industrial wastes were not allowed into the sewage system,an additional distribution system would be needed to get the sludge back to the land.We see,therefore,that additional costs required to make the existing system ecologically sound are prohibitive.

What is needed is not a larger,more centralized and complex system,but a smaller,more decentralized and simpler one.We need a system that is less expensive;and we need a system that is an ecological benefit rather than an ecological drain.

We propose that individual small-scale composting plants begin to replace our present disposal system.Small buildings

工业废水混合在一起。即使工业废水不准向下水道排放，也需建立另一个分配系统来把泥浆输送回地面。因此，我们看到，要使现有的系统成为生态学上健全的污水系统，其附加费用是非常高的。

我们所需要的不是较大、较集中而复杂的系统，而是较小、较分散而又简单的系统。我们需要花钱少的系统；我们需要有利于生态而不是使生态枯竭的系统。

我们建议，用分散的、小规模的堆肥设备来取代我们现今的处理系统。小型楼房可以直接在抽水马桶下面安装它们自己的小型污水处理设备。产生的大块污物便可进入设备。由此形成的腐植质可用来给本楼房周围以及整个社区的土壤补充养分，如同洗澡和洗衣服的废水用来浇地一样。

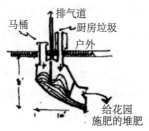

Clivus堆肥装置
Clivus compost chamber

这样的一些小型污水处理设备已作为商品使用，目前应用于瑞士、挪威和芬兰。可供出售的商品名称叫Multrum 或 Clivus，它们甚至能够进口到美国，总价格为1500 美元：比之传统的设备目前的最低价格 2000 美元来要低得多。样品请参阅 Vander Ryn，Anderson and Sawyer，"Composting Privy"，Technical Bulletin#1，Natural Energy Design Center，University of California，Berkeley，Dept.of Architecture，January 1974.

would be equipped with their own miniature sewage plants,located directly under the toilets.All bulky garbage produced on-site would be added to the plants.The resultant humus would be used to replenish the soil surrounding the building and throughout the neighborhood,as would the waste water from bathing and washing.

Such miniature sewage plants are commercially available and are currently in use in Sweden,Norway,and Finland.They are sold under the trade names Multrum or Clivus,and they can even be imported to the United States for a total price of $1500:much lower than the lowest figure of $2000,currently being charged for the conventional system.For a worked example,see Van der Ryn,Anderson and Sawyer,"Composting Privy",Technical Bulletin #1,Natural Energy Design Center,University of California,Berkeley,Dept.of Architecture,January 1974.

These composting plants are so simple that they can be built by amateurs for much less money.An extremely simple homemade composting system is described below:

The privy is built adjoining a larger outbuilding which is built over a root cellar.From overhead joists to floor,the cellar is about 7'deep.And so it was simplest to make the composting chamber beneath the privy 7'deep...

In the composting chamber underneath our privy,we have been using peatmoss—both because it is highly absorbent,and because it comes compactly baled and is convenient to store.We also use some garden dirt and a little lime.

We keep a garbage can full of peat moss in the privy,and dump in about a quart of the moss after each use.The privy is fairly odorless. Whenever there gets to be a smell I add lime,dirt and an extra layer of peat moss.That rakes care of it.I figure that we will use three or four bales of peat moss a year—for a family of four plus a large number of guests.

My privy is of the kind familiarly known as a two-holer,which

这些堆肥设备很简单，可以利用业余时间自己制作，花钱要少得多。一种极为简单的家庭自制的堆肥设备描述如下：

厕所与一较大的外屋相连，外屋建在地窖上方。从上面的托梁到地面，地窖深约7ft。因此很容易在厕所下7ft深造成这个堆肥坑。……

在我们厕所下面的堆肥坑里，我们使用泥炭——既由于它有很强的吸附能力，又由于它是包装好的，易于储藏。我们也用一些菜园的土和少量石灰。

我们在厕所放置一个装满泥炭的垃圾箱，每用一次厕所倒进大约1quant泥炭。厕所基本上没有臭味。每当发出臭味时，我就放点石灰、土，再多加一层泥炭。这就把气味消除了。我估计，我们每年要用三包或四包泥炭——对一个四口之家，外加大批客人，这就够用了。

我的厕所是大家熟知的双坑厕所，看来它很适合于我的堆肥装置。

我们一次只用一个坑。我们使用 A 坑，一直到它堆满18in深。随后我们换用 B 坑，直到它堆积到同 A 坑一样深。然后把 A 坑粪堆铲到 C 坑，·如此做下去。当全部四个坑都填满时，把 C 坑和 D 坑都铲到外面的地上堆起来。我让它在那儿至少放置数周才使用。（*Organic Gardening and Farming*，Emmaus，Pennsylvania：Rodale Press，February 1972.）

因此：

把所有的马桶都放置在干燥的堆肥坑上方。使有机废物顺斜道滑向坑中，然后把合成产物用作肥料。

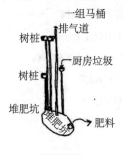

seems necessary for my composting system.

We use only one hole at a time.We use A until there is an accumulation 18 inches deep.We then shift to B and use it until the accumulation there is as great as that of A.Then the heap at A is shoveled to C,and so on.When all four positions are filled,C and D are shoveled into a heap on the ground outside,where I mean to let it stay for at least several weeks before use.(*Organic Gardening and Farming*, Emmaus,Pennsylvania:Rodale Press,February 1972.)

Therefore:

Arrange all toilets over a dry composting chamber. Lead organic garbage chutes to the same chamber,and use the combined products for ferilizer.

                    ဆဩ

Add to the effect of dry composting by reusing waste water;run all water drains into the garden to irrigate the soil;use organic soap—BATHING ROOM(144)...

go back to the inside of the building and attach the necessary minor rooms and alcoves to complete the main rooms;

179.ALCOVES
180.WINDOW PLACE
181.THE FIRE
182.EATING ATMOSPHERE
183.WORKSPACE ENCLOSURE
184.COOKING LAYOUT
185.SITTING CIRCLE
186.COMMUNAL SLEEPING
187.MARRIAGE BED
188.BED ALCOVE
189.DRESSING ROOM

ಬಿ

通过废水的再利用以增强干燥肥堆的肥效；把所有的废水引进花园用来浇灌土壤；使用有机肥皂——**浴室**（144）……

现在回到建筑物内部来，附加一些必需的小房间和凹室将主房间完备起来；

179. 凹室

180. 窗前空间

181. 炉火熊熊

182. 进餐气氛

183. 工作空间的围隔

184. 厨房布置

185. 坐位圈

186. 共宿

187. 夫妻用床

188. 床龛

189. 更衣室

模式179 凹室**

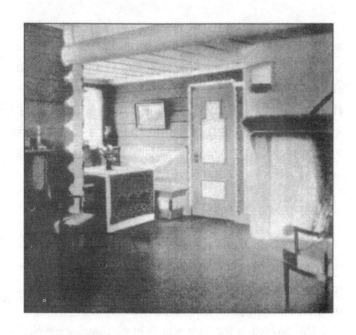

……许多大房间如果没有同它们连通的较小房间和凹室是不完善的。本模式及随后的几个模式规定较小房间和凹室的形状，这些较小房间和凹室有助于完善**中心公用区**（129）、**农家厨房**（139）、**起居空间的序列**（142）、**灵活办公空间**（146）、**等候场所**（150）、**小会议室**（151）以及许多其他模式。

∞∞∞

一个有着统一高度的统一房间无法满足一群人的需要。一个房间应在使一群人有机会都聚集在一起的同时，也要使他们有机会在同一空间里一两个人独处。

179   ALCOVES**

...many large rooms are not complete unless they have smaller rooms and alcoves opening off them.This pattern,and several which follow it,define the form of minor rooms and alcoves which help to complete COMMON AREAS AT THE HEART(129),FARMHOUSE KICHEN(139),SEQUENCE OF SITTING SPACES(142),FLEXIBLE OFFICE SPACE(146),A PLACE TO WAIT(150),SMALL MEETING ROOMS(151),and many others.

⚭

No homogeneous room,of homogeneous height,can serve a group of people well.To give a group a chance to be together,as a group,a room must also give them the chance to be alone,in one's and two's in the same space.

This problem is felt most acutely in the common rooms of a house-the kitchen,the family room,the living room.In fact,it is so critical there,that the house can drive the family apart when it remains unsolved.Therefore,while we believe that the pattern applies equally to workplaces and shops and schools— in fact,to all common rooms wherever they are—we shall focus our discussion on the house,and the use of alcoves around the family common rooms.

In modern life,the main function of the family is emotional;it is a source of security and love.But these qualities will only come into existence if the members of the house are *physically able to be together as a family.*

这个问题表现得最尖锐的是在住宅中的公用房间里——如在厨房、家庭室、起居室里。事实上,这个问题十分关键,如果得不到解决,这样的住宅会导致家庭破裂。因此,虽然我们认为这个模式同样适用于工作空间、商店和学校——实际上适用于任何地方的公共房间——但我们还是把讨论集中在住宅以及如何利用家庭公用房间周围的凹室上。

在现代生活中,家庭主要起感情作用;它是平安和爱的源泉。但这种性质只有当住宅中的人在物质条件上有可能在一起成为一个家庭时,才能表现出来。

这可并不容易。家庭各个成员在一天内进出的时间不同;即使他们都在家,每个人有他自己个人的爱好:做针线活、读书、做作业、干木匠活、搭积木、做游戏。在许多住宅内,这些兴趣使大家各自走进自己的房间,不跟家里人一起待着。产生这样的情况有两个原因。首先,在一般的家庭里,一个人可能很容易被其他人所干的事打扰:这个人想读点什么,其他人在看电视会使他觉得受到干扰。其次,家庭室通常没有任何空间可以让人把东西放在那儿而不被挪动。放在餐桌上的书本在吃饭的时候被拿开了;一场做了一半的游戏不能原封不动留在那里。自然,人们就养成了离开家人到别处去做这些事情的习惯。

为了解决这个问题,必须有某种方法,用这个方法家庭成员即使在做不同的事情也可以在一起。这意味着家庭室需要有许多小空间,在那里人们可以做不同的事情。这些小空间需要跟主要房间有足够的距离,以便小空间中发出的任何喧闹声不会影响大家共同使用主要房间。这些空间需要互相连通,这样大家即使都在自己的空间里待着也还是"在一起"。这意味着,它们需要互相开放。同时,它们需要隐蔽,以便它们里面的人不会受别人打搅。总之,家庭室周围必须有凹室。这种凹室应该有足够的大小,同

This is often difficult.The various members of the family come and go at different times of day;even when they are in the house,each has his own private interests:sewing,reading,homework,carpentry,model-building,games.In many houses,these interests force people to go off to their own rooms,away from the family.This happens for two reasons.First,in a normal family room,one person can easily be disturbed by what the others are doing:the person who wants to read,is disturbed by the fact that the others are watching TV.Second,the family room does not usually have any space where people can leave things and not have them disturbed.Books left on the dining table get cleared away at meal times;a half-finished game cannot be left standing.Naturally,people get into the habit of doing these things somewhere else-away from the family.

To solve the problem,there must be some way in which the members of the family can be together,even when they are doing different things.This means that the family room needs a number of small spaces where people can do different things. The spaces need to be far enough away from the main room,so that any clutter that develops in them does not encroach on the communal uses of the main room.The spaces need to be connected,so that people are still "together" when they are in them:this means they need to be open to each other.At the same time they need to be secluded,so a person in one of them is not disturbed by the others.In short,the family room must be surrounded by small alcoves.The alcoves should be large enough for one or two people at a time:about six feet wide,and between three and six feet deep.To make it clear that they are separate from the main room,so they do not clutter it up,and so that people in them are secluded,they should be narrower than

时可容纳一两个人，宽约 6ft，进深 3 ~ 6ft。为了清楚地标识出它们跟主要房间是分隔的，使它们不致扰乱主房间，同时也使凹室中的人得到安静，它们应比家庭室的墙面窄一些，天花也要比主房间低。

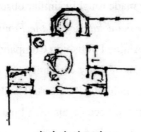

家庭室的凹室
Family room alcoves

由于本模式是如此重要，我们现在引证不同作者的几段话，用以强调许多人都作出过几乎类似的观察这一事实。

《家庭的社会心理内幕》(*Psychosocial Interion of the Family*, Gerald Handel, ed., Chicago, Ⅲ.: Aldine Publishing Company, 1967, p.13) 节选：

家庭生活的这种基本的二重性是意义重大的，因为在一个人竭力发展自己对世界的兴趣，并使自己成为具有独特风格的人的同时，也在设法跟其他成员保持令人满意的联系。同时，其他成员对此人发生他们自己的那种兴趣，也对他们自己发生兴趣。这是家庭生活中相互影响的根源。家庭力图创造一种模式以满足其成员既愿相聚又想独处的要求……

《家庭中的儿童们》(*Children in the Family*, Florence Powdermaker and Louise Grimes, New York : Farrar & Reinhart, Inc., 1940, p.108.) 中提到："即使一个儿童有他自己的房间，但并不喜欢整天待在那儿，而情愿在家中其他地方度过他的大部分时光……"第 112 页上写道："他喜欢和渴望别人的关怀。他喜欢拿出东西让大人看，让大人一起为他的发现感到快乐。他对大人们的活动感到兴趣，

the family room walls,and have lower ceilings than the main room.

Since this pattern is so fundamental,we now present several quotes from various writers to underscore the fact that many people have made roughly similar observations:

From *Psychosocial Interior of the Family*,Gerald Handel, ed.,Chicago, Ⅲ .:Aldine Publishing Company,1967,p.13.

This fundamental duality of family life is of considerable significance,for the individual's efforts to take his own kind of interest in the world,to become his own kind of person,proceed apace with his efforts to find gratifying connection to the other members.At the same time,the other members are engaged in taking their kinds of interest in him,and in themselves.This is the matrix of interaction in which a family develops its life.The family tries to cast itself in a form that satisfies the ways in which its members want to be together and apart...

From *Children in the Family* by Florence Powdermaker and Louise Grimes,New York:Farrar&Reinhart,Inc.,1940, p.108: "Even if a child has a room of his own,he doesn't like being kept there all day long but wants to spend much of his time in other parts of the house..." And p.112: "...he enjoys and craves attention.He likes to show things to adults and have them share in the pleasure of his discoveries.Besides,he is entranced by their activities and would like to have a finger in every pie."

And from Svend Riemer, "Sociological Theory of Home Adjustment," *American Soc.Rev.*,Vol.8,No.3,June 1943, In adjustment to the activities of other members of the family, it will be necessary to "migrate" ...between the different rooms of the family home.Even the same activity may have to be

而且总想插一手。"

《关于家庭调节的社会学理论》(Svend Riemer, *American Soc.Rev.*, Vol.8, No.3, June 1943, p.277.) 节选：

为了与家庭其他成员的活动相适应，需要在家中的不同的房间之间"迁移"。甚至同一活动在一天内不同的时间可能也得从一个房间移到另一个房间。

例如，下午在家自习可能不得不在起居室中进行，因为那时厨房正在准备晚饭；而晚自习可能得在厨房继续进行，因那时起居室被家中其他成员的业余活动占据了。这种在不同房间之间的"迁移"会使智力活动难以集中。它可能引起不安全感。如果儿童是在家中长大，这种可能产生的缺点就得加以严肃考虑。

现在已经很清楚了，在同一时间和同一空间既想独自待着又不想离开大家这种相互矛盾的需要几乎在每个家庭都会出现。不难看到，在所有的公共房间中这样的情形只是大同小异。人们想在一起；但同时他们希望在不脱离大家的情况下有机会享受一点儿隐私。

如果有 10 个人或 5 个人一起在一个房间里，其中两个人想到边上去安静地谈谈，他们需要有一个谈话的地方。只有凹室，或者类似凹室的地方，能够使他们得到所需要的私密性，而又不会迫使他们与大家完全隔离。

因此：

在任何公用房间的边缘设置一些小空间，通常宽不超过 6ft，进深为 3 ~ 6ft，可能还要小得多。这些凹室应该有足够的空间使两个人能坐下来聊天或玩耍，有时还可放一张书桌或工作台。

凹室

moved from one room to the other at different times of the day.

Home studies for example may have to be carried out in the living room during the afternoon,while food is being prepared in the kitchen;they may have to be continued in the kitchen during the evening hours,when the living room is occupied by leisure time activities of other members of the family.This "migration" between different rooms is apt to impair intellectual concentration. It may convey a sense of insecurity.Its possible disadvantages have to be seriously considered whenever children are reared in the family home.

It is clear then,that the opposing needs for some seclusion and some community at the same time in the same space,occur in almost every family.It is not hard to see that only slightly different versions of the very same forces exist in all communal rooms.People want to be together;but at the same time they want the opportunity for some small amount of privacy,without giving up community.

If ten people,or five,are together in a room,and two of them want to pull away to one side to have a quiet talk together,they need a place to do it.Only the alcove,or some version of the alcove,can give them the privacy they need,without forcing them to give the group up altogether.

Therefore:

Make small places at the edge of any common room, usually no more than 6 feet wide and 3 to 6 feet deep and possibly much smaller.These alcoves should be large enough for two people to sit,chat,or play and sometimes large enough to containa desk or a table.

凹室的天花显著低于主要房间的天花板高度——**天花板高度变化**（190）；利用低墙和粗大的柱使凹室和公用房间之间造成局部分隔——**半敞开墙**（193）、**柱旁空间**（226）；当凹室的一边是外墙时，使它成为窗前空间，设有一个漂亮的窗户、低窗台和嵌墙座位——**窗前空间**（180）、**嵌墙坐位**（202）；把它作为**加厚外墙**（211）。关于凹室的形状，详见**室内空间形状**（191）。……

Give the alcove a ceiling which is markedly lower than the ceiling height in the main room—CEILING HEIGHT VARIETY(190);make a partial boundary between the alcove and the common room by using low walls and thick columns-HALF-OPEN WALL(193),COLUMN PLACE (226);when the alcove is on an outside wall,make it into a window place,with a nice window,low sill,and a built-in seat—WINDOW PLACE(180),BUILT-IN SEATS(202);and treat it as THICKENING THE OUTER WALLS(211).For details on the shape of the alcove,see THE SHAPE OF INDOOR SPACE(191)...

模式180　窗前空间**

　　……本模式有助于完善**入口空间**（130）、**禅宗观景**（134）、**两面采光**（159）、**临街窗户**（164）对窗户所作的安排。根据本模式，每个房间至少有一个窗户需要形成一个可应用的窗前空间。

180　WINDOW PLACE**

...this pattern helps complete the arrangement of the windows given by ENTRANCE ROOM(130),ZEN VIEW(134), LIGHT ON TWO SIDES OF EVERY ROOM(159),STREET WINDOWS(164).According to the pattern,at least one of the windows in each room needs to be shaped in such a way as to increase its useulness as a space.

❧

Everybody loves window seats,bay windows,and big windows with low sills and comfortable chairs drawn up to them.

It is easy to think of these kinds of places as luxuries,which can no longer be built,and which we are no longer lucky enough to be able to afford.

In fact, the matter is more urgent.These kinds of windows which create "places" next to them are not simply luxuries; they are necessary.A room which does not have a place like this seldom allows you to feel fully comfortable or perfectly at ease. Indeed,a room without a window place may keep you in a state of perpetual unresolved conflict and tension-slight,perhaps,but definite.

This conflict takes the following form.If the room contains no window which is a "place," a person in the room will be torn between two forces:

1. He wants to sit down and be comfortable.

2.He is drawn toward the light.

人人喜爱窗前座位、凸窗和窗台很低、窗前放置舒适座椅的大窗户。

人们很容易把这样一些地方看作奢侈品，现在再不能去建造它们了，我们也再不会有幸能够享受它们。

事实上，事情很紧要。这类能在它们周围创造出"空间"的窗户不仅仅供人享受，也为人们所必需。的确，如果没有窗前空间，房间会使你处于总也得不到解决的矛盾和紧张的状态中——也许只是稍微有一点儿，但确实是有。

矛盾表现为以下的形式。如果房间没有一个窗前空间，房间中的人会受到两种力量的拉扯。

1. 他想坐下来舒服舒服。

2. 他为光线所吸引。

显然，如果舒适的地方——房间里那些你最想坐的地方——离窗很远，就无法克服这一矛盾。这样你就明白了，我们喜爱窗前空间不是追求奢华，而是机体的直觉，它是由一个人听凭自发的愿望得到满足而产生的。能让你真正感到舒适的房间总会含有某种窗前空间。

当然，现在很难给"窗前空间"以确切的定义。"窗前空间"基本上是房间内一个部分围隔的、可以清楚地辨认出来的地方。在这个意义上，下列结构都可起到窗前空间的作用：凸窗、窗前坐位、前面有一个明显的位置可放舒服的扶手椅的低窗台以及周围有窗户的深的凹室。为使窗前空间的概念更明确一些，对以上每一类型的结构都举例说明，同时讨论它们之中每一种的性能的主要特征。

凸窗 在房间的一端稍微向外突出，周围全是窗户。它起窗前空间的作用，因为这里光线强，从各边的窗户都可以看到景物，你可以端几把椅子或一张沙发坐到那儿去。

Obviously,if the comfortable places-those places in the room where you most want to sit-are away from the windows,there is no way of overcoming this conflict.You see, then,that our love for window "places" is not a luxury but an organic intuition,based on the natural desire a person has to let the forces he experiences run free.A room where you feel truly comfortable will always contain some kind of window place.

Now,of course,it is hard to give an exact definition of a "place". Essentially a "place" is a partly enclosed,distinctly identifiable spot within a room.All of the following can function as "places" in this sense:bay windows,window seats,a low window sill where there is an obvious position for a comfortable armchair,and deep alcoves with windows all around them.To make the concept of a window place more precise,here are some examples of each of these types,together with discussion of the critical features which make each one of them work.

*A bay window.*A shallow bulge at one end of a room,with windows wrapped around it.It works as a window place because of the greater intensity of light,the views through the side windows,and the fact that you can pull chairs or a sofa up into the bay.

*A window seat.*More modest.A niche,just deep enough for the seat.It works best for one person,sitting parallel to the window,back to the window frame,or for two people facing each other in this position.

*A low sill.*The most modest of all.The right sill height for a window place,with a comfortable chair,is very low:12 to 14 inches.The feeling of enclosure comes from the armchair-best of all,one with a high back and sides.

凸窗
A bay uindow

窗前坐位 它比较小，只不过是一个壁龛，进深刚好够一个坐位，供一个人使用最合适，与窗户平行地坐着，背靠着窗框，也可以供两个人以这样的姿势对坐。

窗前坐位
A window seat

低窗台 这是最小的窗前空间。对于一个摆着一把舒适椅子的窗前空间来说，合适的窗台高度是很低的：12～14in。利用扶手椅造成围合的感觉——因此最好放一张靠背和扶手都很高的椅子。

低窗台
A low sill

*A glazed alcove.*The most elaborate kind of window place:almost like a gazebo or a conservatory,windows all around it,a small room,almost part of the garden.

And,of course,there are other possible versions too.In principle,any window with a reasonably pleasant view can be a window place,provided that it is taken seriously as a space,a volume,not merely treated as a hole in the wall.Any room that people use often should have a window place.And window places should even be considered for waiting rooms or as special places along the length of hallways.

Therefore:

In every room where you spend any length of time during the day,make at least one window into a "windowplace".

ಙೋ

Make it low and self-contained if there is room for that ALCOVES (179);keep the sill low—LOW SILL(222);put in the exact positions of frames,and mullions,and seats after the window place is framed,according to the view outside—BUILT-IN SEATS(202),NATURAL DOORS AND WINDOWS(221). And set the window deep into the wall to soften light around the edges—DEEP REVEALS(223).Under a sloping roof,use DORMER WINDOWS(231)to make this pattern...

装有玻璃窗的凹室　这是最精致的窗前空间：很像有玻璃窗的凉亭或温室，周围全是窗户，是一间小房间，几乎是花园的一部分。

装有玻璃窗的凹室
A glazed alcove

当然，还可能有别的花样。原则上，任何一个景色宜人的窗户都可构成窗前空间，只要认真地把它作为一个空间、一个体量，而不仅作为墙里的一个缺口来对待。任何一间人们经常使用的房间都应有一个窗前空间。窗前空间甚至也应被考虑用在等候室或用作过道上的一些特殊场所。

因此：

在每一间你白天度过一些时光的房间里，至少要把一个窗户做成"窗前空间"。

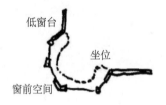

如果有地方可作窗前空间，要使它低矮并自成一统——**凹室**（179）；把窗台做得很低——**低窗台**（222）；根据户外的景色，在窗前空间框定后，把窗框、窗棂和坐位的位置都准确地定下来——**嵌墙坐位**（202）、**借景的门窗**（221）。把窗户深嵌到墙内使边缘周围的光线柔和——**深窗洞**（223）。在坡屋顶下，利用**老虎窗**（231）来建立本模式……

模式181 炉火熊熊*

……本模式有助于创造**中心公共区**（129）的灵魂，甚至有助于安排它的布局和位置，因为它影响到过道和房间之间的相互联系的方法。

❧❧❧

没有别的东西可以取代炉火。

电视机时常成了一个房间的关注中心，但它只不过是房间里一种真正生动和闪光的东西的微不足道的替代品。对火的需要犹如对水的需要一样必不可少。火是激情的试金石，可比得上树木、他人、房屋和天空。但传统的壁炉几乎已经废弃不用，而并新式壁炉常常装在家里作为"奢侈品"。也许这可以说明，为什么这些装点门面的壁炉总是

181　THE FIRE*

...this pattern helps to create the spirit of the COMMON AREAS AT THE HEART(129),and even helps to give its layout and position,because it influences the way that paths and rooms relate to one aother.

৪৩৫৪

There is no substitute for fire.

Television often gives a focus to a room,but it is nothing but a feeble substitute for something which is actually alive and flickering within the room.The need for fire is almost as fundamental as the need for water.Fire is an emotional touchstone,comparable to trees,other people,a house,the sky. But the traditional fireplace is nearly obsolete,and new ones are often added to homes as "luxury items." Perhaps this explains why these showpiece fireplaces are always so badly located. Stripped of the logic of necessity,they seem an after thought,not truly integrated.

The most convincing statement of the need for fire that we have found is in Gaston Bachelard's book,*The Psychoanalysis of Fire*.Here is a long quote from Bachelard to give you some idea of the power of his argument.

The fire confined to the fireplace was no doubt for man the first object of reverie,the symbol of repose,the invitation to repose.One can hardly conceive of a philosophy of repose that would not include a reverie before a flaming logfire.Thus,in our opinion,to be deprived of a reverie before a burning fire is to lose the first use and the truly human

装得不是地方。它们不是非用不可的东西，似乎是计划外的添加物，而非属于整体的一部分。

在加斯顿·巴切拉德所著的《对炉火的心理分析》（*The Psychoanelysis of Fire*）一书中，我们找到了对炉火需求的最有说服力的说明。这里引用一长段巴切拉德的话使你对其论据的力量有一些概念。

壁炉中的火焰无疑是最先招人冥思遐想的东西，它象征着安详，它引人入静。如果一个人得不到在熊熊燃烧着的圆木的火焰前冥思遐想的机会，他就难以体会静默安详的意境。因而，在我们看来，失去在燃烧的火焰前冥思遐想的机会就是失去炉火的最重要的使用价值即它对于人的价值。固然，炉火给我们以温暖和舒适。但人们只有在长时间凝视着火焰之后才会充分意识到这种舒适感；只有当一个人把肘撑在膝盖上，双手托着脑袋，他才能领略火的舒适。这种姿势来源于遥远的过去。坐在炉火旁的儿童自然而然地会这么做的。可思考者采取这种姿势却并非只是出于自然。它会使人的注意力特别集中，与注视和观看不可相提并论。其他的沉思方式很少采取这种姿势。靠近炉火时，你一定得坐下；你要休息但不要睡着；你得面对着一个特定的对象心驰神往。

当然，赞成功利主义心理理论的人不会接受一种理想主义学说。他们会向我们指出炉火的多种用途，以便使我们对它产生的兴趣确定不移：炉火不仅产生热量，它也用来烧肉。似乎合成型壁炉、农家壁炉并不能供人冥思遐想！

……在锯齿状的烟囱挂钩上挂着黑色的锅勺。三只脚的饭锅支在余烬未消的炉火上。我奶奶鼓起腮帮吹着铁管，把快熄灭的炉火重新吹旺。要同时煮好几样东西：给猪吃的土豆和给家人吃的上等土豆。在灰烬底下还为我焙着一个新鲜鸡蛋……在我表现好的那些天，家人们会拿出蛋奶烘饼烤模。这模子是长方形的，它压在葵葵的火焰上，这火焰红得犹如剑形百合的穗状花序。很快奶蛋烘饼就压在我的围嘴布上，它烫手但不烫嘴。于是，当滚

use of fire.To be sure,a fire warms us and gives us comfort.But one only becomes fully aware of this comforting sensation after quite a long period of contemplation of the flames;one only receives comfort from the fire when one leans his elbows on his knees and holds his head in his hands.This attitude comes from the distant past.The child by the fire assumes it naturally.Not for nothing is it the attitude of the Thinker.It leads to a very special kind of attention which has nothing in common with the attention involved in watching or observing.Very rarely is it utilized for any other kind of contemplation.When near the fire,one must be seated;one must rest without sleeping;one must engage in reverie on a specific object...

Of course the supporters of the theory of the utilitarian formation of the mind will not accept a theory so facile in its idealism,and they will point out to us the multiple uses of fire in order to ascertain the exact interest that we have in it:not only does fire give heat,but it also cooks meats.As if the complex hearth,the peasant's hearth,precluded reverie!...

...From the notched teeth of the chimney hook there hung the black cauldron.The three-legged cooking pot projected over the hot embers.Puffing up her cheeks to blow into the steel tube,my grandmother would rekindle the sleeping flames.Everything would be cooking at the same time:the potatoes for the pigs,the choice potatoes for the family.For me there would be a fresh egg cooking under the ashes...on days when I was on my good behavior,they would bring out the waffle iron.Rectangular in form,it would crush down the fire of thorns burning red as the spikes of sword lilies.And soon the gaufre or waffle would be pressed against my pinafore,warmer to the fingers than to the lips.Yes,then indeed I was eating fire,eating its gold,its odor and even its crackling while the burning gaufre was crunching under my teeth...

And it is always like that,through a kind of extra pleasure-like

烫的烤奶蛋烘饼在我的牙齿下嘎吱作响的时候，我真正在品尝着火焰的滋味，它的金黄颜色、它的气味乃至它的酥脆……

炉火每每给予我们一种份外的欢乐——如让我们尝到甜食的滋味——而表明它是人的朋友。它岂止供人煮熟食物而已，它还能把食物烤得酥脆而嘎吱作响。它赋予薄饼一层金黄的外壳；它给人的欢乐以物质的形式。从我们所能回溯到的久远的年代起，烹调得好的食物总是比营养好的食物更加值钱；而正是在欢乐而非痛苦中人们发现自己的才智。得到额外的东西比刚刚满足需要会使我们精神更加振奋。人是欲望的创造者，而不是需求的创造者。

但火炉旁的冥思遐想有着更富哲理的意义。凝视着炉火的人从这里看到一种突然的变化或发展，一种不测的发展。炉火不像流水那样单调和抽象，它甚至比我们在树丛的鸟窝中天天看到的雏鸟生长和变化得更快，它使人联想到，人们渴望变化，加快时间的流逝，使全部生活了结而死后重生。在这种情况下冥思遐想变得真正奇妙而富有戏剧性；它把人的命运放大了；它以小寓大，把炉子变为火山，把圆木头的生命扩大为世态。在奇境中的人能听到火葬柴堆的召唤。此刻对于他而言，变化不如毁灭，毁灭即是新生。……

爱、死亡和火焰在同一瞬间分离了。蜉蝣在火焰的中心献出身躯使我们感受到它的永生。这种不留痕迹的彻底死亡正是我们人的整体离开现世过渡到来生的保证。有得必有失。炉火给予人的启示是明确无误的："你用技能、爱或暴力获取了一切之后，你得全部放弃，你得消灭你自己。"（Gaston Bachelard.*The Psychoanalysis of Fire*，Boston：Beacon Press，1964，pp.14~16. Orignally published as *La Psychanalyse du Feu*，Librarie Gallimard，1938.Reprinted by permission of Beacon Press.）

另一认为炉火是必需的比较现实的观点来自菲尔德夫人，她的话被引用在罗伯特·肯尼迪所著《住宅及其设计艺术》（*The House and the Art of Its Design*）一书中（New

dessert-that fire shows itself a friend of man.It does not confine itself to cooking;it makes things crisp and crunchy.It puts the golden crust on the griddle cake;it gives a material form to man's festivities.As far back in time as we can go,the gastronomic value has always been more highly prized than the nutritive value,and it is in joy and not in sorrow that man discovered his intellect.The conquest of the superfluous gives us a greater spiritual excitement than the conquest of the necessary. Man is a creation of desire,not a creation of need.

But the reverie by the fireside has axes that are more philosophical. Fire is for the man who is contemplating it an example of a sudden change or development and an example of a circumstantial development.Less monotonous and less abstract than flowing water,even more quick to grow and to change than the young bird,we watch every day in its nest in the bushes,fire suggests the desire to change,to speed up the passage of time,to bring all of life to its conclusion,to its hereafter.In these circumstances the reverie becomes truly fascinating and dramatic;it magnifies human destiny;it links the small to the great,the hearth to the volcano,the life of a log to the life of a world.The fascinated individual hears the call of the funeral pyre. For him destruction is more than a change,it is a renewal...

Love,death and fire are untied at the same moment.Through its sacrifice in the heart of the flames,the mayfly gives us a lesson in eternity.This total death which leaves no trace is the guarantee that our whole person has departed for the beyond.To lose everything in order to gain everything.The lesson taught by the fire is clear: "After having gained all through skill,through love or through violence you must give up all,you must annihilate yourself." (Gaston Bachelard,*The Psychoanalysis of Fire*,Boston:Beacon Press,1964,pp.14-16.Originally published as *La Psychanalyse du Feu*,Librarie Gallimard,1938. Reprinted by permission of Beacon Press.)

Another,more down-to-earth,view of the need for fire

York：Reinhold，1953, pp.192~193.）：

在冬月，孩子们常常只能待在屋里做游戏，于是时常发生这种情况：在四点钟左右或稍晚一些时候，他们在游戏室里变得暴躁，爱发脾气，或因无聊而撒野，几乎有点歇斯底里。这时我在起居室的壁炉里生起了火，让孩子们到那儿去观火；如果不把炉火点着，他们会继续吵闹，而且可能把这个安静的房间变成另一个疯人院，但当壁炉上燃烧的火焰一上来，他们立即松弛下来，对火产生兴趣。炉火让他们看得出神，有人讲一个故事逗引大家，于是他们就安静下来了，使我得以脱身去做晚饭给他们吃。炉火无疑具有催眠性质，我们须加以充分利用。

当然，我们必须面对这样的事实，即在世界上许多地方用木材或煤炭烧火是不利于生态的。柴火和煤火污染空气；它们的热力不足；它们会使木材储备枯竭。如果我们想保持在家里烧火的习惯，我们得设法来补救木材燃料的不足。比如，我们可以养成烧一些可燃物的习惯，这些东西可以是家庭四周和整个社区的废料——它们是纸、布、没有加氯的塑料、木屑和锯末。总之，如果我们希望从壁炉上得到感情上的安慰，我们得学会集中使用壁炉，从我们左邻右舍处收集那些不用即作废料处理的材料作为我们自己的燃料。不难想象，人们可以在家中用一个简单的手工操作的夹具把这种废物压成结实的"圆木头"，烧出真正的炉火来。

现在，假定我们会有一种壁炉——也许是一种十分简单的壁炉，但是是明火。我们把它安置在哪里呢？有四点可以考虑。

1. 毫无疑问，主要壁炉应设在住宅的公用区。它会把大家吸引到这个地方来，当壁炉燃烧时，能够给谈话以伴奏。

2. 然而，壁炉应该在经过这个房间的人和在相邻房间特别是在厨房的人的视线之内。炉火会吸引人进来，使一家人更有可能相聚。同时，在进出房间时看见炉火是

comes from Mrs.Field,quoted in Robert Woods Kennedy, *The House and theArt of Its Design*,New York:Reinhold, 1953,pp.192-193:

During the winter months,when the children are often confined indoors for their play,it often happens that around four o-clock or a little after they become cross and grumpy in their playroom,or wild and almost hysterical with boredom.Then I light a fire in the living-room fireplace,and send the children in there to watch it;if the fire were not lighted they would continue their quarreling and perhaps try to turn the quiet room into another bedlam,but with the burning flames on the hearth,they relax into easy interest.They see things in the fire,someone tells a story that interests the whole group,they quiet down,leaving me free to prepare the supper and serve it.It has a definitely hypnotic quality that can be turned to good account.

Of course,we must face the fact that in many parts of the world wood and coal fires are ecologically unsound.They pollute the air;they are inefficient for heating;they are a drain on wood reserves.If we wish to maintain the habit of burning fires in the home,we shall have to find a way of supplementing wood fuel. For example,we can cultivate the habit of burning the inflammable materials that become waste around the house and throughout the community-paper,cloth,non-chlorinated plastics,wood scraps and sawdust.In short,if we want the emotional comfort that can be drawn from a fireplace,we shall have to learn to use the fireplace in a concentrated way,producing our own fuel from materials that would otherwise go to waste in our neighborhoods.It is easy to imagine,a simple hand press which people can use in their homes to press this waste into dense "logs" to make the fire more substantial.

Assume then that we are to have some kind of fireplace—

很有意思的。最合适生火的时间是在傍晚，这时全家聚在一起吃晚饭；这一活动会在厨房和炉火之间起到平衡作用。

3. 还务必要有一个空间使人们能坐在炉火前；而这一空间不能被门和门之间或相邻房间之间的通道所占据。

4. 在炉火不燃时，切记不要将炉前变为一片死寂。熄了火的壁炉，满膛炉灰，黑咕隆咚，会促使人们把它前面的椅子搬走，除非炉子燃烧时椅子向着炉火，炉子熄灭时椅子又可以向着别的东西——一个窗户、一种活动或一处景物。只有这样，不管壁炉燃烧或熄灭，围着炉火摆放的一圈椅子都能始终维持原状而使这个空间富有生气。

白天令人感兴趣的事物
A daytime focus

因此：

要在公用空间——多半在厨房——燃起熊熊炉火，一家人可以围着壁炉谈天说地，也可对着它遐想和沉思。妥善安排壁炉的位置，务使炉火起到沟通其周围公用空间的作用，使身处公共空间中每一处角落的人都能瞥见熊熊的炉火；并置一窗户或布置其他令人感到兴趣的事物，使炉火熄灭时炉前空间依然有着生气。

perhaps something entirely simple,but an open fire nonetheless. Where shall we put it?There are four points to consider:

1.Certainly,the main fireplace should be located in the common area of the house.It will help to draw people together in this area,and when it is burning,it provides a kind of counterpoint to conversation.

2.However,the fireplace should be in view for people passing through the room and people in adjoining rooms, especially the kitchen.The fire will tend to pull people in and make it more likely for the family to gather.And also it is good to view the fire in passing.A welcome time for a fire is in the evening when the family is gathering for the evening meal;and the activity tends to balance between the kitchen and the fire.

3.Make certain,too,that there is a space where people can sit in front of the fire;and that this space is not cut by paths between doors or adjacent rooms.

4.And be sure that the fire is not a dead place when the fire is not burning.A fireplace without a fire,full of ashes and dark,will turn the chairs away,unless the chairs which face the fire when it is lit face something else-a window,or activity,or a view-when it is not lit.Only then will the circle of chairs which forms around the fire be stable and keep the place alive,both when the fire is burning and when it isn't.

Therefore:

Build the fire in a common space-perhaps in the kitchen-where it provides a natural focus for talk and dreams and thought.Adjust the location until it knits together the social spaces and rooms around it,giving them each a glimpse of the fire;and make a window or some other focus to sustain the place during the times whenthe fire is out.

甚至在传统的明火壁炉已无人用其取暖的地方，或在燃料短缺的地方，设法把废料、纸张、木屑和硬纸板变成可燃的芳香的圆木，大多混合以天然树脂，用家制的压钳压制。拿所有这些不能用作**堆肥**（178）的干燥有机物来生火，结果家庭里用剩的所有废料都能派上用场，或作肥料，或作燃料；无疑，炉火的灰烬就可用作堆肥。让坐椅围着炉火排成一圈——**坐位圈**（185）；这些椅子多半可以构成**窗前空间**（180）。

Even where the traditional open fireplace is obsolete for heating or where fuel is scarce,find some way of converting refuse,paper,scraps of wood and cardboard into logs which can be burned,and which smell good—perhaps with some kind of natural resin in a home-made press.Burn all the dry organic materials that do not go to the COMPOST(178),so that the leftovers from the materials which come into the house all serve a useful function,either as fertilizer or as fuel;indeed,the ashes from the fire may go into the compost.Make a circle of chairs around the fire—SITTING CIRCLE(185);perhaps these chairs include a WINDOW PLACE(180).

模式182　进餐气氛

　　……我们已经指出，各种方式的共同进餐对促使维持一群人之间的关系方面起着多么重要的作用——**共同进餐**（147）；我们还提出过，如何能够使共同用餐在厨房内进行——**农家厨房**（139）。本模式讨论进餐气氛的一些细节。

✽✽✽

　　当人们在一起吃饭的时候，他们会在心灵上真正彼此接近——不然的话他们会感到各不相干。某些空间有一种吸引人的力量，使人们觉得在这里进餐悠闲自得，彼此亲近，而另一些空间却迫使人尽快填饱肚子好离开此地到别处去图个痛快。

　　最要紧的是，当桌子上方灯光均匀，桌子周围的墙壁上光线的强度也完全一样时，这样的灯光是不会把人们聚集在一起的；感情的浓度势必消失；人们不会觉得这是一种专门的聚会。但当灯光柔和，灯低挂在餐桌上方，周围墙壁幽暗，使这一点光线照着人们的脸上，使它成为这一群人的中心点，那时这顿饭就非同一般，它联络了众人，沟通了感情。

　　因此：

　　在进餐空间的中心放置一张大餐桌——大到足可供全家或一群人使用。在桌子上方悬置一灯，在进餐的人头顶产生投光区域，这一空间周围是墙壁或明显地比较幽暗。此空间应足够宽大，以便自由地往后挪动椅子，桌子近旁有搁架和台子供放置进餐用的物品。

182   EATING ATMOSPHERE

...we have already pointed out how vitally important all kinds of communal eating are in helping to maintain a bond among a group of people—COMMUNAL EATING(147);and we have given some idea of how the common eating may be placed as part of the kitchen itself—FARMHOUSE KITCHEN (139).This pattern gives some details of the eatig atmosphere.

❧❦

When people eat together,they may actually be together in spirit-or they may be far apart.Some rooms invite people to eat leisurely and comfortably and feel together,while others force people to eat as quickly as possible so they can go somewhere else to relax.

Above all,when the table has the same light all over it,and has the same light level on the walls around it,the light does nothing to hold people together;the intensity of feeling is quite likely to dissolve;there is little sense that there is any special kind of gathering.But when there is a soft light,hung low over the table,with dark walls around so that this one point of light lights up people's faces and is a focal point for the whole group,then a meal can become a special thing indeed,a bond,communion.

Therefore:

Put a heavy table in the center of the eating space—large enough for the whole family or the group of people using it. Put a light over the table to create a pool of light over the

中心的灯光
light in the middle

∞⟡∞

关于灯光的详细情况请参阅**投光区域**（252）；选好色彩使这个空间在夜间也能够显得温暖、幽暗、舒适——**暖色**（250）；在附近放置若干软椅——**各式坐椅**（251）；或设**嵌墙坐位**（202），坐位上有一些大的软垫靠在一面墙上；在供存放东西的地方，设置**敞开的搁架**（200）和半人高的**搁架**（201）……

group, and enclose the space with walls or with contrasting darkness.Make the space large enough so the chairs can be pulled back comfortably,and provide shelves and counters close at hand for things relate to the meal.

∞∞∞

Get the details of the light from POOLS OF LIGHT (252);and choose the colors to make the place warm and dark and comfortable at night—WARM COLORS(250);put a few soft chairs nearby—DIFFERENT CHAIRS(251);or put BUILT-IN SEATS(202)with big cushions against one wall;and for the storage space—OPEN SHELVES(200)and WAIST-HIGH SHELF(201)...

模式183　工作空间的围隔**

　　……本模式在帮助创造一种使人们能够有效地进行工作的气氛方面起着极为重要的作用。你可以一点一点地用它来为工作空间创造一些较大的模式，如**灵活办公空间**（146）、**半私密办公室**（152）和**家庭工作间**（157）。当然，它也可用以帮助完善这些较大的模式，如果你已经把它们列入设计之中。甚至在与家庭公用区隔开的凹室里——**凹室**（179），你也可以依据本模式，将周围加以围隔，使之更适合于工作。

183   WORKSPACE ENCLOSURE**

...this pattern plays a vital role in helping to create an atmosphere in which people can work effectively.You can use it piecemeal to generate the larger patterns for workspace like FLEXIBLE OFFICE SPACE(146),HALF-PRIVATE OFFICE(152),and HOME WORKSHOP(157).Or,of course,it can be used to help complete these larger patterns,if you have already built them into your design.Even in an alcove off the family commons-ALCOVES(179),you can make the workspace more suitable for work,by placing and shaping the enclosure immediately around it according to this pattern.

᛭ᛒᚩᚳᛋ

People cannot work effectively if their workspace is too enclosed or too exposed.A good workspace strikes the balance.

In many offices,people are either completely enclosed and feel too isolated,or they are in a completely open area as in the office landscape and feel too exposed.It is hard for a person to work well at either of these two extremes—the problem is to find the right balance between the two.

To find the proper balance,we conducted a simple experiment.We first defined 13 variables,which we thought might influence a person's sense of enclosure in his workspace.

These 13 variables are:

1.Presence or absence of a wall immediately behind you.

2.Presence or absence of a wall immediately beside you.

如果人们的工作空间过于封闭或过于开敞，他们都不能有效地工作。一个良好的工作空间要做到使这两者兼顾。

在许多办公室，人们或者完全封闭而感到与世隔绝；或者完全开敞，如同在景观办公室，感到过于暴露。这两种极端的情况中无论哪一种，都难以使人安心地工作——问题就在于如何找到这两者之间恰当的平衡。

为找到适当的平衡，我们做了一个简单的实验。首先我们规定 13 种可变因素，我们认为这些因素会影响一个人在其工作空间中对封闭程度的感觉。

这 13 种可变因素是：

1. 靠近你的后背处有没有墙。

2. 靠近你的旁边处有没有墙。

3. 你的前方有多大的空地。

4. 工作空间的面积。

5. 工作空间的周围围隔的总量。

6. 向外的视野。

7. 同最近的人的距离。

8. 在你的工作地点你能感觉到的人数。

9. 噪声：级别和类型。

10. 有无直接面对着你的人。

11. 有多少不同的位置你可以坐下来。

12. 在你的工作地点你能看到的人数。

13. 你不用提高嗓门就可与之谈话的人的数目。

然后我们作出 13 种假设来把这些可变因素跟工作地点的舒适程度联系起来。这些假设在下面列表说明。我们走访了 17 个人，有男有女，他们在不同的办公室工作。在访问的时候，我们首先请每个人想一想他（或她）曾经工作过的最好的工作空间和最差的工作空间。然后请他（或她）

3.Amount of open space in front of you.

4.Area of the workspace.

5.Total amount of enclosure around the immediate work space.

6.View to the outside.

7.Distance to nearest person.

8.Number of people you are aware of from your workplace.

9.Noise:level and type.

10.Presence or absence of a person facing you directly.

11.Number of different positions you can sit in.

12.Number of people you can see from your workspace.

13.The number of people you can talk to without raising your voice.

We then formulated thirteen hypotheses which connect these variables with the comfort of the work space.The hypotheses are listed below.We interviewed 17 men and women who had all worked in several different offices.In the interview,we first asked each person to think of the very best workspace he(or she)had ever worked in and the very worst;and then asked him(or her)to make a sketch plan of both spaces. Then we asked questions to identify the value of each of these 13 variables in the "best" and "worst" workspaces.Thus,for instance,we might point to one of the sketches a person had drawn,and say "How far away was that wall" to establish the value of the third variable.The values of the variables for the 17 best and worst workspaces are given in the following table.

On the basis of this table,we then calculated the probable significance of our hypotheses,according to the chisquared test.Nine of the hypotheses appear to be significant

画出这两种空间的草图。然后我们提出问题以确定在这些"最好"和"最差"的工作空间里那 13 个可变因素中的每一个的价值。例如，我们可能指向某人所画的草图中的一张问道："那道墙离得多远？"以确定第三项可变因素的价值。这 17 个最好和最差的工作空间的可变因素的价值可通过一个列表来表示。

在这个表的基础上，根据 X 平方检定法，我们于是估算出这些假设可能具有的意义。根据 X 平方检定法，假设中有 9 个看来是有意义的，有 4 个没有多大意义。现在我们将这 9 个有意义的假设罗列出来，对其中每一个都在括号内试图解释它成功的原因。

1. 如果你后背处有墙，你在一个工作空间会感到舒服一些。（如果你的背部是暴露的，你会感到没有保护——说不准有谁在看着你，或有谁从后面往你走来。）数据表明这一假设有 1％的意义。

2. 如果一旁有墙，你在一个工作空间会感到舒服一些。（如果你的工作空间前面和两边都是开敞的，你会感到过于暴露。原因也许跟这样的事实有关，即你可能对你周围 180°范围内发生的事情大致有一个印象，但你如果不时转动你的脑袋，你就不能真正控制这样大的角度。如果你的旁边有一面是墙，你只需管住一个 90°的角，这就容易多了，因此，你会感到更安全一些。）数据表明这个假设的意义为 5％。

3. 在你前面 8ft 之内不应有无窗墙。（在你工作的时候，有时你想抬起头来看看，让你的眼光落在离工作台稍远的某处，使眼睛得到休息。如果 8ft 内有一堵无窗墙，你的视线没有变化，眼睛就得不到放松。在此情况下你会感到环境过于封闭。）数据表明这一假设的意义为 5％。

according to the chi-squared test,and four are not.We now list the nine "significant" hypotheses,and with each one,in parentheses,we venture an explanation for its success.

1.You feel more comfortable in a workspace if there is a wall behind you.(If your back is exposed you feel vulnerable—you can never tell if someone is looking at you,or if someone is coming toward you from behind.)The data support this hypothesis at the 1 percent level of significance.

2.You feel more comfortable in a workspace if there is a wall to one side.(If your workspace is open in front and on both sides,you feel too exposed.This is probably due to the fact that though it is possible to be vaguely aware of everything that goes on 180 degrees around you,you cannot feel in real control of such a wide angle without moving your head all the time.If you have a wall on one side,you only have to manage an angle of 90 degrees,which is much easier,so you feel more secure.)The data support this hypothesis at the 5 percent level of signifiCance.

3.There should be no blank wall closer than 8 feet in front of you.(As you work you want to occasionally look up and rest your eyes by focusing them on something farther away than the desk.If there is a blank wall closer than 8 feet your eyes will not change focus,and they get no relief.In this case you feel too enclosed.)The data support this hypothesis at the 5 percent level of significance.

4.Workspaces where you spend most of the day should be at least 60 square feet in area.(If your workspace is any smaller than 60 square feet you feel cramped and claustrophobic.)The data support this hypothesis at the 5 percent level of significance.

对每种假设的可变因素的估价

问题序号	工作空间背后有墙吗 1		工作空间旁边有墙吗 2		前面8ft之内有墙吗 3		工作空间的面积 4		围隔程度 5		能否看到外界 6		私人谈话会不会被人听到 7		感觉得到的人数 8		是否被各种噪声打扰 9		引起别人注意吗 10		工作时能看到几种景致 11		看得见的人数 12		可与之谈话的人数 13	
	B	W	B	W	B	W	B	W	B	W	B	W	B	W	B	W	B	W	B	W	B	W	B	W	B	W
托尼	Y	N	Y	N	N	N	35	20	50	37	Y	N	—	—	2	1	Y	Y	N	N	1	1	2	1	2	1
艾琳	Y	N	Y	N	Y	N	135	35	60	0	Y	Y	Y	N	3	4	Y	Y	N	N	3	2	2	2	0	1
埃菲	Y	N	Y	N	Y	Y	150	35	86	25	Y	Y	Y	N	4	5	Y	Y	N	N	2	2	0	5	0	3
佩奇	Y	N	Y	N	N	N	90	36	50	2	Y	N	Y	N	8	16	Y	N	Y	Y	4	2	8	16	0	1
罗恩	Y	N	Y	N	Y	Y	63	32	57	44	Y	N	Y	N	8	1	Y	N	N	N	2	1	4	1	4	1
琼	Y	N	N	N	Y	N	120	20	50	0	Y	N	Y	N	3	9	Y	N	N	Y	2	1	3	9	3	2
莱斯利	Y	N	Y	N	N	N	50	20	25	0	Y	Y	Y	Y	10	0	Y	N	N	N	3	1	9	0	7	0
弗吉尼亚	Y	N	N	N	N	N	80	20	50	25	Y	N	Y	N	3	50	Y	N	N	N	1	4	3	8	1	4
弗朗	N	N	N	Y	Y	Y	100	50	55	50	Y	Y	N	N	2	2	N	N	N	N	1	1	1	2	1	1
登多尔	Y	Y	Y	Y	Y	N	20	20	20	50	Y	Y	N	N	4	150	Y	Y	N	N	1	1	4	2	4	1
菲艾斯	Y	Y	Y	Y	N	N	70	150	50	0	Y	Y	Y	Y	3	1	N	N	N	N	3	3	0	15	0	1
玛丽娜	Y	N	Y	Y	Y	Y	20	20	37	25	Y	N	Y	N	3	1	Y	Y	N	Y	2	2	2	0	2	0
弗雷德	N	Y	Y	N	N	N	400	20	75	25	Y	Y	Y	N	21	2	N	N	N	N	1	1	1	1	1	1
杰里	N	N	Y	N	N	Y	200	100	37	43	N	Y	N	Y	1	5	Y	N	N	Y	2	1	1	15	1	3
格里	N	N	N	Y	N	N	20	40	25	31	Y	N	N	Y	3	3	N	N	N	N	2	1	3	2	2	2
莱尔	Y	Y	Y	Y	N	Y	50	64	50	25	Y	Y	N	Y	2	60	Y	Y	N	N	2	1	2	60	2	1
（未标明）							100	112	75	95	Y	Y	Y	N	20	16	Y	Y	N	N	1	1	20	2	0	0

注：B—最好；W—最差；Y—是；N—否

Values of the variables for each hypothesis.

Question Number	1 Wall Behind in Workspace?		2 Wall to Side in Workspace?		3 Wall in Front Within 8'?		4 Area of Workspace?		5 Enclosure		6 View to Outside?		7 Aural Privacy?		8 No of People in Awareness		9 Bother by Noise of a Different Kind?		10 Catch Somebody's Eye?		11 No of Different Views While Working		12 No of People in Sight		13 No of People Available for Chat	
	B	W	B	W	B	W	B	W	B	W	B	W	B	W	B	W	B	W	B	W	B	W	B	W	B	W
Tony	Y	Y	Y	Y	Y	N	35	20	50	37	Z	Y	–	Y	2	1	Y	Y	Z	Y	1	1	2	1	2	1
Irene	Y	Y	Y	Y	N	Z	135	35	60	0	Y	Y	Z	Y	3	4	Z	Y	Z	Z	3	2	2	2	0	1
Effie	Y	Y	Y	Y	Y	Y	150	35	86	25	Y	Y	Z	Y	4	5	Y	Y	Z	Z	2	2	0	5	0	3
Peggy	Y	Y	Y	Y	N	Z	90	36	50	2	Y	Y	Z	Y	8	16	Y	Y	Z	Y	4	2	8	16	0	1
Ron	Y	Y	Y	Y	Y	Y	63	32	57	44	Z	Y	Z	Y	8	1	Y	Y	Z	Z	2	1	4	1	3	1
Joan	Y	Y	Y	Y	Y	Y	120	20	50	0	Y	Y	Y	Y	3	9	Z	Y	Z	Z	2	1	3	9	3	2
Leslie	Y	Y	Y	Z	Z	Z	50	20	25	0	Y	Y	Z	Y	10	0	Y	Y	Y	Z	3	1	3	0	7	0
Virginia	Y	Z	Z	Y	Z	Z	80	20	50	25	Z	Y	Z	Y	3	50	Y	Y	Z	Z	2	4	3	8	1	4
Fran	N	Z	Y	Y	Z	Y	100	50	55	50	Y	Y	Z	Y	2	2	Z	Y	Z	Z	1	1	1	2	1	1
Dendal	N	Y	Y	Y	Y	Z	20	20	25	50	Y	Y	Y	Z	4	150	Y	Z	Z	Z	3	3	4	2	4	1
Phyllis	Y	Y	Y	Y	Z	Z	70	150	50	0	Y	Y	Y	Y	3	1	Y	Y	Z	Z	2	2	0	15	0	1
Ina	Y	N	Y	Y	Z	Z	20	20	37	25	Y	Y	Z	Y	3	1	Y	Y	Z	Z	2	1	2	0	2	0
Mary	N	Z	Y	Y	Z	Z	400	20	75	25	Y	Y	Z	Y	21	2	N	Y	Z	Z	1	2	1	1	1	3
Fred	Y	N	Y	Y	Y	Y	200	100	37	43	N	Z	Y	Y	1	5	Y	Z	Y	Z	2	1	1	5	2	2
Jerry	N	N	Y	Y	N	Z	20	40	25	31	Y	Y	Y	Y	3	3	Z	Y	Z	Z	1	1	3	2	2	1
Gerry	Z	Z	Y	Y	Z	Y	50	64	50	25	N	Y	Y	Z	2	60	Y	N	Z	Z	2	1	2	60	0	0
Lyle	Y	Y	Y	Y	Y	N	100	112	75	95	Y	Y	Z	Y	20	16	Y	Y	Z	Z	1	1	20	2	0	0

* Best & Worst

4. 你一天度过大部分时间的工作空间面积至少要有 $60ft^2$。（如果你的工作空间小于 $60ft^2$，你会觉得空间太小而产生幽闭感。）数据表明这一假设的意义为 5%。

5. 每个工作空间应有 50%～75% 被墙或窗围隔。（我们估计，用窗户围隔会产生相当于用实心墙围隔一半的感觉，因此一个工作空间周围一半有墙，一半有窗，可以算作有 75% 的围隔，而一个周围全是半高墙，上面开敞的工作空间，其围隔程度被认为是 50%。）数据表明这一假设的意义为 1%。

6. 每个工作空间都应使人看到外面。[如果你看不到外面，即使你在一个大的开敞的办公室内工作，你也会感到过于闭塞和受到建筑物的压抑。**参阅俯视外界生活之窗** （192）。] 数据表明这个假设的意义为 0.1%。

7. 在你工作的空间 8ft 之内应该没有别人在工作。（你应当可以和别人通电话或跟人当面谈话而不会感到好像有人能听到你说的每句话。每个办公室的噪声平均水平为 45dB。噪声在 45dB 时，同你的距离在 8ft 之内的人实际上不能不听到你的谈话，见《噪声测量手册》（*Handbook of Noise Measurement* by Peterson and Gross，Sixth Edition，West Concord，Mass.：General Radio Company，1967）。数据表明这个假设的意义为 5%。

8. 在你工作的时候，若你感觉不到周围至少有其他两个人，这是令人不舒服的。另外，你也不愿意有 8 个以上的人在你感觉得到的范围之内。（如果你感觉得到周围有 8 个以上的人，你就不知道自己是在整个机构内的什么地方。你感到自己只是一架大机器上的一个齿轮。你接触的人太多了。另外，如果你感觉不到有任何人在你周围，你会感到孤独，似乎无人关心你和你的工作。在这种情况下，你太闭塞了。）数据表明这一假设的意义为 5%。

9. 你应该听不到同你的工作空间出来的噪声完全不同

5. Each workspace should be 50 to 75 percent enclosed by walls or windows.(We guess that enclosure by windows creates about half the feeling of enclosure that solid walls have,so that a workspace which is surrounded half by wall and half by window is considered to have 75 percent enclosure,while a workspace completely surrounded by a half-height wall and otherwise open,is considered to have 50 percent enclosure.)The data support this hypothesis at the 1 percent level of significance.

6.Every workspace should have a view to the outside.[If you do not have a view to the outside,you feel too enclosed and oppressed by the building,even if you are working in a large open office.See WINDOWS OVERLOOKING LIFE(192).]The data support this hypothesis at the 0.1 percent level of significance.

7.No other person should work closer than 8 feet to your workspace.(You should be able to hold conversations either on the phone or in person with someone,without feeling as though someone else can hear every word you are saying. The noise level in an average office is 45 dB.At 45 dB people closer than 8 feet to you are virtually forced to overhear your conversations.From the *Handbook of Noise Measurement* by Peterson and Gross,Sixth Edition,West Concord,Mass:General Radio Company,1967.)The data support this hypothesis at the 5 percent level of significance.

8.It is uncomfortable if you are not aware of at least two other persons while you work.On the other hand,you do not want to be aware of more than eight people.(If you are aware of more than eight people,you lose a sense of where you are in the whole organization.You feel like a cog in a huge machine. You are exposed to too many people.On the other hand,if you are not aware of anyone else around you,you feel isolated and

的噪声。（你的工作空间应该围隔得很好使它足以隔断跟你制造的噪声种类不同的噪声。有证据说明，如果一个人的周围都跟他干相同的工作，而不是干别的事情，那么此人能够更好地集中精力干一件工作。）数据表明这一假设的意义为5%。

我们检验过的假设中有四个通过统计资料表明不具有重要意义，它们是：

10. 不应该有人坐在你的正对面面向着你。

11. 工作空间应该使你有可能面向各个方向。

12. 从你的工作空间，你应至少能看到其他两个人，但不超过四个人。

13. 至少应该有另外一个人跟你坐得很近，可以同他说话。

因此：

每一个工作空间的面积至少应有 60ft^2。使每一个工作空间周围的墙和窗的总面积（把窗的面积折半计算）相当于假定这 60ft^2 四面全用实心墙围隔的50%～75%。使工作空间的前部至少有 8ft 的空地，向一个比它大的空间敞开。办公桌的摆放应使坐在桌前工作的人从前面或侧面看见室外风光。如有其他的人在附近工作，要进行适当的围隔，使一个工作人员感觉到自己同其他两三个人有着联系；但绝对不要在看得见和听得见的范围内布置 8 个以上的工作空间。

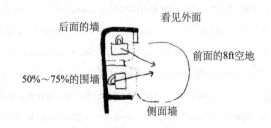

后面的墙　看见外面

前面的8ft空地

50%～75%的围墙

侧面墙

as though no one cares about you or your work.In this case,you are too enclosed.)The data support this hypothesis at the 5 percent level of significance.

9.You should not be able to hear noises very different from the kind of noise you make,from your workplace.(Your workplace should be sufficiently enclosed to cut out noises which are of a different kind from the ones you make.There is some evidence that one can concentrate on a task better if people around him are doing the same thing,not something else.)The data support this hypothesis at the 5 percent level of significance.

Four of the hypotheses we tested were not supported to a statistically significant extent by the data.They are the following:

10.No one should be sitting directly opposite you and facing you.

11.Workspaces should allow you to face in different directions.

12.From your workspace,you should be able to see at least two other persons;but no more than four.

13.There should be at least one other person close enough to talk to.

Therefore:

Give each workspace an area of at least 60 square feet. Build walls and windows round each workspace to such an extent that their total area(counting windows at one-half)is 50 to 75 percent of the full enclosure that would be there if all four walls around the 60 square feet were solid.Let the front of the workspace be open for at least 8 feet in front,always into a larger space.Place the desk so that the person working at it has a view out,either to the front or to the side. If there are other people working nearby,arrange the enclosure so that the

∞∞∞

　为了看见风景，使每一个工作空间都有一个开向外界的窗户——**俯视外界生活之窗**（192）；用内有搁架和储藏间的厚墙围住这个空间——**半敞开墙**（193）、**厚墙**（197）、**敞开的搁架**（200）、**半人高的搁架**（201）；在办公桌上方布置白炽灯光池给它照明——**投光区域**（252）；设法在工作空间旁安排一个休息处，这样可使一天的工作有节奏地进行，谈话有个方便的地方——**坐位圈**（185）。关于工作空间的形状，详见**室内空间形状**（191）……

person has a sense of connection to two or three others;but never put more than eight workspaces within view or earshot of one another.

For the view,give each workspace a window to the outside—WINDOWS OVERLOOKING LIFE(192);surround the space with thick walls which contain shelves and storage space—HALF-OPEN WALL (193),THICK WALLS(197), OPEN SHELVES(200),WAIS-THIGH SHELF (201);arrange a pool of incandescent light over the work table to set it off— POOLS OF LIGHT(252);and try to make a sitting place,next to the workspace,so that the pulse of work,and talk can happen easily throughout the day—SITTING CIRCLE(185).For details on the shape of the workspace,see THE SHAPE OF INDOOR SPACE(191)...

模式184　厨房布置*

184   COOKING LAYOUT*

...within the FARMHOUSE KITCHEN(139),or any other kind of kitchen,it is essential that the cooking area be fashioned as a workshop for the preparation of food,and not as some kind of magazine kitchen with built-in counters and decorator colors. This down-to-earth and working character of a good kitchen comes in large part from the arrangement of the stove and food and counter.

൝൝

Cooking is uncomfortable if the kitchen counter is too short and also if it is too long.

Efficiency kitchens never live up to their name.They are based on the notion that the best arrangement is one that saves the most steps;and this has led to tiny,compact kitchens. These compact layouts do save steps,but they usually don't have enough counter space.Preparing dinner for a family is a complex operation;several things must go on at once,and this calls for the simultaneous use of counter space for different projects.If there isn't enough counter space,then the ingredients and utensils for one thing must be moved,washed,or put away before the next thing can be prepared;or else things become so jumbled that extra time and effort must be taken to find what's needed at the proper moment.On the other hand,if the counter is too long or too spread out,the various points along its length are too far apart—and cooking is again uncomfortable,because your movements as you cook are so inefficient and slow.

Empirical support for the notion that there is insufficient

……在**农家厨房**（139）或任何其他一种厨房内部，重要的是使烹调区成为适合于制作食物的工作间，而不是某种带有固定工作台和色彩协调的画报中见到的厨房。一个好的厨房之所以实用和工作效率高，多半因为对炉灶、食物和工作台进行恰当的布置。

❧❧

厨房工作台太短或太长都会使烹调工作做起来不舒服。

所谓实效厨房从来名不符实。它们所依据的想法是，最好的安排就是使步骤尽量简便；结果就使厨房变得小巧而紧凑。这些紧凑的布置确实节省步骤，但它们通常没有足够的工作台空间。给一家人做顿正式的晚饭是一项复杂的工作；几样事情要一起做，而这要求同时使用工作台来完成不同的工序。如果工作台空间不够，用来制作某种食品的原料和用具都得挪动、清洗或放在一旁才能准备下一道食物；不然，事情就会搞乱，结果须在适当的时候另花时间和精力去找出所需要的东西。另外，如果工作台过长或铺得很开，台面上各点的距离过大——做起饭菜来又会感到不舒服，因为你在做饭菜时的动作效率低而且缓慢。

伊利诺伊大学的"小家庭协会"在其最新发表的作品中，根据实验认定，提出许多厨房工作台太小这一意见是正确的。这个协会发现，在一百多个住宅区里，67%的厨房工作台空间被认为太小。但没人抱怨说他们的厨房太大。

在《自己建造家园》（*The Owner Built Home* Yellow Springs, Ohio, 1961, Volume IV, p.30）中，肯·克恩指出，在厨房设计中一个基本概念是在厨房的每一个主要的食品制作中心提供储藏处所和工作空间。他根据康奈尔大学的一项研究，把主要的中心定为水槽、炉灶、冰箱、调制食品和端送饭菜的地方。给每一中心提供储藏处所要求有

counter space in many kitchens comes from a recent work by the Small Homes Council,University of Illinois.The Council found that in over a hundred housing developments,67 per cent had too little counter space.No one complained that their kitchens were too large.

In The Owner Built Home(Yellow Springs, Ohio, 1961,Volume IV,p.30),Ken Kern notes that a principal concept in cooking design is to provide for storage and workspace at each of the major cooking centers in the kitchen.Drawing on a Cornell University study he identifies the major cooking centers as the sink,the stove,the refrigerator,the mixing,and the serving areas.To provide storage for each center requires 12 to 15 feet of free counter space,excluding the sink,drainboards,and stove. (*The Cornell Kitchen*,Glenn Beyer,Cornell University,1952.)

As far as the limits on the distance between these major cooking centers are concerned,there is less empirical evidence. Estimates vary.The rule of thumb we postulate is that no two of them should be more than three or four steps,or about 10 feet,apart.

Therefore:

To strike the balance between the kitchen which is too small,and the kitchen which is too spread out,place the stove,sink,and food storage and counter in such a way that:

1. No two of the four are more than 10 feet apart.

2. The total length of counter-excluding sink,stove,and refrigerator-is at least 12 feet.

3.No one section of the counter is less than 4 feet long.

There is no need for the counter to be continuous or entirely "built-in"as it is in many modern kitchens-it can even consist of free-standing tables or counter tops.Only the three functional relationships described above are critical.

12～15ft 的自由工作台空间，不包括水池、滴水板和炉灶。（*The Cornell Kitchen*，Glenn Beyer，Cornell University，1952.）

关于厨房内几个主要的中心之间应有多大距离限度，实验证据不多。大家估算不一。我们所采用的经验公式是其中任何两个之间的距离不应超过三步至四步，或 10ft 左右。

一个真正实效的厨房：很大，又很实用
A kitchen that really works:huge,but great.

因此：

为了使厨房不致太小，也不致过分扩大，应把炉灶、水槽、食物储藏处和工作台以这样的方式来布置：

1. 这四个物体中任意两个之间不应相距 10ft 以上。

2. 工作台的总长度——不包括水槽、炉灶和冰箱——不应小于 12ft。

3. 工作台的任何一部分不短于 4ft。

工作台没有必要连成一片或如在许多现代厨房中那样完全"固定"——它甚至可以是独立的桌子或台面。唯有上述这三种功能关系才是最为重要的。

꙰

Place the most important part of the working surface in the sunlight—SUNNY COUNTER(199);put all the kitchen tools and plates and saucepans and nonperishable food around the walls,one deep,so all of it is visible,and all of it directly open to reach—THICK WALLS(197),OPEN SHELVES(200)...

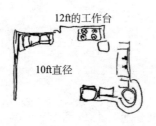

12ft的工作台

10ft直径

&C&

　　要让工作台台面上的最重要部分照到阳光——有阳光的**厨房工作台**（199）；把所有的厨房用具、盘子、小平底锅和不易腐烂的食品放在墙壁周围，一排进深，这样所有的东西都看得见，所有的东西都可直接取用——**厚墙**（197）、**敞开的搁架**（200）……

模式185 坐位圈*

　　……根据**起居空间的序列**（142），在整个办公楼、住宅或工作间会有一系列各种不同的起居空间——一些是正式的，一些是非正式的，一些大，一些小，这部分地是根据**私密性层次**（127）来安排的。本模式研究这些起居空间中任何一个空间的实际的形状布置。当然，可以利用它以有助于创造起居空间的序列，一次一个空间，逐个地进行。

✄

　　一批椅子、一张沙发或一张椅子、一摞垫子——这是

185   SITTING CIRCLE*

...according to the SEQUENCE OF SITTING SPACES
(142),there will be a variety of different kinds of sitting space
throughout an office building or a house or workshop—some
formal,some informal,some large,some small,laid out in part
according to the INTIMACY GRADIENT(127).This pattern
deals with the actual physical layout of any one of these sitting
spaces.And of course,it can be used to help create the sequence
of sitting spaces,piecemeal,one space at a time.

&⊃⊂&

**A group of chairs,a sofa and a chair,a pile of cushions-
these are the most obvious things in everybody's life-and yet
to make them work,so people become animated and alive in
them,is a very subtle business.Most seating arrangements
are sterile,people avoid them,nothing ever happens there.
Others seem somehow to gather life around them,to
concentrate and liberate energy.What is the difference
between the two?**

Most important of all,perhaps,is their position.A sitting
circle needs essentially the same position as a COMMON
AREA AT THE HEART(129),but in miniature:a well defined
area,with paths running past it,not cutting through it,and
placed so that people naturally pass by it,stop and talk,lean on
the backs of chairs,gradually sit down,move position,get up
again.These characteristics are vital.The reasons are exactly
the same as those given in COMMON AREAS AT THE
HEART(129);only the scale is different.

平常生活中司空见惯的东西——但要使它们发挥作用，让人们坐在椅子上兴高采烈，生气勃勃，那是一件巧妙的工作。大多数的座位布置得令人感到枯燥无味，人们避开它们，那里无人问津。另外一些则看来能够吸引人，使人精神集中，精力充沛。这两者之间的区别何在？

首要的也许是它们的位置。一个坐位圈基本上需要与**中心公共区**（129）具有相同的位置，但地方小一点：一个物色得合适的地区，有过道从这里经过，但并不穿过它，而是使人们自然地绕着它走，停下来说说话，在椅背上靠一会儿，逐渐地坐了下来，挪挪位子，又站了起来。这些特征是极为重要的。原因同**中心公共区**（129）里提出的那些完全一样，唯一区别是规模不同。

其次是坐位圈的大体形状。当人们坐下来一起谈话时，他们总想大致把他们自己围成圆圈。玛格丽特·米德对此提出过实验证据（"Conference Behavior," *Columbia University Forum*, Summer, 1967, pp.20~25）。围成圆圈而不是其他形状的一个理由也许是人们坐着的时候喜欢彼此形成一个角度，而不是并排而坐（Robert Sommer, "Studies in Personal Space," *Sociometry*, 22 September 1959, pp.247~260）。在一个圆圈里，甚至相邻的人彼此也略有角度。这跟第一点结合起来表明，大致围成圆圈是最合适的。

但光是椅子围成圆圈是不够的。只有当实际的建筑——柱子、墙壁、壁炉、窗户——形成一个大致圆圈形的局部闭合和固定的地区时，椅子才能维持这样的状态。壁炉特别有助于形成一个坐位圈。其他一些东西差不多也同样可起这种作用。

第三，我们注意到，坐位需要安排得稍微松动一点——不要太规整。使许多各式各样的沙发、坐垫和椅子都可以随便挪动这种比较松动的安排会使坐位圈显得生动活泼。

Second,the rough shape of a circle.When people sit down to talk together they try to arrange themselves roughly in a circle.Empirical evidence for this has been presented by Margaret Mead ("Conference Behavior," *Columbia University Forum*,Summer,1967,pp.20-25.).Perhaps one reason for the circle,as opposed to other forms,is the fact that people like to sit at an angle to one another,not side by side(Robert Sommer, "Studies in Personal Space," *Sociometry*,22 September 1959,pp.247-260.)In a circle,even neighbors are at a slight angle to one another.This,together with the first point,suggests that a rough circle is best.

But it is not enough for the chairs to be in a circle.The chairs themselves will only hold this position if the actual architecture-the columns,walls,fire,windows—subtly suggest a partly contained,defined area,which is roughly a circle.The fire especially helps to anchor a sitting circle.Other things can do it almost as well.

Third,we have observed that the seating arrangement needs to be slightly loose—not too formal.Relatively loose arrangements,where there are many different sofas,cushions,and chairs,all free to move,work to bring a sitting circle to life. The chairs can be adjusted slightly,they can be turned at slight angles;and if there are one or two too many,all the better:this seems to animate the group.People get up and walk around,then sometimes sit back down in a new chair.

Therefore:

Place each sitting space in a position which is protected, not cut by paths or movement,roughly circular,made so that the room itself helps to suggest the circle-not too strongly-with paths and activities around it,so that people naturally

椅子的位置可以稍稍调整，它们可以转动一个小的角度；如果有一两张多余的，反而更好：看来这会使气氛更为活跃。人们站起来踱踱步，又可以不时地在一张新的椅子上重新坐下来。

因此：

使每一个座位空间都处于不受干扰的状态，它不在过道上，人们来回走动不会穿过这个地方，座位大体摆成圆圈形，还要使这个房间本身略呈圆形，周围有走道和人们的各种活动，这样当人们想要坐一会儿的时候，他们自然地会受这些椅子的吸引。把椅子和垫子松散地摆放成一圈，而且要放些备用的。

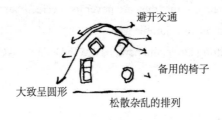

避开交通

备用的椅子

大致呈圆形

松散杂乱的排列

୫୦୯୧

利用壁炉、柱子和半敞开墙造成圆圈形——**炉火熊熊**（181）、**室内空间形状**（191）、**半敞开墙**（193）；但不要使它们过分正式或围合得过紧——**中心公共区**（129）、**起居空间的序列**（142）。使用**各式坐椅**（251），有大的、小的，有椅垫，还有些备用的，这样它们就不会排列得太整齐，而总有一点杂乱。设置一处**投光区域**（252）来显示座位圈的位置，它多半是**窗前空间**（180）……

gravitate toward the chairs when they get into the mood to sit.Place the chairs and cushions loosely in the circle,and have a few too man.

སུཇིཪྠ

Use a fire,and columns,and half-open walls to form thc shape of the circle—THE FIRE(181),THE SHAPE OF INDOOR SPACE (191),HALF-OPEN WALL(193);but do not make it too formal or too enclosed—COMMON AREAS AT THE HEART(129),SEQUENCE OF SITTING SPACES(142).Use DIFFERENT CHAIRS(251),big ones,small ones,cushions,and a few too many,so that they are never too perfectly arranged,but always in a bit of a jumble.Make a POOL OF LIGHT (252)to mark the sitting circle,and perhaps a WINDOW PLACE (180)...

模式186 共宿

……到现在为止供睡眠的区域已经规定好了——**夫妻的领域**（136）、**儿童的领域**（137）、**朝东的卧室**（138）、**多床龛卧室**（143）。剩下的只是固定具体的放置床位的地方——**夫妻用床**（187）、**床龛**（188）。然而，在我们考虑这些模式之前，我们想提出一个可能影响床的具体位置的更为普遍一些的模式。

✴✴✴

在许多传统的和原始的文化中，睡觉是公共性的活动，不像今天在西方世界那样会引起性方面的联想。我们认为它可能是非常重要的社会功能，如同共同进餐那样，是人们的基本需要，为人们所不可缺少。

例如，在印度农村，在干旱的季节里，男人们在黄昏时把他们的床搬到大院里，在那里一块儿聊天和抽烟，然后慢慢地睡着。这是社区中社交生活很重要的一部分。西方人的篝火晚会与此颇为相似：人们爱好野营说明那种要求还是人人都有的。

睡觉作为一项公共活动很可能是健康的社会生活的重要部分，不仅适用于儿童，也适用于所有的成年人。我们如何使这种需要同跟睡觉有联系的私密性和性活动这一明显的事实协调起来呢？

当然，在黎明和夜晚时分，夫妻在一起窃窃私语，一起入睡，共同醒来，充满柔情蜜意。但我们认为，有时候也有可能创造一种环境，使几家人可以在一起睡觉。

186   COMMUNAL SLEEPING

...by this time the sleeping areas have been defined—COUPLE'S REALM(136),CHILDREN'S REALM (137),SLEEPING TO THE EAST(138),BED CLUSTER(143). It remains only to build in the actual detailed space which forms the beds themselves-MARRIAGE BED(187),BED ALCOVE(188).However,before we consider these patterns,we wish to draw attention to a slightly more general pattern which may affect their detailed position.

❧❦

In many traditional and primitive cultures,sleep is a communal activity without the sexual overtones it has in the West today.We believe that it may be a vital social function, which plays a role as fundamental and as necessary to people as communal eating.

For instance,in Indian villages during the dry season the men pull their beds into the compound at sundown and talk and smoke together,then drift off to sleep.It is a vital part of the social life of the community.The experience of the campfire is the closest westem equivalent:people's love of camping suggests that the urge is stilla common one.

It is possible that sleep as a communal activity may be a vital part of healthy social life,not only for children,but for all adults.How might we harmonize this need with the obvious facts of privacy and sexuality that are linked with sleeping?

Of course,it is a beautifully intimate thing-the moment in the morning and at night when a couple are together,in

特别是，我们可以为大都市文化设想这种特殊的活动方式，因为在大都市朋友们经常各住一方，彼此距离遥远。下面这种情况你经历过多少次了呢？你和你的朋友晚上一起出去，然后回到他们的家中喝酒、谈话、拨火。最后，夜深了，该回家了。他们常常会说："请在这里过夜吧"——但实际上你不大会在那儿过夜。你谢绝了，然后拖着疲乏的身子，带着几分醉意驱车回家，到"你自己的床"上去睡觉。

在我们看来，特别在这样的情况下，同朋友一家共宿是入情入理的。它会有助于加强我们去看望住在远处的友人这样的社交活动。

但必须有提供共宿的环境，不然的话，我们就不会欣然同意这么做。人们在别处过夜总感到不方便，因为这通常意味着得准备客床，或睡在地毯上，或在沙发上拘束地过一夜。如果在住宅的主要卧室周围或公用区周围有些凹室，配备些床铺被褥，在夜阑人静时，供一两个客人在这里打个盹儿，想想，这有多吸引人。

从实用观点出发，这种凹室有两种可供选择的位置：

1. 可设置在公用区——不在每个人的私密空间——在晚会刚散、炉火将残的深夜，这里可供大家在一起睡个觉——让孩子们和父母一块儿睡，在这里过一个不平常的夜晚。这样做可能很简单：只需一个大垫子和一些毯子。

2. 另一解决办法是本模式的一个考虑得更加周到的做法：把住宅中的夫妻领域设计得比平常的稍大一些，有一两个凹室或可兼作床位的窗前座位。比如，一个嵌墙坐位，其宽度和长度都足可容人躺下，有一个薄的垫子铺在它上面，就可作床使用。有了若干这样的空间，夫妻卧室便随时可成为共宿的环境。

private,falling asleep or waking up together.But we believe that it is also possible to create a situation where,occasionally,people can sleep together in big,family-size groups.

In particular,we can imagine a special version of this activity for metropolitan culture,where so often friends live many miles away from each other.How many times have you experienced this situation:You have been out for the night with your friends and end up back at their house for drinks,to talk,to build a fire.Finally,late into the night,it is time to leave.Often they will say,"Please,spend the night"-but this rarely happens.You decline,and make the weary,half-drunken drive home to "your own bed."

It seems to us that under these conditions especially, communal sleeping makes sense.It would help to intensify the social occasions when we do see our friends who live far away.

But the environment must invite it,or we shall never overcome our reluctance.People are uneasy about spending the night because it usually means having to make up a guest bed,or sleeping on the rug,or cramped on the sofa.Think how much more inviting it would be if,at the end of the night,people simply dozed off,in ones and twos,in alcoves,and on mats with quilts,around the main sleeping area of the house,or around the commons.

From a practical point of view,there are two alternative positions for the alcoves:

1.There might be a place in the commons—not in any one person's private space—a place where late at night after people have been together for the evening and the fire is dying out,it is simple to draw together and sleep—a place where children and parents can sleep together on special nights.It could be very simple:one large mat and some blankets.

2.The other solution is a more deliberate version of the

在上述无论哪一种情况下，解决问题的办法必须是简单的，要做的事情不过是拿出毯子和垫子。如果得铺设专门的座位，需要重新布置房间，共宿是绝对做不得的。当然，放置客人用床的空间在不用于睡觉的时候也不要使它成为一个死角。要使它两种功能兼备——这个地方还可以放摇篮、当坐位、放衣服——**凹室**（179）、**窗前空间**（180）、**更衣室**（189）。

本模式初看起来令人觉得奇怪，但我们的打字员读了之后却被它迷住了，她决定同她一家人在一个星期六晚上试一试。他们在整个起居室铺了一个大垫子。全家人一块儿起床，帮小儿子去送报，然后吃点早点。编辑：他们还在这么干吗？作者：不，两周后他们被捕了。

现在言归正传：

把睡觉的地方布置成有可能使儿童和成年人睡在同一空间，彼此看得见和听得见，这至少可以令他们平常的睡觉习惯偶尔起些变化。

共宿可以安排在壁炉附近的公用区，一家人和客人们都可以一起睡在那儿——在每一个凹室里放一个大的垫子和一些毯子。也有可能在一个扩大的夫妻卧室里设些床龛供客人在这里过夜。

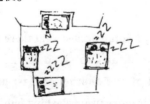

彼此看得见和听得见的床龛
beds within sight and sound of other beds

pattern:the couple's realm in a family house could be slightly larger than normal,with one or two alcoves or window seats that could double as beds.A built-in seat,for example,that is wide enough and long enough to lay down on,with a thin mat spread across it,becomes a bed.A few places like this,and,at a moment's notice the couple's bedroom becomes a setting for communal sleeping.

In either case,the solution must be simple and must involve nothing more than reaching for a blanket and a mat. If special beds must be made and the room rearranged,it will never happen.And,of course,the space for guest's beds must be made so that it is not dead when it is not used for sleeping. It needs a compatible double function—a place to put a crib,a scat,a place to lay out clothes—ALCOVE(179),WINDOW PLACE(180),DRESSING ROOM(189).

This pattern may seem strange at first,but when our typist,read it,she was fascinated and decided to try it one Saturday night with her family.They spread a big mat across the living room.They all got up together and helped the youngest son on his paper route;then they had some breakfast.Ed:Are they still doing it??Au:No,after 2 weeks they were arrested.

Seriously though:

Arrange the sleeping area so that there is the possibility for children and adults to sleep in the same space,in sight and sound of one another,at least as an occasional alternative to their more usual sleeping habits.

This can be done in the common area near the fireplace,where the entire household and guests can sleep together—one large mat and some blankets in an alcove.It is also possible to build bed alcoves for overnight guests,in an extended couple's ralm.

∽∞∾

　　相应地设置**凹室**（179）、**夫妻用床**（187）、**床龛**（188）
和**更衣室**（189）。儿童们自己已经有了本模式——如果他
们的卧室是多床龛的——**多床龛卧室**（143）……

   Place the ALCOVES(179)and MARRIAGE BED(187) and the BED ALCOVES(188)and DRESSING ROOMS(189) accordingly.The children have this pattern for themselves already-if bed alcoves are placed in a cluster—BED CLUSTER (143)...

模式187　夫妻用床

……**夫妻的领域**（136）这一模式强调家庭中夫妻私生活的重要性。在夫妻的领域内，床的布置和特征自然是最重要的事情。

⊱⊰

床是夫妻共同生活的中心：在这儿他们谈情说爱，共度良宵，有病时互相照顾。但床和卧室的布置方式常常无法显示它们的重要性，因而起不到这些作用。

187   MARRIAGE BED

...the pattern COUPLE'S REALM(136)gives emphasis
to the importance of the couple's private life together within a
household.Within that couple's realm,the placing and nature of
the bed is naturally the most important ting.

❧❧

**The bed is the center of a couple's life together:the
place where they lie together,talk,make love,sleep,sleep
late,take care of each other during illness.But beds and
bedrooms are not often made in ways which intensify their
meaning,and these experiences cannot take hold.**

It is true that there are extra wide beds,special bedspreads
and frames,water beds,soft lighting,and all kinds of accessories
on the night table.But these are all essentially gadgets.They still
don't make a bed which nourishes intimacy and love.

There are three far more basic points which go to establish
the marriage bed.

1.The space around the bed is shaped around the bed.There
is a low ceiling,or a partial ceiling,over the bed.The walls and
windows are made to contain the bed.See BED ALCOVE(188).

2.It is crucial that the couple choose the right time to build
the bed,and not buy one at the drop of the hat.It is unlikely that
the bed can come to have the right feeling until a couple has
weathered some hard times together and there is some depth to
their experience.

不错，卧室中有宽敞的床、特别考究的床单、特制的床框、电热温水褥、柔和的灯光，床头柜上夜间用品一应俱全。但这些从根本上说都是小东西。它们还不能构成体现亲热和爱情的夫妻用床。

构成夫妻用床须有三个远比这些重要的条件。

1. 在床的周围形成床旁空间。在床的上方天花板要低，或只有部分天花板。使床旁有墙壁和窗户。参阅**床龛**（188）。

2. 重要的是使夫妻俩选择适当的时间把床建造起来，而不要随便买一张了事。如果夫妻不共同度过一些艰难时光，使他们有着深切的体验，床是无法使他们建立起真正的感情的。

3. 设法给床和床旁空间增添点什么，使它随着岁月的流逝更加富有独特的个性；例如，一块雕上花饰，漆了又漆的床头板，一顶可以更换的绣花布帐。

床作为夫妻生活中的依靠的重要性从《荷马史诗》的以下段落里得到证明。奥德修经过 20 年的浪迹和灾祸之后回到家中。他的妻子潘奈洛佩不敢认他——冒名顶替的人太多了，而他离家的时间也太长了。他请求她相信这是他，但她拿不定主意。奥德修心灰意冷，只得转身离开她。潘奈洛佩开口了：

"你才奇怪呢。我并不是自高自大，也不是轻视你，也不是惊讶得不知所措。不过我还记得很清楚，你乘着长橹的船离开伊大嘉时是什么模样。尤吕克累，你去给他铺好那个结实的床，就是他自己手制的那个，把它放在我的卧房外面；你再在结实的床架上铺上褥子，再加上羊皮、外套和光滑的被单。"

她用这个话来试探她的丈夫。奥德修动了火：

"夫人，你说的话真叫人伤心，谁把我的床搬开了？就是有天大本领的人，也不容易做到这件事。只有一位天神才能随意把床搬走。一个凡人，不管他多么有力气，也不可能把它轻易移动，因为那个巧制的床藏着一个机关，那完全是我一个人设计的。院

3.Find a way of adding to the bed and the space around it, so that it will become more personal and unique over the years;for example,a headboard that can be carved, painted,repainted,or a cloth ceiling that can be changed, embroidered.

The importance of the bed as an anchor point in a couple's life is brought home in this passage from Homer.Odysseus is home after 20 years of wandering and misadventure.His wife,Penelope,does not recognize him—there have been so many imposters,and he has been away so long.He pleads with her to believe it is him,but she is unsure.Frustrated,Odysseus turns away from her.Penelope speaks:

"Strange man,I am not proud,or contemptuous,or offended,but I know what manner of man you were when you sailed away from Ithaca.Come Eurycleia,make the bed outside the room which he built himself;put the fine bedstead outside,and lay out the rugs and blankets and fleeces."

This was a little trap for her husband.He burst into a rage:

"Wife,that has cut me to the heart!Who has moved my bed?That would be a difficult job for the best workman,unless God himself should come down and move it.It would be easy for God,but no man could easily prize it up,not the strongest man living!There is a great secret in that bed.I made it myself,and no one else touched it.There was a strong young olive tree in full leaf growing in an enclosure,the trunk as thick as a pillar.Round this I built our bridal chamber;I did the whole thing myself,laid the stones and built a good roof over it,jointed the doors and fitted them in their places.After that I cut off the branches and trimmed the trunk from the root up,smoothed it carefully with the adze and made it straight to the line.This tree I made the bedpost.That was the beginning of my bed;I bored

子里曾经长了一株枝叶修长的橄榄树，健壮而茂盛，粗得像一根柱子；我在树的周围修建了卧室，用石头紧密筑成，又精巧地加上屋顶，装上紧密无缝的两扇门；然后我把这株枝叶修长的橄榄树从根砍断，去掉上面的枝条，用铜锛把树干细细锛平修直，作为床柱，又用钻子穿了孔；这样我做好床基，上面镶了金银和象牙，又铺上漂亮的紫色牛皮；这就是我的秘密设计。夫人，那个床还好好的在原处吗？还是有人把橄榄树根砍断，把床移到别处去了？"

他这样说；潘奈洛佩认出奥德修所说的可靠标志，坐在那里心神无主，四肢无力；她流着泪一直跑过去，用双手抱着奥德修的头颈，吻他的头，向他说道：

"奥德修，不要生我的气；你是比任何人都明白道理的。天神给我们许多痛苦；他们不愿意让我们在一起享受青春时光，直到老年临近。你现在不要生气，不要责备我吧，虽然我最初看见你的时候，我没有欢迎你。这是因为我总是不放心，生怕什么人到这里来说假话欺骗我；很多人是会作出邪恶的计策的。现在你说出了我们婚床的明显标记，你使我真正相信了。"（*The Odyssey*，W.H.D.Rouse 译。以上引文见中文本《奥德修记》，杨宪益译。）

英译者给这一事件作了以下的脚注："这是这个情节复杂的故事中奥德修第一次感情冲动的讲话；他对各式各样的结局早有准备，但这一未卜的小事却解开了他心中的疙瘩。"

老实说，这个模式是否有意义，我们没有把握。一方面，我们觉得它是有意义的：因为这是一个美妙的想法；几乎是牧歌式的。但是，面对冷酷严峻的事实和我们周围夫妻关系中的破裂和矛盾，看来很难希望这一模式会完全实现。我们决定把它保留，只不过因为这是一个美妙的想法。但我们请你把它当作奥勃洛摩夫的梦，一幅超现实的图画，一个完美的、田园诗般环境的不能实现的幻想，这样也许有助于使我们乱糟糟的生活现实稍具新意——但我们不可对此全信。

holes through it,and fitted the other posts about it,and inlaid the framework with gold and silver and ivory,and I ran through it leather straps coloured purple.Now I have told you my secret. And I don't know if it is still there,wife,or if some one has cut the olive at the root and moved my bed!"

She was conquered,she could hold out no longer when Odysseus told the secret she knew so well.She burst into tears and ran straight to him,throwing her arms about his neck.She kissed his head,and cried:

"Don't be cross with me,my husband,you were always a most understanding man!The gods brought affliction upon us because they grudged us the joy of being young and growing old together!Don't be angry,don't be hurt because I did not take you in my arms as soon as I saw you!My heart has been frozen all this time with a fear that some one would come and deceive me with a false tale;there were so many imposters!But now you have told me the secret of our bed,that settles it." (From *The Odyssey*,translated by W.H.D.Rouse.Reprinted by arrangement with The New American Library,Inc.,New York,New York.)

The translator footnotes this incident as follows: "This is the first time in all the eventful tale when Odysseus speaks on impulse;he has been prepared for everything,but this unexpected trifle unlocks his heart."

Quite honestly,we are not certain whether or not this pattern makes sense.On the one hand,it does:it is a beautiful idea;idyllic almost.Yet,face to face with cold hard fact and with the dissolution and struggles in the marriages around us,it seems hard to hope that it could ever be quite real.We have decided to leave it in,just because it is a beautiful idea.But we ask you to treat it like Oblomov's dream,a picture more

因此:

在夫妻生活中的适当时候,重要的是由他们自己来制造一张独特的床——用它维持他们生活中的亲密感情;床旁要稍有围隔,有一个低的天花或帐幔,一个属于它的空间;多半可以用许多小窗在床周围造成一个小空间。使床有它自己的某种独特形状,也许会是一张四柱卧床,床头架有多年手工雕刻的花纹或油漆的光泽。

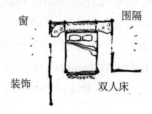

在夫妻用床附近设两个独立的更衣室——**更衣室**(189);关于床周围的空间,详见**床龛**(188);床上方的天花板稍低——**天花板高度变化**(190),给床的周围加以某种特殊的装饰——**装饰**(249)。关于床周围空间的形状,详见**室内空间形状**(191)⋯⋯

real than reality,an impossible dream of perfect and idyllic circumstances,which may help perhaps,to make a little more sense of our muddled everyday reality—but only if we take it with a pinch of salt.

Therefore:

At the right moment in a couple's life,it is important that they make for themselves a special bed-an intimate anchor point for their lives;slightly enclosed,with a low ceiling or a canopy,with the room shaped to it;perhaps a tiny room built around the bed with many windows.Give the bed some shape of its own,perhaps as a fourposter with head board that can be hand carved or painted over the years.

❧❧❧

Make two separate dressing rooms or alcoves near the bed—DRESSING ROOMS(189);for more details on the space around the bed,see BED ALCOVE(188);lower the ceiling over the bed—CEILING HEIGHT VARIETY(190),and provide some way of creating special ornament all around it—ORNAMENT(249).For the detailed shape of the space around the bed,see THE SHAPE OF INDOOR SPACE(191)...

模式188 床龛**

......床龛有助于形成**多床龛卧室**（143）、**共宿**（186）和**夫妻用床**（187）。对于儿童，每个床龛也可作为他们的**个人居室**（141），这样，即使在最小的住宅里，不仅成年人，甚至是每个儿童，至少都有一个可算作他自己的小处所。

188   BED ALCOVE**

...bed alcoves help to generate the form of BED CLUSTERS(143),COMMUNAL SLEEPING(186)and MARRIAGE BED(187).For children,each alcove also functions as AROOM OFONESOWN(141),so that even in the smallest house,not only the adults,but every child can have at least a small place to call his own.

⊗⊗⊗

Bedrooms make no sense.

The valuable space around the bed is good for nothing except access to the bed.And all the other functions-dressing,working,and storage of personal belongings which people stuff uncomfortably into the comers of their bedrooms-in fact,need their own space,and are not at all well met by the left over areas around a bed.

In BED CLUSTERS(143),we have already argued that each child in a family should have a bed alcove of his own,opening off a common play-space.This is based purely on the balance between community and privacy.We shall now try to establish the fact that,for everyone in the house,isolated beds,not only those in clusters,are better off in alcoves than in bedrooms.There are two reasons.

First,the bed in a bedroom creates awkward spaces around it:dressing,working,watching television,sitting,are all rather foreign to the side spaces left over around a bed.We have found that people have a hard time adapting the space around the bed to their needs for bedroom space.

卧室没有多大的意义。

床周围宝贵的空间除了上下床使用以外别无用处。所有其他的功能——穿衣服、工作、储放个人用品（人们总是把它们胡乱堆放在自己卧室的角落里）——事实上都需要有它们各自的空间，床周围空出的地方完全不能满足这些要求。

在**多床龛卧室**（143）中，我们已经论证过，一个家庭中的每个孩子都应有他自己的床龛，通向共同的游戏室。这完全是为了既满足共同性的需要，也满足私密性的需要。现在我们打算证实下面的论据：对于住宅中每个人来说，不仅那些在多床龛卧室中的床，而且单独的床，最好也放在凹室，而不放在卧室。理由有二。

首先，卧室中放着床，床周围的空间会让人感到别扭：穿衣服、工作、看电视、坐在椅子上，人们会觉得在床旁空间做这些事情都不合适。我们发现，人们很难使床周围的空间适应卧室空间的需要。

其次，看来把床安置在一个适合于它的空间，会使人感到更加舒服。在我们作设计试验时，请外行人利用这些模式来设计他们自己的住宅，我们注意到，他们都有一种很强烈的愿望要设置一个放床的凹角，使床有一些围隔。显然，本模式打动人们的心弦。

一旦把床放进了适合于它的空间，就可以根据需要随心所欲地安排卧室中的其余空间，供起居、游玩、更衣和储放东西之用。

要使一个床龛适用，什么是关键所在呢？

宽敞 不要把它设计得太窄小。应能很自如地进出和铺床。如果床龛用来给儿童作为**个人居室**（141），它应该是一个很小的房间，有一面墙是开敞的。

Second,the bed itself seems more comfortable in a space that is adjusted to it.In our design experiments,where lay people have used these patterns to design their own houses,we have noticed a rather strong urge to give the bed a nook of its own,some kind of enclosure.Apparently this particular pattern strikes a chord in people.

Once the bed has been built into a space that is right for it,then the rest of the bedrom space is free to shape itself around the needs for sitting space,play areas,dressing,and storage.

What are the issues at stake in making a good bed alcove?

Spaciousness.Don't make it too tight.It must be comfortable to get in and out and to make the bed.If the alcove is going to function as A ROOM OF ONE'S OWN(141)for a child,then it needs to be almost a tiny room,with one wall missing.

Ventilation.Bed alcoves need fresh air;at least a vent of some kind that is adjustable,and better still a window.

Privacy.People will want to draw into the alcove and be private.The opening of the alcove needs a curtain or some other kind of enclosure.

Ceiling.According to the arguments developed with the pattern CEILING HEIGHT VARIETY(190),the bed,as an intimate social space for one or two,needs a ceiling height somewhat lower than the room beside it.

Therefore:

Don't put single beds in empty rooms called bedrooms,but instead put individual bed alcoves off rooms with other nonsleeping functions,so the bed itself becomes a tiny private haven.

If you are building a very small house no more than 300 or 400 square feet-perhaps with the idea of adding to

我们在秘鲁设计的一个住宅中的六个床龛
Six bed alcoves in one of our houses in Peru

通风 床龛需要新鲜空气；至少要有某种可调整的通风孔，最好有一个窗户。

私密性 人们走到床龛中来是愿意独处的。床龛的开口处需要挂一个门帘或其他用来遮挡的东西。

天花板 根据**天花板高度变化**（190）这个模式所持的论据表明，床作为仅供一两个人使用的一种私密的社会空间，需要的天花板高度要比房间内其他地方略低。

从家庭室中分隔出来的床龛
Bed alcoves off a family room

因此：

别把单人床放在叫作卧室的房间里，而是要从不供睡觉用的房间分隔出一些独立的床龛，使床成为一个小小的私密的安息所。

如果你在建造一幢很小的住房，不大于 300ft^2 或 400ft^2——也许你想给它逐渐加点什么——本模式将起到重要作用。在这种情况下，从家庭室分隔出一些床龛来也许是最合适的。

it gradually—this pattern plays an essential role.It will probably be best then to put the alcoves off the family room.

∞ↄ

Build the ceiling low—CEILING HEIGHT VARIETY (190); add some storage in the walls around the alcove—THICK WALLS(197),OPEN SHELVES(200),and a window,in a natural position—NATURAL DOORSAND WINDOWS(221).Perhaps HALF-OPEN WALL(193)will help to give the alcove the right enclosure.Where space is very tight, combine the bed alcove with DRESSING ROOM(189).And finally,give each alcove,no matter how small,the characteristics of any indoor space—THE SHAPE OF INDOOR SPACE(191)...

床
凹室
看见更大的公共空间

❦❧

　　把天花板做得低一点——**天花板高度变化**（190）；在床龛周围的墙内安装一些储藏柜——**厚墙**（197）、**敞开的搁架**（200），对着自然景物开出一个窗户——**借景的门窗**（221）。也许**半敞开墙**（193）会有助于使床龛得到适当的围隔。在空间十分窄小的地方，把床龛同**更衣室**（189）合在一起。最后，使每一床龛，不管它多小，都具有室内空间的特征——**室内空间形状**（191）……

模式189 更衣室*

189   DRESSING ROOMS*

...if the beds are in position—MARRIAGE BED(187), BED ALCOVES (188)—we can give detailed attention to the dressing spaces-both to the closets where people keep their clothes and to the space they use for dressing.These dressing spaces may also help to form the BATHING ROOM(144).

∞∞∞

Dressing and undressing,storing clothes,having clothes lying around,have no reason to be part of any larger complex of activities.Indeed they disturb other activities: they are so self-contained that they themselves need concentrated space which has no other function.

We have argued,in BED ALCOVES(188),that the concept of the bedroom leads to wasted space around the bed.This pattern lends further support to the idea that "bedrooms" in their present form are not valuable entities to have in a house.

The arguments are:

1.Clothes lying around are messy;they can take over a great deal of space;they need some kind of individual space.A dressing space can be for one person or shared by a couple.The important thing is that it be organized as a small space where it is comfortable to store clothes and to dress.When such a space is not provided, *the whole bedroom* is potentially the dressing room;and this can destroy its integrity as a room.It becomes more a big closet to "keep neat," than a room to stay in and relax.

2.People tend to take up a private position while they dress,even where they are relatively intimate with the people

……如果床位已经安排妥当——**夫妻用床**（187）、**床龛**（188）——我们就可以详细考虑更衣空间了——包括人们放衣服的壁柜和他们穿衣服时使用的空间。这些更衣空间也可能有助于形成**浴室**（144）。

୨୦୧୫

穿衣服和脱衣服，储藏衣服和在周围放衣服，这些最好都不要在一个有多种用途的大空间里进行。它们势必会干扰其他活动：它们有着自己的独特性，需要一个不作其他用处的专用空间。

在**床龛**（188）中，我们已经论证了，卧室这一概念容易导致卧床周围空间的浪费。本模式将进一步证明，目前这种"卧室"的形式在住宅中没有多少实用价值。

论据是：

1. 室内到处放衣服，显得很不整齐；它们会占据很大一片地方，需要给它们单独的一种空间。一个更衣空间可供一个人使用，也可让夫妻共用。重要的事情是应使它成为一个放衣服和穿衣服都很方便的小空间。如果不设置这么一个空间，整个卧室实际上就是更衣室；而这就会破坏它作为一个房间的完整性。它会变得更像一个用来"保持整洁"的大的衣柜，而不像一个供人居住和休憩的房间。

2. 人们习惯于找一个私密的地方换衣服，即使在同他们生活在一起的亲人面前也是如此。甚至在一个公共更衣室内，人们在换衣时也会半转过身去避开他人。这表明更衣空间是比较私密的。演员休息室或闺房中老式的立式屏风就起这种作用；它们能形成半私密的更衣空间。

they live with.Even in a locker room,people will make a half-turn away from others as they dress.This suggests that the space for dressing be relatively private.The old fashioned standing screens in a green room or a boudoir worked this way;they created a half-private dressing space.

3.The time of dressing,the activity,is a natural moment of transition in the day.It is a time when people think about the day ahead,or unwind at the end of the day and get ready for bed.If you dwell,for a moment,on this transitional quality of dressing,it seems clear that the dressing space can be made to help support it.For example,a good place to dress will have beautiful natural light;this requires as much thought in your design as any room—see,for example,LIGHT ON TWO SIDES OF EVERY ROOM(159).

4.The dressing space should be large enough,with room to stretch your arms and turn around.This means six or seven feet of open area.It must also have about six feet of clothes hanging space,another six feet of open shelves,and a few drawers for each person.These figures are rough.Check your own closet and shelves,think about what you really need,and make an estimate.

Therefore:

Give everyone a dressing room—either private or shared—between their bed and the bathing room.Make this dressing room big enough so there is an open area in it at least six feet in diameter;about six linear feet of clothes hanging space;and another six feet of open shelves;two or three drawers;and a mirror.

೮೦೦೪

Place each dressing room so that it gets plenty of

3. 人们换衣服的时间是自然界中昼夜交替的时刻。在这个时刻，人们想到白昼将临，或一天过去了松弛下来准备就寝。如果你对更衣具有的这一时间上的过渡性质细想片刻，看来就会明白，建立更衣空间对此是有好处的。例如，一个合适的更衣室会有明亮的自然光；这要求你在设计更衣室时要同设计任何房间一样周到地考虑问题——例见**两面采光**（159）。

4. 更衣空间应该有足够的地方使你能够伸开胳膊和转身。这表明要有 6ft 或 7ft 宽的空地。还必须有大约 6ft 宽的挂衣服用的空间，另有 6ft 宽的敞开的搁架及几个供每人使用的橱柜。这些数字是个约数。查看一下你自己的衣柜和搁架，考虑一下你真正需要什么，然后作一个估算。

因此：

每个人在其床和浴室之间都应有一个更衣室——供一个人专用，或几个人共用。使这个更衣室至少有一个直径为 6ft 的空地；大约 6ft 宽的挂衣空间；另有 6ft 宽的敞开的搁架；两三个橱柜；一面镜子。

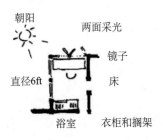

在设计每一处更衣室时要使它得到充足的天然的**两面采光**（159）。以**厚墙**（197）、**居室间的壁橱**（198）及**敞开的搁架**（200）作为它的墙；包括一个围着边缘的很宽的搁

natural LIGHT ON TWO SIDES(159).Use THICK WALLS (197),CLOSETS BETWEEN ROOMS(198),and OPEN SHELVES(200)to form its walls;include a wide shelf around the edge—WAIST-HIGH SHELF(201);and for the detailed shape of the room,see THE SHAPE OF INDOOR SPACE(191)...

fine tune the shape and size of rooms and alcoves to make them precise and buildable;

架——半人高的搁架（201）；关于这一空间的形状，详见**室内空间形状**（191）……

　　仔细规划好各种房间和凹室的形状和大小，使它们精确和便于建造；

　　190. 天花板高度变化
　　191. 室内空间形状
　　192. 俯视外界生活之窗
　　193. 半敞开墙
　　194. 内窗
　　195. 楼梯体量
　　196. 墙角的房门

模式190　天花板高度变化**

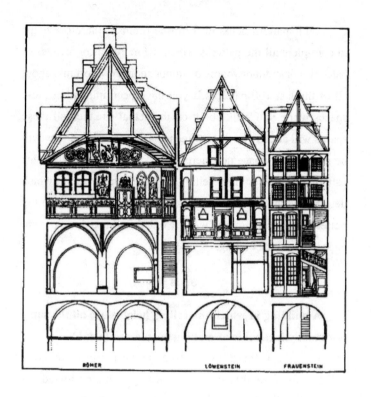

ROMER　　　　LÖWENSTEIN　　　FRAUENSTEIN

　　……本模式有助于构成空间的形状。因而它有助于完善确定房间、拱廊、阳台、户外空间或小房间的所有模式。简言之，有助于完善确定大约这之前的 100 个模式。当你在实际的地基上漫步的时候，如果你设想过这些空间，所有这些空间都会立体地呈现在你的心中：它们是具有空间的体量，而不仅是图纸上的面积。现在，有了决定天花板高度的本模式，决定每个空间的精确形状的下一个模式，以及本模式语言中的其他模式，我们就充实了关于建筑物的这一空间的概念。

190   CEILING HEIGHT VARIETY**

...this pattern helps to form the rooms.It therefore helps to com-plete all the patterns which define rooms,or arcades,or balconies,or outdoor rooms or minor rooms:in short,just about all of the last 100 patterns.If you have been imagining these spaces while you walk about on the actual site,then all these spaces will already be three-dimensional in your mind:they will be volumes of space,not merely areas on plan.Now,with this pattern,which determines ceiling heights,the next pattern which determines the exact shape of each room,and the remaining patterns in the language,we fill out this three dimensional conception of the buildig.

≫✧≪

A building in which the ceiling heights are all the same is virtually incapable of making people comfortable.

In some fashion,low ceilings make for intimacy,high ceilings for formality.In older buildings which allowed the ceiling heights to vary,this was almost taken for granted. However,in buildings which are governed by standard components,it is very hard to make the ceiling height vary from room to room,so it tends to be forgotten.And people are willing to let it go,because they have forgotten what an important psychological reason there is for making the heights vary.

We have presented three different theories over the years in our attempts to explain the significance of ceiling height variety,and we shall present the evolution of all three theories

൦൭

天花板高度千篇一律的建筑物是绝无可能让人感到舒适的。

通过某种样式，低天花板令人感到私密，高天花板则表现庄重。在天花高低不一的较古老的建筑中，这被认为是理所当然的事。然而，在用标准构件建成的房屋中，很难使各个空间的天花板高度彼此不同，因此人们已将此忘诸脑后。人们也无心去管这些，因为他们忘却了使天花板高度有所变化的重要的心理学原因。

这些年我们提出了三种不同的理论，力图解释天花板高度变化的重要性，因为它能使你对这件事的重要性有一个正确的理解，从而也许使你自己条理分明地形成本模式。

理论 1　天花板高度应与空间的长度和宽度有关，因为这是一个比例问题，人们是根据空间的比例感到舒适或不舒适的。

人们作了许多努力来制定一些规则，以确定空间是否"合乎比例"。例如，帕拉第奥创立三条关于比例的规则：这些规则都具有这样的特点，即一个空间的高度应介乎它的长度和宽度之间。

在传统的日本建筑学里，这一思想是由一个经验公式形成的。一个房间的天花高度是 6ft 3in+（3.7× 房间里塌塌米的数目）in。这使地面面积同天花板高度直接联系了起来。一个很小的房间（三席地面）其天花板高度为 7ft 2in。一个大房间（12 席地面）的天花板高度为 9ft 11in。(参阅 Heinrich Engle，*The Japanese House*，Rutland Vermont：Charles E.Tuttle Company，1964，pp.68~71.)

尽管这一方法在某些情况下看起来可能很有用处，但它显然不是全面有效的几何原理。有许多空间天花板很低，特别是那些农舍式别墅和非正式住宅，但它们十分宜人——即使它们完全违背了帕拉第奥的原则和日本人的经

here,because it puts the matter in perspective and will perhaps allow you to formulate the pattern most coherently for yourself.

*Theory one.*The ceiling height should be related to the length and breadth of the room,because the problem is one of proportion,and people feel comfortable or uncomfortable according to the room's proportions.

Many efforts have been made to establish rules which will make sure that rooms are "well proportioned." Thus,for instance,Palladio laid down three rules of proportion:all of them shared the feature that the height of a room should be intermediate between its length and its breadth.

In traditional Japanese architecture,this idea is captured by a simple rule of thumb.The ceiling height of a room is 6 feet 3 inches $+(3.7 \times$ the number of tatami in the room)inches. This creates a direct relationship between floor area and ceiling height.A very small room(3 mats)has a ceiling height of 7 feet 2 inches.A large room(12 mats)has a ceiling height of 9 feet 11 inches.(See Heinrich Engle,*The Japanese House*,Rutland Vermont:Charles E.Tuttle Company,1964,pp.68-71.)

However sound this approach may seem in certain cases,it is clearly not a completely valid geometric principle.There are many rooms with extremely low ceilings,especially in cottages and informal houses,which are extremely pleasant-even though they violate Palladio's principle and the Japanese rule of thumb utterly.

*Theory two.*The ceiling height is related to the *social distance* between people in the room,and is therefore directly related to their relative intimacy or nonintimacy.

This theory makes it clear what is wrong with badly proportioned rooms,and gives the beginning of a functional basis for establishing the right height for different spaces.

验公式。

理论 2　天花板高度关系到房间内人们之间的社交距离，从而直接关系到他们彼此间的亲疏程度。

这一理论说明比例失调的房间有什么偏差，并从功能出发确立不同空间的天花板合适高度。这个问题的关键是看社交距离是否适当。众所周知，在各种社交场合在人与人之间有着适当的和不适当的距离（参阅 Edward Hall, *The Silent Language*, NewYork : Doubleday, 1959, pp.163~164 ; and Robert Sommer, "The Distance for Comfortable Conversation," *Sociometry*, 25, 1962, pp.111~116）。看来，一个空间的天花板高度在两个方面关系到人们的社交距离：

A. 天花板高度看来会影响听者对声源距离的感觉。因此，在低天花板下面声源看来比其实际距离要近；在高天花板下面它们看来比其实际距离要远。

由于声音（说话声、脚步声、沙沙作响等）是判别人们彼此间有多大距离的重要依据，这就表明天花板高度会改变人们之间的距离感。在一个高天花板下面，人们感觉到比他们实际的距离远。

根据天花板的这一功效，毫无疑问，私密性强的地方要求天花板很低，如双人床上的布帐；私密性不太强的地方要求较高的天花板，如壁炉旁的凹角；正规的场所要求高天花板，如正规的接待室；纯属公共性质的场所要求最高的天花板，如中央车站。

B. 以立体的"气泡"为媒介。我们知道，每一社交场所都有一定的水平尺寸或直径。我们可以把这看作一种围合这个场所的膜或气泡。可能这种气泡需要一个垂直组件——高度与它的直径相等。如果这样，为了舒适，天花板高度应该相当于房间里的主要社交距离。由于在中央车站里都是陌生人，有效社交距离长达 100ft，这就可以解释，为什么天花板应该很高；同样，在一个私密性很强的凹角

The problem hinges on the question of appropriate social distance.It is known that in various kinds of social situations there are appropriate and inappropriate distances between people.(See Edward Hall,*The Silent Language*,New York:Doubleday,1959,pp.163-164;and Robert Sommer, "The Distance for Comfortable Conversation," *Sociometry*,25,1962, pp.111-116.)Now,the ceiling height in a room has a bearing on social distance in two ways:

A.The height of a ceiling appears to affect the *apparent distance* of sound sources from a hearer.Thus,under a low ceiling sound sources seem nearer than they really are;under a high ceiling they seem further than they really are.

Since the sound is an important cue in the perception of distance between people(voice,footstep,rustle,and so on),this means that the ceiling height will alter the apparent distance between people.Under a high ceiling people seem further apart than they actually are.

On the basis of this effect,it is clear that intimate situations require very low ceilings,less intimate situations require higher ceilings,formal places require high ceilings,and the most public situations require the highest ceilings:for example,the canopy over the double bed,a fireside nook,high-ceilinged formal reception room,Grand Central Station.

B.Through the medium of three-dimensional "bubbles". We know that each social situation has a certain horizontal dimension or diameter.We may think of this as a kind of membrane or bubble which encloses the situation.It is likely that this bubble needs a vertical component—equal in height to its diameter.If so,the height of the ceiling should,for comfort,be equal to the dominant social distance in the room.Since people

里，或在双人床的上方，这些地方社交距离不过 5ft 或 6ft，天花板应该很低。

理论 3 虽然以上两种理论都包含有价值的观点，但它们至少还有些偏差，因为它们假定，任何一个空间的绝对天花板高度都具有重大的功能效果。实际上，绝对天花板高度不像人们从理论 1 和理论 2 中所料想的起那么大的作用。

例如，爱斯基摩人圆顶小屋中的私密性最强的空间可能高不及 5ft；但在热带地区，即使最私密的房间也可能有 9ft 高。这说明，空间的绝对高度受其他因素——气候和文化——的制约。那么，显然，任何对某一社交场所或空间大小规定一个绝对高度的理论都不可能是正确的。那是什么因素在起作用呢？为什么天花板高度会有变化？它们的变化会产生什么样的功能效果？

最后，我们得出结论，重要的是变化本身，而不仅是某一特定空间的绝对高度。因为，如果一幢建筑物内部有几种不同的天花板高度，而天花板高度影响社交关系（由于上述理由），那么只要做到各处天花高度不同，人们就能根据他们所寻求的私密性程度，或者从高的空间搬到低的空间去，或者相反——因为他们知道，每个人都把私密性跟天花板高度联系起来。

根据这一理论，天花板高度不能直接产生效果；因为在人与空间之间存在着一种复杂的相互作用，在这种情况下，人们从房屋中不同的天花板高度看出一些信息，他们根据这些信息采取自己的态度。他们是否感到舒适，要看他们是否能够参与这个过程，以及是否能够因为选择了一个私密性合适的空间而感到心安理得。

最后，在贯彻这一模式时要特别注意某些特殊事项。在单层建筑物中没有问题；因为天花板高度可以随意变化。然而，在多层楼房中，就很难使天花板高低分明了。上层

in Grand Central are strangers,and have an effective social distance of as much as 100 feet,this would explain why the ceiling has to be very high;similarly,in an intimate nook,or over a double bed,where the social distance is no more than five or six feet,the ceiling has to be very low.

*Theory three.*Although both of the previous theories contain valuable insights,they must be at least slightly wrong because they assume that the absolute ceiling height in any one room has a critical functional effect.In fact,the absolute ceiling height does not matter as much as one would expect from theories one and two.

For example,the most intimate room in an igloo may be no more than five feet high;yet in a very hot climate even the most intimate rooms may be nine feet high.This makes it clear that the absolute height of rooms is governed by other factors too—climate and culture.Obviously,then,no theory which prescribes an absolute height for any given social situation,or room size,can be correct.What then,is going on?Why do ceiling heights vary?What functional effect does their variation have?

We have been led,finally,to the conclusion that it is the *variation itself* which matters,not merely the absolute height in any given room.For if a building contains rooms with several different ceiling heights in it and the height has an effect on social relationships(for the reasons given),then the mere fact that the ceiling heights vary,allows people to move from high rooms to low rooms,and vice versa,according to the degree of intimacy they seek—because they know that everyone correlates intimacy with ceiling height.

According to this theory,the effect of the ceiling height is not direct;there is instead a complex interaction between people

的楼板总应该是平的；如你要使楼下的天花板高度有所变化，就明显地是个难题。下面一些意见可能有助于解决这些问题：

1. 你要想降低天花板高度，在天花板和楼板之间造一储藏间——至少 2ft 高。

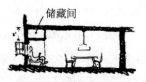

低天花上面的储藏间
Storage over a low ceiling

2. 在一个凹室之上再建一凹室。如果每个凹室的高度为 6ft 3in，主要的天花板高度便是 13ft，这对于公共性很强的空间是很合适的。

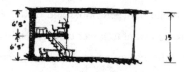

双层凹室
Stacked alcoves

3. 用台阶来提高地面的高度，而不去降低天花板高度。

提高地面使空间高度发生变化
The floor does it

4. 使某些房间的天花板低到 7ft 或 7ft 6in，这是一个好办法——这样的房间很美观。

5. 除了在单层建筑以外，低天花板的空间放在上层比较合理，实际情况也是这样，平均天花板高度随着层数的

and space,in which people read the different ceiling heights in a building as messages,and take up positions according to these messages.They are comfortable or uncomfortable according to whether they can take part in this process,and can then feel secure in the knowledge that they have chosen a place of appropriate intimacy.

Finally,some special notes are required on the implementation of this pattern.In a one story structure there is no problem;the ceiling heights may vary freely.In buildings with several stories however,it is not so clear cut.The floors of the upper stories must be more or less flat;and this obviously creates problems as you try to vary the ceiling heights underneath.Here are some notes which may help you to solve this problem:

1.Build storage between floors and ceilings—at least two feet deep-where you want to lower ceiling heights.

2.Put two alcoves over each other.If each is 6 feet 3 inches,this gives a main ceiling of 13 feet,which is good for very public spaces.

3.Raise the floor level with steps,instead of lowering the ceiling.

4.It is very important to have some rooms with ceilings as low as 7 feet or 7 feet 6 inches—these are very beautiful.

5.Except in one-story buildings,the low ceilinged rooms will make most sense on upper stories;indeed,the average ceiling height will probably get lower and lower with successive stories-the most public rooms,for the largest gatherings,are typically on the ground,and rooms get progressively more intimate the further they are from the ground.

升高会越来越低——供最大集会用的公共性质最强的空间多半在底层，从底层开始，楼层越高，空间的私密性越强。

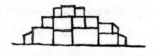

上层的天花板较低
Louer ceilings upstairs

因此：

在整幢建筑物中，特别是在互相连通的空间之间，天花板高度应不断变化，让人感到不同空间的私密性程度不一。特别是，在那些公用的或供大型集会使用的空间天花要很高（10~12ft），使用的人较少的空间天花板可以低一些（7~9ft），只供一两个人使用的房间或凹室天花板要做得很低（6~7ft）。

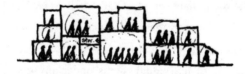

❧❦

建造地面的筒拱结构几乎自然会造成天花板高度的变化，因为筒拱结构最低处高度约为 6ft 6in，还要升高相当于它空间直径的五分之一——**楼面天花拱结构**（219）。在同一楼层里天花高度发生变化的地方，在不同高度的空间之间建造储藏间——**大储藏室**（145）。从**室内空间形状**（191）和**结构服从社会空间的需要**（205）中获得在一定的天花板高度下各个空间的形状；使每一楼层天花板高度都不一样——最底下一层天花板最高，最高一层天花板最低——参阅**柱的最后分布**（213）中的数据表……

Therefore:

Vary the ceiling heights continuously throughout the building,especially between rooms which open into each other,so that the relative intimacy of different spaces can be felt.In particular,make ceilings high in rooms which are public or meant for large gatherings(10 to 12 feet),lower in rooms for smaller gatherings (7 to 9 feet),and very low in rooms or alcoves for one or two people(6 to7 feet).

❧❦

The construction of floor vaults will create variations in ceiling height almost automatically since the vault starts about 6 feet 6 inches high and rises a further distance which is one-fifth of the room diameter—FLOOR-CEILING VAULTS(219). Where ceiling height varies within one story,put storage in the spaces between the different heights-BULK STORAGE(145). Get the shape of individual rooms under any given ceiling height from THE SHAPE OF INDOOR SPACE(191)and STRUCTURE FOLLOWS SOCIAL SPACES(205);and vary ceiling heights from story to story-the highest ceilings on the ground floor and the lowest on the top floor—see the table in FINAL COLUMN DISTRIBUTION(213)...

模式191 室内空间形状**

……从**天花板高度变化**（190）中，你对楼房中每一层高度的变化情况已有一个全面的概念，一般是处于中间位置的最大空间所在的天花最高，靠近边缘的小空间所在的天花较低，而且高度随着楼层的不同而有所变化，低楼层的平均天花高度比高楼层高。本模式研究高低不同的各个空间的情况，并给它们以更加确定的形状。

⊱⊰

完全透明的正方形和长方形的超现代建筑对人或对结构来说都没有特别的意义。它们只表达了人们过分注重他们的生产方法和手段时产生的固执的欲望和幻想。

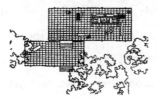

透明建筑
…*Crystalline*…

为了避免这种狂热，一种新思潮干脆完全抛弃直角。许多新的有机技术所建造的房屋和房间颇像子宫和洞穴。

仿生学建筑
…*Pseudo biological*…

但这些仿生学空间正如它们力图取代的僵化的透明建

191 THE SHAPE OF INDOOR SPACE**

...from CEILING HEIGHT VARIETY(190)you have an overall conception of each floor in the building as a cascade of heights,typically highest in the middle where the largest rooms are,lower toward the edge where the small rooms are,and varying with floor also,so that the lower floors will tend to have a higher average ceiling height than upper floors.This pattern takes each individual space,within this overall cascade,and gives it a more definite shape.

❧❦❧

The perfectly crystalline squares and rectangles of ultramodern architecture make no special sense in human or in structural terms.They only express the rigid desires and fantasies which people have when they get too preoccupied with systems and the means of their production.

To get away from this madness a new wave of thought has thrown the right angle away completely.Many of the new organic technologies create buildings and rooms shaped more or less like wombs and holes and caves.

But these biological rooms are as irrational,as much based on images and fantasies as the rigid crystals they are trying to replace.When we think about the human forces acting on rooms,we see that they need a shape which lies between the two.There are reasons why their sides should be more or

筑一样不合理，同样建立在虚构和幻想的基础之上。如果我们考虑作用于空间的人的因素，我们就会看到这些空间需要一种介乎这二者之间的形状。有理由使它们的四边多少有点直线；它们的角，或其中的许多角，应该大体上是直角。但没有充分的理由使它们的各边完全一样，但也没有充分理由使它们的角都是准确的直角。它们只要大体上是一个不规则、不完全的长方形就可以了。

我们争论的焦点就在于此。我们主张，每个空间，除非墙很厚，只要人们能够辨认出来并有明确墙界，就必须有大体上是直线的墙，墙的两面都应凹进去。

理由很简单。每道墙的两面都有社交空间。因为社交空间总是凸形的——请参阅**户外正空间**（106）中的大量论据——它必须或者有凹墙（这样可以形成凸空间），或者有一完全笔直的墙。但任何向一面凹进去的"薄"墙一定会向另一面凸出来，从而至少在一面留下一个凹的空间。

两个凸空间彼此紧靠在一起，在它们之间形成一道直墙
Two convex spaces pressed up each other, form a straight wall between them

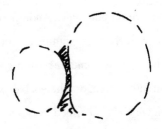

墙很厚，使两面都能凹进去
A wall thick enough to be concave on both sides

less straight;and there are reasons why their angles,or many of them anyway,should be rough right angles.Yet their sides have no good reason to be perfectly equal,their angles have no good reason to be perfectly right angles.They only need to be irregular,rough,imperfect rectangles.

The core of our argument is this.We postulate that every space,which is recognizable and walled enough to be distinct,must have walls which are roughly straight,except when the walls are thick enough to be concave in both directions.

The reason is simple.Every wall has social spaces on both sides of it.Since a social space is convex—see the extensive argument in POSITIVE OUTDOOR SPACE(106)—it must either have a wall which is concave(thus forming a convex space)or a wall which is perfectly straight.But any "thin" wall which is concave toward one side,will be convex toward the other and will,therefore,leave a concave space on at least one side.

Essentially then,every wall with social spaces on both sides of it,must have straight walls,except where it is thick enough to be concave on both sides.And,of course,a wall may be curved whenever there is no significant social space on the outside of it.This happens sometimes in a position where an entrance butts out into a street,or where a bay window stands in a part of a garden which is unharmed by it.

So much for the walls.They must most often be roughly straight.Now for the angles between walls.Acute angles are

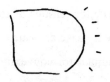

墙很薄，在一面造成一个凸空间，却破坏了另一面

A thin wall, makes a convex space on one side, and destroys the other side

那么，基本上两边皆有社会空间的墙必须是直墙，使两边都可凹进去的厚墙除外。当然，墙也可以是曲线的，只要它的外面没有重要的社会空间。这种情况有时见于入口临街的地方，或有凸窗的地方，该凸窗位于花园而又不会对花园造成破坏。

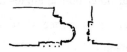

在这个地方可以把墙做成曲线，因为这里的墙只同外部空间发生联系

A place where a wall can be curved, because it works with the outside

以上讲的是墙。它们在大多数情况下都应当大体上是直线的。现在讲讲墙之间的夹角。锐角几乎总不合适，理由还是为了保持社会空间的完整性。要使房间中有一个实用的锐角是十分艰难的。由于主张使房间向外突出，角度就不能超过180°，这意味着空间的角必须总是介于80°和180°之间的钝角。（我们说80°，是因为小于直角几度关系不大。）

否 ← 可 → 否

夹角的各种可能的形状

The range of possible corners

hardly ever appropriate,for reasons of social integrity again. It is an uphill struggle to make an acute angle in a room,which works.Since the argument for convexity rules out angles of more than 180 degrees,this means that the comers of spaces must almost always be obtuse angles between 80 and 180 degrees.(We say 80 degrees,because a few degrees less than a right angle makes no difference.)

And one further word about the angles.Most often rooms will pack in such a way that angles somewhere near right angles(say between 80 and 100 degrees)make most sense.The reason,simply,is that other obtuse angles do not pack well at comers where several rooms meet.Here are the most likely typical kinds of corners:

This means that the majority of spaces in a building must be polygons,in plan,with roughly straight walls and obtuse-angled corners.Most often they will probably be irregular, squashed,rough rectangles.Indeed,respect for the site and the subtleties of the plan will inevitably lead to slightly irregular shapes.And occasionally they may have curved walls—either if the wall is thick enough to be concave on both sides or,on an exterior wall,where there is no important social space outside.

A final point.Our experience has led us to an even stronger version of this pattern—which constrains the shape of ceilings too.Specifically,we believe that people feel uncomfortable in spaces like these:

关于角的问题还要说几句话。在大多数情况下，房间相邻时，以近似直角的角（例如在80°和100°之间）最为合理。理由很简单，当几个空间相交在一起时，它们的角很难形成钝角。最可能形成的常见的角有以下几种（各图）：

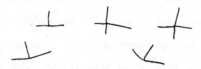

只有近似直角的角才能成功地组合在一起
Only angles that are nearly right angles pack successfully

这说明，在一个建筑物里，大部分空间在平面图上必须是墙壁略直、角落呈钝角的多边形。大多数情况下它们多半是不规则的、拉长的、近似的长方形。的确，考虑地基情况和设计中小的细节都难免会使形状稍有不规则。偶尔这些空间可能出现曲线墙——如果墙很厚使两边都能凹进去，或向墙外没有重要的社会空间的外墙凹进。

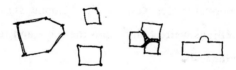

多角形、近似的长方形、厚的曲线墙、曲线外墙
Polygon, rough rectangle, thick curved wall, exterior curved wall

最后一点。我们的经验让本模式发挥更大的作用——使它也能限制天花板的形状。具体地讲，我们认为人们在这样的空间中会感到不舒服的：

天花板会使你感到不舒服的房间
Rooms whose ceilings can make you uncomfortable

We can only speculate on the possible reasons for these feelings.It seems just possible that they originate from some kind of desire for a person to be surrounded by a spherical bubble roughly related to the human axis.Room shapes which are more or less versions of this bubble are comfortable;while those which depart from it strongly are uncomfortable.*Perhaps when the space around us is too sharply different from the imaginary social bubble around us,we do not feel quite like persons.*

A ceiling that is flat,vaulted in one direction or vaulted in two directions,has the necessary character.A ceiling sloping to one side does not.We must emphasize that this conjecture is not intended as an argument in favor of rigidly simple or symmetric spaces.It only speaks against those rather abnormal spaces with one-sided sloping ceilings,high apexed ceilings,weird bulges into the room,and re-entrant angles in the wall.

Therefore:

With occasional exceptions,make each indoor space or each position of a space,a rough rectangle,with roughly straight walls,near right angles in the corners,and a roughly symmetrical vault over each room.

我们对这种感觉的原因只能做些推测。看来它们可能源于人的某种欲望，他总想使自己处于一个大致与人的轴线相当的球形泡的包围之中。房间的形状多少与这种气泡相类似会令人感到舒服；而与它相差甚远的那些则令人感到不快。也许当我们周围的空间跟我们想象中我们周围的社交气泡相差甚远时，我们会感到这不像人住的地方。

空间气泡的形状
The shape of the space bubble

平的天花板和单向或双向呈筒拱状的天花板都具备这种必需的特征。而一面倾斜的天花板不具备这种特征。我们必须予以强调，我们并不想把这一推测作为一种论据，来证明应当构造十分简单的或完全对称的空间。我们只不过是反对那些很不正常的空间，比如天花板倾斜或高高耸起，房间很不自然地一面鼓起和墙上形成许多凹角。

因此：

除了偶然的例外情况，每一室内空间或每一空间位置大致都应形成长方形，它们的墙大体上是直的，墙角近乎直角，每一空间的上方为基本对称的筒拱结构。

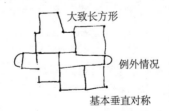

大致长方形

例外情况

基本垂直对称

You can define the room with columns,one at each corner—COLUMNS AT THE CORNERS(212);and the shape of the ceiling can be given exactly by the ceiling vault—FLOOR AND CEILING LAYOUT(210),FLOOR-CEILING VAULT(219).Avoid curved walls except where they are strictly necessary—WALL MEMBRANES(218).Where occasional curved walls like bay windows do jut out into the outside,place them to help create POSITIVE OUTDOOR SPACES(106). Make the walls of each room generous and deep—THICK WALLS(197),CLOSETS BETWEEN ROOMS(198);and where it is appropriate,make them HALF-OPEN WALLS(193).For the patterns on the load-bearing structure,engineering,and construction,begin with STRUCTURE FOLLOWS SOCIAL SPACES(205)...

అ८౪

　　你可在每一个墙角立一根柱来界定空间——**角柱**（212）：利用天花拱结构来精确地构成天花形状——**楼面和天花板布置**（210）、**楼面天花拱结构**（219）。避免用曲线墙，除非非用它不可——**墙体**（218）。如采用像凸窗这样不常见的曲线墙，可利用它向户外空间伸出的特点来创造**户外正空间**（106）。使每一空间都有宽厚的墙——**厚墙**（197）、**居室间的壁橱**（198）；在合适的地方，把它们做成**半敞开墙**（193）。关于承重结构、土木工程和施工等模式，参看**结构服从社会空间的需要**（205）及之后的几个模式……

模式192　俯视外界生活之窗*

　　……本模式有助于完善前面几个使每一空间具备自己形状的模式——**两面采光**（159）、**天花板高度变化**（190）、**室内空间形状**（191）。一旦对这些模式了解清楚了，就可借助于本模式更精确地决定窗在墙上的位置。它规定墙上应有多少窗户，窗与窗之间的距离应多大，窗的总面积应是多少。

192 WINDOWS OVERLOOKING LIFE*

...this pattern hdps to complete the earlier patterns which give each room its shape:LIGHT ON TWO SIDES OF EVERY ROOM(159),CEILING HEIGHT VARIETY(190),and THE SHAPE OF INDOOR SPACE(191).Once these patterns are clear,this pattern helps to place the windows rather more precisely in the walls.It defines just how many windows there should be,how far apart,and what their total area should be.

❧❧

Rooms without a view are prisons for the people who have to stay in them.

When people are in a place for any length of time they need to be able to refresh themselves by looking at a world different from the one they arc in,and with enough of its own variety and life to provide refreshment.

Amos Rapoport gives written descriptions of three windowless seminar rooms at the University of California. The descriptions-by teachers and students of English who were asked to write descriptions of the rooms as part of a writing exercise-are heavily negative,even though they were not asked to be,and in many cases refer directly to the windowless,boxed-in,or isolated-from-the-world character of the rooms.

 Here are two examples:

Room 5646 is an unpleasant room in which to attend class because in it one feels detached and isolated from the rest of the world under the buzzing fluorescent lights and the high sound-proofed ceilings,amid the sinks,cabinets,and pipes,surrounded by empty space.

看不到外界的房间对于其中的居者来说无异于一个监狱。

不管人们在一个地方停留多久，他们都需要看看外界环境，这环境跟他们所在的环境不同，它以自己的特色和生活气息使人心情舒畅，人们需要借此来调剂自己的精神。

拉波波特在他的文章里描写过加利福尼亚大学的三间没有窗户的教室。该校英语专业师生在进行一种写作练习时对这些教室进行描写，尽管谁也没有让他们描写缺点，可在他们的笔下这些教室太糟糕了，而问题大部分在于这些教室没有窗户，坐在教室里面如同关在箱子里，与世隔绝。

请看以下两例：

在5646号教室上课令人很不愉快，因为在那里，在丝丝作响的荧光灯下，在隔音良好的天花下，在水池、柜子和管道之间，周围一片空寂，与世隔绝。

这巨大、空空如也、没有窗户的教室，四面是坚硬、闭合、单调的灰墙，叫人讨厌不得、喜欢不得、人们也许容易忘记，他们是如何禁闭在这个地方的。(Amos Rapoport, "Some Consumer Comments on a Designed Environment," *Arena-The Architectural Association Journal*, January 1967, pp.176~178.)

布赖恩·韦尔斯在研究办公人员对工作地点的选择时发现，81％的人选择靠近窗口的位置。(*Office Design：A Study of Environment*, Peter Manning, ed., Pilkington Research Unit, Department of Building Science, University of Liverpool, 1965, pp.118~121.) 许多被调查的人说他们选择靠窗的理由是为了"阳光"而不是"风景"。但在同一篇报告的另一处表明，离窗户很远的被调查者认为他们接

The large and almost empty,windowless room with its sturdy,enclosing,and barren grey walls inspired neither disgust nor liking;one might easily have forgotten how trapped one was. (Amos Rapoport, "Some Consumer Comments on a Designed Environment," *Arena-The Architectural Association Journal*,January 1967,pp.176-78.)

Brian Wells,studying office workers'choice of working positions,found that 81 percent of all subjects chose positions next to a window.(*Office Design:A Study of Environment*,Peter Manning,ed.,Pilkington Research Unit,Department of Building Science,University of Liverpool,1965,pp.118-121.)Many of the subjects gave "daylight" rather than "view" as a reason for their choice.But it is shown elsewhere in the same report that subjects who are far from windows grossly overestimate the amount of daylight they receive as compared with artificial light.(*Office Design* p.58.) This suggests that people want to be near windows for other reasons over and above the daylight.Our conjecture that it is the view which is critical is given more weight by the fact that people are less interested in sitting near windows which open onto light wells,which admit daylight,but present no view.

And Thomas Markus presents evidence which shows clearly that office workers prefer windows with meaningful views-views of city life,nature-as against views which also take in large areas,but contain uninteresting and less meaningful elements.(Thomas A.Markus, "The Function of Windows:A Reappraisal," *Building Science*,2,1967,pp.97-121;see especially p.109.)

Assume then that people do need to be able to look out of windows,at some world different from their immediate surroundings.We now give very rough figures for the total area of the windows in a room.The area of window needed will depend to a large extent on climate,latitude,and the amount

受到的自然光比人造光多得多。（*Office Design* p.58.）这表明，人们想坐在窗边除了阳光以外，还有其他更重要的理由。我们猜想，主要是为了风景，因为人们对于坐在面向天井的窗边并无多大兴趣，尽管这些窗户一样接受阳光，只不过窗外无景可赏。

托马斯·马库斯提供的证据清楚地表明，办公人员喜欢的窗户要能看到引人入胜的景致——城市生活和大自然的风光——而不喜欢同样占有很大面积，但令人兴味索然又无多大意义的景物。（Thomas A.Markus，"The Function of Windows：A Reappraisal，" *Building Science*，2，1967，pp.97~121；see especially p.109.）

那么可以设想，人们确实需要能够向窗外张望，看到不同于他们周围的另一种环境。我们估算了一下一个房间其窗户总面积的大约数字。窗户所需要的面积在很大程度上取决于建筑物所在地的气候、纬度以及建筑物外部周围反射表面的量。然而，比较合理的看法是，地面同窗户的比例关系虽然在不同地区不同，但在一个特定区域内多少应该固定。

因此我们建议，在你居住的城市到处转一转，选择你觉得光线确实很好的五六个房间。在每一个房间都量一下窗户面积，算一下它同地面面积的百分比；然后得出这些不同百分比的平均数。

在我们这个地方——加利福尼亚州伯克利市——我们发现最令人愉快的房间其窗户面积大约占 25%——有时多达 50%——（这就是说，每 100ft^2 地面对应 25 ~ 50ft^2 的窗户）。但我们重申，这个数字显然会因地域的不同而有很大的变化。请你想象一下这些地方：拉巴特、廷巴克图、南极洲、挪威北部、意大利、巴西丛林……

of reflecting surfaces around the outside of the building. However,it is fairly reasonable to believe that the floor/window ratio,though different in different regions,may be more or less constant within any given region.

We suggest,therefore,that you go round the town where you live,and choose half a dozen rooms in which you really like the light.In each case,measure the window area as a percentage of the floor area;then take the average of the different percentages.

In our part of the world—Berkeley,California—we find that rooms are most pleasant when they have about 25 percent window—sometimes as much as 50 percent(that is, 25~50 square feet of window for every 100 square feet of floor).But we repeat,obviously this figure will vary enormously from one part of the world to another.Imagine:Rabat,Timbuctoo,Antarctica, Northern Norway,Italy,Brazilian jungle...

Therefore:

In each room,place the windows in such a way that their total area conforms roughly to the appropriate figures for your region (25 percent or more of floor area,in the San Francisco Bay Area),and place them in positions which give the best possible views out over life:activities in streets,quiet gardens,anything different from the indoo scene.

ഇൻവ

Fine tune the exact positions of the windows at the time that you build them—NATURAL DOORS AND WINDOWS (221);break the area of each window into SMALL PANES (239);give each window a very LOW SILL(222)to improve the view and DEEP REVEALS(223)to make the light as soft as possible inside...

因此:

使每个房间窗户的总面积基本符合你所在地区的合适的比例数值(在旧金山湾区为地面面积的 25% 或多一些),并能从窗户中尽可能多地看到外界的生活:街上的活动、幽静的花园,任何不同于室内环境的景物。

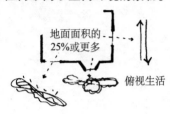

地面面积的 25% 或更多

俯视生活

෨෬

在你建造窗户的时候,要选择好窗户的准确位置——**借景的门窗**(221);把每个窗户的面积分成**小窗格**(239);使每个窗户都有一个**矮窗台**(222)以开阔视野和造成**深窗洞**(223),使室内光线尽可能柔和⋯⋯

模式193　半敞开墙*

……**室内空间形状**（191）规定大小房间的形状。本模式将详述这些房间之间的墙。凡是有**半私密办公室**（152）、**六英尺深阳台**（167）、**凹室**（179）、**坐位圈**（185）、**床龛**（188）、**有顶街道**（101）、**拱廊**（119）、**穿越空间**（131）的地方，必须使墙局部敞开或保持半敞开状态，以便使封闭和敞开之间得到巧妙的平衡。

৪০৫৪

过分闭塞的空间不能使社交活动畅通无阻，也妨碍人们从一个社交场合顺利地过渡到另一社交场合。而过分敞开的空间则不能满足社会生活的不同要求。

以一个四面皆墙的封闭的房间为例，它显然能够满足完全不同于邻近房间所进行的那些活动。在这个意义上它是很理想的。但人们很难随时参与或退出这些活动。人们想要这样做，就只有在门上装上玻璃，墙上开个窗户，或使房间有一个通路，这样他们才可以在谈话间歇时，从容地走上前来，自然而然地参与到活动中来。

反之，一个周围没有墙的开敞的空间，仅靠地上的一块地毯和摆放一些椅子作为标志，完全暴露在人们面前，无论如何都不会令人感到舒畅。在这样的地方安排什么活动都困难，因为谁都看得见；这样的地方只适合于干些不动感情的琐事，如喝酒、读报、看电视、观赏风景、无所事事地待着——你不会发现有生气的交谈、争论、激动，不会有人在这里干活、画画、打牌、猜谜、拉小提琴。人们要到那些周围有些围合的地方去参加这些丰富多彩的活

193   HALF-OPEN WALL*

...THE SHAPE OF INDOOR SPACE(191)defines the shapes of rooms and minor rooms.This pattern gives more detail to the walls between these rooms.Wherever there are HALF-PRIVATE OFFICES(152),SIX-FOOT BALCONIES(167),ALCOVES(179),SITTING CIRCLES(185),BED ALCOVES(188),BUILDING THOROUGHFARES(101),ARCADES(119),or THE FLOW THROUGH ROOMS(131),the spaces must be given a subtle balance of enclosure and openness by partly opening up the walls or keeping them half-open.

ଡ଼୦ୠ

Rooms which are too closed prevent the natural flow of social occasions,and the natural process of transition from one social moment to another.And rooms which are too open will not support the differentiation of events which social life requires.

A solid room,for instance,with four walls around it can obviously sustain activities which are quite different from the activities in the next room.In this sense it is excellent.But it is very hard for people to join in these activities or leave them naturally.This is only possible if the door is glazed,or if there is a window in the wall,or if there is an opening,so that people can gradually come forward,just when there is a lull in the conversation,and naturally become a part of what is happening.

On the other hand,an open space with no walls around it,just a place marked by a carpet on the floor and a chair arrangement, but entirely open to the spaces all around it,is so exposed that people never feel entirely comfortable there.No

动，那些地方至少一半有墙，或有栏杆、柱子，同邻近空间有些分隔。

总之，在开敞和封闭之间微妙的矛盾中自然要求有一种平衡。但由于某种原因，现代有关房间和室内空间的形象把人们引向这两种极端，而没有引向所需要的平衡。

一种空间，如要既能满足多种活动的需要，又便于从一种活动转向另一种活动，就需要比全部围合的空间开敞些，同时又要比全部开敞的空间有较多的围合。

一种半敞开、半封闭的墙——拱门、棚架、带有装饰性柱子的与柜台一般高的墙、由缩小通路或在角落增大柱子所形成的墙、墙内列柱——有助于使封闭和开敞得到适当的平衡；结果在这些地方人们感觉舒适。

半敞开墙数例
Examples

从**工作空间的围隔**（183）中我们曾经证明，多大的围隔才算合适。那一模式告诉我们，一个人感到最舒适的是他大约"一半"受到围隔——他大约两面受到围合，或他的四面大约都是半开半合。

因此我们推想，任何半敞开墙都应该半虚半实。这并不是说它应该是一个屏风。例如，粗大的柱、深的梁、拱门结合在一起也能造成敞开和封闭的平衡。栏杆则开敞得太大了。但有粗大支座的回栏常常适合于作为半敞开墙。

one activity can establish itself because it is too vulnerable;and so the things that happen there tend to be rather blaand—a drink,reading the paper,watching television,staring at the view, "sitting around": you will not find animated conversations,arguments,excitement,people making things,painting,card games,charades,or some one practicing the violin.People let themselves go into these more highly diferentiated activities,when there is some degree of enclosure around them—at least a half-wall,a railing,columns,some separation from the other nearby spaces.

In short,the subtle conflict between exposure and enclosure naturally requires a balance.But for some reason the modem images of rooms and indoor space lead people to the two extremes,and hardly ever to the balance which is needed.

The kind of space which most easily supports both diferentiation of activities and the transition between different activities has less enclosure than a solid room,and more enclosure—far more—than a space inside an open plan.

A wall which is half-open,half-enclosed—an arch,a trellised wall,a wall that is counter height with ornamented columns,a wall suggested by the reduction of the opening or the enlargement of the columns at the corners,a colonnade of columns in the wall—all these help get the balance of enclosure and openness right;and in these places people feel comfortable as a result.

From WORKSPACE ENCLOSURE(183)we have some evidence for the amount of enclosure required. We found there that a person is comfortable when he is about "half" enclosed—when he has material around him on about two sides,or the four sides around him are about half solid and half-open.

这很适用于户外小空间和阳台；同样也适用于同更大的房间相连而又同它们部分分开的室内空间——凹室、工作空间、厨房、床。在以上情况下，用以围隔空间、把较小空间同较大空间分开的墙需要部分开敞和部分闭合。

在我们自己和我们的许多朋友中间，我们发现，需要将一幢房屋加以改造实际上就是需要在这房子的各部分之间创造一些半敞开墙。看来，即使不给本模式取名，人们也会本能地使一个空间"敞开"；或使另外一个空间"更多地围合起来"。

因此：

调整每一室内空间的墙、通路和窗，要让敞开的、流动的空间和像船舱一样封闭的空间之间取得合适的平衡。不要想当然地认为每一空间都是一个房间；另一方面也不要认为所有的空间都应彼此流通。恰当的平衡总是介于这两种极端之间：没有一个房间是完全隔断的；也没有一个空间全部跟另一空间相连。把柱、半敞开墙、门廊、室内窗户、推拉门、矮窗台、法式双扇玻璃门、坐墙等结合起来加以利用来达到恰当的平衡。

50%敞开

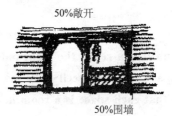

50%围墙

We therefore guess that the enclosure of any half-open wall should itself consist of about 50 percent void and 50 percent solid.This does not mean that it has to be a screen.For example,a combination of thick columns,deep beams,arched openings,also creates this balance of openings and enclosures. A railing is too open.But a balustrade with thick supports will often be just right.

This applies very strongly to outdoor rooms and balconies;and equally to all those indoor spaces which are connected to larger rooms but partly separate from them—an alcove,workspace,kitchen,bed.In all these cases the wall which forms the enclosure and separates the smaller space from the larger one,needs to be partially open and partially closed.

Among ourselves and many of our friends,we have found that the urge to remodel a house is virtually one and the same with the urge to create half-open walls between various parts of the house.It seems that without ever naming this pattern,people have the instinct to "open up" a room;or to give "more enclosure" to some other space.

Therefore:

Adjust the walls,openings,and windows in each indoor space until you reach the right balance between open,flowing space and closed cell—like space.Do not take it for granted that each space is a room;nor,on the other hand,that all spaces must flow into each other.The right balance will always lie between these extremes:no one room entirely enclosed;and no space totally connected to another. Use combinations of columns,half-open walls,porches,indoor windows,sliding doors,low sills,frenchdoors,sitting walls,and so on,to hit te right balance.

ಹಿಂ

当一个小空间处在一个较大空间之内但又与它稍有分隔时，这两个空间之间的墙应做到半开半合——**凹室**（179）、**工作空间的围隔**（183）。使围隔的部分和敞开的部分集中在一起，这样就会形成许多小的通路，每一通路都由粗大的柱子、半人高的搁架、深拱腹和角落上的拱架或支柱而形成，将围隔的地方和敞开的地方相交处加以装饰——**内窗**（194）、**角柱**（212）、**柱旁空间**（226）、**柱的连接**（227）、**小窗格**（239）、**装饰**（249）……

❧❧❧

Wherever a small space is in a larger space, yet slightly separate from it, make the wall between the two about half-open and half-solid—ALCOVES(179), WORKSPACE ENCLOSURE(183).Concentrate the solids and the openings, so that there are essentially a large number of smallish openings, each framed by thick columns, waist high shelves, deep soffits, and arches or braces in the corners, with ornament where solids and openings meet—INTERIOR WINDOWS (194), COLUMNS AT THE CORNERS(212), COLUMN PLACE(226), COLUMN CONNECTIONS(227), SMALL PANES(239), ORNAMENT(249)...

模式194　内窗

　　……在建筑物内各个地方，在空间和空间之间的墙上开窗会使这些空间更富有活力，因为窗能扩大人们的视野，并使更多的光线能够照射进黑暗的角落。例如，在过道和房间之间或相邻的各个起居室之间，或相邻的各个工作室之间都可以有内窗——**有顶街道**（101）、**入口空间**（130）、**穿越空间**（131）、**短过道**（132）、**明暗交织**（135）、**起居空间的序列**（142）、**半敞开墙**（193）。

❧❧

　　窗在大多数情况下用于连接室内和室外。但也有许多情况，一个室内空间与另一个室内空间需要用窗来连接。

194   INTERIOR WINDOWS*

...at various places in the building,there are walls between rooms where windows would help the rooms to be more alive by creating more views of people and by letting extra light into the darkest corners.For instance,between passages and rooms or between adjacent living rooms,or between adjacent work rooms—BUILDING THOROUGHFARE(101),ENTRANCE ROOM(130),THE FLOW THROUGH ROOMS(131),SHORT PASSAGES(132),TAPESTRY OF LIGHT AND DARK (135),SEQUENCE OFSITTING SPACES(142),HALF-OPEN WALL(193).

ক্রে৪

Windows are most often used to create connections between the indoor and the outdoors.But there are many cases when an indoor space needs a connecting window to another indoor space.

This is most often true for corridors and passages.These places can easily seem deserted.People feel more connected to one another by interior windows,and the passages in the building become less deserted.

The same may hold for certain rooms,especially small rooms.Three bare walls and a window can seem like a prison. Windows placed between rooms,or between a passage and a room,will help to solve these problems and will make both the passages and the rooms more lively.

Furthermore,when rooms and passages are visibly connected to one another,it is possible to grasp the overall

这种情况在走廊和过道最为常见。这些地方很容易让人感到凄凉冷清。有了内窗人们会感到彼此之间有了联系,也给建筑物内的过道增添生气。

对于某些房间,特别是小房间,情况可能也一样。三堵空墙和一个窗户看起来颇似牢房。在房间与房间之间,或在过道和房间之间,开些窗户会有助于解决这些问题,并会使过道和房间都充满生气。

再则,当人们看到房间和过道彼此相连时,有可能更好地看清建筑物的总体布局,而在一个全部空间之间有墙无窗的建筑物里却远远做不到这一点。

这些窗只要使人能够看见它外面的东西就可以了;它们无须打开,也无须做成可打开的窗户。普通的、便宜的、固定的玻璃便能解决问题。

因此:

在那些由于很少有活动而显得死气沉沉的空间之间,或空间内部光线昏暗的地方,可以装上固定的玻璃窗。

普通的固定窗户

෴

这些窗同任何其他窗户一样,可以使用小窗格玻璃——**小窗格**(239)。在某种情况下,在门上装上内窗可能是合适的——**镶玻璃板门**(237)……

arrangement of a building far more clearly than in a building with blank walls between all the rooms.

It is enough if these windows allow people to see through them;they do not need to be open nor the kind which can be opened.Ordinary,cheap,fixed glazing will do all that is required.

Therefore:

Put in fully glazed fixed windows between rooms which tend to be dead because they have too little action in them or where inside rooms ar unusually dark.

∞∞∞

Make the windows the same as any other windows,with small panes of glass—SMALL PANES(239).In some case it may be right to build interior windows in the doors—SOLID DOORS WITH GLASS (237)...

模式195　楼梯体量*

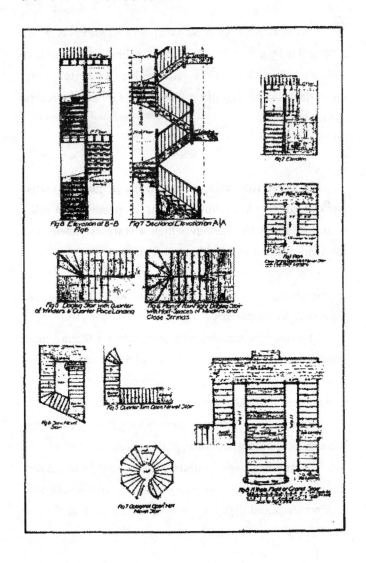

195   STAIRCASE VOLUME*

...STAIRCASE AS A STAGE(133)and OPEN STAIRS (158)will tell you roughly where to place the various stairs,both indoors and outdoors.This pattern gives each stair exact dimensions and treats it like a room so that it becomes realitic in the plan.

❧❧❧

We are putting this pattern in the language because our experiments have shown us that lay people often make mistakes about the volume which a staircase needs and therefore make their plans unbuildable.

Here are some examples of the stairs which people who are not used to building,draw,or think of,when they try to lay out houses for themselves.

Obviously,these stairs will not work;and the misunder standings of the nature of the stair are so basic,that it is hard to correct these plans without destroying them.In order to put in a realistic stair,it would be necessary to rethink the plan entirely. To avoid this kind of mental backtracking,it is essential that stairs be more or less realistic *from the very start.*

The simplest way to understand a stair is this.*Every staircase occupies a volume,two stories high.*If this volume is the right shape,and large enough to give the stair its rise,then it will be possible to fill it later,with a stair which works.

……**有舞台感的楼梯**（133）和**室外楼梯**（158）会使你大致知道，各种楼梯，无论是室内的还是室外的，应该建造在何处。本模式给出每一楼梯的精确尺寸，并把它看作一个空间，使它成为图纸上的实体。

৪৩৫৪

我们之所以把本模式列入本语言中，因为我们的实验表明，外行人在楼梯需要多大体量这个问题上常犯错误，结果用他们的设计图纸盖不成楼房。

下面几个楼梯的平面图是不熟悉建筑的人在设法自己设计住宅时画出来或想象出来的。

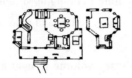

楼梯间有问题——太短
Staircae problems—too short

楼梯间有问题——太短
Staircae problems—too short

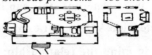

楼梯间有问题——太短
Staircae problems—too short

……没有上层体量
…no upstairs volume

There are several possible layouts for this volume:any one of them will work,provided that the length of run is long enough for the slope of the stair,and the floor to floor height.We urge you to be as free as possible when you decide the slope of the stair.Unfortunately,the search for perfect safety in housing laws,insurance standards,and bank policies,has exaggerated the standardisation of slopes.For example,Federal Housing Authority regulations specify that stairs should be between 30 and 35 degrees in slope.But in some cases—a very small house,a stair to the roof—such a shallow stair is a waste of space;a steep stair is far more appropriate.And in other cases— a main stair in a public building,or an outdoor stair—a much shallower stair is more generous,and more appropriate.

Therefore:

Make a two story volume to contain the stairs.It may be straight,L-shaped,U-shaped,or C-shaped.The stair may be 2 feet wide(for a very steep stair)or 5 feet wide for a generous shallow stair.But,in all cases,the entire stairwell must form one complete structural bay,two stories high.

Do not assume that all stairs have to have the "standard"angle of 30 degrees.The steepest stair may almost be a ladder.The most generous stair can be as shallow as a ramp and quite wide.As you work out the exact slope of your stair,bear in mind the relationship: riser+trea=$17\frac{1}{2}$ inches.

显然，这些楼梯都没有用处；由于它们根本上弄错了楼梯的性质，很难把这些设计图改正过来，只好毁弃不用。为了建造一个可用的楼梯，需要完全重新考虑这个设计图。与其改头换面，不如从一开始就把楼梯设计好。

了解楼梯的最简单的方法是：**每一楼梯间都占两层楼高的体量。如这一体量形状合适，有足够的大小可作楼梯踏步，就能建造一条可供使用的楼梯。**

利用这个体量可以作出几种可能的布置：每一种都会是实用的，只要楼梯踏步

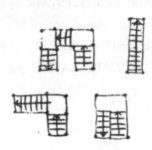

两层楼高的空间
Two-story space

的长度符合楼梯的斜度和楼层到楼层的高度。当你在决定楼梯的斜度时，我们希望尽量自由选择。令人遗憾的是，在房屋规范、保险标准和银行政策中，为了保证绝对安全，过分强调斜度的标准化。如联邦房管局的条例规定，楼梯斜度应该是 30° ～ 35°。但在某些情况下——一幢很小的住宅、一条通往房顶的楼梯——建造这样平缓的楼梯是一种空间的浪费；陡峭的楼梯反而会合适得多。而在另一些情况下——公共建筑物的主楼梯或室外楼梯——比较平缓的楼梯显得更加大方，也更加适用。

不同的斜度
Different slopes

Construct the staircase as a vault,within a space defined by columns,just like every other room—COLUMNS AT THE CORNERS(212),STAIR VAULT(228).And make the most of the staircase;underneath it is a place where the children can play and hide—CHILD CAVES(203);and it is a place to sit and talk—STAIR SEATS(125)...

因此：

楼梯应有两层楼高的体量。它可以是直的、L 形的、U 形的或 C 形的。楼梯可以是 2ft 宽（用于很陡的楼梯）或 5ft 宽（用于宽大平缓的楼梯）。但不论什么情况，整个楼梯井一定要形成一个完全的结构开间，两层楼高。

不要认为所有的楼梯都须有 30° 的标准角度。最陡的楼梯差不多可以是一架梯子。最平缓的楼梯可以像坡道那样平缓，而且踏步很宽。在你设计楼梯的准确斜度时，记住这个关系式：楼梯踏步竖板 + 级宽 =17 $\frac{1}{2}$ in。

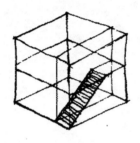

8003

把楼梯建造成由角柱限定的空间内部的穹窿结构，这与其他每个空间一样——**角柱**（212）、**楼梯拱**（228）。尽量把楼梯间设计好；它下面是儿童们可以游玩和躲藏的地方——**儿童猫耳洞**（203）；楼梯也是一个可供坐着聊天的地方——**能坐的台阶**（125）……

模式196　墙角的房门*

……本模式有助于你确定门的位置。借助于它可以创造更大的**穿越空间**（131）。你也可以用它来形成**起居空间的序列**（142），把小角落留出作坐位，不受门的干扰；你也可以用它来造成**明暗交织**（135），因为每扇门，如果装有玻璃，离窗又近，会创造出一种吸引人的天然投光区域。

❧❧❧

一个房间是否设计得好，在很大程度上取决于门的位置。如果由门造成的一种活动方式使房内的空间无法利用，这个房间绝对不会使人感到舒适。

首先是单门房间的情况。一般地说，这个门最好位于角落。如它位于墙的中部，它造成的活动方式势必把房间分成两半，使中心区无法使用，并造成房间内每一块地方都小到派不上用场。但狭长的房间一般不受这一规律的影响。对于狭长的房间，在长边的中间开门是合情合理的，因为这样就会造成两块地方，每一块差不多都呈正方形，这样大的面积足可供人利用。这种开在中间的门，当房间两部分有各自独立的功能时，特别有用，这样可使房间自然地一分为二。

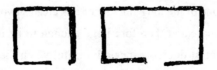

只有一扇门的房间
Rooms with one door

196   CORNER DOORS*

...this pattern helps you place doors exactly.Use it to help create the larger FLOW THROUGH ROOMS(131).You can use it too,to generate a SEQUENCE OF SITTING SPACES (142),by leaving small corners for sitting,uninterrupted by the doors;and you can use it to create TAPESTRY OF LIGHT AND DARK(135),since every door,if glazed and near a window,will create a natural pool of light which people gravitate toward.

⊰⊱

The success of a room depends to a great extent on the position of the doors.If the doors create a pattern of movement which destroys the places in the room,the room will never allow people to be comfortable.

First there is the case of a room with a single door.In general,it is best if this door is in a corner.When it is in the middle of a wall,it almost always creates a pattern of movement which breaks the room in two,destroys the center,and leaves no single area which is large enough to use.The one common exception to this rule is the case of a room which is rather long and narrow.In this case it makes good sense to enter from the middle of one of the long sides,since this creates two areas,both roughly square,and therefore large enough to be useful.This kind of central door is especially useful when the room has two partly separate functions,which fall naturally into its two halves.

Now,the case of a room with two or more doors:the

现在讲一下一个房间有两扇或几扇门的情况。根据上述理由，每扇门应该仍旧开在角落。但我们现在必须考虑的不仅是各扇门的位置，还有门和门之间的关系。只要做得到，应该将它们开在同一边，使房间的其余地方不受人们进出的影响。

作为普遍规律，如果我们画一条各门之间的连线，连线之外的空间应该有足够的大小可供利用，而且还应具备完整的有效体形——在活动的走道之间留下三角形空间没有什么使用价值。

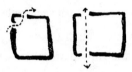

多门的房间
Rooms with more than one door

最后请注意，不能将本模式应用于很大的房间。在一个很大的房间里，或在一个当中放有一张大桌子的房间里，门可以开在中间，仍能给人一种特别正规的、宽敞的感觉。事实上，在此情况下，为了能够让人产生这种感觉，把门开在中间会更好。但这种办法只能在房间大到能从中获益的情况下才可行。

因此：

除了在很大的房间以外，把门开在墙的中部不会有多少好处。比如，入口空间的门要开在中部，因为这种空间主要是通过门表现其特征的。但对大多数房间，特别是小房间，要尽可能地把门开在靠近房间角落的地方。如果这个房间有两扇门，人们又要在这个房间穿行，则应使两扇门都开在同一侧。

individual doors should still be in the corners for the reasons given above.But we must now consider not only the position of the individual doors,but the relation between the doors. If possible,they should be placed more or less along the same side,so as to leave the rest of the room untouched by movement.

More generally,if we draw lines which connect the doors,then the spaces which are left uncut by these lines,should be large enough to be useful, and should have a strong positive shape—a triangular space left between paths of circulation will hardly ever be used.

Finally, note that this pattern does not apply to very large rooms.In a very large room,or in a room with a big table in the middle,the doors can be in the middle,and still create a special formal,spacious feeling.In fact,in this case,it may even be better to put them in the middle,just to create this feeling.But this only works when the room is large enough to benefit from it.

Therefore:

Except in very large rooms,a door only rarely makes sense in the middle of a wall.It does in an entrance room,for instance,because this room gets its character essentially from the door.But in most rooms,especially small ones,put the doors as near the corners of the room as possible.If the room has two doors,and people move through it,keep both doors at one end of the room.

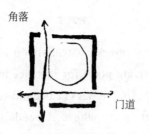

角落

门道

&C&

　　当门标志过渡时，如通过门进入一个卧室或一个私密的地方，要大胆地把它做得尽量低——**低门道**（224）；如一个空间需要特别私密，用壁橱加厚门道——**居室间的壁橱**（198）。以后，当你做门框时，使它跟墙连成一个整体，可以随意装饰它——**门窗边缘加厚**（225）、**装饰**（249）；除房间非常私密以外，要在门上装窗——**镶玻璃板门**（237）……

　　凡是有凹室、窗户、搁架、壁橱或坐位的地方，所有的墙都要有一定的深度。

　　197. 厚墙

　　198. 居室间的壁橱

　　199. 有阳光的厨房工作台

　　200. 敞开的搁架

　　201. 半人高的搁架

　　202. 嵌墙坐位

　　203. 儿童猫耳洞

　　204. 密室

80C3

When a door marks a transition,as it does into a bedroom or a private place,for instance,make it as low as you dare—LOW DOORWAY(224);and thicken the entry way with closet space where it needs to be especially private—CLOSETS BETWEEN ROOMS(198).Later,when you make the door frame,make it integral with the wall,and decorate it freely—FRAMES AS THICKENED EDGES(225),ORNAMENT(249);except when rooms are very private,put windows in the door—SOLID DOORS WITH GLASS(237)...

give all the walls some depth,wherever there are to be al-coves,windows,shelves,closets or seats.

197. THICK WALLS
198. CLOSETS BETWEEN ROOMS
199. SUNNY COUNTER
200. OPEN SHELVES
201. WAIST-HIGH SHELF
202. BUILT-IN SEATS
203. CHILD CAVES
204. SECRET PLACE

模式197 厚墙**

197　THICK WALLS**

　　...once the plan is accurate to the nearest 5 or 6 feet,there is a final process in which the smallest spaces—niches,built-in seats,counters,closets and shelves—get built to form the walls.Or of course,you can build this pattern into an existing house.In either case,use the pattern so that it helps to create the proper shapes for rooms—THE SHAPE OF INDOOR SPACE(191),the ceiling heights—ALCOVES(179),WINDOW PLACES(180),and CEILING HEIGHT VARIETY (190),and,on the outside of the rooms,the nooks and crannies of the BUILDING EDGE(160).

୫୦୧୫

　　Houses with smooth hard walls made of prefabricated panels,concrete,gypsum,steel,aluminum,or glass always stay impersonal and dead.

　　In the world we live in today,newly built houses and apartments are more and more standardized.People no longer have a chance to make them personal and individual.A personal house tells us about the people who live there.A child's swing hanging in a doorway reflects the attitude of parents to their children.A window seat overlooking a favorite bush supports a contemplative,dreamy nature.Open counters between kitchen and living space are specific to informal family life;small closable hatches between the two are specific to more formal styles.An open shelf around a room should be seen at one height to display a collector's porcelain,best seen from

……当设计进行到靠墙最外面的五六英尺时，要做最后的一道工序，即把最小的空间——壁龛、嵌墙坐位、厨房工作台、壁橱和搁架——建造起来以形成墙体。或者，你当然也可以把这一模式应用到现有的住宅中。无论哪一种情况下，都可利用本模式使之有助于创造出房间的合适形状——**室内空间形状**（191）、天花板高度——**凹室**（179）、**窗前空间**（180）和**天花板高度变化**（190），而在房间外面，创造出**建筑物边缘**（160）的凹角和缝隙。

☙☙☙❧

住宅如果采用预制板、混凝土、石膏、钢筋、铝材或玻璃制成平滑的硬墙，则毫无个性，且无变动的余地。

在我们今天所生活的世界上，新盖的住宅和公寓越来越标准化。人们已不再能够使它们具有个性和特点。一幢有个性的住宅能够表达出住宅主人的个性。一架挂在门口的儿童秋千反映了父母对他们孩子的态度。一个能俯视一片树丛的窗前坐位引人冥思遐想。摆在厨房和起居空间之间的厨房工作台表现了不拘礼节的家庭生活的特点；在厨房和起居空间之间装上小门则具有较为庄重的风格。房间周围敞开的搁架，为了展示主人收藏的瓷器，应有一定的高度，让人居高临下看得清楚；如用来摆放一个摄影师最新摄制的照片，就要有另一高度和深度；用于放置主人宴请宾客的饮料，则又要有另一高度。一个足够大的装有壁炉的凹室，要有足够多的嵌墙坐位，以吸引六口之家围坐团聚。

这些器物中的每一件都可以使我们了解住在这个房子里的人，因为它们都表达了某种特殊的个人需要。每个人都需要有机会使其周围环境适应自己的生活方式。

above;at another height and depth if it is to be used to support a photographer's latest pictures;at another height again for setting down drinks in the house of a perennial party-giver.A large enough fireplace nook,with enough built-in seats,invites a family of six to sit together.

Each of these things gives us a sense about the people living in the house because each expresses some special personal need.And everyone needs the opportunity to adapt his surroundings to his own way of life.

In traditional societies this personal adaptation came about very easily.People lived in the same place for very long periods,often for whole lifetimes.And houses were made of hand-processed materials like wood,brick,mud,straw,plaster, which are easily modified by hand by the inhabitants themselves. Under these conditions,the personal character of the houses came about almost automatically from the fact of occupancy.

However,in a modern technological society,neither of these two conditions holds good.People move frequently,and houses are increasingly built of factory-made,factory-finished materials,like 4×8 foot sheets of finished plaster board,aluminum windows,prefabricated baked enamel steel kitchens,glass,concrete,steel—these materials do not lend themselves at all to the gradual modification which personal adaptation requires.Indeed,the processes of mass production are almost directly incompatible with the possibility of personal adaptation.

The crux of the matter lies in the walls.Smooth hard flat industrialized walls make it impossible for people to express their own identity,because most of the identity of a dwelling lies in or near its surfaces—in the 3 or 4 feet near the walls.

在传统的社会里这种适合个性的布置很容易做到。人们长时间住在一个地方，常常一住一辈子。这些住宅又都是用一些经手工处理的材料建造的，如木头、砖块、泥浆、麦秆、抹灰，这些材料都很容易由住宅主人自己用手来进行改造。在这样的情况下，住宅自然而然地表现出它主人的个性。

然而，在现代技术社会里这两种状况都已不复存在。人们流动频繁，房子又越来越多地由工厂预制和装修的材料建造，如 4ft×8ft 预制抹灰板块、铝制窗框、预制厂焙制的搪瓷钢灶具、玻璃、混凝土、钢筋——这些材料完全不能使人加以逐步改造，因此难以满足个人的要求。毫无疑问，大规模生产的过程几乎直接同满足个人要求相矛盾。

问题的症结在于墙。工业化制作的光滑、坚硬、平整的墙使人无法表现他们个人的特点，因为住宅的特点大多体现在墙的内部或周围——即在墙附近 3ft 或 4ft 处。在这些地方人们放置他们的大部分家产；在这些地方装有特制的灯座；在这些地方摆设特制的嵌墙家具；在这些地方有着特殊且舒适的凹室和角落，供家庭某些成员作为个人居室；在这些地方时常会发生小范围的变化；在这些地方人们最容易改换环境，欣赏到自己的手工成绩。

住宅的特点通过它的墙表现出来
The identity of a house comes from its walls

This is where people keep most of their belongings;this is where special lighting fixtures are;this is where special built-in furniture is placed;this is where the special cosy nooks and corners are that individual family members make their own;this is where the identifiable small—scale variation is;this is the place where people can most easily make changes and see the product of their own craftsmanship.

The house will become personal only if the walls are so constructed that each new family can leave its mark on them—they must,in other words,invite incremental fine adjustments,so that the variety of the inhabitants who live in it rubs off on them.And the walls must be so constructed that these fine adjustments are permanent—so that they do accumulate over time and so that the stock of available dwellings becomes progressively more and more differentiated.

All this means that the walls must be extremely deep. To contain shelves,cabinets,displays,special lights,special surfaces,deep window reveals,individual niches,built in seats and nooks,the walls must be at least a foot deep;perhaps even three or four feet deep.

And the walls must be made of some material which is inherently structural—so that however much of it gets carved out,the whole remains rigid and the surface remains continuous almost no matter how much is removed or added.

Then,as time goes on,each family will be able to work the wall surfaces in a very gradual,piecemeal,incremental manner. After a year or two of occupancy,each dwelling will begin to show its own characteristic pattern of niches,bay windows,breakfast nooks,seats built into the walls,shelves, closets,lighting arrangements,sunken parts of the floor,raised parts of the ceiling.

住宅是否具有个性，要看墙的构造是否能使每个新住进去的家庭都在它上面留下自己的标记——换句话说，要看墙是否可以让人不断作些细微的调整，使住户的特色在它上面体现出来。而且墙的构造还应使这些细微的调整永久地保留下来——这样它们能够长期积累起来，而使各种住宅逐渐都变得具有自己的特色。

所有这一切都说明，墙必须很厚。为了能容纳搁架、柜子、陈设品、特制的灯具、有特色的表面装饰、深的窗洞、各自独立的壁龛、嵌墙坐位和凹角，墙至少要有 1ft 厚，多半甚至要有 3ft 或 4ft 厚。

墙必须由某些固有的结构材料制成，不管它被挖去多少，其整体始终坚固，其表面无论去掉或增加多少，始终连续。

于是，随着岁月的流逝，每个家庭都将能够以很缓慢的、一点一滴的、不断增加的方式完善自家的墙。住了一两年之后，每个住家会开始使自己的壁龛、凸窗、小餐室、嵌墙坐位、搁架、壁橱、灯具、下沉式地面、隆起的天花板，都呈现出自己的独特品味。

每幢住宅都会有自己的纪念物；不同人的特点和个性都可以记载在这厚墙之中；住宅越古老，其特征越突出。每个人都可通过选择和进行点滴修改来营造适合自己的环境。本模式全文最初发表在亚历山大的《厚墙》一文中。(*Architectural Design*，July 1968，pp.324~326.)

因此：

可以大胆考虑把你住宅的墙做得很厚，使它具有很大的体量——甚至具有实际上可利用的空间——而不应只是一层没有多少厚度的薄膜。要确定应在哪里建造这些厚墙。

Each house will have a memory;the characteristics and personalities of different human individuals can be written in the thickness of the walls;the houses will become progressively more and more differentiated as they grow older,and the process of personal adaptation—both by choice and by piecemeal modification—has room to breathe.The full version of this pattern was originally published by Christopher Alexander: "Thick Walls," *Architectural Design*,July 1968, pp.324-326.

Therefore:

Open your mind to the possibility that the walls of your building can be thick,can occupy a substantial volume— even actual usable space—and need not be merely thin membranes which have no depth.Decide where these thick walls ought to be

⁊⁋

Where the thickness is 3 or 4 feet,build the thickness and the volume of the walls according to the process described in THICKENING THE OUTER WALLS(211);where it is less,a foot or 18 inches,build it from open shelves stretched between deep vertical columns—OPEN SHELVES(200),COLUMNS AT THE CORNERS(212).Get the detailed position of the various things within the wall from the patterns which define them:WINDOW PLACE(180),CLOSETS BETWEEN ROOMS(198),SUNNY COUNTER(199),WAIST HIGH SHELF(201),BUILTIN SEATS (202),CHILD CAVES (203),SECRET PLACE(204)...

1~4ft

可用手工挖开

❀❀❀

　　根据**加厚外墙**（211）中所描述的过程，在厚度为 3～4ft 的地方，建造墙的厚度和体量；在薄一些的地方，1ft 或 18in，根据深垂直柱之间的敞开搁架的宽度造墙——**敞开的搁架**（200）、**角柱**（212）。墙内各种东西的摆法详见规定它们的各个模式：**窗前空间**（180）、**居室间的壁橱**（198）、**有阳光的厨房工作台**（199）、**半人高的搁架**（201）、**嵌墙坐位**（202）、**儿童猫耳洞**（203）、**密室**（204）……

模式198 居室间的壁橱*

198   CLOSETS BETWEEN ROOMS*

...given the layout of rooms,it is now necessary to decide exactly where to put the built-in cupboards and closets. Use them,especially,to help form the enclosure around a workspace—WORKSPACE ENCLOSURE(183),around a dressing space—DRESSING ROOM(189),and around the doors of rather private rooms so that the doorway itself gets some depth—CORNER DOOR(196)

ഇൿരൟ

The provision of storage and closets usually comes as an afterthought.

But when they are correctly placed,they can contribute greatly to the layout of the building.

Perhaps the most important secondary feature of storage space is its sound insulating quality.The extra wall sections,and the doors enclosing the closet,as well as the clothes,boxes,and so on,that are being stored,all work to create substantial acoustical barriers.You can take advantage of this feature of closet space by locating all required storage areas within the walls separating rooms rather than in exterior walls,where they cut off natural light.

In addition,when storage is placed in the interior walls of a room,around the doorway,the resulting thickness will make the transitions between rooms and corridors more distinct.For the person entering such a room,the thickness of the wall creates a subtle "entry" space,which makes the room more private.This

……各居室的布局已经完成，现在需要确定，在哪里建造嵌墙碗橱和壁橱。特别是要利用它们来形成工作空间周围的围隔——**工作空间的围隔**（183），更衣空间周围的围隔——**更衣室**（189）以及很私密的房间的门周围的围隔，使门道有一定的深度——**墙角的房门**（196）。

❧❧❧

建造储藏间和壁橱通常要三思而行。

但一旦它们的位置选得合适了，它们对建筑物的布局起很大作用。

储藏空间第二个最重要的特征多半是它的隔音性能。附设的墙、关住壁橱的门，以及藏在橱内的衣服、箱子等杂物，都能起很好的隔音作用。你可以利用壁橱空间的这一特点，但要使所有必需的储藏区位于分隔居室的内墙，而不在外墙，在外墙它们会遮挡自然光。

此外，当储藏间设在居室的内墙和门道周围时，所形成的厚度会使居室和过道之间的过渡更加分明。对于进入这样的居室的人，墙的厚度产生一个小的"入口"空间，它使居室显得更加私密。这种在入口周围构成壁橱"厚度"的方法因而适合于这样一些空间，如**夫妻的领域**（136）和各种私密的房间——**个人居室**（141）。

壁橱形成居室的入口
Closets form the entrance to the rooms

way of making the closet "thickness" around an entrance is therefore appropriate for spaces like the COUPLE'S REALM (136)and the various private rooms—A ROOM OF ONES OWN(141).

Therefore:

Mark all the rooms where you want closets.Then place the closets themselves on those interior walls which lie between two rooms and between rooms and passages where you need acoustic insulation.Place them so as to create transition spaces for the doors into the rooms.On no account put closets on exterior walls.It wastes the opportunity for good acoustic insulation and cuts off precious ight.

৪৩৩

Later,include the closets as part of the overall building structure—THICK WALLS(197)...

因此：

把你想要有壁橱的所有居室都标志出来。然后把壁橱做在内墙，使它们处于需要隔声的两个居室之间或居室与过道之间。利用壁橱的位置来营造出由门进入居室的过渡空间。绝不可把壁橱设在外墙，否则会白白浪费了营造良好隔声效果的机会并阻挡了宝贵的阳光。

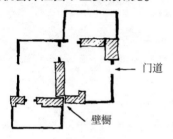

随后，把壁橱作为全部建筑结构的一部分——**厚墙**（197）……

模式199　有阳光的厨房工作台*

　　……农家厨房（139）和厨房布置（184）讲述了厨房
的全面设计及其工作空间的情况。室内阳光（128）要使厨
房内一定有阳光。但为了有助于创造这些较大的模式，并
使厨房尽可能地暖和与美观，如何选定厨房工作台和它的
窗户的位置是十分值得注意的。

⊰⊱

　　阴暗的厨房是令人不愉快的。厨房比起其他房间需要
更多而非更少的阳光。

199   SUNNY COUNTER*

...FARMHOUSE KICHEN(139)and COOKING
LAYOUT(184)give the overall design of the kitchen,and its
workspace.INDOOR SUNLIGHT(128)makes sure of sunshine
in the kitchen.But to help create these larger patterns,and to
make the kitchen as warm and beautiful as possible,it is worth
taking a great deal of care placing the counter and its widows.

ജൗൽ

**Dark gloomy kitchens are depressing.The kitchen
needs the sun more than the other rooms,not less.**

Look how beautiful the workspace in our main picture
is.Nearly the whole counter is lined with windows.The work
surface is bathed in light,and there is a sense of spaciousness all
around.There is a view out,an air of calm.

Compare it with this gloomy kitchen.There is no natural
light on the work counter,the cabinets are a clutter;it is a shabby
experience to work there-to work below a cabinet,facing a wall
with artificial light in the middle of the day.

This gloomy kitchen is typical of many thousands
of kitchens in modern houses.It happens for two reasons.
First,people often place kitchens to the north,because they
reserve the south for living rooms and then put the kitchen in
the left over areas.And it happens,secondly,when the kitchen
is thought of as an "efficient" place,only meant for the
mechanical cooking operations.In many apartments,efficiency
kitchens are even in positions where they get no natural

瞧，本模式的主要插图中的这个工作间多美。几乎整个厨房工作台紧靠窗户。工作面沐浴着阳光，周围多么宽敞。窗外风光，一片宁静气象。

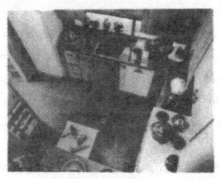

阴暗的厨房
A groomy kitchen

把它跟这个阴暗的厨房比一比。厨房工作台上不见阳光，食品柜杂乱无章；在这儿干活要站在碗柜下方，面对着大白天都要靠灯光照亮的墙壁。这真叫人难堪。

这种阴暗的厨房就是在现代住宅里最常见到的成千上万的标准厨房。它之所以这样，原因有二：首先，人们常常把厨房设在北边，因为他们要把南边留下来作起居室，而把厨房放在被挑剩下的区域；其次，厨房被认为是"讲究工作效率"的地方，只是用来进行机械的烹调操作的地方。在许多公寓，简易厨房里甚至看不到一点自然光。但是，我们在**农家厨房**（139）中提出的要把厨房变成起居室而不仅是机械工作间的论点自然把这一切都改变了。

因此：

把厨房工作台的主要部分放在厨房的南边或东南边，其周围有宽大的窗户，使阳光能够直射进来，上下午厨房都沐浴着金黄色的光辉。

light at all.But,of course,the arguments we have presented in FARMHOUSE KITCHEN(139)for making the kitchen a living room,not merely a machine-shop,change all this.

Therefore:

Place the main part of the kitchen counter on the south and southeast side of the kitchen,with big windows around it,so that sun can flood in and fill the kitchen with yellow light both morning and afernoon.

৪০৫৪

Give the windows a view toward a garden or the area where children play—WINDOWS OVERLOOKING LIFE (192).If storage space is tight,you can build open shelves for bowls and plates and plants right across the windows and still let in the sun—OPEN SHELVES (200).Build the counter as a special part of the room,integral with the building structure,able to take many modifications later—THICKENING THE OUTER WALLS(211).Use WARM COLORS(250)around the window to soften and warm the sunlight...

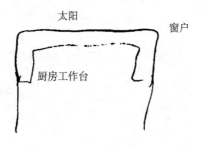

太阳　　　　　　　　　窗户

厨房工作台

⳾〇⳽

使厨房的窗户对着花园或儿童们玩耍的地方——**俯视外界生活之窗**（192）。如果储藏空间窄小，你可以正对窗户做些敞开的搁架放置碗盘，还可以布置一些装置，但仍然能让阳光照射进来——**敞开的搁架**（200）。使厨房工作台成为该空间的特殊部分，同建筑结构连成一体，以后可以进行许多修改——**加厚外墙**（211）。在窗周围使用**暖色**（250）使阳光柔和而温暖……

模式200　敞开的搁架*

200   OPEN SHELVES*

...within the THICK WALLS(197),especially around the FARMHOUSE KITCHEN(139)and WORKSPACE ENCLOSURE(183),but possibly throughout the building,there is a need for shelves.This pattern helps you decide exactly where you want them and how they shall be organized.Mary Louise Rogers first made the pattern explicitfor us.

৪০০৪

Cupboards that are too deep waste valuable space,and it always seems that what you want is behind something else.

It is easy to think that you have good storage in a room or in a building just because you have enough closets, cupboards,and shelves.But the value of storage depends as much on the ease of access as on the amount.An enormous amount of cupboard space in a place where no one can get to it is not very useful.It is useful when you can find the things which you have put away at a glance.

This means,essentially,that except for BULK STORAGE (145),things should be stored on open shelves, "one deep." Then you can see them all.It means,in effect,that you are flattening out the total storage all over the walls—instead of having it in solid lumps,hidden,and hard to reach.

The need for open storage is most obvious in kitchens.In badly planned kitchens,the shelves are filled with things three or four items deep,sometimes stacked on top of each other,and something is always in the way of what you need.But in well-

……在**厚墙**（197）内，特别是在**农家厨房**（139）和**工作空间的围隔**（183）周围，还可能在整个建筑物中，都需要搁架。本模式帮你确定，在什么地方你需要搁架及如何把它们安排好。玛丽·罗杰斯首先替我们解释了本模式。

୫୦୯ଓ

太深的柜子浪费宝贵的空间，而且你总觉得，你要拿的东西在其他东西的后面。

只要你有足够的壁橱、碗柜和搁架，人们就会认为你的居室或住房储藏条件好。但储藏空间的价值不仅在于大小，还在于存取东西是否容易。如果柜子的空间有很大一部分谁也够不着，那是没有多大用处的。只有在你一眼就能找到自己所存放的东西时，它才有用处。

这表明，除了**大储藏室**（145）外，东西应该以"一个深度"放在敞开的搁架上。这样所有东西你都能看得见。这就等于说，你把储藏的全部东西都摊开在墙上——而不是整堆地放在那儿，看不见，够不着。

储藏的东西需要敞开，这在厨房中表现得最明显。在规划糟糕的厨房里，搁架上放了形形色色的东西，里三层，外三层，有时候还摞起来，以致总有东西碍你手脚，使你找不到要找的东西。但在规划整齐的厨房里，每一类东西都摆成一排。搁架上罐头放成一排，玻璃杯也放成一排，锅盆在墙上挂成一排；有专门的调料架放小瓶子和各种调料，使这些东西摆成一排。

我们认为，这样做，储藏起东西来就方便了。家中珍藏的器物和礼品，不管是厨房用品或在别处使用的物品，只要它们放在柜子里面或壁橱的后架上，就谁也看不见它们。若一排一排地把它们放在敞开的搁架上，这些美器家珍会使满屋生辉。

planned kitchens,all storage is one item deep.Shelves are one can deep,glasses are stored one row deep,pots and pans are hung one deep on the wall;for small jars and spices there are special spice shelves that hold the items just one deep.

We think this property is common to all convenient storage.A family's most prized possessions,gifts,whether for the kitchen or any place else in the house,are hidden away when they are stored in cupboards and the back shelves of closets. Openly stored,one deep,these things are beautiful around the house.

Many forms of storage can be one-deep:swinging cabinets that have shelves inside the doors;pegboards for pots and pans;tool racks.It is even possible to create narrow open shelves in front of windows.When things are just one deep,there is still enough light coming in to make the window useful.

Therefore:

Cover the walls with narrow shelves of varying depth but always shallow enough so that things can be placed on them one deep—nothing hiding behind anyhing else.

❧✧❧

At waist height put in an extra deep shelf for plates, phonograph,TV,boxes,displays,treasures—WAIST-HIGH SHELF(201).Mark the open shelves along with all the other deep spaces in the walls—THICKENING THE OUTER WALL(211)...

许多种储藏设备都可以让东西摆成一排：门内有搁架的吊橱；搁锅盆的木栓板；工具架。甚至在窗前也有可能布置一些狭窄的敞开的搁架。当东西摆成一排时，足够的光线射进来使窗户更具有使用价值。

立在窗对面的敞开的搁架
Open shelves across a window

因此：

可以在墙的周围布置深浅不一的狭窄的搁架，但无论如何要使搁架的深度只能放一排物品——不要在它的后面再藏着东西。

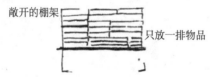

敞开的棚架　　　　　　　　　　只放一排物品

✿

在半人高的地方设置另一较深的搁架用来放盘子、唱机、电视机、盒子、摆设品、珍藏品——**半人高的搁架**（201）。将敞开的搁架同墙内所有其他深的空间一起标记出来——**加厚外墙**（211）……

模式201 半人高的搁架*

……在任何有敞开的搁架的地方和在任何一个房间周围，只要这些地方有着盆栽花木、书本、盘子、纸张、漂亮的花瓶和你旅行时采购来的小玩意儿，都需要有空间把这些东西放置妥当，而不致把房间弄得乱七八糟——**厚墙**（197）、**敞开的搁架**（200）。

8003

在每一幢住宅和每一个工作空间，每天都有一些常用物品带进携出。除非这些东西能够多随手拿到，否则生活中流通不畅，错误百出；东西不是被遗忘，就是放错地方。

这个问题的关键在于"随手拿到"这句话。事情正是这样，也应当这样。当一个人在拿一样东西的时候，他的手基本上同腰部一样高。在房间四周、在过道两旁、在门边，凡是有半人高的表面，人们自然就把东西放在这儿，以后又从这儿拿走。零钱、画片、打开的书本、一个苹果、一个包裹、一张报纸、日常的信件、一张条子，这些东西都可以在半人高的搁架上随手拿到。如果没有这样一些表面，东西要不就得收拾起来，然后被遗忘和丢失，要不就放着碍事，得不断地清理掉。

其次，收集在半人高搁架上的东西自然是一些日积月累的最普通的东西——一个人生活中最常用的东西。由于这些东西因人而异，半人高的搁架就很容易使房间变得独特而具有个性。

201   WAIST-HIGH SHELF*

...anywhere where there are open shelves,and around any room which tends to accumulate potted plants,books,plates,bits of paper,boxes,beautiful vases,and little things you have picked up along your travels,there is a need for space where these things can lie undisturbed,without making the room a mess—THICK WALLS (197),OPEN SHLVES(200).

❧❦

In every house and every workplace there is a daily"traffic"of the objects which are handled most.Unless such things are immediately at hand,the flow of life is awkward,full of mistakes;things are forgotten,misplaced.

The essence of this problem lies in the phrase "at hand." This is literally true and needs to be interpreted as such. When a person reaches for something,his hands are roughly at waist height.When there are surfaces here and there,around the rooms and passages and doors,which are at waist height they become natural places to leave things and later pick them up.Pocket change,pictures,open books,an apple,a package,a newspaper,the day's mail,a reminder note:these things are at hand on a waist high shelf.When there are no such surfaces,then things either get put away and are then forgotten and lost,or they are in the way and must continually be cleared aside.

Furthermore,the things that tend to collect on waist high shelves become a natural,evolving kind of display of the most ordinary things—the things that are most immediately a

因此：

至少可以在人们生活和工作的主要房间的某部分周围设置一些半人高的搁架。把它们做成长条形，9~15in 深，在底下有架子或柜子。使搁架挨近坐位和门窗。

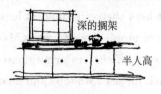

深的搁架

半人高

∞∞∞

把搁架直接造到建筑物结构中去——**加厚外墙**（211）。这是放置你个人珍藏品的好地方——**生活中的纪念品**（253）……

part of one's life.And since for each person these things will vary,the waist high shelf helps a room become unique and personal,effortlessly.

Therefore:

Build waist-high shelves around at least a part of the main rooms where people live and work.Make them long,9 to 15 inches deep,with shelves or cupboard underneath. Interrupt the shelf for seats,wndows,and doors.

ଅଠକ୍ଷ

Build the shelf right into the structure of the building— THICKENING THE OUTER WALL(211).It is a good place to put your personal treasures—THINGS FROM YOUR LIFE(253)...

模式202　嵌墙坐位*

　　……在整幢房屋里——**起居空间的序列**（142）——凹室、入口处、角落和窗户这些地方都可很自然地设置一些嵌墙坐位——**入口空间**（130）、**凹室**（179）、**窗前空间**（180）。本模式有助于完善它们。

❧❦❧

　　嵌墙坐位妙极了。人人喜爱它们。房子里有嵌墙坐位，人们感到舒适，得到享受。可惜它们常常不能真正发挥作用，因为地点不合适，或地方太窄，或椅背没有斜度，或景致不佳，或坐位太硬。本模式告诉你，怎样才能使一个嵌墙坐位真正有用。

202   BUILT-IN SEATS*

...throughout the building—SEQUENCE OF SITTING SPACES(142)—there are alcoves,entrances,corners,and windows where it is natural to make built-in seats—ENTRANCE ROOM(130),ALCOVES (179),WINDOW PLACE(180).This pattern helps complete them.

❧

Built-in seats are great.Everybody loves them.They make a building feel comfortable and luxurious.But most often they do not actually work.They are placed wrong,or too narrow,or the back does not slope,or the view is wrong,or the seat is too hard.This pattern tells you what to do to make a built-in seat that really works.

Why do built-in seats so often not work properly?The reasons are simple and fairly easy to correct.But the problems are critical.If the seats are wrongly made,they just will not be used,and they will be a waste of space,a waste of money,and a wasted golden opportunity.What are the critical considerations?

Position:It is natural to put the built-in seat into an unobtrusive corner—that is where it melts most easily into the structure and the wall.But,as a result,it is often out of the way. If you want to build a seat,ask yourself where you would place a sofa or a comfortable armchair—and build the seat *there*,not tucked into some hopeless corner.

Width and comfort:Built-in seats are often too hard,too narrow,and too stiff-backed.No one wants to sit on a shelf, especially not for any length of time.Make the seat as wide as

为什么嵌墙坐位常常不能发挥应有的作用呢？理由很简单，也容易纠正。但这些问题却很重要。如果嵌墙坐位设置得不好，人们就不爱使用它们，它们白白浪费空间，浪费钱财，而且使人坐失良机。那么最要紧的应当考虑些什么问题呢？

位置：人们自然而然地把嵌墙坐位设置在一个不惹人注目的角落，在这些地方它容易同结构和墙融为一体。但这样一来，它便十分偏僻了。如你想安排一个坐位，问问你自己，想在什么地方放一张沙发或一张舒适的扶手椅，你就会把坐位定在那儿，而不会把它塞到某个遥远的角落。

宽度和舒适：嵌墙坐位常常做得太硬、太窄、靠背太直。谁也不愿意坐在一个架子上，更别提要坐得很久。要把坐位做得跟舒适的椅子一样宽（至少18in），靠背稍斜（不是直的），在坐位和靠背上各放一个暖和的软垫，使坐位确实让人感到舒服。

视野：大多数人坐着的时候总想看看什么东西——或者看人，或者看景。嵌墙坐位常常使人既无景可看，也看不到屋里其他的人。要使人坐在坐位上能观看使他感到兴趣的东西。

因此：

在你建造嵌墙坐位之前，拿一张旧的扶手椅或一张沙发，放在你打算做坐位的位置上。来回地搬动椅子，直到你确实感到满意为止。把椅子摆在那儿放几天，看看你是否喜欢坐在那儿。如你不喜欢，再挪动它。当椅子放到了你喜欢的位置，这是你常常爱坐的地方，你就知道这个位置准合适。这时你就可以建造一个同原来的位置一样宽、铺垫得一样舒适的坐位了——你的嵌墙坐位就一定能合用。

a really comfortable chair(at least 18 inches),with a back that slopes gently (not upright),and put a warm soft cushion on it and on the back,so that it is really comfortable.

View:Most people want to look at something when they sit—either at other people or a view.Built-in seats often place you so that you are facing away from the view or away from the other people in the room.Place the seat so that a person sitting down is looking at something interesting.

Therefore:

Before you build the seat,get hold of an old arm chair or a sofa,and put it into the position where you intend to build a seat.Move it until you really like it.Leave it there for a few days.See if you enjoy sitting in it.Move it if you don't. When you have got it into a position which you like,and where you often find yourself sitting,you know it is a good position.Now build a seat that is just as wide,and just as well padded—and your built-in seat will work.

೮೦೦೪

Once you decide where to put the seat,make it part of the THICK WALLS(197),so that it is a part of the structure,not just an addition—THICKENING THE OUTER WALL(211)...

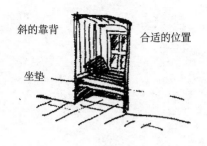

斜的靠背　　　　　合适的位置

坐垫

❧❦❧

　　一旦你把建造坐位的地方确定下来，就把它做在**厚墙**
（197）里面，使坐位成为结构的一部分，而不是后加进去
的——**加厚外墙**（211）……

模式203　儿童猫耳洞

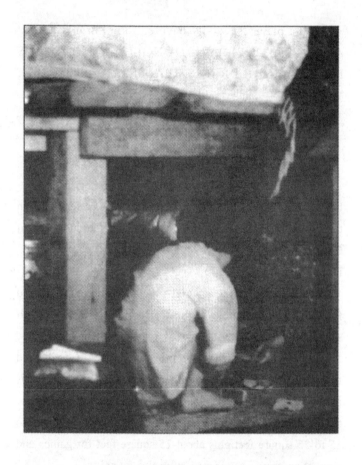

　　……专门给儿童玩耍的一些地方——**冒险性的游戏场地**（73）、**儿童之家**（86）、**儿童的领域**（137）和**厚墙**（197）——可以特别仔细地加以修饰。

203   CHILD CAVES

...the places specially devoted to children's play—
ADVENTURE PLAYGROUND(73),CHILDREN'S
HOME(86),CHILDREN'S REALM(137)—and THICK
WALLS(197)—can be embdlished with a special detail.

8003

Children love to be in tiny,cave-like places.

In the course of their play,young children seek out cave—
like spaces to get into and under—old crates,under tables,in
tents,etc.(For evidence see L.E.White, "The Outdoor Play of
Children Living in Flats," *Living in Towns*,Leo Kuper,ed.,
London,1953,pp.235-264.)

They try to make special places for themselves and
for their friends—most of the world about them is "adult
space" and they are trying to carve out a place that is kid size.

When children are playing in such a "cave" —each child
takes up about 5 square feet;furthermore,children like to do this
in groups,so the caves should be large enough to accommodate
this:these sorts of groups range in size from three to five—so
15 to 25 square feet,plus about 15 square feet for games and
circulation,gives a rough maximum size for caves.

Therefore:

**Wherever children play,around the house,in the
neighborhood,in schools,make small"caves"for them.Tuck
these caves away in natural left over spaces,under stairs,
under kitchen counters.Keep the ceiling heights low—2 feet
6 inches to 4 feet—and the entranc tiny.**

儿童们喜欢待在一些小的、洞穴般的地方。

年龄稍小的儿童在游戏的时候，总想找些洞穴般的空间钻进去或躲在它下面，他们钻进旧木箱，藏在桌子底下或帐篷里。(例见：L.E.White, "The Outdoor Play of Children Living in Flats," *Living in Towns*,Leo Kuper,ed.,London,1953,pp. 235~264.)

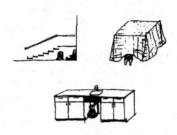

他们想方设法替他们自己和朋友创造一些他们专用的地方，因为他们周围的世界大部分都是"成年人的空间"，他们想办法设计出一个适合于儿童的小空间。

当儿童们在这样一个猫耳洞里玩耍的时候，每个儿童占有大约 5ft^2；而且，儿童们喜欢结伴儿玩这种游戏，因此猫耳洞的大小应与此相适应：这些成群结队的人数由三至五人不等——这样他们要占 15~25ft^2，加上供做游戏和走动用的大约 15ft^2，合起来就是猫耳洞大致的最大尺寸。

因此：

无论在住宅周围、在邻里或在学校，凡是儿童玩耍的地方，都要为他们建造一些猫耳洞。把这些猫耳洞设置在天然剩下来的空间：在楼梯和在厨房工作台底下。使猫耳洞顶部很低——2ft 6in~4ft——出入口很小。

೮ﻼೞ

Build the caves right into the fabric of the walls—
THICKENING THE OUTER WALLS(211).Make the doors
very tiny to match the caves—an extreme version of LOW
DOORWAY(224)...

顶部3~4ft

⊰ৎ⊱

　　把猫耳洞直接造到墙的结构中去——**加厚外墙**（211）。
配合猫耳洞的门也要做得很小——这是一种特别的**低门道**
（224）……

模式204　密室

……本模式最后涉及厚墙，或许甚至低天花——**厚墙**（197）、**天花板高度的变化**（190）。

⊱⋇⊰

何处可以珍藏宝贵物品，使之可以消失而后复现？

我们认为，人们在自己的住家中需要有一个密室：此室非比寻常，不遇特殊情况不打开。

家设密室会丰富你的生活。它邀你藏入某些珍品，封存起来，只让某些人知道此秘密，他人不得而知。它使你有可能完全以自己的方式保管你的珍品，谁也不会发现它，有朝一日你可以告诉你的友人说："现在我要让你看一样珍奇的东西"——然后把它的来龙去脉统统说出来。

加斯顿·巴切拉德在他的《空间的诗意》(*The Poetics of Space*，New York：The Omen Press，1964) 中强调生活中有这种需要。让我们引用该书第三章中的一段话：

说到抽屉、柜子、锁和衣橱，我们不禁又想起那些潜藏心底的私密的幻梦。

带有搁板的衣橱、带有抽屉的书桌、带有假底的柜子都体现秘密的心理活动。的确，如果没有这些"物件"以及人们同样非常喜爱的另外一些东西，我们私密的生活就难以表现私密。这些东西具有混杂和主观的性质。如同我们自己一样，它们具有私密性，这是借由我们和为了我们表现出来的。

如果我们给物品一种它们应得的友谊，我们打开柜子就不会毫不动心。在它的赤褐色的木材底下，柜子像杏仁一样雪白。打开它，即经历一次雪白色的洗礼。

一本献给如银箱和珠宝盒这种小盒子的选集，会组成心理学

204 SECRET PLACE

...and here is a finishing touch to the thick walls,perhaps even to the low ceilings—THICK WALLS(197),CEILING HEIGHT VARIET(190).

❧❧❧

Where can the need for concealment be expressed;the need to hide;the need for something precious to be lost,and then revealed?

We believe that there is a need in people to live with a secret place in their homes:a place that is used in special ways,and revealed only at very special moments.

To live in a home where there is such a place alters your experience.It invites you to put something precious there,to conceal,to let only some in on the secret and not others.It allows you to keep something that is precious in an entirely personal way,so that no one may ever find it,until the moment you say to your friend, "Now I am going to show you something special" —and tell the story behind it.

There is strong support for the reality of this need in Gaston Bachelard's *The Poetics of Space*(New York:The Omen Press,1964).We quote from Chapter 3:

With the theme of drawers,chests,lock s and wardrobes,we shall resume contact with the unfathomable store of daydreams of intimacy.

Wardrobes with their shelves,desks with their drawers,and chests with their false bottoms are veritable organs of the secret psychological life.Indeed,without these "objects" and a few others in equally high favor,our intimate life would lack a model of intimacy.They are hybrid

上重要的一章。这些由工匠创造的精细的物品本身就说明它们需要秘密，使人自然想到要收藏起来。事情不仅在于要保护好个人的私人用品。能抵抗绝对暴力的锁是不存在的，所有的锁都会招惹盗贼。锁是心理的门槛……

因此：

在住宅中设置一个地方，可能只需几平方英尺，把它上锁保密；一个绝无可能被发现的地方——除非人家告诉你它在何处；一个保存该住宅档案或其他更大秘密的地方。

密室

家庭的生活史　　　　　　　　珍贵的物品

住宅的历史

⋙⋘

传统的密室设有可向后滑动的嵌板，暴露出墙内的洞口，地毯下面有活动的木板，还有活板门——**居室间的壁橱**（198）、**加厚外墙**（211）、**楼面天花拱结构**（219）……

objects,subject objects.Like us,through us and for us,they have a quality of intimacy...

If we give objects the friendship they should have,we do not open a wardrobe without a slight start.Beneath its russet wood,a wardrobe is a very white almond.To open it,is to experience an event of whiteness.

An anthology devoted to small boxes,such as chests and caskets,would constitute an important chapter in psychology. These complex pieces that a craftsman creates are very evident witnesses of the need for secrecy,of an intuitive sense of hiding places.It is not merely a matter of keeping a possession well guarded.The lock doesn't exist that could resist absolute violence,and all locks are a n invitation to thieves.A lock is a psychological threshold...

Therefore:

Make a place in the house,perhaps only a few feet square,which is kept locked and secret;a place which is virtually impossible to discover—until you have been shown where it is;a place where the archives of the house,or other more potent secrets,mght be kept.

⚘

Classic types of secret places are the panel that slides back,revealing the cavity in the wall,the loose board beneath the rug,the trap door—CLOSETS BETWEEN ROOMS(198),THICKENING THE OUTER WALLS(211), FLOOR-CEILING VAULTS(219)...

CONSTRUCTION

构

造

在这一阶段，你会有个体建筑的完整设计。如果你遵循所提出的模式，则会有一个空间设计方案，既可以用桩在地面上标出，又可以在图纸上示出，准确性达到1ft上下。你会知道房间的高度、门窗的大致尺寸和位置，也会粗略地知道住宅的屋顶和花园的布局。

本语言如下的最后一部分会告诉你，如何直接根据这一粗糙的空间设计方案来建造一幢真实的住宅以及如何建成它的详尽的细节。

∞∞∞

这最后一部分中列出的模式，对本语言第二部分所述的各种住宅建筑有密切关系的构造施工，表达了我们的实际观点。这些构造施工模式是特意为施工人员设计的——不论是职业施工人员，还是房主兼业余施工人员均适用。

每一模式都说明有关结构的原理和材料的性质。这些原理在实际施工中可用各种方式来加以充实。我们力图说明这些原理赖以形成的各种途径。但是，一部分原因是这些模式未得到最充分的发展，另一部分原因是由于建筑模式本身的性质，很可能，读者将对这些模式作许多补充。例如，为了实现这些模式而实际使用的建筑材料会是各种各样的……

当你阅读这部分内容时，可以记住以下要点：在本部分中我们的意图是为那些已经成为机器时代和现代建筑遗产的专家主宰一切刻板的施工方式提供一种可供选择的途径。

At this stage,you have a complete design for an individual building.If you have followed the patterns given,you have a scheme of spaces,either marked on the ground,with stakes,or on a piece of paper,accurate to the nearest foot or so.You know the height of rooms,the rough size and position of windows and doors,and you know roughly how the roofs of the building,and the gardens are laid out.

The next,and last part of the language,tells you how to make a buildable building directly from this rough scheme of spaces,and tells you how to build it,in detail.

കൗൽ

The patterns in this last section present a physical attitude to construction that works together with the kinds of buildings which the second part of the pattern language generates.These construction patterns are intended for builders—whether professional builders,or amateur owner-builders.

Each pattern states a principle about structure and materials. These principles can be implemented in any number of ways when it comes time for actual building.We have tried to state various ways in which the principles can be built.But,partly because these patterns are the least developed,and partly because of the nature of building patterns,the reader will very likely have much to add to these patterns.For example,the actual materials used to implement them will vary greatly from region to region...

Perhaps the main thing to bear in mind,as you look over this material,is this:Our intention in this section has been to provide an alternative to the technocratic and rigid ways of building that have become the legacy of the machine age and modern architecture.

这里所描述的施工方法会使建筑因地制宜而且具有特色。这取决于施工人员对建筑工程的负责态度；取决于他们施工所依据的详尽的施工图——先将入口、窗户和空间尺寸做成大模型进行实验，再按所得结果直接施工。

本部分中的模式在以下几方面具有鲜明特色。

首先，这些模式的顺序比本语言中前面的任何一组都更为具体。这不仅符合用户对设计在理性上深思熟虑的顺序，而且也符合实际施工的顺序。这就是说，除了开头4个处理结构原理的模式外，其余模式均可按所列的顺序实际应用于建筑施工。本语言的顺序几乎完全符合在地皮上实际施工的顺序。除此之外，这部分的模式本身比本语言的任何其他模式更具体而又更抽象。

它们之所以更具体，是因为我们对每一模式都作出至少一种可以直接用于施工的说明。例如，**柱基**这一模式，我们提出了一种特殊的说明，以表明这是可以办到的，同时也给读者提供一种直接的、切实可行的施工方法。

而且同时它们也更为抽象。我们对每一模式所作的特别具体而又系统的阐述，均可用其他成百上千种的方式来解释和修改。因此，就有可能把握**柱基**这一模式的一般概念，即柱基像一棵树的树根那样起作用，它把建筑锚固在地面上——从而创造出许多截然不同的符合这一基本施工法的物理系统。从这种意义上来说，这些模式比本书中的其他模式更为抽象，因为可以说明它们的范围更宽了。

为了说明这一事实，即各种不同的实际施工系统，以这些模式为基础，是能够发展的，现在我们根据不同的具体情况，提出我们已经开发出来的3种方案。

The way of building described here leads to buildings that are unique and tailored to their sites.It depends on builders taking responsibility for their work;and working out the details of the building as they go—mocking up entrances and windows and the dimensions of spaces,making experiments,and building directly according to the results.

The patterns in this section are unique in several ways.

First,the sequence of the patterns is more concrete than in any of the earlier portions of the language.It not only corresponds to the order in which a design matures conceptually,in the user's mind,but also corresponds to the actual physical order of construction.That is,except for the first four patterns,which deal with structural philosophy,the remaining patterns can actually be used,in the sequence given, to build a building.The sequence of the language corresponds almost exactly,to the actual sequence of operations on the building site.In addition,the patterns themselves in this section are both more concrete,and more abstract,than any other patterns in the language.

They are more concrete because,with each pattern,we have always given at least one interpretation which can be built directly.For instance,with the pattern ROOT FOUNDATION,we have given one particular interpretation,to show that it can be done,and also to give the reader an immediate,and practical, buildable approach to construction.

Yet at the same time,they are also more abstract.The particular concrete formulation which we have given for each pattern, can also be interpreted,and remade in a thousand ways. Thus,it is also possible to take the general idea of the pattern,the idea that the foundation functions like a tree root,in

在墨西哥：带有搭接筋的混凝土基础构件块；墙和柱采用的以竹筋加固的空心自调土模构件块；用粗麻布制成的混凝土梁；用土和沥青覆面的陡筒拱——每一构件都可粉刷。

在秘鲁：地面和墙基整体浇筑；用烧制的软质砖铺砌地面；用硬木（diablo fuerte）作柱和梁；竹条抹灰后用作柱间的剪力墙；斜铺木质天花板和地板；还有竹编隔墙。

在伯克利：有彩色蜡饰面的混凝土平板；墙外皮为一层木板，墙里皮为一层石膏板，中间填轻质混凝土；箱形柱也如此浇筑；还有木板条和粗麻布支模筑成的 2in 厚的混凝土天花板和地面。

从这些例子中可以看到我们在系统地阐明这些模式的同时，对造价是十分注意的。在这些模式实例中我们力求使用最便宜的和最易得到的材料。我们设计的住宅可由非专业人员来施工（因此，他们可以完全不计劳务费用）；而且我们也设计出了这样一种建筑：如果由专业人员来施工，其劳动的费用也是低廉的。

本语言的三部分中，这第三部分是发挥得最不充分的。《城镇》和《建筑》这两部分内容都已经过实践的检验，前者是部分地而后者是全面、彻底地经过实践的检验。这第三部分中至今只有少量比较小型的建筑进行过实践的检验。很明显，这意味着第三部分内容还须不断加以充实和完善。

然而，我们还是试图尽可能对所有这些模式在各种不同的建筑物——住宅、公共建筑、细部和各种附加建筑——中进行充分试验。我们一旦有足够的例子值得进行评论的话，我们将再次发表另一卷来论述这些案例，并叙述我们的发现。

the way that it anchors the building in the ground—and invent a dozen entirely different physical systems,which all work in this fundamental way.In this sense,these patterns are more abstract than any others in the book,since they have a wider range of possible interpretations.

To illustrate the fact that a great variety of actual building systems can be developed,based on these patterns,we present three versions that we have developed,in response to different contexts.

In Mexico:Concrete block foundations with re-bar connectors;hollow self-aligning molded earth blocks reinforced with bamboo for walls and columns;burlap formed concrete beams; steep barrel vaults with earth and asphalt covering— everything whitewashed.

In Peru:Slab floors poured integrally with wall foundations; finished with soft baked tiles;hard wood(diablo fuerte)columns and beams;plaster on bamboo lath acting as shear walls between columns;diagonal wood plank ceiling/floors; bamboo lattice partitions.

In Berkeley:Concrete slab finished with colored wax;walls of exterior skin of I x boards and interior skin of gypboard filled with light weight concrete;box columns made of I x boards, filled with lightweight concrete;2-inch concrete ceiling/ floor vaults formed with wood lattice and burlap forms.

As you can see from these examples,we have formulated these patterns with very careful attention to cost.We have tried to give examples of these patterns which use the cheapest,and most easily available,materials;we have designed them in such a way that such buildings can be built by lay people(who can therefore avoid the cost of labor altogether);and we have designed it so that the cost of labor,if done professionally,is also low.

这一部分虽在许多方面还嫌粗糙，但仍然是本语言中最激动人心的。因为正是在这里，在这些为数不多的模式中，我们才能够生动地看到在其影响下呈现在我们面前的名副其实的建筑。

有关构造施工的实际过程的描述可参阅《建筑的永恒之道》第 23 章。这些模式序列在构造施工的实际过程中创造出建筑。

在设计出构造施工的细部之前，建立起一种可以将结构直接从你的建筑平面图和你对于该建筑的概念中发展起来的结构原理。

205. 结构服从社会空间的需要

206. 有效结构

207. 好材料

208. 逐步加固

Of the three parts of the language,this third part is the least developed.Both the part on Towns and the part on *Buildings* have been tested,one partially,the other very thoroughly, in practice.This third part has so far only been tested in a small number of relatively minor buildings.That means, obviously,that this material needs a good deal of improvement.

However, we intend,as soon as possible,to test all these patterns thoroughly in various different buildings—houses, public buildings,details,and additions.Once again,as soon as we have enough examples to make it worth reporting on them,we shall publish another volume which describes them,and our findings.

In many ways,rough though it is,this is the most exciting part of the language,because it is here,in these few patterns,that we can most vividly see a building literally grow before our eyes,under the impact of the patterns.

The actual process of construction,in which the sequence of their patterns creates a building,is described in chapter 23 of *The Timeless Way*.

Before you lay out construction details,establish a philosophy of structure which will let the structure grow directly from your plans and your conception of the buildings.

205. STRUCTURE FOLLOWS SOCIAL SPACES
206. EFFICIENT STRUCTURE
207. GOOD MATERIALS
208. GRADUAL STIFFENING

模式205 结构服从社会空间的需要**

  ……如果你已经利用本语言前面的模式，你的平面图就是以社会空间的精巧多变的安排为基础了。当你开始建造时，除非你找出一种施工方法，它既能服从社会空间的需要，而又不致歪曲空间或由于工程上的原因而重新安排空间，否则所有这些社会空间的优美和精致就会遭到破坏。

  本模式提供这种施工方法的初步入门。它是专门处理结构和施工的 49 个模式中的第一个；它是所有语言，包括从房间和建筑设计的较大模式到详细说明构造施工过程的较小模式，全部需要通过的一个难关。它不仅包括自己固

205 STRUCTURE FOLLOWS
SOCIAL SPACES**

...if you have used the earlier patterns in the language,your plans are based on subtle arrangements of social spaces. But the beauty and subtlety of all these social spaces will be destroyed,when you start building,unless you find a way of building which is able to follow the social spaces without distorting or rearranging them for engineering reasons.

This pattern gives you the beginning of such a way of building. It is the first of the 49 patterns which deal specifically with structure and construction;it is the bottleneck through which all languages pass from the larger patterns for rooms and building layout to the smaller ones which specify the process of construction.It not only has its own intrinsic arguments about the relation between social spaces and load-bearing structure—it also contains,at the end,a list of all the connections which you need for patterns on structure,columns,walls, floors,roofs,and all the details of construction.

&ΟΕ&

No building ever feels right to the people in it unless the physical spaces(defined by columns,walls,and ceilings) are congruent with the social spaces(defined by activities and human groups).

And yet this congruence is hardly ever present in modern construction.Most often the physical and social spaces are incongruent.Modern construction—that is,the form of construction most commonly practiced in the mid-twentieth

有的关于社会空间和承重结构之间关系的各种论据；而且，最后，还包含与结构、柱、墙、地板、屋顶和施工的全部细部等有关模式之间种种联系的一览表。

⋙⋘

除非物理空间（由柱、墙、天花所限定）与社会空间（由各种活动和人群限定）相协调，否则就没有任何建筑能够让其内部的人们感到舒适。

可是，这种协调在现代建筑施工中几乎是不存在的。最为常见的恰是物理空间和社会空间不协调。现代构造施工——即20世纪中叶最广泛的施工形式——通常硬把社会空间塞入建筑的构架，而构架的形状取决于工程上的各种考虑。

这种不协调有两种截然不同的情况：

一方面，有一些建筑的结构形状确实是所需的，实际上迫使社会空间服从构造施工的形式——例如巴克明斯特·富勒的穹顶、双曲抛物面和受拉结构等。

另一方面，有些建筑的结构构件很少——只是几根巨型立柱而已，别无他物。在这样的建筑中社会空间是由轻质的非承重的隔墙所限定，而这种隔墙在工程所提供的"灰色的"物理结构空间内可以自由移动。如密斯·凡·德·罗、斯基德莫尔、奥因斯和梅里尔的建筑就是例证。

短程穹顶
Geodesic dome

钢架和玻璃
Steel and glass

century usually forces social spaces into the framework of a building whose shape is given by engineering considerations.

There are two different versions of this incongruence.

On the one hand,there are those buildings whose structural form is very demanding indeed and actually forces the social space to follow the shape of the construction—Buckminster Fuller domes,hyperbolic paraboloids,tension structures are examples.

On the other hand,there are those buildings in which there are very few structural elements—a few giant columns and no more. In these buildings the social spaces are defined by lightweight nonstructural partitions floating free within the "neutral" physical structure given by the engineering.The buildings of Mies van der Robe and Skidmore Owings and Merrill are examples.

We shall now argue that both these kinds of incongruence do fundamental damage—for entirely different reasons.

In the first case the structure does damage simply because it constrains the social space and makes it different from what it naturally wants to be.To be specific:we know from our experiments that people are able to use this pattern language to design buildings for themselves;and that the plans they create,unhampered by other considerations,have an astonishing range of free arrangements,always finely tuned to the details of their lives and habits.

Any form of construction which makes it impossible to implement these plans and forces them into the strait jacket of an alien geometry,simply for structural reasons,is doing social damage.

Of course,it could be argued that the structural needs of a building are as much a part of its nature as the social and psychological needs of its inhabitants.This argument might perhaps,perhaps,hold water if there were indeed no way of building buildings which conform more exactly to the loose

现在我们来探讨一下这两种不协调所造成的严重破坏——其原因全然不同。

在第一种情况下，结构遭到损坏仅仅因为它限制社会空间，并使社会空间与它应有的空间格格不入。明确地说，根据我们的实验结果可以知道，人们能利用本模式语言为自己设计建筑，并且他们所绘制的平面图因不受其他种种考虑的牵制，所以能有一个惊人的、可自由布置的空间，结果始终能完全和他们生活和习惯上的细微要求相协调。

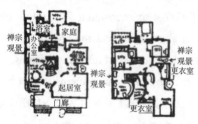

用户的住宅平面图
User's house plan

任何一种构造施工方式，如果它无法实现平面图，而仅仅出于结构的原因，把平面硬塞到几何形状迥异的外壳里，那么它将有损于社会。

当然，建筑结构上的需要，如同其居住者的社会和心理上的需要一样，都是建筑的本质的一部分。这种看法是可以争论的。如果实在没有别的施工方法可以建造出更加准确符合松散的、以活动为基础的平面图的建筑，这种争论也许是站得住脚的。

但是，本书中下面列举的一些模式会使人确信无疑：存在着既在结构上合理而又完全与社会空间一致的施工方法，而且不必采取任何折中措施。由此可见，我们可以合理地摒弃任何一种不能完全适应社会活动所需的空间形状的构造施工方法。

plans based on activities alone.

But the next few patterns in this book make it very clear that there do exist ways of building which are structurally sound and yet perfectly congruent with social space,without any compromise whatever. It is therefore clear that we may legitimately reject any form of construction which cannot adapt itself perfectly to the forms of space required by social action.

What of the second kind of incongruence between social space and building form—the kind where the structure creates huge areas of almost uninterrupted "flexible" space,punctuated by occasional columns,and the social spaces are created inside this framework by nonst uctural partitions.

Once again,many important patterns cannot be incorporated into the design—LIGHT ON TWO SIDES OF EVERY ROOM(159),for example simply cannot be included in a giant rectangle.But in this type of building,there is an additional kind of incongruence between social space and engineering structure which comes from the fact that the two are virtually independent of each other.The engineering follows its own laws,the social space follows its laws—and they do not match.

This mismatch is perceived and felt not merely as a mismatch,but as a fundamental and disturbing incoherence in the fabric of the building,which makes people feel uneasy and unsure of themselves and their relation to the world.We offer four possible explanations.

First:the spaces called for by the patterns dealing with social and psychological needs are critical.If the spaces are not right,the needs are not met and problems are not solved. Since these spaces are so critical,it stands to reason that they must be felt as real spaces,not flimsily or haphazardly

在社会空间和建筑形式之间的第二种不协调就是结构形成大面积的、几乎不间断的、不时被为数不多的立柱所阻隔的"灵活"空间；并且社会空间在这构架范围内是由非承重的隔墙所形成的。

许多重要的模式再也不能被掺合进这样的设计——每室**两面采光**（159）中去了，例如，它们再也不能被套用在大的矩形之中。但是在这种施工方式中，在社会空间和工程结构之间存在着一种额外的不协调，这是因为社会空间和工程结构两者实际上是相互独立的。工程结构遵循其自身的规律，社会空间也遵循其自身的规律——结果它们互不匹配。

这种明显察觉到的失配现象，不能仅限于就事论事而已，而且要把它视为建筑结构中具有破坏性的、严重的不协调。这种情况使人们感到不快，并使其自身以及自己和周围世界的关系感到不安和无把握。现在我们提出 4 种可能的解释。

第一，涉及社会需求和心理需求的那些模式所要求的空间是非常重要的。如果空间不合适，这些需求就不会得到满足，而问题就会悬而未决。因为这些空间如此重要，所以有理由认为，它们必须被人们确定是实在的空间，而不是不足信的或任意分割的空间。而任意分割的空间只是口惠（而实不至）而已，满足不了人们的社会需求和心理需求。例如，一个入口空间如果是由轻而薄的隔墙所形成的，那是不会得到认可的；人们是不会认真对待它的。只有当最结实的建筑构件形成的空间，人们才会切实感觉到，从而他们对空间的需要才会得到充分的满足。

partitioned spaces,which only pay lip-service to the needs people experience.For instance,if an entrance room is created with flimsy partitions,it will not take hold;people won't take it seriously.Only when the most solid elements of the building form the spaces will the spaces be fully felt and the needs which call for the space then fully be satisfied.

Second:a building will also seem alien unless it gives to its users a direct and intuitive sense of its structure—how it is put together.Buildings where the structure is hidden leave yet another gap in people's understanding of the environment around them.We know this is important to children and suspect it must be important to adults too.

Third:when the social space has,as its own surrounding,the fabric of the load-bearing structure which supports that space,then the forces of gravity are integrated with the social forces,and one feels the resolution of all the forces which are acting in this one space.The experience of being in a place where the forces are resolved together at once is completely restful and whole.It is like sitting under an oak tree:things in nature resolve all the forces acting on them together:they are,in this sense,whole and balanced.

Fourth:it is a psychological fact that a space is defined by its corners.Just as four dots define a rectangle to your eye,so four posts(or more)define an imaginary space between them.

This is the most fundamental way in which solids define space.Unless the actual solids which make up the building lie at the corners of its social spaces,they must,instead,be creating other virtual spaces at odds with the intended ones.The building will only be at rest psychologically if the corners of its rooms are clearly marked and coincide,at least in the majority of cases,with its most solid elements.

第二，如果建筑物不能使用户对它的结构具有一个直接而又直观的了解，即它是怎样组合而成的，那么它也似乎是异类的了。结构隐蔽的建筑会造成人们对周围环境理解上的分歧。我们知道，这种情况对孩子们来说是重要的，但我们怀疑，它对成年人来说是否也必定重要。

第三，当社会空间，与其周围环境一样，具备支持这种空间的承重结构的建筑物时，重力就和社会力量结合在一起，结果人们会感觉到作用于这一空间的所有的力在分解。对某一处各力立即一起分解的感知是完全静态的、整体的。这仿佛我们大家坐在一棵大橡树下感觉到：各物本质上都在分解同时作用于它们的所有的力。从这种意义来讲，各种力既是个整体，而又处于平衡之中。

第四，空间由四角规定这是人的一种心理现象。正如4个点对你的眼睛来说会形成一个矩形一样，4根（或更多的）柱子就会形成一种想象的空间。

四点成一矩形
Four points make a rectangle

这是实体构件规定空间的最基本的方法。如果构成建筑的实体构件不是定位于社会空间的四角，则这些实体必定会形成违背原来意图的其他的虚拟空间。如果一幢建筑的各个房间的角隅有清晰的标志，并且在大多数情况下，至少和大多数实体构件重合，该建筑才会在人们的心理上形成静止的感觉。

Therefore:

A first principle of construction:on no account allow the engineering to dictate the building's form.Place the load bearing elements—the columns and the walls and floors—according to the social spaces of the building;never modify the social spaces to conform to the engineering structue of the building.

৪০‍৫৪

You will be able to guarantee that structure follows social spaces by placing columns at the corner of every social space—COLUMNS AT THE CORNERS(212);and by building a distinct and separate vault over each room and social space—FLOOR-CEILING VAULTS(219).

For the principles of structure which will make it possible to build your building according to this pattern,begin with EFFICIENT STRUCTURE(206);for the class of compatible materials,see GOOD MATERIALS(207);for the fundamentals of the process of construction,see GRADUAL STIFFENING(208)...

因此：

构造施工的首要原则是：绝不能由工程结构来决定建筑的形式。将承重构件——柱、墙和地面——根据建筑的社会空间来进行布置；千万不要试图通过修改社会空间去适应建筑的工程结构。

结构

社会空间

❧❧❧

通过在每一社会空间的角隅布置立柱，你就能保证结构服从社会空间的需要——**角柱**（212）；并要把每一房间和社会空间建成不同的、分离的圆拱——**楼面天花拱结构**（219）。

至于说到结构原理——它们将使你按本模式建造住宅成为可能——先从**有效结构**（206）开始；关于这类可以相容的材料，参阅**好材料**（207）；关于构造施工过程的基本原理，参阅**逐步加固**（208）……

模式206 有效结构*

    ……本模式对**结构服从社会空间的需要**（205）是个补充。前一模式限定社会空间和结构之间的关系，而本模式仅规定纯粹由工程要求的结构类型。正如你所看到的，本模式与**结构服从社会空间的需要**（205）是并行不悖的，并且有助于后者的形成。

206   EFFICIENT STRUCTURE*

...this pattern complements the pattern STRUCTURE
FOLLOWS SOCIAL SPACES(205).Where that pattern defines
the relationship between the social spaces and the structure,this
pattern lays down the kind of structure which is dictated
by pure engineering.As you will see,it is compatible with
STRUCTURE FOLLOWS SOCIAL SPACES,and wil help to
create it.

❧❧❧

**Some buildings have column and beam structures;
others have load-bearing walls with slab floors; others are
vaulted structures,or domes,or tents.But which of these, or
what mixture of them,is actually the most efficient?What is
the best way to distribute materials throughout a building,
so as to enclose the space,strongly and well,with the least
amount of material?**

Engineers usually say that there is no answer to this
question.According to current engineering practice it is first
necessary to make an arbitrary choice among the basic possible
systems—and only then possible to use theory and calculation
to fix the size of members within the chosen system.But,the
basic choice itself—at least according to prevailing dogma—
cannot be made by theory.

To anyone with an enquiring mind,this seems quite unlikely.
That such a fundamental choice,as the choice between column and
beams systems and load-bearing wall systems and vaulted systems,

一些建筑有柱结构和梁结构；另一些建筑则有带楼板的承重墙；还有一些则是圆拱结构，或是穹顶，或是帐篷结构。但是，到底哪一种结构或合用哪几种结构才是实际上最有效的呢？什么是整幢建筑物中分配使用材料的最佳方案呢？而这样节省用料所围拢的空间能使人感到舒适安全吗？

　　工程师们通常会说：对这个问题还没有答案。根据现在的工程实践的情况来看，头等重要的是在各种基本合理的系统中作任意选择，而且也只有在那时，在所选择的系统内才有可能应用理论和计算方法去确定构件的尺寸。但是，基本选择本身——至少根据流行的教条——是不能由理论来作出的。

　　对于任何一个爱刨根问底的人来说，这似乎是不可能的。像这样的基本选择，诸如在柱系统和梁系统之间的选择以及在承重墙系统和圆拱系统之间的选择纯粹是凭借一时的灵感——在这些原型之间可能存在着大量的混合系统，甚至也一概不予以考虑——所有这一切表明对可以利用的理论现状比对任何重要的见解更需进行深入的研究。

　　确实，正如我们现在在力图表明的那样，要解决建筑有效结构这一问题就要利用原型，最佳的解决方案就在3个最著名的原型之中。它们就是承重墙、按通常间距加厚的壁柱和采用圆拱的楼面天花结构，这三者构成一个完整系统。

　　我们将分3个步骤来分析出最有效结构的特性。首先，我们将说明建筑的房间和空间典型体系的三维空间特性。其次，我们将说明有效结构。这种结构使用最小量、最廉价、仅布置在房间之间间隔处的稳定材料。它们能支撑自身和房间的荷载。最后，我们将获得有效结构的种种细部。关于类似的论述，参阅克利斯托弗·亚历山大的著

should lie purely in the realm of whim—and that the possible myriad of mixed systems,which lie between these archetypes, cannot even be considered—all this has more to do with the status of available theory than with any fundamental insight.

Indeed,as we shall now try to show,the archetypal,best solution to the problem of efficient structure in a building is one which does lie in between the three most famous archetypes.It is a system of load-bearing walls,supported at frequent intervals by thickened stiffeners like columns,and floored and roofed by a system of vaults.

We shall derive the character of the most efficient structure in three steps.First,we shall define the three-dimensional character of a typical system of rooms and spaces in a building. We shall then define an efficient structure as the smallest cheapest amount of stable material,placed only in the interstices between the rooms,which can support itself and the loads which the rooms generate.Finally,we shall obtain the details of an efficient structure.For a similar discussion,see Christopher Alexander, "An attempt to derive the nature of a human building system from first principles," in Edward Allen,*The Responsive House,M.I.T.Press*,1974.

I.*The three-dimensional character of a typical building based purely on the social spaces and the character of rooms.*

In order to obtain this from fundamental considerations,let us first review the typical shape of rooms—see THE SHAPE OF INDOOR SPACE(191)—and then go on to derive the most efficient structure for a building made up of these kinds of rooms:

1.The boundary of any space,seen in plan,is formed by segments which are essentially straight lines—though they need not be perfectly straight.

作。(Christopher Alexander，"An attempt to derive the nature of a human building system from first principles，"in Edward Allen，*The Responsive House*，*M.I.T.Press*，1974.)

一、纯粹以社会空间和房间性质为基础的典型建筑的三维空间特性

为了从种种基本考虑中获得这种性质，首先让我们来考察一下房间的典型形状——参阅**室内空间形状**（191），然后再从这些不同类型的房间所构成的建筑中导出最有效的结构。

1. 在平面图上所见到的任何空间边缘基本上是由扇形体的直线段构成——虽然它们并不一定全部是直线。

2. 各空间的天花板高度随其社会功能而变化。概括地说，天花板高度随地面面积的大小而变化——大空间具有较高的天花板、小空间具有较低的天花板——**天花板高度变化**（190）。

3. 空间的边缘基本上与头高垂直——即头高约 6ft。超过该高度，空间的边缘和空间就是一回事了。一个标准房间的墙壁和天花板之间的上部角隅是没有功能作用的，所以把它们视为空间的重要部分是毫无用处的。

4. 每一空间都具有一水平地面。

5. 建筑就是许多个多边形空间的集合体。每一多边形空间均有一个蜂窝状截面和一个随其规模大小而变化的高度。

一个多边形蜂窝状空间的集合体
A packing of polygonal beehive spaces

如果我们遵循**结构服从社会空间的需要**（205）的原理，我们可以设想，这种三维空间排列必须保持完整，并不被结构的构件所中断。这意味着有效结构一定是这样一

2.The ceiling heights of spaces vary according to their social functions.Roughly speaking,the ceiling heights vary with floor areas—large spaces have higher ceilings,small ones lower—CEILING HEIGHT VARIETY(190).

3.The edges of the space are essentially vertical up to head height—that is,about 6 feet.Above head height,the boundaries of the space may come in toward the space.The upper corners between wall and ceiling of a normal room serve no function,and it is therefore not useful to consider them as an essential part of the space.

4.Each space has a horizontal floor.

5.A building then is a packing of polygonal spaces in which each polygon has a beehive cross section,and a height which varies according to its size.

If we follow the principle of STRUCTURE FOLLOWS SOCIAL SPACES (205),we may assume that this three-dimensional array of spaces must remain intact,and not be interrupted by structural elements.This means that an efficient structure must be one of the arrangements of material which occupies only the interstices between the spaces.

We may visualize the crudest of these possible structures by means of a simple imaginary process.Make a lump of wax for each of the spaces which appears in the building,and construct a three-dimensional array of these lumps of wax, leaving gaps between all adjacent lumps.Now,take a generalized "structure fluid," and pour it all over this arrangements of lumps,so that it completely covers the whole thing, and fills all the gaps.Let this fluid hard-en.Now dissolve out the wax lumps that represent spaces.The stuff which remains is the most generalized building structure.

种材料的排列：材料只占据空间之间的空隙位置。

我们只要简单地想象一下，就可以想象出这些可能被采用的结构的大致情况。试按建筑中出现的每一空间做一蜡模，用这些蜡模构成一个三维空间组合，在所有相邻的蜡模之间留下空隙。现在再用广义的"结构流体"注入全部蜡模组合的空隙中去，以便注满所有的空隙和覆盖住全部蜡模。让这种流体硬化。然后再把代表各种空间的蜡模熔化掉。而保留下来的物体就是具有这种三维空间组合基本特征的建筑结构了。

二、某特定空间体系最有效的结构

很明显，由结构流体形成的想象结构并不是现实的结构。此外，它的效率相当低。要是真的那样做，就会用掉大量的材料。现在我们必须提出这样的问题：怎样才能形成一个相似于这一想象结构的结构而又用料最省？正如我们将会看到的那样，这种最有效的结构将是受压结构和连续结构。前者的弯曲和拉力减少到最小值，后者的全部构件均为刚性连接。每一构件至少承担任何一种荷载所引起的部分应力。

1. 受压结构。在有效结构中，我们要使每一点细微的材料都物尽其用。更确切地说，我们要使应力遍布于各种材料，使得每立方英寸的材料所受的应力都相等。这种情况，例如，在简支木梁中是不会发生的。材料受最大应力的地方是在梁的顶部和底部；而梁的中部只有很小的应力，因为在梁的中间部位，相对于应力分布来说，材料用量大。

通常我们可以说，处于弯曲中的构件总是承受不均匀的应力分布，因此，如果结构要完全不弯曲，我们只能将应力均匀地分布到各种材料上去。总而言之，一个完全有效的结构必须是不弯曲的。

有两种结构可以完全避免弯曲：纯受拉结构和纯受压结构。虽然纯受拉结构在理论上令人感到兴趣，并且适合

II.*The most efficient structure for a given system of spaces.*

Obviously,the imaginary structure made from the structure fluid is not real.And besides,it is rather inefficient:it would,if actually carried out,use a great deal of material.We must now ask how to make a structure,similar to this imaginary one,but one which uses the smallest amount of material.As we shall see,this most efficient structure will be *a compression structure*,in which bending and tension are reduced to a minimum and *a continuous structure*,in which all members are rigidly connected in such a way that each member carries at least some part of the stresses caused by any pattern of loading.

1.A *compression structure*.In an efficient structure,we want every ounce of material to be working to its capacity.In more precise terms,we want the stress distributed throughout the materials in such a way that every cubic inch is stressed to the same degree.This is not happening,for example,in a simple wooden beam.The material is most stressed at the top and bottom of the beam;the middle of the beam has only very low stresses,because there is too much material there relative to the stress distribution.

As a general rule,we may say that members which are in bending always have uneven stress distributions and that we can therefore only distribute stresses evenly throughout the materials if the structure is entirely free of bending. In short,then,a perfectly efficient structure must be free of bending.

There are two possible structures which avoid bending altogether:pure tension structures and pure compression structures.Although pure tension structures are theoretically interesting and suitable for occasional special purposes,the

于少数的特殊用途中，但是在**好材料**（207）中所提到的种种考虑排除了纯受拉结构占优势的地位，这是因为抗拉材料很难获得，而且价格昂贵，而几乎一切材料都能抵抗压缩力。尤其要注意，建筑的两种主要受拉材料——钢材和木材——一般都很缺乏，由于生态的原因，再也不能被大量使用了——再次参阅**好材料**（207）。

2. 连续结构。在有效结构中，单独的构件承重时，它们都有均匀的应力分布；结构作为一个整体而起作用。这两点都是正确的。

考虑一下，例如，一只柳条筐的情况。柳条筐的单根柳条强度不大。没有一根柳条能独自承受大量荷载。但是柳条筐编得如此精巧，以致它的所有柳条都能一齐承受哪怕是最小的荷载。如果你用手指加压于筐的一部分，它上面的所有柳条，包括离手指最远的柳条，都会同时受力。当然，因为该柳条筐的结构是作为整体在起作用、在受力，所以，筐的任何一部分都不单独具有很高的强度。

这一原理对于像建筑这样的结构尤为重要。目前建筑结构已面临着范围十分广泛的、各种不同的荷载条件。天有不测风云。忽而狂风从一个方向吹来；忽而地震来袭，使建筑物发生摇摆；近几年来，因为一些地基下沉得比另外一些慢，所以建筑物的不均衡下陷会造成荷载的重新分布；还有，只要建筑物存在，在它里面的人和家具始终在不断地运动。如果每一构件的强度大到足以单独承受可能施加到它上面的最大荷载，它就一定是一个巨大的构件。

但建筑物是连续的，就像那只柳条筐，它的每一部分都有助于承受最小的荷载，当然，荷载的不可预言的特性丝毫不会造成任何困难。构件可能是很小的。因为不管如何受力，建筑物的连续性将把荷载分布在作为一个整体的所有构件，而建筑物也将作为一个整体来承受荷载。

建筑物的连续性取决它的连接：即材料和形状的实际

considerations described in GOOD MATERIALS(207) rule them out over-whelmingly on the grounds that tension materials are hard to obtain,and expensive, while almost all materials can resist compression.Note especially that wood and steel,the two principle tension materials in buildings,are both scarce,and can—on ecological grounds—no longer be used in bulk-again,see GOOD MATERIALS(207).

2.*A continuous structure*.In an efficient structure,it is not only true that individual elements have even stress distributions in them when they are loaded.It is also true that the structure acts as a whole.

Consider,for example,the case of a basket.The individual strands of the basket are weak.By itself no one strand can resist much load.But the basket is so cunningly made,that all the strands work together to resist even the smallest load. If you press on one part of the basket with your finger,all the strands in the basket—even those in the part furthest from your finger—work together to resist the load.And of course,since the whole structure works as one,to resist the load,no one part has,individually,to be very strong.

This principle is particularly important in a structure like a building,which faces a vast range of different loading conditions.At one minute,the wind is blowing very strong in one direction;at another moment an earthquake shakes the building;in later years,uneven settlement redistributes dead loads because some foundations sink lower than others;and,of course,throughout its life the people and furniture in the building are moving all the time.If each element is to be strong enough,by itself,to resist the maximum load it can be subjected to,it will have to be enormous.

连续性。要在不同的材料之间形成像连续材料一样有效地转移荷载的连续连接是十分困难的，几乎是不可能的；由此可见，重要的是建筑物应当由一种材料构成，它从一个构件到另一个构件实际上是连续的。构件之间的连接形状也是十分重要的。直角往往造成非连续性：只有当墙和天花板、墙和墙以及柱和梁接合的地方有呈对角的圆角时，各种力才能被分布到整个建筑物。

三、有效结构的种种细部

如果我们现在假设有两种有效建筑即受压的和连续的，我们就能通过直接推理获得有效建筑结构的主要形态特征。

1. 有效结构的天花板、地面和房间都必须是拱结构。这是直接推理得出的结论。拱顶或穹顶的形状是纯粹受压的唯一形状。如果地面和屋顶在它们的边缘处呈曲线向下，它们才能和墙壁一起成为连续的。而且社会空间的形状也直接要求这样。因为墙和天花板之间的三角形空间不便于利用，所以这一三角形空间就自然而然地成为布置结构材料的空间。

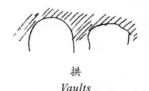

拱
Vaults

2. 墙都必须是承重的。很明显，任何一种非承重隔墙都是和连续性原理相矛盾的。连续性原理指明，建筑物的任何一小部分均有助于承重。此外，带有柱间非承重隔墙的柱子需要剪力支承。承重墙自然能提供这种剪力；墙、地面和天花板的连续性的形成只能靠连接地面和天花板的承重墙的作用。

But when the building is continuous,like a basket,so that each part of the building helps to carry the smallest load,then,of course,the unpredictable nature of the loads creates no difficulties at all.Members can be quite small,because no matter what the loads are,the continuity of the building will distribute them among the members as a whole,and the building will act as a whole against them.

The continuity of a building depends on its connections: actual continuity of material and shape.It is very hard,almost imposible,to make continuous connections between different materials,which transfer load as efficiently as a continuous material;and it is therefore essential that the building be made of one material,which is actually continuous from member to member.And the shape of the connections between elements is vital too.Right angles tend to create discontinuities:forces can be distributed throughout the building only if there are diagonal fillets wherever walls meet ceilings,walls meet walls,and columns meet beams.

III.*The details of an efficient structure.*

If we assume now that an efficient building will be both compressive and continuous,we can obtain the main morphological features of its structure by direct inference.

1.*Its ceilings,floors,and rooms must all be vaulted.*This follows directly.The dome or vault shape is the only shape which works in pure compression.Floors and roofs can only be continuous with walls,if they curve downward at their edges. And the shape of social spaces also invites it directly—since the triangle of space between the wall and ceiling serves no useful purpose,it is a natural place for structural material.

2.*Walls must all be load-bearing.*Any non load-bearing

承重墙
Load-bearing walls

　3.墙必须用壁柱按其长度作等间距加固。如果墙含有一定量的材料，当墙料不均匀地重新分布而形成壁柱时，它才能最有效地承重。这种墙在抗压曲方面最为有效——确实，这种大部分以厚度来增强的加固法就是实际上要求墙充分发挥其抗压能力——参阅**柱的最后分布**（213）。结果，墙有助于抗水平荷载，因为加劲柱像梁一样是能抗水平力的。

立式加劲杆
Vertical stiffners

　4.墙和地面之间的连接以及墙和墙之间的连接，都必须使用那些能沿缝形成圆抹角的附加材料。连接处是连续性最弱的点。直角连接最差。可是，我们从**室内空间形状**（191）知道，我们无法避免墙和墙接合处的近似直角。为了抵消直角的不良影响，有必要用材料把该角"填实"。有关这一原理的论述，参阅**柱的连接**（227）。

增厚连接
Thickened connections

　5.墙的开口必须具有厚实的边框，并在上部两角处呈圆形。这是从连续性原理直接得出的结论。这一结论我们将在**门窗边缘加厚**（225）中详加论述。

partition evidently contradicts the principle of continuity which says that every particle of the building is helping to resist loads. Furthermore,columns with non load-bearing partitions between them need shear support.The wall provides it naturally;and the continuity of the walls,floor,and ceiling can only be created by the action of a wall that ties them together.

3.*Walls must be stiffened at intervals along their length by columnar ribs.*If a wall is to contain a given amount of material,then the wall acts most efficiently when its material is redistributed,nonhomogeneously,to form vertical ribs.This wall is most efficient in resisting buckling—indeed,at most thicknesses this kind of stiffening is actually required to let the wall act at its full compressive capacity—see FINAL COLUMN DISTRIBUTION(213).And it helps to resist horizontal loads,because the stiffeners act as beams against the horizontal forces.

4.*Connections between walls and floors,and between walls and walls,must all be thickened by extra material that forms a fillet along the seam.*Connections are the weakest points for continuity,and right-angled connections are the worst.However,we know from THE SHAPE OF INDOOR SPACE(191)that we cannot avoid rough right angles where walls meet walls;and of course,there must be rough right angles where walls meet floors.To counteract the effect of the right angle,it is necessary to "fill" the angle with material.This principle is discussed under COLUMN CONNECTIONS (227).

5.*Openings in walls must have thickened frames,and rounding in the upper corners.*This follows directly from the principle of continuity and is fully discussed in FRAMES AS THICKENED EDGES(225).

开口
Openings

因此：

把建筑物想象为由受压材料构成的连续体。就其几何形状而言，可把它想象为一个由若干独立拱空间组成的三维系统，其中大多数空间呈矩形；每一面薄承重墙必须用壁柱按其长度作等间距加固；墙和墙的接合处以及墙和拱顶的接合处，还有开口的周边一律都要增厚加固。

抗压材料　　　　材料的连续性

ജ○ൠ

内拱的布局可参阅**楼面和天花板布置**（210）和**楼面天花拱结构**（219）；屋顶的外拱布局可参阅**屋顶布置**（209）和**拱式屋顶**（220）；墙的壁柱布置可参阅**柱的最后分布**（213）；墙和墙接合处的增厚布置可参阅**角柱**（212）；墙和拱顶连接处的增厚可参阅**圈梁**（217）；柱和墙的构造施工可参阅**箱形柱**（216）和**墙体**（218）；门框和窗框的增厚可参阅**门窗边缘加厚**（225）；柱和梁之间的非直角连接可参阅**柱的连接**（227）……

Therefore:

Conceive the building as a building made from one continuous body of compressive material.In its geometry, conceive it as a three-dimensional system of individually vaulted spaces,most of them roughly rectangular;with thin load-bearing walls,each stiffened by columns at intervals along its length,thickened where walls meet walls and where walls meet vaults and stiffened aroun the openings.

�✧ᘓ

The layout of the inner vaults is given in FLOOR AND CEILING LAYOUT(210)and FLOOR-CEILING VAULTS(219);the layout of the outer vaults which form the roof is given in ROOF LAYOUT(209)and ROOF VAULTS(220).The layout of the stiffeners which make the walls is given in FINAL COLUMN DISTRIBUTION(213);the layout of the thickening where walls meet walls is given by COLUMNS AT THE CORNERS(212);the thickening where walls meet vaults is given by PERIMETER BEAMS(217);the construction of the columns and thewalls is given by BOX COLUMNS(216)and WALL MEMBRANES(218);the thickening of doors and window frames is given by FRAMES AS THICKENED EDGES(225);and the non-right-angled connection between columns and beams by COLUMN CONNECTION(227)...

模式207　好材料**

　　……结构原理允许你想象出一幢建筑物，它的材料是以最有效的方式分布的，并且它与平面图所规定的社会空间协调一致——**结构服从社会空间的需要**（205）和**有效结构**（206）。但是，不必讳言，结构的概念仍然是图解式的、含混不清的。只有当你知道该建筑将由什么材料构成时，你对它的结构才会一清二楚、了如指掌。本模式可供你选用建筑材料时参考。

207 GOOD MATERIALS**

...the principles of structure allow you to imagine a building in which materials are distributed in the most efficient way,congruent with the social spaces given by the plan—STRUCTURE FOLLOWS SOCIAL SPACES(205), EFFICIENT STRUCTURE(206).But of course the structural conception is still only schematic.It can only become firm and cogent in your mind when you know what materials the building will be made of.This pattern helps you sette on materials.

❧❦

There is a fundamental conflict in the nature of materials for building in industrial society.

On the one hand,an organic building requires materials which consist of hundreds of small pieces,put together,each one of them hand cut,each one shaped to be unique according to its position. On the other hand,the high cost of labor,and the ease of mass production,tend to create materials which are large,identical,not cuttable or modifiable,and not adaptable to idiosyncracies of plan.These "modern" materials tend to destroy the organic quality of natural buildings and,indeed,to make it impossible. In addition,modern materials tend to be flimsy and hard to maintain—so that buildings deteriorate more rapidly than in a pre-industrial society where a building can be maintained and improved for hundreds of years by patient attention.

The central problem of materials,then,is to find a collection of materials which are small in scale,easy to cut on site, easy to work on site without the aid of huge and expensive

在工业社会中，在有关选用何种建筑材料方面，存在着尖锐的意见冲突。

一方面，一座有机的建筑物需要用成百上千种小块材料组合拼装，每一块材料都可进行手工切割，每一块材料均按其位置加工成独特形状。另一方面，劳动力费用高和成批生产简便易行就容易导致去生产大型的、统一规格的、不可切割的或不可更改的材料，而且不符合于平面图所要求的特有风格。这些"现代的"建筑材料会使天然成趣的建筑物的有机性质遭到破坏，而实际情况也确实如此。此外，现代的材料趋向于薄而轻，而且难以保养——结果就是建筑物会很快变质损坏，其损坏的速度远比前工业社会更为迅速，那时的建筑物，由于细心而周到的保养维修，经历数百年而完整无损。

由此可见，建材的中心问题是去收集各种材料，它们尺寸小，在地基上容易切割和施工，用不着庞大而又昂贵的施工机械的帮助。这些建材容易变化和适应，重量适中，牢固耐久，容易养护，而且容易建造。不需要受过专门训练的、成本高昂的劳动力，只要四处招聘廉价的劳动力就可以了。

此外，这类好材料必须对于生态环境是无害的：可还原处理的、能量消耗低的，而且是取之不尽用之不竭的材料。

当我们将这些要求汇总起来时，我们就会想到一个璀璨夺目的美名"好材料"——这是一类几乎完全不同于我们今天普遍使用的材料。下面我们试图说明它。这种说明肯定是不全面的；但也许有助于你更加审慎地考虑建材问题。

machinery,easy to vary and adapt,heavy enough to be solid, longlasting or easy to maintain,and yet easy to build,not needing specialized labor,not expensive in labor,and universally obtainable and cheap.

Furthermore,this class of good materials must be ecologically sound:biodegradable,low in energy consumption, and not based on depletable resources.

When we take all these requirements together,they suggest a rather startling class of "good materials"—quite different from the materials in common use today.The following discussion is our attempt to begin to define this class of materials.It is certainly incomplete;but perhaps it can help you to think through the problem of materials more carefully.

We start with what we call "bulk materials"—the materials that occur in the greatest volume in a given building.They may account for as much as 80 percent of the total volume of materials used in a building.Traditionally,bulk materials have been earth,concrete,wood,brick,stone,snow...Today the bulk materials are essentially wood and concrete and,in the very large buildings,steel.

When we analyze these materials strictly,according to our criteria,we find that stone and brick meet most of the requirements,but are often out of the question where labor is expensive,because they are labor intensive.

Wood is excellent in many ways.Where it is available people use it in great quantities,and where it is not available people are trying to get hold of it.Unfortunately the forests have been terribly managed;many have been devastated;and the price of heavy lumber has skyrocketed.From today's paper: "Since the end of federal economic controls the price of

我们先从所谓的"常用材料"开始说起吧。常用材料在建筑中大量使用，占用料总量的 80%。从传统上说，常用材料就是泥土、混凝土、木材、砖、石等……今天常用材料基本上就是木料、混凝土，在大型建筑物中还有钢材。

当我们根据自己的标准对这些材料进行严格分析的时候，我们发现，石头和砖能满足大部分的需要，但是有一点，工人的劳动强度大和劳动报酬昂贵。

木材在许多方面都是无与伦比的。凡在容易获得木材的地方，人们都大量使用，凡在不容易获得木材的地方，人们都尽量设法搞到它。不幸的是，森林由于滥砍滥伐而遭到了严重的破坏；许多地区已成为一片荒芜；厚的优质板料价格飞涨。据今天的报纸所载："因为联邦经济失去控制，木材价格在一个月内猛涨约 15%，现在已超高去年同期约 55%"（*San Francisco Chronicle*, February 11, 1973）。因此，我们将把木材视为贵重材料，而不是当作常用材料或用于结构目的。

钢作为常用材料似乎是不可能的。我们不需要把钢材用于高层建筑物，因为高层建筑物没有什么社会意义——**不高于四层楼**（21）。而对于较小的建筑物来说钢材太昂贵，使用过程中其数值也无法修正，它在生产中耗能大。

泥土是一种有趣的常用材料。泥土很难稳定，土墙重得难以置信，因为它非常厚实。凡是容易得到泥土的地方就用得多。无论如何，泥土无疑是"好材料"之一。

标准混凝土太细密。它很重且难以使用。当用混凝土浇筑成一定形状的整体材料后，施工人员就无法切入它或钉入钉子，它外表粗糙，如不覆盖上一层与结构非一体化的、昂贵的饰面材料，人们会感到它的外表粗陋、冰冷而又坚硬。

lumber has been jumping about 15percent a month and is now about 55 percent above what it was a year ago." *San Francisco Chronicle*, February 11,1973.We shall therefore look upon wood as a precious material,which should not be used as a bulk material or for structural purposes.

Steel as a bulk material seems out of the question.We do not need it for high buildings since they do not make social sense—FOUR-STORY LIMIT(21).And for smaller buildings it is expensive,impossible to modify,high energy in production.

Earth is an interesting bulk material.But it is hard to stabilize,and it makes incredibly heavy walls because it has to be so thick.Where this is appropriate,and where the earth is available,however,it is certainly one of the "good materials."

Regular concrete is too dense.It is heavy and hard to work. After it sets one cannot cut into it,or nail into it.And its surface is ugly,cold,and hard in feeling,unless covered by expensive finishes not integral to the structure.

And yet concrete,in some form,is a fascinating material.It is fluid,strong,and relatively cheap.It is available in almost every part of the world.A University of California professor of engineering sciences,P.Kumar Mehta,has even just recently found a way of converting abandoned rice husks into Portland cement.

Is there any way of combining all these good qualities of concrete and also having a material which is light in weight, easy to work,with a pleasant finish?*There is.It is possible to use a whole range of ultra-lightweight concretes which have a density and compressive strength very similar to that of wood.They are easy to work with,can be nailed with ordinary nails,cut with a saw,drilled with wood-working tools,easily repaired.*

We believe that ultra-lightweight concrete is one of the

然而，某些成形的混凝土是很有吸引力的。它是流体，强度大，且价格比较便宜。它几乎在世界上任何地方都可以获得。加利福尼亚大学工科教授比·库马·梅塔最近已找到一种方法：把废弃的稻壳变成硅酸盐水泥。

　　有没有一种方法可将混凝土的所有这些优点结合而成一种重量轻、易操作、外观光泽的新型材料呢？有。利用各种和木材具有相似密度和抗压强度的超轻质混凝土可以生产出这种材料。使用它们极为方便，可钉入普通的钉子，可用锯切割，可用木工加工工具钻孔，且容易修复。

　　我们认为超轻质混凝土是未来最基本的常用材料之一。

　　为了尽可能说清楚这点，我们现在来论述轻质混凝土的承力范围。经过反复的实验，我们得出结论，最好的轻质混凝土，施工中最有用的混凝土，就是那些密度范围为每立方英尺受力 40～60lbs 的混凝土，它们在抗压时每平方英寸受力 600～1000lbs。

　　说来奇怪，这种特殊规格的混凝土是目前可以使用的混凝土系列中开发最少的。正如我们从下面的图表中所能看到的那样，所谓"结构混凝土"通常密度更大（至少每立方英尺受力 90lbs），而且强度也大得多。最普通的"轻质"混凝土使用蛭石作为骨料，用于地面底层或用作隔热层。轻质混凝土很轻，但强度不够，通常不能成为有用的结构材料，抗压时每平方英寸受力约 300lbs 最为常见。然而，一系列混合轻质骨料，包括蛭石、珍珠岩、浮石和按不同比例掺和的膨胀页岩，在世界任何地方都能够很容易地生产出容重为每立方英尺 40～60lbs、抗压强度为每平方英寸 600lbs 的轻质混凝土。很幸运，我们已掌握了 1-2-3 的混合比例，即水泥、辉橄霞斜岩和蛭石（按 1：2：3）混合制成混凝土。

most fundamental bulk materials of the future.

To make this as clear as possible,we shall now discuss the range of lightweight concretes.Our experiments lead us to believe that the best lightweight concretes,the ones most useful for building,are those whose densities lie in the range of 40 to 60 pounds per cubic foot and which develop some 600 to 1000 psi in compression.

Oddly enough,this particular specification lies in the least developed part of the presently available range of concretes.As we can see from the following diagram,the so-called "structural" concretes are usually more dense(at least 90 pounds per cubic foot)and much stronger.The most common "lightweight" concretes use vermiculite as an aggregate, are used for underflooring and insulation,and are very light,but they do not usually develop enough strength to be structurally useful—most often about 300 psi in compression. However,a range of mixed lightweight aggregates,containing vermiculite,perlite,pumice,and expanded shale in different proportions,can easily generate 40-60 pound,600 psi concretes anywhere in the world.We have had very good luck with a mix of 1-2-3:cement-kylite-vermiculite.

Beyond the bulk materials,there are the materials used in relatively smaller quantities for framework,surfaces,and finishes.These are the "secondary" materials.

When buildings are built with manageable secondary materials,they can be repaired with the same materials:repair becomes continuous with the original building.And the buildings are more apt to be repaired if it is easy to do so and if the user can do it himself bit by bit without having to rely on skilled workers or special equipment.With prefabricated materials this is impossible,the materials are inherently

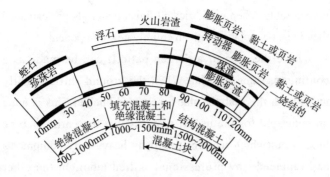

目前使用的混凝土混合料
Currently available concrete mixes

　　除了常用材料外，还有使用数量比较少的、用于框架、表面和装饰的材料。它们被称为"第二类"材料。

　　当住宅用可行的第二类材料建成之后，也用同样的材料来修缮：住宅一旦建成，保养修缮就要连续不断地进行。如果修缮工作容易进行，如果住户能自己逐步修缮而不依赖熟练的工人或专门设备，那么住宅就更容易修复。应用预制构件建成的住宅，修复就困难了，因为这些材料是内在地不可修复的。预制装饰材料遭到破坏后必须用全新的构件来替换。

　　试以花园式庭院为例。它可以被浇筑成连续的混凝土地面。当地面下的地基稍有移动时，平板就出现裂缝和扭曲隆起。这对住户来说是无法修理的，必须要粉碎整块地面和完全替换它（这需要专门的熟练工人操作比较大型的设备才能实现。）另外，花园式庭院也可以一开始就用许多小砖、软砖和石头来建造。当地基发生移动时，住户就能取出破碎的软砖，加点泥土，换上新的软砖——所有这一切就不需要昂贵的机械或专门的熟练工人的帮助了。如果一块软砖或砖坏了，就能很容易地被替换。

　　第二类好材料都有哪些呢？木料（我们不想把它列为常用材料）是制作门、窗、饰面和家具的最好的第二类材料。

unrepairable.When prefabricated finish materials are damaged they must be replaced with an entirely new component.

Take the case of a garden patio.It can be made as a continuous concrete slab.When the ground shifts slightly underneath this slab,the slab cracks and buckles.This is quite unrepairable for the user.It requires that the entire slab be broken out(which requires relatively heavy-duty equipment) and replaced—by professional skilled labor.On the other hand,it would have been possible to build the patio initially out of many small bricks,tiles,or stones.When the ground shifts,the user is then able to lift up the broken tiles,add some more earth,and replace the tile—all without the aid of expensive machinery or professional help.And if one of the tiles or bricks becomes damaged,it can be easily replaced.

What are the good secondary materials? Wood,which we want to avoid as a bulk material,is excellent as a secondary material for doors,finishes,windows,furniture.Plywood,particle board,and gypsum board can all be cut,nailed,trimmed,and are relatively cheap.Bamboo,thatch,plaster,paper,corrugated metals,chicken wire,canvas,cloth,vinyl,rope,slate,fiberglass, nonchlorinated plastics are all examples of secondary materials which do rather well against our criteria.Some are dubious ecologically—that is,the fiberglass and the corrugated metals—but again,these sheet materials need only be used in moderation,to form and finish and trim the bulk materials.

Finally,there are some materials which our criteria exclude entirely—either as bulk or secondary materials.They are expensive,hard to adapt to idiosyncratic plans,they require high energy production techniques,they are in limited reserves....For example:steel panels and rolled steel sections;aluminum;hard

胶合板、木屑板和石膏板都是第二类材料，它们都能被切割、钉入钉子或被整饰，它们都比较便宜。还有竹子、茅草、灰泥、纸、波纹金属板、钢丝网、帆布、织物、乙烯树脂、绳索、石板、纤维玻璃和非氯化塑料都属于第二类材料，所有这些材料都颇符合我们所提出的标准。从生态学的角度来看，有些材料还是把握不大的，如纤维玻璃和波纹金属板，但是还要在此提一下，这些板材只要使用得当，就能使常用材料具备一定的外形，并成为它们的饰面和装饰材料。

最后，还有一些材料不是常用材料就是第二类材料，但完全不符合我们的标准。它们价格昂贵，难以适合平面图的特殊要求，它们需要高能生产技术，且储量有限……例如，钢板条和轧钢型材、铝、质地坚硬的预应力混凝土构件、氯化泡沫材料、结构用木材、水泥砂浆、各种型号规格的平板玻璃……

还要请那些认为能永远继续使用钢筋的乐观主义者考虑一下如下的事实吧：即使是在整个地球表面储量丰富的铁资源有朝一日也会耗尽的。如果铁的消耗量以现在的增长速率继续发展下去（如果世界的广大地区还没有达到美国及其他西方发达国家消耗铁资源的水平，也许这是一件幸运的事），铁的资源将于 2050 年消耗殆尽。

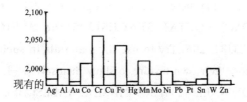

假定目前的消耗率像1960年至1968年那样
增长，各种金属将被用尽的年份如上图所列
Years at which various metals will be depleted assuming current
usage rate continues to increase as it did between 1960 and 1968

and prestressed concrete;chlorinated foams;structural lumber;cement plaster;immense sections of plate glass....

And,for any optimist who thinks he can go on using steel reinforcing bars forever—consider the following fact.Even iron,abundant as it is all over the earth's surface,is a depletable resource.If consumption keeps growing at its present rate of increase (as it very well may,given the vast parts of the world not yet using resources at American and western consumption levels),the resources of iron will run out in 2050.

Therefore:

Use only biodegradable,low energy consuming materials, which are easy to cut and modify on site.For bulk materials we suggest ultra-lightweight 40-60 lbs.concrete and earth-based materials like tamped earth,brick,and tile. For secondary materials,use wood planks,gypsum, plywood, cloth,chickenwire,paper,cardboard,particle board, corrugated iron,lime plasters,bamboo,rope,and tile.

In GRADUAL STIFFENING(208),we shall work out the way of using these materials that goes with STRUCTURE FOLLOWS SOCIAL SPACES(205)and EFFICIENT STRUCTURE(206).Try to use the materials in such a way as to allow their own texture to show themselves—LAPPED OUTSIDE WALLS(234),SOFT INSIDE WALLS(235)...

因此：

只利用可还原处理的、能量消耗少的，并在建筑工地上易于切割和修正的材料。所谓常用材料，我们指的是每立方英尺承力 40 ~ 60lbs 的超轻质混凝土和土基材料，如夯实的泥土、砖和软砖。至于第二类材料，可使用木板条、石膏、胶合板、织物、钢丝网、纸、硬纸板、木屑板、波纹铁板、石灰砂浆、竹子、绳子和瓦。

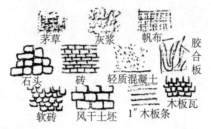

超轻质混凝土或有机的或土基的材料
Ultra-lightweight concrete or organic or earth-based materials

                                                                                        ଚ୦ଔ

在**逐步加固**（208）中，我们将说明如何使用这些材料的方法。这种方法符合**结构服从社会空间的需要**（205）和**有效结构**（206）。设法使用那些能显示其质感纹理的材料——**鱼鳞板墙**（234）和**有柔和感的墙内表面**（235）……

模式208 逐步加固**

　　……我们在**结构服从社会空间的需要**（205）和**有效结构**（206）中，解释了施工的初步原理和方法。**好材料**（207）已告诉我们有关一些材料的情况，为了满足人和生态的要求，我们应当使用这些材料。现在，在我们开始制定出建筑结构布局的切实方案之前，很有必要对这个更有哲理意义的模式进行考虑。本模式说明这样一种施工过程：它一方面采用恰当的建材，另一方面又要使结构的总体概念正确无误。

208   GRADUAL STIFFENING**

...in STRUCTURE FOLLOWS SOCIAL SPACES(205) and EFFICIENT STRUCTURE(206)we have set down the beginnings of a philosophy,an approach,to construction.GOOD MATERIALS(207)tells us something about the materials we ought to use in order to meet human and ecological demands.Now,before we start the practical task of making a structural layout for a building,it is necessary to consider one more philosophical pattern:one which defines the process of construction that will make it possible to use the right materials and get the overall conception of the structure right.

৪৩৫৩

The fundamental philosophy behind the use of pattern languages is that buildings should be uniquely adapted to individual needs and sites;and that the plans of buildings should be rather loose and fluid,in order to accommodate these subtleties.

This requires an entirely new attitude toward the process of construction.We may define this attitude by saying that it is desirable to build a building in such a way that it starts out loose and flimsy while final adaptations in plan are made,and then gets stiffened gradually during the process of construction,so that each additional act of construction makes the structure sounder.

To understand this philosophy properly,it is helpful to imagine a building being made like a basket.A few strands

在使用本模式语言时，应注意更深层次的重要原理：建筑应当尽可能和特定的具体要求和基地实际环境相适应；为了适应这些细微的变化，建筑平面应当是流动、可变的。

这就要求对构造施工过程持有一种完全崭新的观点，我们可以说明如下：建造一幢住宅的理想步骤是：开始时用常用的、薄而轻的材料而结束时却能够和图纸相符。而且，在结构施工过程中对住宅进行逐步加固，使得每一附加的施工活动都能令结构更加牢固。

为了正确理解这一基本原理，把住宅想象成一个柳条筐是大有好处的。筐上几根孤零零的柳条是不结实的、容易折断的。但它们和其他的柳条一起被编成一只筐，情况就不同了，筐逐渐变得越来越牢固了。建筑的最后结构强度只是由于它的所有构件同时受力才达到的，在它未完全竣工之前是达不到的。从这种意义上来说，这样一个构造施工过程就建成一幢建筑，它的各个部分在结构上共同受力——参阅**有效结构**（206）。

为什么逐步加固这一原理看来是一个非常合理的构造施工过程呢？

首先，这样一种结构允许实际的施工过程成为有创造性的活动。它允许将建筑逐步建成。建筑的构件在牢固定位前可以移动。所有这些细部设计的决定绝无可能在图纸上预先作出，而只能在施工过程中作出。而且这一施工过程允许你把三维空间视为一个整体，并把施工的每一步骤视为要增加更多的材料。

这意味着在施工过程中所增添的每一种新材料必须完美无缺地适应建筑的构架，所以每一种新材料必须比前一种材料具有更大的适应性、更多的灵活性且更能应付各种变化。这样，虽然建筑作为一个整体是从使用脆弱材料到

are put in place.They are very flimsy.Other strands are woven in.Gradually the basket gets stiffer and stiffer.Its final structural strength is only reached from the cooperation of all the members,and is not reached until the building is completely finished.In this sense,such a process produces a building in which all parts of it are working structurally—see EFFICIENT STRUCTURE(206).

Why does the principle of gradual stiffening seem so sensible as a *process* of building?

To begin with,such a structure allows the actual building process to be a creative act.It allows the building to be built up gradually.Members can be moved around before they are firmly in place.All those detailed design decisions which can never be worked out in advance on paper,can be made during the building process.And it allows you to see the space in three dimensions as a whole,each step of the way,as more material is added.

This means that since each new material that is added in the process must adapt perfectly to the framework that is there, each new material must be more adaptable,more flexible,more capable of coping with variation,than the last.Thus,though the building as a whole goes from flimsy to strong,the actual materials that are added go from the strongest and stiffest,to the gradually less stiff,until finally fluid materials are added.

The essence of this process is very fundamental indeed. We may understand it best by comparing the work of a fifty-year-old carpenter with the work of a novice.The experienced carpenter keeps going.He doesn't have to keep stopping, because every action he performs,is calculated in such a way

使用坚固材料的过渡，而所增加的实际材料是从强度最大的、最牢固的材料到牢固性逐渐减小最后直至流体材料。

这种施工过程的本质确实非常重要。这一本质我们可以通过比较一个年过半百的老木工和一个新木工的工作情况就能深刻理解。经验丰富的老木工可连续作业而不中断，因为他完成的每一动作都是计算好的，后一个动作恰好与前一个动作紧密衔接，并在某种程度上把不完善的地方加以改进。这里重要的是动作的连续性。他不会做出令他后来无法改正的动作，所以他可以连续不断、信心十足、从容不迫地进行作业。

而新木工就相形见绌了。他要花大量时间去揣摩做什么。之所以如此，根本原因是他知道他现在的操作稍一不慎就会酿成无法挽回的错误；如果他不小心翼翼，就会发生这样的事：在某一接合处要求缩短某一关键性部件，可是已为时太晚了。他对这类错误时时提心吊胆，这就迫使他花许多时间去提前动脑筋想办法。结果他不得不尽量按精确图纸施工，因为图纸能保证他避免这类错误。

新老木工之间的差别是显而易见的：新手还未学会施工中只出些小差错，而老手却知道其动作的先后顺序，这使他的操作得心应手，完全不会出大差错。正是老木工的这种朴素而又重要的认识才使他自己感到他的工作是轻松愉快的、出色稳妥的，而几乎毫不介意其工作的简单性。

我们在构造施工中有着与此完全相同的问题，只不过是明显放大了。重要的是，大多数现代构造施工都具有新木工干活的性质，而不具有老木工干活的性质。施工人员不知道如何放松紧张情绪，如何利用后面的细部处理前面的失误；他们不知道正确的施工顺序；通常也没有一种施工体系或施工过程可以令他们自己感到轻松愉快的同时不时有奇思妙想迸发出来。相反，他们都像这位新木工这般，完全按精心设计的图纸施工；建筑竣工时与图纸的要求完

that some later action can put it right to the extent that it is imperfect now.What is critical here,is the sequence of events. The carpenter never takes a step which he cannot correct later;so he can keep working,confidently,steadily.

The novice,by comparison,spends a great deal of his time trying to figure out what to do.He does this essentially because he knows that an action he takes now may cause unretractable problems a little further down the line;and if he is not careful,he will find himself with a joint that requires the shortening of some crucial member—at a stage when it is too late to shorten that member.The fear of these kinds of mistakes forces him to spend hours trying to figure ahead:and it forces him to work as far as possible to exact drawings because they will guarantee that he avoids these kinds of mistakes.

The difference between the novice and the master is simply that the novice has not learnt,yet,how to do things in such a way that he can afford to make small mistakes.The master knows that the sequence of his actions will always allow him to cover his mistakes a little further down the line. It is this simple but essential knowledge which gives the work of a master carpenter its wonderful,smooth,relaxed,and almost unconcerned simplicity.

In a building we have exactly the same problem,only greatly magnified.Essentially,most modern construction has the character of the novice's work,not of the master's. The builders do not know how to be relaxed,how to deal with earlier mistakes by later detailing;they do not know the proper sequence of events;and they do not,usually,have a building system,or a construction process,which allows them to develop this kind of relaxed and casual wisdom.Instead,like

全一致；建筑与图纸的要求稍有偏离就容易引起严重的后果，也许会不可避免地使部分工程返工。

就对待细部来说，这种如新木工一般的、惊慌失措的态度引起的严重后果有二。第一，和新木工一样，建筑师们不惜花费大量时间设法提前设计出所有细部，而不是将功夫用在扎实稳妥的施工上。很明显，这样做要花很多钱；只顾设计细部而不顾具体施工，就会助长"完美无缺的"建筑被造得像机器一样呆板。第二，更为严重的后果是：细部支配整体。在平面图中由于模式自由支配设计所形成的美和精巧别致突然遭到扼杀和破坏，这是因为人们担心各种细部设计不周全，连接的细部和构件就会统治该平面图了。结果，房间的形状就有点不正确，窗户错位，门和墙之间的空间变得面目全非、毫无用处了。总而言之，现代建筑的全部特性，即在构造施工中微不足道的细部支配较大的空间已占上风。

现在所需要的恰恰是其反面——一种细部适应整体的施工过程。这就是老木工的诀窍。在《建筑的永恒之道》（*The Timeless Way of Building*）一书中我们把这一点作为全部有机形式和所有成功的建筑的基础加以描述。我们在这里所论述的逐步加固的施工过程是这一重要原理在实际上和顺序上的具体化。现在我们试问一下，怎样才能根据**好材料**（207）所规定的含义实际上形成一个逐步加固的结构。

有关材料的各种事实就是我们赖以形成本模式的出发点。

1. *板材易于生产并能够形成最佳的连接*。在传统社会中，板材数量甚少。可是现在的工厂却大量生产板材，因为板材比其他形状的材料更易于生产。随着我们进入成批生产的时代，板材的品种日益多样丰富。不言而喻，板材强度大、重量轻而又便宜。石膏板、胶合板、织物、乙烯树脂、帆布、纤维玻璃、木屑板、木板条、波纹金属板和钢丝网均属板材。

the novice,they work exactly to finely detailed drawings;the building is extremely uptight as it gets made;any departure from the exact drawings is liable to cause severe problems,may perhaps make it necessary to pull out whole sections of the work.

This novice-like and panic-stricken attention to detail has two very serious results.First,like the novice,the architects spend a great deal of time trying to work things out ahead of time,not smoothly building.Obviously,this costs money;and helps create these machine-like "perfect" buildings.Second,a vastly more serious consequence:the details control the whole. The beauty and subtlety of the plan in which patterns have held free sway over the design suddenly becomes tightened and destroyed because,in fear that details won't work out,the details of connections,and components,are allowed to control the plan. As a result,rooms get to be slightly the wrong shape, windows go out of position,spaces between doors and walls get altered just enough to make them useless.In a word,the whole character of modem architecture,namely the control of larger space by piddling details of construction,takes over.

What is needed is the opposite—a process in which details are fitted to the whole.This is the secret of the master carpenter; it is described in detail in *The Timeless Way of Building* as the foundation of all organic form and all successful building.The process of gradual stiffening,which we describe here,is the physical and procedural embodiment of this essential principle. We now ask how,in practice,it is possible to create a gradually stiffened structure within the context defined by the pattern GOOD MATERIALS(207).

Facts about materials give us the starting point we need.

1.*Sheet materials are easy to produce and make the best*

而且，板材是强度最大的连接材料。连接点是结构中的薄弱环节。板材易于连接，因为板材连接处能相互搭接。凡由板材构成的一切物体天生比块材或棒料所构成的物体强度大。

　　2. 超轻质混凝土是一种极好的填料。它的密度和木材一样，强度大、重量轻，易于切割，易于修补和易于钉钉子——而且到处都容易取得。这一点已在**好材料**（207）中详细说明了。

　　3. 可是，任何一种混凝土都需要制模：制模费用巨大。的确，使混凝土成为形状复杂的构件费用十分昂贵；在常规的建筑系统中制模工序或多或少会排除我们所描述的这种"有机的"结构。此外，在制造标准的混凝土构件中，模板最后总要报废而被扔掉。

　　我们认为，饰面材料在任何合理的建筑系统中应和施工过程以及结构本身（正如它们存在于几乎所有传统建筑中一样）融为一体。对任何建筑系统中的建筑物不得不另外"增加"饰面材料，确是一种浪费，而且也无法做到自然协调。

　　4. 因此，我们建议超轻质混凝土应当使用于浇筑由容易获得的板材所制成的板模中去，并使它们留在原处形成饰面材料。板材可以和织物、帆布、木板条、石膏板、纤维玻璃、胶合板、纸、抹上灰泥的钢丝网、波纹金属板等混合使用。如有可能，它也可与空心砖、砖和石头混合使用——参阅**好材料**（207）。我们推荐使用珍珠岩、膨胀页岩或浮石作为超轻质混凝土的骨料。如果荷载允许，经过夯实的泥土、风干土坯和非氯化泡沫材料都可以代替混凝土。

connections.

In traditional society there are few sheet materials. However,factory production tends to make sheets more easily than other forms of material.As we move into an age of mass production,sheet materials become plentiful and are naturally strong,light,and cheap.Gypsum board,plywood,cloth,vinyl,canvas,fiberglass,particle board,wood planks,corrugated metals,chicken wire,are all examples.

And sheet materials are the strongest for connections. Connections are the weak points in a structure.Sheet materials are easy to connect,because connections can join surfaces to one another.Anything made out of sheets is inherently stronger than something made of lumps or sticks.

2.*Ultra-lightweight concrete is an excellent fill material— it has the density of wood,is strong,light,easy to cut,easy to repair,easy to nail into—and is available everywhere.*This is discussed fully in GOOD MATERIALS(207).

3.*However,any kind of concrete needs formwork:and the cost of formwork is enormous.*

This makes it very expensive indeed to build any complex form;and within conventional building systems,it more or less rules out the kind of "organic" structure which we have described.Furthermore,in regular concrete work,the formwork is eventually wasted,thrown away.

We believe that the finishes in any sensible building system should be integral with the process of construction and the structure itself(as they are in almost all traditional buildings)and that any building system in which finishes have to be "added" to the building are wasteful,and unnatural.

4.*We therefore propose that ultra-lightweight concrete be*

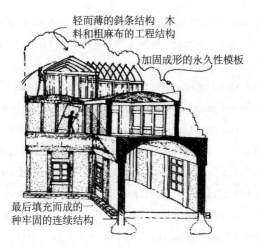

轻而薄的斜条结构　木料和粗麻布的工程结构

加固成形的永久性模板

最后填充而成的一种牢固的连续结构

一种逐步加固方案——利用1in厚的木板条、石膏板和粗麻布作为板材，并用超轻质混凝土作为填料
One version of gradual stiffening,using one inch planks,gypsum board and burlap as sheets,with ultra-lightweight concrete as fill

上图显示了实现这种逐步加固模式的一种特殊方法。但是逐步加固的原理远比这一特殊的实施方法更具有普遍的意义。的确，这种情况以某种方式发生于几乎所有的传统的建筑形式之中。爱斯基摩人的雪块圆顶小屋结构和非洲人的柳筐式结构两者都是逐步加固的结构。这些结构的每一施工步骤都要妥善处理现存的构架，要对它增添材料并加固它。意大利南部的阿尔贝罗贝洛地区的石头建筑就是例证。在伊丽莎白地区的半木结构建筑也是例证。

因此：

确定你们不是像吊装机那样正在用构件装配建筑，而是正在编制一种开始呈球形的、完整但脆弱的结构；继而对它进行逐步加固，但它仍然是相当不结实的；只在最后才能做到完全加固它，并使它成为坚硬的、受力大的结构。

我们认为，在我们的时代，这种施工过程所应用最自然的方案是放置由板材构成的薄壳结构，用抗压填料填充，使它变得非常坚固。

CONSTRUCTION
构　造

poured into forms which are made of the easily available sheet materials:and that these materials are then left in place to form the finish.

The sheet materials can be any combination of cloth, canvas,wood planks,gypsum boards,fiberboards,plywood,paper,plastered chickenwire,corrugated metals,and where it is possible,tile,brick,or stone—see GOOD MATERIALS(207). For the ultra-lightweight concrete we recommend a perlite, expanded shale,or pumice aggregate.Tamped earth, adobe,nonchlorinated foams,may also do instead of the concrete,if loads allow it.

The drawing above,shows one particular realisation of this kind of gradual stiffening.But the principle is far more general than this particular use of it.Indeed,it occurs,in one way or another,in almost all traditional forms of building.Eskimo igloo construction and African basket structures are both gradually stiffened structures,where each next step copes with the existing framework,adds to it,and stiffens it.The stone buildings of Alberobello in southern Italy are examples.So is Elizabethan halftimber construction.

Therefore:

Recognize that you are not assembling a building from components like an erector set,but that you are instead weaving a structure which starts out globally complete,but flimsy; then gradually making it stiffer but still rather flimsy; and only finally making it completely stiff and strong.

We believe that in our own time,the most natural version of this process is to put up a shell of sheet materials, and then make it fully strong by filling it with a compressive fill.

软外壳模框　　抗压填料

∂০cs

　　尽量选用天然材料作为外层薄壳材料——木板条供柱用，帆布或粗麻布供拱用，灰泥板或木板条或砖或空心砖供墙用——**好材料**（207）。

　　使用每立方英尺容重为 40 ～ 60lbs 的超轻质珍珠岩混凝土以充当抗压填料——它和木材的密度一样，可以切割并钉入钉子。无论在施工期间，还是在今后的使用过程中，如果此料破损已非修不可，就可随时修复——**好材料**（207）。

　　先竖立空心柱，然后用超轻质混凝土灌注其内；接着架梁，同样用超轻质混凝土浇筑；接下去就建造拱，用一层薄薄的混凝土覆盖其上，硬化而成薄壳；再用各种更加轻的材料充填薄壳而形成楼面；下一步就是筑墙和窗框，用混凝土填实；最后才建造屋顶，也是一个拱，它是由一层薄的织物和覆盖于其上的一层混凝土构成的一种薄壳——**箱形柱**（216）、**圈梁**（217）、**墙体**（218）、**楼面天花拱结构**（219）和**拱式屋顶**（220）……

　　在这一结构原理的范围内，以你所画的平面图为基础，进行完整的结构布置；这就是你在开始实际施工之前，在纸面上工作的最后一件事；

　　209. 屋顶布置

　　210. 楼面和天花板布置

　　211. 加厚外墙

　　212. 角柱

　　213. 柱的最后分布

Choose the most natural materials you can,for the outer shell itself—thin wood planks for columns,canvas or burlap for the vaults,plaster board or plank or bricks or hollow tiles for walls—GOOD MATERIALS(207).

Use ultra-lightweight 40 to 60 pounds perlite concrete for the compressive fill—it has the same density as wood and can be cut and nailed like wood,both during the construction and in later years when repairs become necessary—GOOD MATERIALS(207).

Build up the columns first,then fill them with the ultra-lightweight concrete;then build up the beams and fill them;then the vaults,and cover them with a thin coat of concrete which hardens to form a shell;then fill that shell with even lighter weight materials to form the floors;then make the walls and window frames,and fill them;and finally,the roof,again a thin cloth vault covered with a coat of concrete to form a shell— BOX COLUMNS(216),PERIMETER BEAM(217),WALL MEMBRANE(218),FLOOR-CEILING VAULTS (219),ROOF VAULTS(220)...

within this philosophy of structure,on the basis of the plans which you have made,work out the complete structural layout;this is the last thing youdo on paper,before you actually start to build;

209.ROOF LAYOUT

210.FLOOR AND CEILING LAYOUT

211.THICKENING THE OUTER WALLS

212.COLUMNS AT THE CORNERS

213.FINAL COLUMN DISTRIBUTION

模式209　屋顶布置*

　　……现在假定你已有一幅按楼房每一层楼的比例而绘制的大致平面图。在此情况下,根据**重叠交错的屋顶**（116）**和带阁楼的坡屋顶**（117），你大概早已知道屋顶的形状了；而且你确切知道何处屋顶是平的，以便在紧接不同楼层的空间形成屋顶花园——**屋顶花园**（118）。本模式给你指明如何为住宅设计出详尽的屋顶平面图。这将有助于实现你在每一平面图中所画的那些模式。

୫୦୯୫

　　哪一种屋顶平面图是与你的建筑的特性有机地联系在一起的呢？

209   ROOF LAYOUT*

...assume now that you have a rough plan,to scale,for each floor of the building.In this case you already know roughly how the roofs will go,from CASCADE OF ROOFS(116)and SHELTERING ROOF (117);and you know exactly where the roof is flat to form roof gardens next to rooms at different floors—ROOF GARDEN(118).This pattern shows you how to get a detailed roof plan for the building,which helps those patterns come to life,for any plan which you have drawn.

଼ଷଔ

What kind of roof plan is organically related to the nature of your building?

We know,from arguments presented in THE SHAPE OF INDOOR SPACE(191),that the majority of spaces in an organic building will have roughly—not necessarily perfectly—straight walls because it is only then that the space on both sides of the walls can be positive,or convex in shape.

And we know,from similar arguments,that the majority of the angles in the building will be roughly—again,not exactly—right angles,that is,in the general range of 80 to 100 degrees.

We know,therefore,that the class of natural plans may contain a variety of shapes like half circles,octagons,and so on—but that for the most part,it will be made of very

根据**室内空间形状**（191）中所提出的论点，我们知道，在有机材料住宅中大部分空间都一定具有粗糙的——而不必是完美无缺的——直墙，因为唯有如此，墙两侧空间的形状才是正向的，即凸形的。

根据相似的论点，我们还知道，建筑的大多数的角都一定近似于——再一次强调，不是精确的——直角，即一般在 80°～100° 的范围内。

由此可见，这类正常的平面图可包含各种图形，如半圆形、八角形等——但绝大部分的图形一定是十分近似的、不精确的矩形。

根据**带阁楼的坡屋顶**（117），我们还知道，无论何时，所有的厢房都应尽可能地和正房处在同一个屋顶之下；建筑的屋顶应是多种混合的，即平顶、坡顶和圆顶等，而重点应放在不是平顶的那些屋顶上。

因此，我们可以在规定屋顶的布置中对这一问题作如下说明：如果已有一任意平面图，如上所述，我们怎样做才能使那些符合于**重叠交错的屋顶**（116）、**带阁楼的坡屋顶**（117）**和屋顶花园**（118）的屋顶布置与该平面图相一致呢？

在详细解释屋顶布置的过程之前，我们要强调为这一过程提供基础的 5 种假设。

1. "陡斜的"屋顶实际上可以是倾斜的，或是呈斜弧线的拱顶，或筒拱——正如**拱式屋顶**（220）中所描述的那样。在所有这 3 种情况下，总过程是相同的。（关于呈斜弧线的拱顶，把斜率定义为高宽比。）

一个拱顶的"斜高"
The "pitch" of a vaulted roof

rough,sloppy rectangles.

We also know,from SHELTERING ROOF(117),that entire wings should be under one roof whenever possible and that the building is to be roofed with a mixture of flat roofs and sloping or domical roofs,with the accent on those which are *not* flat.

We may therefore state the problem of defining a roof layout as follows:*Given an arbitrary plan of the type described above,how can we fit to it an arrangement of roofs which conforms to the* CASCADE OF ROOFS(116)and SHELTERING ROOF(117)and ROOF GARDENS (118)?

Before explaining the procedure for laying out roofs in detail,we underline five assumptions which provide the basis for the procedure.

1.The "pitched" roofs may actually be pitched,or they may be vaults with a curved pitch,or barrel vaults—as described in ROOF VAULTS(220).The general procedure,in all three cases,is the same.(For curved vaults,define slope as height-to-width ratio.)

2.Assume that all roofs in the building,which are not flat,have roughly the same slope.For a given climate and roof construction,one slope is usually best;and this greatly simplifies construction.

3.Since all roofs have the same slope,the roofs which cover the widest wings and/or rooms will have the highest peaks;those covering smaller wings and rooms will be relatively lower. This is consistent with MAIN BUILDING(99),CASCADE OF ROOFS(116),and CEILING HEIGHT VARIETY(190).

2. 假设建筑的所有屋顶都不是平顶的，都有大致相同的坡度。对于某一给定的气候条件和屋顶构造来说，同一坡度通常是最佳的；并且这可大大简化施工过程。

全部屋顶坡度相同
The same slope throughout

3. 因为所有的屋顶坡度相同，所以覆盖最宽侧翼和（或）最宽房间的屋顶一定有最高的屋脊；覆盖较窄的侧翼和较窄的房间的屋顶，相对说来，一定是较低的。这种情况是和**主要建筑**（99）、**重叠交错的屋顶**（116）以及**天花板高度变化**（190）等模式完全符合的。

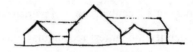

宽屋顶是最高的
Wide roofs are highest

4. 住宅有助于围成户外亭榭或庭院的所有地方都需要有一条整齐的屋檐线，以便它具有一个"房间"的空间形式。不规则的、带山墙的屋檐线通常会破坏小庭院的空间。因此，屋顶四周边缘在同一水平线上，屋顶应是四坡的。这是十分必要的。

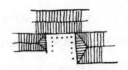

一庭院四周低矮的屋檐
Low roof edge round a courtyard

5. 在其他所有位置上，使建筑的端部和侧翼的端部成为山墙的端部。

4.Any place where the building helps to enclose an outdoor room or courtyard needs an even eave line so that it has the space of a "room." An irregular roof line,with gable ends,will usually destroy the space of a small courtyard.It is necessary,therefore,that roofs be hipped in these positions to make the roof edge horizontal.

5.In all other positions,leave the ends of buildings and wings as gable ends.

We shall now discuss the rules for roofing a building by using an example of a house designed by a layman using the pattern language.This building plan is shown below.It is a single-story house and it contains no roof gardens or balconies.

We first identify the largest rectangular cluster of rooms and roof it with a peaked roof,the ridge line of which runs the long direction:

Then we do the same with smaller clusters,until all the major spaces are roofed.

Then we roof remaining small rooms,alcoves,and thick walls with shed roofs sloping outward.These roofs should spring from the base of the main roofs to help relieve them of outward thrusts;their outside walls should be as low as possible.

Finally,we identify the outdoor spaces(shown as A,B,and C),and hip the roofs around them to preserve a more continuous eave line around the spaces.

We shall now discuss a slightly more complicated example,a two story building.

一种屋顶布置方案——使用超轻质混凝土拱结构作为屋顶

One version of a roof layout,using ultra-lightweight concrete vaults as roofs

我们现在开始论述建筑的屋顶布置的细则。我们想通过非专业人员利用本模式语言所设计的一幢住宅的实例来加以说明。该建筑的平面图如下图所示。这是一幢平房,它既不含屋顶花园,也不含阳台。

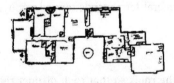

我们首先鉴定一下一个最大的矩形图,它由若干房间组成,并有一个有纵向屋脊的屋顶。

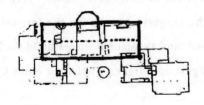

We begin with the top story,roofing the entire master bedroom and bath under one peaked roof with the ridge running lengthwise:

Next we move to the lower story,roofing the children's wing under a flat roof to form a ROOF GARDEN(118)for the master bedroom,and the larger living room under a pitched roof,again with the ridge running lengthwise.

Then we bring the roof over the master bedroom down over the interior loft.

Finally,we smooth the living room roof ridge line into the side of the roof over the loft.This completes the roof layout.

It is very helpful,when you are laying out roofs,to remember the structural principle outlined in CASCADE OF ROOFS(116).When you have finished,the overall arrangement of the roofs should form a self-buttressing cascade in which each lower roof helps to take up the horizonal thrust generated by the higher roofs—and the overall section of the roofs,taken in very very general terms,tends toward a rough upside down catenary.

Therefore:

Arrange the roofs so that each distinct roof corresponds to an identifiable social entity in the building or building complex.Place the largest roofs—those which are highest and have the largest span—over the largest and most important and most communal spaces;build the lesser roofs off these largest and highest roofs;and build the smallest

接着，我们对较小的房间团组进行同样的鉴定，直到所有主要空间都设置了屋顶为止。

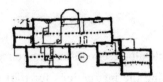

接下来，我们把剩下来的小房间、凹室和带有向外倾斜的单坡屋顶的厚墙都一一盖上屋顶。这些屋顶应当从主屋顶的底面抬升，以利于减少它们的外推力；它们的外墙应当尽量低矮。

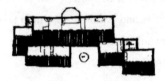

最后，我们来鉴定户外空间（如图中A、B和C所示），沿空间四周盖上四坡屋顶，以便环绕空间周边能够保持一条更连续的屋檐线。

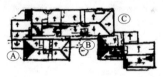

现在我们来论述更为复杂的例子，一幢两层楼的建筑。

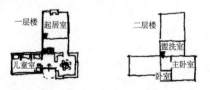

我们先从顶层开始吧。其主卧室和盥洗室在同一坡屋顶之下，屋脊呈纵向走向。

roofs of all off these lesser roofs,in the form of halfvaults and sheds over alcoves and thck walls.

⊗⊘⊗

You can build all these roofs,and the connections between them,by following the instructions for roof vaults—ROOF VAULTS (220).When a wing ends in the open,leave the gable end at full height;when a wing ends in a courtyard,hip the gable,so that the horizontal roof edge makes the courtyard like a room—COURTYARDS WHICH LIVE(115).

Treat the smallest shed roofs,which cover thick walls and alcoves,as buttresses,and build them to help take the horizontal thrust from floor vaults and higher roof vaults—THICKENING THE OUTER WALLS(211)...

下面我们就把话题转到底层。儿童的卧室上布置平屋顶，以便形成主卧室的**屋顶花园**（118），而较大的起居室上则布置斜屋顶，也是纵向屋脊的。

接下来我们把主卧室的屋顶延伸下去作为内阁楼的屋顶。

最后，我们调整起居室的屋脊线，使其插入阁楼屋顶的一侧。屋顶布置到此就告完成。

在布置屋顶时，回忆在**重叠交错的屋顶**（116）中所勾画结构原理是十分有益的。当你结束屋顶布置后，屋顶的总体布置应当形成一种自我支撑式的重叠交错，其中每一个下层屋顶都有助于承受上层屋顶所传递的水平推力——并且，用最一般的角度来看，所有屋顶的总截面近似一条颠倒的悬链线。

因此：

布置屋顶时务必应使每一个不同的屋顶和建筑或建筑群体中的易识别的社会实体相符合。将最大的屋顶——即具有最高和最大跨度的屋顶——放置于最大的、最重要的和最公共的空间之上；将较小的屋顶偏置于这些最大的和最高的屋顶之下；将最小的屋顶偏置于这些较小的屋顶之下，在凹室和厚墙之上做成半拱或单坡屋顶。

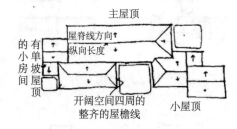

根据拱式屋顶的说明——**拱式屋顶**（220），就能建成所有这些屋顶和屋顶之间的连接。当建筑某翼端部外露时，务必使山墙顶端具有充分的高度；当某翼端部外露于庭院时，将山墙筑成四坡屋顶，以便水平屋檐将庭院围成类似房间的空间——**有生气的庭院**（115）。

将覆盖厚墙和凹室的最小的单坡屋顶作为墙垛来处理，并将它们建造成有助于承受来自楼面拱和较高拱屋顶的水平推力——**加厚外墙**（211）……

模式210 楼面和天花板布置

……**有效结构**（206）告诉我们，建筑空间应当是拱式的，以便楼面和天花板几乎全用抗压材料来建造。为了布置拱结构楼面和天花板，我们必须使拱结构适合于各单个房间的天花板高度变化——**天花板高度变化**（190），并在顶层使它们适合于拱屋顶的布置——**屋顶布置**（209）。

✂✃✄

重申：基本的问题是保持平面图中的社会空间的完整性。

根据**结构服从社会空间的需要**（205），我们知道，楼面和天花拱结构必须和平面图中的重要社会空间相符合。但是，存在着大量的社会空间，就其大小来说，从如**窗前空间**（180），也许直径为5ft，到如**农家厨房**（139），也许直径为15ft，直到如**中心公共区**（129），也许直径为35ft的各种空间的组合。

在不同宽度的拱相互接近的地方，你必须记住，要注意上面楼层的标高。既可以通过使较小的拱按比例加高拱顶去找该楼层的标高，也可以在低拱顶之上填以附加材料从而使小拱顶继续保持其低的位置——参阅**天花板高度变化**（190），你还可以在上面的楼层地面上修筑台阶，使之符合下面的拱尺寸的变化。

在不同楼层上的拱结构无须上下完全对齐。从这种意义来说，它们远比柱梁结构更具有灵活性，同理，也就更能适应于**结构服从社会空间的需要**（205）。可是，有着种种限制。如果上面拱结构的位置刚好使其荷载传送到下面拱结构的拱顶中部，那么它就会把过量的应力传给下面的拱结构。而我们则利用通过连续抗压介质的垂直力以45°

210 FLOOR AND CEILING LAYOUT

...EFFICIENT STRUCTURE(206)tells us that the spaces in the building should be vaulted so that the floors and ceilings can be made almost entirely of compression materials.To lay out the floor and ceiling vaults,we must fit them to the variety of ceiling heights over individual rooms—CEILING HEIGHT VARIETY(190)and,on the top story,to the layout of the roof vaults—ROOF LAYOU(209).

❧❧❧

Again,the basic problem is to maintain the integrity of the social spaces in the plan.

We know,from STRUCTURE FOLLOWS SOCIAL SPACES(205),that floor and ceiling vaults must correspond to the important social spaces in the plan.But there are a great number of social spaces,and they range in size from spaces like WINDOW PLACE(180),perhaps five feet across,to spaces like FARMHOUSE KITCHEN(139),perhaps 15 feet across,to collections of spaces,like COMMON AREAS AT THE HEART(129),perhaps 35 feet across.

Where vaults of different width are near each other,you must remember to pay attention to the level of the floor above.Either you can level out the floor by making the smaller vaults have proportionately higher arches,or you can put extra material in between to keep the small vaults low— see CEILING HEIGHT VARIETY(190),or you can make steps in the floor above to correspond to changes in the vault

的角锥往下扩散开去。如果下面的柱始终在该角锥内，那么上面拱结构不会对下面拱结构造成结构上的破坏。

垂直力往下扩散的角度
The angle at which a vertical force spreads downward

为了把拱结构体系的合理结构的完整性作为一个整体来保持，所以我们建议，每一拱体的位置应使其荷载沿45°的斜线把力传到下面受力的拱体的柱上。

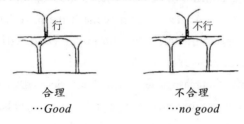

合理　　　　　　　　不合理
…Good　　　　　　　…no good

记住这一切之后，就可为你的建筑设计出一个拱结构的平面图。我们建议，要尽量使拱体和各房间对准，并在必要时予以调整，以便可以适应一个很大的房间或一个很小的凹角处或凹室。下附各图表示一幢简单建筑的楼面和天花板的布置。

为拱结构所选择的每一空间既可以是双向拱（底面为矩形的球形天花拱），又可以是单向拱（即筒拱）。双向拱在结构上是最有效的；但空间呈狭长时，球形拱承载性质就和筒拱一样了。因此我们建议，在长边不超过短边一倍的那些空间可采用球形拱，而在那些更狭窄的空间采用筒拱。

sizes below.

Vaults on different floors do not have to line up perfectly with one another.In this sense they are far more flexible than columnbeam structures,and for this reason also better adapted to STRUCTURE FOLLOWS SOCIAL SPACES(205).However,there are limits.If one vault is placed so that its loads come down over the arch of the vault below,this will put undue stress on the lower vault. Instead,we make use of the fact that vertical forces,passing through a continuous compressive medium,spread out downward in a 45 degree angle cone.If the lower columns are always within this cone,the upper vault will do no structural damage to the vault below it.

To maintain reasonable structural integrity in the system of vaults as a whole,we therefore suggest that every vault be placed so that its loads come down in a position from which the forces can go to the columns which support the next vault down, by following a 45 degree diagonal.

With all this in mind then,work out a vault plan for your building.We suggest that you try to keep the vaults aligned with the rooms,with occasional adjustments to suit a very big room, or a very small nook or alcove.The drawing on the next page shows a floor and ceiling layout for a simple building.

Each space that you single out for a vault may have either a two-way vault(a domical ceiling on a rectangular base)or a one-way vault(a barrel vault).The two-way vaults are the most efficient structurally;but when a space is long and narrow,the domical shape begins to act like a barrel vault.We therefore suggest domical vaults for spaces where the long side is not more than twice the short side and barrel vaults for the spaces

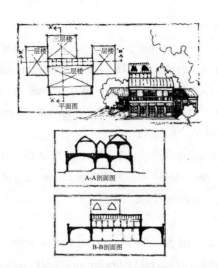

一幢简单的超轻质混凝土建筑的楼面和天花板布
置方案的平面图和剖面图
A version of floor-ceiling layout, shown in plan and section,
for a simple ultralightweight concrete building

我们同时还建议，对直接位于屋顶之下的房间采
用筒拱。这种屋顶本身一般都是筒拱——参阅**拱式屋顶**
（220）——所以，它很自然地恰好使屋顶下面的空间的天
花板形成筒拱。

在**楼面天花拱结构**（219）中所描述的拱结构跨度可达
5～30ft，并且它们需要起拱的高度至少为短跨的13%。

因此：

**为每一楼层画出一拱结构平面图。最常见的手法是使
用双向拱；在长度超过宽度一倍以上的那些空间采用单向
筒拱。在绘制拱结构平面图时，应画出建筑的剖面图，并
要牢记下列要点：**

1. 一般来说，拱体应当与房间相一致。

2. 在每个拱结构的各侧都必须有支承物：这通常是墙
的顶部。在特殊的环境下，这可以是梁或拱券。

3. 拱结构的跨度小至5ft，大至30ft。但其拱高不能小

which are narrower.

We also suggest that you use barrel vaults for the rooms immediately under the roof.The roof itself is generally a barrel vault—see ROOF VAULT(220)—so it is most natural to give the ceiling of the space just under the roof a barrel vault as well.

The vaults described in FLOOR-CEILING VAULTS(219) may span from 5 to 30 feet.And they require a rise of at least 13 per cent of the short span.

Therefore:

Draw a vault plan,for every floor.Use two-way vaults most often;and one-way barrel vaults for any spaces which are more than twice as long as they are wide.Draw sections through the building as you plan the vaults,and bear the following facts in mind:

1.Generally speaking,the vaults should correspond to rooms.

2.There will have to be a support under the sides of each vault:this will usually be the top of a wall.Under exceptional circumstances,it can be a beam or arch.

3.A vault may span as little as 5 feet and as much as 30 feet.However,it must have a rise equal to at least 13 per cent of its shorter span.

4.If the edge of one vault is more than a couple of feet(in plan)from the edge of the vault below it—then the lower vault will have to contain an arch to support the load from the upper vault.

于短跨的 13%。

　　4. 如果一个拱结构的边缘长度超过它下面的拱结构好几英尺（在平面图中）——那么下面拱结构必须有一拱券去支承来自上拱体的荷载。

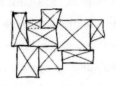

房间上面的拱结构

上下拱结构相一致

❧❧❧

　　沿着承重墙的顶部或有跨度的开口处，在每一拱结构的四周放置一圈梁（217）。根据**楼面天花拱结构**（219），设计出拱结构的形状，并且在设计拱结构的剖面时，要切记上面楼层的圈梁位置要越来越低，因为上层的柱必须是较短的（顶层柱高约 4ft，顶层下面一层的柱高约 6ft，顶层下面二层的柱高为 6 ～ 7ft，顶层下面 3 层的柱高为 8ft）——**柱的最后分布**（213）。楼层标高的变化应和安静区及公共区两者不同的特点相吻合——**地面面层**（233）。最后确定由拱结构和**角柱**（212）所形成的个别空间。把沿建筑四周的所有最小的拱体都包括在**加厚外墙**（211）之内……

Put a PERIMETER BEAM(217)on all four sides of every vault,along the top of the bearing wall,or spanning openings. Get the shape of the vaults from FLOOR-CEILING VAULTS(219)and as you lay out the sections through the vaults,bear in mind that the perimeter beams get lower and lower on higher floors,because the columns on upper stories must be shorter(top floor columns about 4 feet,one below top 6 feet,two below top 6 to 7 feet,three below top 8feet)—FINAL COLUMN DISTRIBUTION(213).Make sure that variations in floor level coincide with the distinctions between quiet and more Public areas—FLOOR SURFACE(233).Complete the definition of the individual spaces which the vaults create with COLUMNS AT THE CORNERS(212).Include the smallest vaults of all,around the building edge,in THICKENING THE OUTER WALLS(211)...

模式211   加厚外墙*

……屋顶和楼面拱结构的布置会产生水平方向的外推力，因而需要做扶壁加强——**重叠交错的屋顶**（116）。但也有这样的情况：在合理构造的楼房内，每一层四周，在不同的地方，都有小凹室、窗座、壁龛以及在房间外缘形成"厚墙"的扶壁——**窗前空间**（180）、**厚墙**（197）、**有阳光的厨房工作台**（199）、**嵌墙坐位**（202）、**儿童猫耳洞**（203）和**密室**（204）。一幢不矫揉造作的楼房其美妙之处就在于这些厚墙具有扶壁的功能，因为厚墙始终需要比自身所构成的房间较低的天花板。

211   THICKENING THE OUTER WALLS*

...the arrangement of roof and floor vaults will generate horizontal outward thrust,which needs to be buttressed— CASCADE OF ROOFS (116).It also happens,that in a sensibly made building—every floor is surrounded,at various places,by small alcoves,window seats,niches,and counters which form "thick walls" around the outside edge of rooms— WINDOW PLACE(180),THICK WALLS(197),SUNNY COUNTER(199),BUILT-IN SEATS(202),CHILD CAVES(203),SECRET PLACE(204).The beauty of a natural building is that these thick walls—since they need lower ceilings,always,than the rooms they come from—can work as buttresses.

Once the ROOF LAYOUT(209),and the FLOOR AND CEILING LAYOUT(210)are clear these thick walls can be laid out in such a way as to form the most effective butresses, against the horizontal thrust developed by the vaults.

❧❧❧

We have established in THICK WALLS(197),how important it is for the walls of a building to have "depth" and "volume,"so that character accumulates in them,with time. But when it comes to laying out a building and constructing it, this turns out to be quite hard to do.

The walls will not usually be thick in the literal sense,except in certain special cases where mud construction, for example, lends itself to the making of walls.More

一旦对**屋顶布置**（209）以及**楼面和天花板布置**（210）一清二楚了，这些厚墙就可如此布置，使它们形成最有效的扶壁以抗衡拱结构的水平推力。

❦❧

我们在厚墙（197）中已经阐明建筑的墙具有"厚度"和"体积"是何等的重要。随着时间的推移，墙的特性逐渐形成。但当实际的设计和施工时，墙是很难处理的。

墙就其字面意思讲通常是不厚的，除像土筑的一些特殊情况之外。更为常见的是，墙的厚度取决于筑墙的材料，如泡沫材料、灰泥、立柱、支撑和薄隔板。在此情况下，首先是立柱，起着主要的作用，因为人们筑墙主要靠立柱。例如，如果墙的构架是由独立于墙背面的柱来构成的话，那么墙就需要进行修正——墙就成为自然的，并且极易把木板钉到柱上，从而做成坐位、搁架，和其他许多有用的东西。但是，对于一垛平平的、纯粹的无窗墙，这种做法通常是不予鼓励的。即使在理论上讲，人们总能把粘不上墙的东西加到墙上去，但平整度十分良好的墙就不大可能发生这种事。于是我们就可以假定，厚墙在立柱限定其体积时才是最有效的。

由柱限定的有效厚墙
Thick walls made effective by columns

怎样才能证明这种墙有利于建筑结构而增加费用是合理的呢？事实上，建筑被想象成抗压结构，其楼面和屋顶均为拱结构——**有效结构**（206）——意味着在建筑外围有水平推力的地方，拱并不能相互抵消推力。

often,the thickness of the wall has to be built up from foam,plaster,columns,struts,and membranes.In this case columns,above all,play the major role,because they do the most to encourage people to develop the walls.For instance,if the framework of a wall is made of columns standing away from the back face of the wall,then the wall invites modification—it becomes natural and easy to nail planks to the columns,and so make seats,and shelves,and changes there.But a pure,flat,blank wall does not give this kind of encouragement.Even though,theoretically,a person can always add things which stick out from the wall,the very smoothness of the wall makes it much less likely to happen.Let us assume then,that a thick wall becomes effective when it is a volume defined by columns.

How is it possible for a wall of this kind to justify its expense by helping the structure of the building?The fact that the building is conceived as a compressive structure, whose floors and roofs are vaults—EFFICIENT STRUCTURE(206),means that there are horizontal thrusts developed on the outside of the building,where the vaults do not counterbalance one another.

To some extent this horizontal thrust can be avoided by arranging the overall shape of the building as an upside down catenary—see CASCADE OF ROOFS(116).If it were a perfect catenary,there would be no outward thrust at all. Obviously,though,most buildings are narrower and steeper than the ideal structural catenary,so there are horizontal thrusts remaining.Although these thrusts can be resolved by tensile reinforcing in the perimeter beams—see PERIMETER BEAMS(217)—it is simplest,and most natural,and stable to use the building itself to buttress the horizontal thrusts.

在某种程度上，这种水平推力是可以避免的，办法是把建筑的总体形状设计成倒置的悬链线——参阅**重叠交错的屋顶**（116）。假如这是一条完整的悬链线，则就根本不存在任何外向推力了。不过，很明显，大多数建筑比理想结构的悬链线更窄和更陡，所以仍然有水平推力。虽然这些推力可以通过圈梁内的抗拉钢筋分解掉——参阅**圈梁**（217）——但是利用建筑本身去承受水平推力是最简单、最合乎自然的，而且也是最稳定的。

凡有"厚墙"的地方，很自然地就可能出现凹室、窗座或房间外围的任何其他小空间。这些空间的天花可能比主要房间的天花低，因此，它们的屋顶的形状可做成室内拱式天花的延续。这就要求把厚墙布置在主要房间的结构之外，结果它们的屋顶和墙就近似形成一条和主要拱体相连的悬链线。

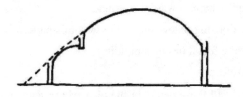

在悬链线内的凹室
Alcoves within the catenary

当然，基本上不太可能使凹室或厚墙接近真正的悬链线截面——我们几乎从不要求凹室或厚墙那样深或那样低。但是，即使当厚墙或凹室在悬链线以内时，它们仍然有助于抗衡外向推力。而且，如果把它们的屋顶做得重些，它们的扶壁支撑效果将会有更加明显的提高。该附加的重量将会把来自主要拱体的力的方向重新稍微向垂直方向改变传至地面。

下图表示本模式的工作方式以及它对建筑所起到的效果。

This possibility occurs naturally wherever there are "thick walls"—alcoves,window seats,or any other small spaces at the outside edge of rooms,which can have lower ceilings than the main room and can therefore have their roofs shaped as continuations of the ceiling vault inside.This requires that thick walls be outside the structure of the main room,so that their roofs and walls come close to forming a catenary with the main vault.

It is of course rare to be able to have the alcove or thick walls approach a true catenary section—we hardly ever want them that deep or that low.But even when the thick walls and alcoves are inside the line of the catenary,they are still helping to counter outward thrusts.And their buttressing effect can be improved still more by making their roofs heavy.The extra weight will tend to redirect the forces coming from the main vault slightly more toward the ground.

The drawing below shows the way this pattern works,and the kind of effect it has on a building.

Therefore:

Mark all those places in the plan where seats and closets are to be.These places are given individually by ALCOVES (179),WINDOW PLACES(180),THICK WALLS(197), SUNNY COUNTER(199),WAIST-HIGH SHELF(201), BUILT-IN SEATS(202),and so on.Lay out a wide swath on the plan to correspond to these positions. Make it two or three feet deep;recognize that it will be outside the main space of the room;your seats,niches,shelves,will feel attached to the main space of rooms but not inside them.Then,when you lay out columns and minor columns,place the columns in such a way that they surround and define these thick volumes of wall,as if they were rooms or alcoves.

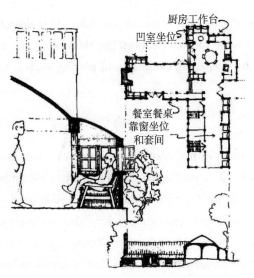

厨房工作台
凹室坐位

餐室餐桌
靠窗坐位
和套间

平面图和剖面图所示的加厚外墙的效果
The effect of thickening the outer walls,shown in plan and section

因此:

在平面图中标明哪里是坐位,哪里是盥洗室。凹室
(179)、窗前空间(180)、厚墙(197)、有阳光的厨房工
作台(199)、半人高的搁架(201)、嵌墙坐位(202)等
均逐一标出。在平面图上按相应位置画出一条宽带。其深
度为 2ft 或 3ft;考虑到它将位于房间的主空间之外;你的
坐位、壁龛、搁架将附属于各房间的主空间,但不在主空
间之内。随后,当你布置立柱和次要支柱时,要使立柱放
置在这些体积大的厚墙四周,并限定它们,仿佛这些立柱
就形成了房间或凹室。

对于不到 2ft 宽的搁架和工作台,没有必要再加宽了。
加厚外墙的简单办法就是加大立柱的厚度,并将搁架布置
于立柱之间。

For shelves and counters less than 2 feet deep,there is no need to go to these lengths.The thickening can be built simply by deepening columns and placing shelves between hem.

୫◦©◈

In order to make an alcove or thick wall work as a buttress,build its roof as near as possible to a continuation of the curve of the floor vault immediately inside.Load the roof of the buttress with extra mass to help change the direction of the forces—ROOF VAULTS (220).Recognize that these thick walls must be outside the main space of the room,below the main vault of the room—FLOOR-CEILING VAULTS(219),so that they help to buttress the horizontal forces generated by the main vault of the ceiling.When you lay out columns and minor colunms,put a column at the corner of every thick wall,so that the wall space,like other social spaces,becomes a recognizable part of the building structure-COLUMNS AT THE CORNERS(212)...

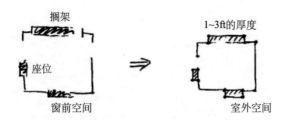

搁架

座位

窗前空间

1~3ft的厚度

室外空间

⚜

　　为了使凹室或厚墙起到扶壁的功能，就要尽量使屋顶
完全成为内部楼面拱体曲线的延续。扶壁的屋顶所增加的
荷载有助于改变传力的方向——**拱式屋顶**（220）。考虑到
这些厚墙应位于房间的主空间之外，房间的主要拱体之
下——**楼面天花拱结构**（219），所以，厚墙有助于抗衡天
花的主要拱体所产生的水平（推）力。当你布置立柱和次
要支柱时，将立柱放置在各个厚墙的角隅上，使得墙的
空间像其他社会空间一样，成为建筑结构易识别的一部
分——**角柱**（212）……

模式212　角柱**

　　……假定你已设计出屋顶平面图，并布置完毕各层每个房间的天花拱体——**屋顶布置**（209）、**楼面和天花板布置**（210），这些拱体不仅是结构的基础，而且也限定了其下的社会空间；现在该是布置拱体角柱的时候了，我们如同之前清楚地说明社会空间——**结构服从社会空间的需要**（205）——那样，来完整地阐明角柱。同时角柱也是建筑安装在结构方面的第一个步骤——**逐步加固**（208）。

❧☙

　　我们早已明确这样的观念：建筑的结构构件应和建筑的社会空间相符合。

212   COLUMNS AT THE CORNERS**

...assume that you have worked out the roof plan,and laid out ceiling vaults for every room on every floor—ROOF LAYOUT(209),FLOOR AND CEILING LAYOUT(210).These vaults are not only the basis of the structure,but also define the social spaces underneath them.Now it is time to put columns at the corners of the vaults.This will both complete them as clearly defined social spaces—STRUCTURE FOLLOWS SOCIAL SPACES(205)—and also be the first constructive step inthe erection of the building—GRADUAL STIFFENING(208).

⟡⟡⟡

We have already established the idea that the structural components of a building should be congruent with its social spaces.

In STRUCTURE FOLLOWS SOCIAL SPACES(205) we have established that the columns need to be at corners of social spaces for psychological reasons.In EFFICIENT STRUCTURE(206)we have established that there needs to be a thickening of material at the corners of a space for purely structural reasons.

Now we give yet a third still different derivation of the same pattern—not based on psychological arguments or structural arguments,but on the process by which a person can communicate a complex design to the builder,and ensure that it can be built in an organic manner.

在**结构服从社会空间的需要**（205）中，我们已经提出了，由于心理原因，柱应放置在社会空间的角隅处。在**有效结构**（206）中，我们提出了，由于纯粹结构上的原因，在空间角落处的构件应当增厚。

现在我们还要提出本模式第三个不同的论据——它不是心理学方面或结构方面的论据，而是施工过程的论据，即人们在施工过程中能把复杂的设计清楚地交待给施工人员，并确保这一设计能以有机的方式加以实现。

让我们先从测量和施工详图的问题谈起吧。最近数十年来，通常实际工作中都是利用施工详图来规定建筑的平面图。这些精确的详图要被带往施工现场；建筑工人在工地根据该详图的每一细部放大尺寸进行精心施工。

这一施工过程会使建筑受到损害。画出这种施工详图不用丁字尺是不行的。绘制施工详图的过程就不可避免地会改变平面图，使它不会轻易变更，使它成为一种能画、能测量的平面图。

但是，利用本模式语言所能绘制出的平面图相较于上述的那种平面图自由灵活得多——而且不那么容易绘制和测量。不管这些平面图是你在施工现场上构想出来的，并用棒、石头或石灰标志好的，还是你在信封背面或描图纸的碎片上粗略画出的——如果施工人员能建造出一幢生气蓬勃的建筑，即使平面图的所有线条稍欠整齐匀称和角度不准确，你想体现在平面图中的丰富多样性才能够被保存下来。

地面上的石灰标志
Chalk marks on the ground

为此，建筑必须以完全不同的方式来建造。按施工详图依样画葫芦是行不通的。主要做的应是确定构成各种空

We begin with the problem of measurement and working drawings.For the last few decades it has been common practice to specify a building plan by means of working drawings. These measured drawings are then taken to the site;the builder transfers the measurements to the site,and every detail of the drawings is built in the flesh,on site.

*This process cripples buildings.*It is not possible to make such a drawing without a T-square.The necessities of the drawing itself change the plan,make it more rigid,turn it into the kind of plan which can be drawn and can be measured.

But the kind of plans which you can make by using the pattern language are much freer than that—and not so easy to draw and measure.Whether you conceive these plans out on the site—and mark them on the site with sticks and stones and chalk marks—or draw them roughly on the back of envelopes or scraps of tracing paper—in all events,the richness which you want to build into the plan can only be preserved if the builder is able to generate a living building,with all its slightly uneven lines and imperfect angles.

In order to achieve this aim,the building must be generated in an entirely different manner.It cannot be made by following a working drawing slavishly.What must be done,essentially,is to fix those points which generate the spaces—*as few of them as possible*—and then let these points generate the walls,right out on the building site,during the very process of construction.

You may proceed like this:first fix the corner of every major space by putting a stake in the ground.There are no more than a few dozen of these comers in a building,so this is possible,even if the measurements are intricate and irregular. Place these comer markers where they seem right,without

间的各点的位置——这些点要尽量少——接着，就让这些点完全在建筑工地现场上，在施工过程中形成墙。

可以这样来进行施工：在地基上打桩，首先把每一主要空间的转角点固定下来。在一幢房屋中这样的角不会超过几十个。所以，即使尺寸复杂且不规则的，这不是能办到的。把这些转角点的标志确定在大致正确的位置，而不必考虑它们之间的精确距离。没有任何理由来硬性规定它们之间的模数距离。如果角度稍有偏差，正如它们经常发生的那样，间距模数化无论如何是不可能的。

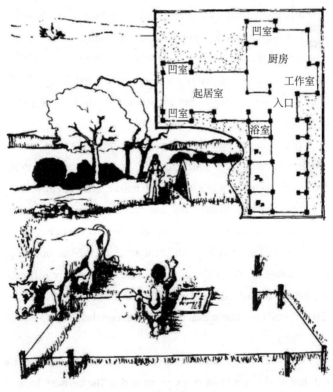

凹室
凹室
厨房
工作室
起居室
入口
凹室
浴室

"打桩"
"Staking out"

regard for the exact distances between them.There is no reason whatever to try and make modular distances between them. If angles are slightly off,as they often will be,the modular dimensions are impossible anyway.

These simple marks are all you need to build the building.Once construction starts,you can start very simply,by building a column,over each of these marks.These columns will then generate the rest of the building,by their mere presence,without any further need for detailed measurements or drawings,because the walls will simply be built along the lines which connect adjacent columns:and everything else follows.

For the upper storys,you can make drawings of the column positions and once again transfer them to the actual building while it is being built.As you will see from FINAL COLUMN DISTRIBUTION (213),upper story columns do not need to line up perfectly with downstairs columns.

With this procedure,it becomes possible to transfer a rather complex building from your mind,or from a scrap of paper,to the site—and regenerate it in a way which makes it live out there.

The method hinges on the fact that you can fix the corners of the spaces first—and that these corners may then play a significant role in the construction of the building. It is interesting that although it is based on entirely different arguments from STRUCTURE FOLLOWS SOCIAL SPACES (205),it leads to almost exactly the same conclusion.

Therefore:

On your rough building plan,draw a dot to represent a column at the corner of every room and in the corners formed by lesser spaces like thick walls and alcoves.Then transfer these dots onto the ground out on the site withstakes.

这些简单的标志都是在建房时所需要的。一旦施工开始，就可以直接在每一标志上立柱。这些柱将逐渐形成房屋的其余部分。仅仅由于这些柱的存在，就无需任何进一步的详细测量或施工详图，因为沿着连结相邻的柱之间的线可以轻而易举地把墙建造起来。其他步骤如下：

对于上部楼层来说，你可以绘出柱位置图，在施工建造过程时，按图纸上的位置立上真实的柱。正如你将在**柱的最后分布**（213）中所看到的那样，上层的柱不必和楼下的柱分毫不差地排成一条线。

随着这一施工过程的进展，就有可能把你头脑中的或画在图纸碎片上的一幢相当复杂的房屋在施工现场上变成现实——结果就会在那里生动地再现出来。

这种方法取决于你首先要能确定这些空间的转角点这一事实——这些转角点在这幢房屋的施工过程中极其重要。有趣的是，虽然这一方法是以完全不同于**结构服从社会空间的需要**（205）的论据为基础，但是它可以得出几乎完全相同的结论。

因此：

在粗略的建筑平面图上画上点以示每一房间的角柱以及由较小的空间如厚墙和凹室所形成的各个角隅。然后用桩将这些点反映在施工现场上。

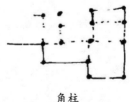

角柱
columns at comers

Once you have the columns for each floor on your vault plan, reconcile them from floor to floor and put in intermediate columns—FINAL COLUMN DISTRIBUTION(213).Note, especially,that it is not necessary for the corner columns to fall on a grid.The floor vaults and roof vaults can be made to fit any arrangement of columns,and still make a coherent structure—thus allowing the social spaces to determine the building shape without undue constraint from purely structural considerations—FLOORCEILING VAULTS(219),ROOF VAULTS (220).

These columns will not only guide your mental image of the building,they will also guide construction:first put the columns and the column foundations in place;then,to make the frame complete,tie the columns together around each room with the perimeter beam—ROOT FOUNDATIONS(214),BOX COLUMNS(216),PERIMETER BEAMS(217).Give special emphasis to all free-standing columns with the idea that when you build them,you will make them very thick—COLUMN PLACE(226)...

一旦在拱结构平面图上画上各楼层的立柱，就使它们在层与层之间协调一致，并放置中间柱——**柱的最后分布**（213）。尤其要注意的是，没有必要把角柱画在坐标方格上。楼面拱和屋顶拱的位置可以和柱的任何一种排列相适应，并且还能形成一种相互连贯的结构——因此，从纯粹的结构角度考虑，就能使社会空间决定建筑的形状而避免不适当的限制——**楼面天花拱结构**（219）和**拱式屋顶**（220）。

这些柱不仅指导有关建筑的形象思维，也指导施工实践：首先把柱和柱基就位；然后架立起全部框架结构，并用圈梁把每一房间四周的柱连接起来——**柱基**（214）、**箱形柱**（216）和**圈梁**（217）。对于所有的独立柱来说，要特别强调的是：在建造它们时，应将它们造得十分粗大——**柱旁空间**（226）……

模式213 柱的最后分布**

　　……假定你已把限定空间的角柱布置完毕——**角柱**（212）。现在正如**有效结构**（206）所要求的那样，有必要用中间加劲杆来填充柱之间的空隙。本模式提出关于这些中间加劲杆的空间分布方案。它有助于形成**有效结构**（206）所要求的那种墙，同样也有助于形成**天花板高度变化**（190）。

め◯♂

　　如何根据天花板高度、楼层数量和房间大小的不同，对加强墙的那些次要柱子进行配置？

213   FINAL COLUMN DISTRIBUTION**

...assume that you have placed the comer columns which define the spaces—COLUMNS AT THE CORNERS(212).It is now necessary to fill in the gaps between the columns with intermediate stiffener columns as required by EFFICIENT STRUCTURE(206).This pattern gives the spacing of these intermediate stiffener columns,and helps to generate the kind of walls which EFFICIENT STRUCTURE(206)requires.It also helps to generate CEILING HEIGHT VARIETY(190).

ଞୠଔଔ

How should the spacing of the secondary columns which stiffen the walls,vary with ceiling height,number of stories and the size of rooms?

In some very gross intuitive way we know the answer to this question.Roughly,if we imagine a building with the walls stiffened at intervals along their length,we can see that the texture of these stiffeners needs to be largest near the ground,where social spaces are largest and where loads are largest,and smallest near the roof,where rooms are smallest and where loads are least.In its gross intuitive form this is the same as the intuition which tells us to expect the finest texture in the ribbing at the fine end of a leaf where everything is smallest,and to expect the grosser,cruder structure to be near the large part of the leaf.

These intuitions are borne out by many traditional building forms where columns,or frames,or stiffeners are larger and further apart near the ground,and finer and closer together

借助某种非常明显的直观方式我们就会知道这一问题的答案。粗略地说，如果我们想象一幢楼房的墙在沿它的长度上以等间距加固，我们就会看到，这些加劲杆的结构在靠近地面处最大，那里社会空间最大，荷载也最大；而靠近屋顶处加劲杆最小，那里房间最小，荷载也最小。这正同我们从叶子得到的直观印象是一模一样的：一片叶子最精细的结构是其端部的叶脉，那里一切都是最小的，而在靠近叶子的根部，它的结构比较粗大。

叶子
Leaf

这些直观印象来自许多传统的建筑形式，这些建筑的柱子或框架或加劲杆在接近地面处较大，间距也较大，而越往上，则加劲杆越细，间距也越小。本模式首页上插图说明了这些例子。可是，什么东西是这些直观印象的结构基础呢？

弹性薄板理论给我们提供了合理的解释。

设想有一片未加固的薄墙承受着轴向荷载。这片墙因为薄，在纯压力使其损坏之前就往往被压曲了。这意味着，墙的材料未得到有效使用。墙的抗压强度本来可能承受的压力荷载，它却不能承受，因为它太薄了。

因此很自然，设计墙时，既要有足够的厚度，又要有足够的加强，以便使它能充分发挥抗压能力来承受荷载而不是受弯而被破坏。这样的墙在利用材料的抗压能力方面已达到极限，而且也可以满足**有效结构**（206）的要求。

这种临界因素就是墙的细长度：即它的高厚比。对于一片未加固的混凝土墙的简单情况，美国混凝土协会法规告诉我们，如果这片墙的细长比为 10 或不到 10，它就只能发挥 93％ 的效率（即承受它的潜在抗压能力的 93％ 而无弯曲）。因此，从这种意义上讲，一片 10ft 高、1ft 厚的墙

higher up.Our key picture shows examples.But what is the structural basis for these intuitions?

Elastic plate theory gives us a formal explanation.

Consider an unstiffened thin wall carrying an axial load. This wall will usually fail in buckling before it fails in pure compression because it is thin.And this means that the material in the wall is not being used efficiently.It is not able to carry the compressive loads which its compressive strength makes possible because it is too thin.

It is therefore natural to design a wall which is either thick enough or stiffened enough so that it can carry loads up to its full compressive capacity without buckling.Such a wall,which uses its material to the limits of its compressive capacity,will then also satisfy the demands of EFFICIENT STRUCTURE(206).

The critical factor is the slenderness of the wall:the ratio of its height to its thickness.For the simple case of an unstiffened concrete wall,the ACI code tells us that the wall will be able to work at 93 per cent efficiency(that is,carry 93 per cent of its potential compressive load without buckling),if it has a slenderness ratio of 10 or less.A wall 10 feet high and 1 foot thick is therefore efficient in this sense.

Suppose now,that we extrapolate to the case of a stiffened wall using elastic plate theory.By using the equation which relates allowable stress to the spacing of stiffeners,we can obtain similar figures for various walls with stiffeners.These figures are presented in the curve below.For example,a wall with a slenderness of 20needs stiffeners at 0.5H apart(where H is the height)thus creating panels half as wide as they are high. In general,obviously,the thinner the wall is,in relation to its

是有效的。

现在假定，我们利用弹性薄板理论来推断加固墙的情况。利用容许应力与加劲杆间距的关系方程式，我们就能获得具有加劲杆的各种墙的近似数据。这些数据由下面的曲线示出。例如，一片细长比为20的墙加劲杆间距需要0.5H（此处H是高度），从而就形成许多宽度为其高度一半的镶嵌板。很明显，一般来说，与高度成比例，墙越薄，则沿其长度上就越要频繁加固。

在所有情况下，曲线都表示使墙发挥其93％抗压强度所必需的加劲杆的间距。总而言之，根据**有效结构**（206）原理所建的墙必须按该曲线加强。

不同楼层上的柱间距增减率可直接从该曲线得出。我们从下例中可以看到这一点。一幢四层楼房其各个楼层的墙荷载比例大致如下：4 : 3 : 2 : 1（只是非常粗略）。

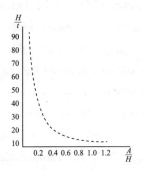

此曲线表示墙的细长度与加劲杆间距的关系
The curve which relates wall slenderness to the spacing of stiffeners.
图内说明：这条曲线由下式求出：f_c=+5.0E（t/H）^{2}+3.6（t/A）^{2}轻质混凝土容许抗压应力标准定为f_c=93％，并采用美国混凝土协会对非加固情况的值t/H=1/10，式中t/A=0。

在所有情况下，我们越往楼房的高处走，墙所承受的荷载就越小。如果所有的墙即将达到它们的极限抗压能力，这就意味着，我们越往楼的高处走，墙就会渐渐地变得越

height,the more often it needs to be stiffened along its length.

In every case,the curve gives the spacing of stiffeners which is needed to make the wall work at 93 per cent of its compressive strength.In short,we may say that a wall built according to the principle of EFFICIENT STRUCTURE(206) ought to be stiffened in accordance with this curve.

The gradient of column spacing over different floors follows directly from this curve.We may see this in the following manner. The walls in a four story building carry loads which are very roughly in the ratio 4 ∶ 3 ∶ 2 ∶ 1(only very roughly).In any case,the loads the walls carry get less and less the higher we go in the building.If all the walls are reaching their full compressive capacity,this means that they must be getting steadily thinner too,the higher one goes in the building.If we assume that the walls all have the same height,then the four walls will therefore have progressively greater and greater slenderness ratios,and *will therefore fall further and further to the left on the curve,and will therefore need to be stiff-ened at closer and closer intervals.*

For example,suppose a four story building has 8 foot high walls on all floors and has wall thicknesses of 12 inches,9 inches,6 inches,and 3 inches on its four floors.The slenderness ratios are 8,11,17,and 33.In this case,reading off the curve,we find the ground floor has no stiffeners at all(they are infinitely far apart),the second floor has stiffeners at about 8 feet apart,the third floor has them about 5 feet apart,and the top floor has them about 2 feet apart.

In another case,where the walls are thinner(because materials are lighter and loads smaller),the spacing will be closer.Suppose,for example,that the necessary wall thicknesses are 8,6,4,and 2 inches. Then the slenderness ratios are 12,16,24,and 48,and the stiffeners

来越薄。如果我们假定，墙都具有同一高度，则四片墙的细长比将越来越大，因此，将越来越向曲线的左侧移动，而需要加固的间距也就越来越小。

例如，假设一幢四层楼的建筑各层都有8ft高的墙，墙厚分别为12in、9in、6in和3in。其细长比为8、11、17和33。在这种情况下，当我们读出曲线的值时，就会发现一层完全没有加劲杆（它们的间距无限），二层有加劲杆，间距约为8ft，三层有加劲杆，间距为5ft，顶层有加劲杆，间距约为2ft。

在另一种情况下，墙较薄（因为材料较轻和荷载较小）的地方，则间隔将会更小。例如假设所需的墙厚分别为8in、6in、4in和2in。那么，细长比则为12、16、24和48，同时加劲杆需要的间距就比以前更小了：在底层相隔9ft，在二层相隔5ft，在三层相隔3ft，在顶层相隔15in。

正如你从这些例子中所看到的那样，柱间隔的变化大得惊人，实际上，比直觉所允许的大。但变化之所以如此大是因为我们已经假设，每层天花板的高度都是一样的。事实上，在一幢正确设计的楼房内，天花板的高度将随楼层的不同而发生变化；在这样的条件下，正如我们将要看到的那样，柱间距的变化显得更为合理了。天花板高度随楼层的不同而变化的原因有二：一是社会的原因，另一是结构的原因。

在大多数的楼房中，底层的空间和房间应更大一些——因为公用房间、会议室等房间，一般说来，位置靠近楼房的入口处较好，而私密的和较小的房间宜设在更高一些的楼层上，位于更深入建筑的地方。因为天花板的高度随社会空间的大

小房间低墙

大房间高墙

房间大小的变化
Variation of room sizes

need to be spaced closer together than before:nine feet apart on the ground story,5 feet apart on the second story,3 feet apart on the third,and 15 inches apart on the top.

As you can see from these examples,the variation in column spacing is surprisingly great;greater,in fact,than intuition would allow.But the variation is so extreme because we have assumed that ceiling heights are the same on every floor.In fact,in a correctly designed building,the ceiling height will vary from floor to floor;and under these circumstances,as we shall see,the variation in column spacing becomes more reasonable.There are two reasons why the ceiling height needs to vary from floor to floor,one social and one structural.

In most buildings,the spaces and rooms on the first floor will tend to be larger—since communal rooms,meeting rooms,and so on,are generally better located near the entrance to buildings,while private and smaller rooms will be on upper stories,deeper into the building.Since the ceiling heights vary with the size of social spaces—see CEILING HEIGHT VARIETY(190)—this means that theceiling heights are higher on thc ground floor,getting lower as one goes up.And the roof floor has either very short walls or no wall at all—see SHELTERING ROOF(117).

And there is a second,purely structural explanation of the fact that ceilings need to be lower on upper stories.It is embodied in the drawing of the granary shown below.Suppose that a system of columns is calculated for pure structure.The columns on upper stories will be thinner,because they carry less load than those on lower stories.But because they are thinner,they have less capacity to resist buckling,and must therefore be shorter if we are to avoid wasting material.As a

小而变化——参阅**天花板高度变化**（190）——这意味着，底层天花板高度较高，而越往高处，天花板高度就越低。而且，顶层或是有很矮的墙，或是根本没有墙——参阅**带阁楼的坡屋顶**（117）。

对于上面楼层的天花板必须是较低的这一事实还有第二种，即纯结构的解释。这一点体现在下列的粮仓图中。假设柱体系的计算只是出于纯结构的目的。上面楼层的柱将较细，因为它们比下面楼层的柱承受较小的荷载。但是，正因为它们较细，所以它们抗压曲的能力也较小。如果我们想避免浪费材料，柱就必须是较短的。结果，即使在粮仓里，天花板高度的变化与任何社会原因无关，但出于纯属结构上的考虑，就有必要使下面楼层的柱粗大，天花板要高，越往高处柱就越细，天花板就越低。

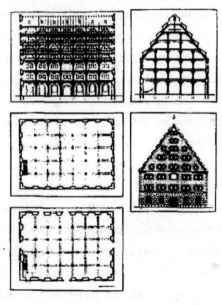

德国粮仓
German granary

result,even in a granary,where there are no social reasons for variation in ceiling height,purely structural considerations create the necessity for thick columns and high ceilings on the lower stories and for thinner and thinner columns and lower and lower ceilings the higher one gets in the building.

The same conclusion comes from consideration of our curve.We have used the curve,so far,to tell us that stiffeners need to be closer together on upper stories,because the walls are more slender.We may also use the curve to tell us that,for a given load,we should try to keep the slenderness ratio as low as possible.On the upper stories,where walls are most apt to be thin,we should therefore make the walls as low as possible,in order to keep the slenderness ratios low.

Let us assume now,that the wall heights do vary in a building,in a manner consistent with these arguments.A four story building,with an attic story on top,might then have these wall heights(remember that the vault height,in a vaulted room,is higher than the wall height):9 feet on the ground floor,7 feet on the second,6 feet on the third,and 4 feet on the fourth,where the pitched roof comes down low over the eaves.And let us assume that the wall thicknesses are 12 inches,6 inches,5 inches,and 3inches,respectively.In this case,the slenderness ratios will be 9,14,14,15.The ground floor needs no stiffeners at all;the second has them 6 feet apart;the third has them 5 feet apart;and the fourth has them 3 feet apart.We show a similar distribution in the drawing opposite.

When you try to apply this pattern to floor plan,you will find a certain type of difficulty.Since the corners of rooms may already be fixed by COLUMNS AT THE CORNERS(212),it is not always possible to space the stiffeners correctly within

同样的结论也能从我们的曲线中得出。直到现在我们一直使用这一曲线是为了告知读者，上面楼层的加劲杆的间距应小些，因为墙更为细长。我们也可以利用曲线来告知读者，当荷载固定时，我们应使细长比尽可能减小。为了保持小的细长比，在上面楼层的大多适合的薄墙，我们应使它们尽可能矮些。

我们现在假定，楼房的墙高变化在某种意义上是符合这些论据的。一幢带有顶部阁楼的四层楼房或许会有如下这些墙高（请记住，在一个拱结构的房间内，拱的高度大于墙高）：底层为 9ft，二层为 7ft，三层为 6ft，四层为 4ft。四层的斜屋顶从上往下盖在屋檐上。我们还假设，墙厚分别为 12in、6in、5in 和 3in。在此情况下，细长比将为 9、14、14、15。底层完全不需要加劲杆；二层加劲杆间距为 6ft；三层加劲杆间距为 5ft；四层加劲杆间距为 3ft。我们在下页的图中表示出了近似的分布。

在设法运用本模式于楼层平面图时，你会发现有某种困难。因为房间的各个角隅可能早已被**角柱**（212）所固定了，并非都可以将加劲杆正确地分布在任何一个指定的房间的墙内。当然这无关紧要；加劲杆的间距只需大致正确；间距可以随不同的房间而灵活变化，以便适应墙的长短。可是，从整体上说，房间小时应设法使加劲杆的间距缩短，房间大时，使加劲杆的间距增大。如果不这样做，楼房的外观就会显得稀奇古怪，因为它违背了结构的直观性原则。

试考虑一下同一层楼的两个房间，一个为另一个的两倍大。大房间有两倍的周长，但是它的天花板却有 4 倍的荷载；因此，墙的每一单位长度将承受更大的荷载。在理想的有效结构中，这意味着墙必须是较厚的；因此，按照早已提出的论据，大房间需要的加劲杆的间距应比小的房间（小房间承受较小的荷载，并且它的墙较薄）更大些。

我们承认，在楼房的同一层，为数不多的施工人员在

the wall of any given room.Naturally this does not matter a great deal;the stiffeners only need to be about right;the spacing can comfortably vary from room to room to fit the dimensions of the walls.However,on the whole,you must try and put the stiffeners closer together where the rooms are small and further apart where rooms are large.If you do not,the building will seem odd,because it defies one's structural intuitions.

Consider two rooms on the same floor,one twice as large as the other.The larger room has twice the perimeter,but its ceiling generates four times the load;it therefore carries a greater load per unit length of wall.In an ideal efficient structure,this means that the wall must be thicker;and therefore,by the arguments already given,it will need stiffeners spaced further apart than the smaller room which carries less load and has thinner walls.

We recognize that few builders will take the trouble to make wall thicknesses vary from room to room on one floor of the building.However,even if the wall is uniformly thick,we believe that the stiffeners must at least not contradict this rule. If,for reasons of layout,it is necessary that the spacing of stiffeners varies from room to room,then it is essential that the larger spacings of the stiffeners fall on those walls which enclose the larger rooms.If the greater spacing of stiffeners were to coincide with smaller rooms,the eye would be so deceived that people might misunderstand the building.

One important note.All of the preceding analysis is based on the assumption that walls and stiffeners are behaving as elastic plates.This is roughly true,and helps to explain the general phenomenon we are trying to describe.However,no wall behaves perfectly as an elastic plate—least of all the kind of lightweight concrete walls we are advocating in the rest of

确保墙厚随不同的房间而变化时会不辞辛劳。可是，我们认为，即使墙的厚度完全一致，加劲杆至少必须和这一规则不相矛盾。如果由于布置的原因，加劲杆的间距随房间的不同而变化是必需的话，那么重要的是，加劲杆大间距应布置在大房间的墙上。假如加劲杆的大间距用在小房间，人们的眼睛将会受到欺骗，以致对楼房产生误解。

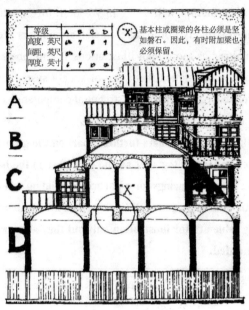

一幢四层楼房内柱的最后分布，根据我们的有关柱、墙和拱结构的模式建造的

The final column distribution in a four story building,
built according to our patterns for cloumns, walls and vaults

有一点值得重视。前面的全部分析都是以如下的假设为基础的：墙和加劲杆都起着弹性薄板的作用。粗略地说这是对的，并有助于解释我们正试图描述的一般现象。可是，没有一面墙真正起着弹性薄板的作用，这种弹性薄板是轻质混凝土墙中最轻的，我们在其余的构造施工模式中正在大力提倡使用它。由此可见，我们一直在使用按照美国混

the construction patterns.We have therefore used a modified form of the elastic plate theory,calibrated according to the ACI code,so that the numbers in our analysis are based on the elastic behavior of concrete(and fall within the limits of its tension and compression).However,when the plate goes out of the elastic range and cracks,as it almost certainly will in a concrete design,other factors will enter in.We therefore caution the reader most strongly not to take the actual numbers presented in our analysis as more than illustrations.The numbers reflect the general mathematical behavior of such a system,but they are not reliable enough to use in structural computations.

Therefore:

Make column stiffeners furthest apart on the ground floor and closer and closer together as you go higher in the building. The exact column spacings for a particular building will depend on heights and loads and wall thicknesses.The numbers in the following table are for illustration only,but they show roughly what is needed.

building height in stories	ground floor	2nd floor	3rd floor	4th floor
1	$2'\sim5'$			
2	$3'\sim6'$	$1'\sim3'$		
3	$4'\sim8'$	$3'\sim6'$	$1'\sim3'$	
4	$5'\sim8'$	$4'\sim8'$	$3'\sim6'$	$1'\sim3'$

Mark in these extra stiffening columns as dots between the corner columns on the drawings you have made for different floors.Adjust them so they are evenly spaced between each pair of corner columns;but on any one floor,make sure that they are closer together along the walls of small rooms and further apart along the walls of large rooms.

凝土协会法规所标定的、弹性薄板理论的修正公式，所以，我们分析的数值是以混凝土的弹性状态为基础的（并在它的拉力和压力的极限之内）。但是，薄板超出弹性范围就会破裂，因为混凝土设计中几乎肯定会发生破裂，所以还应考虑其他因素。因此，我们郑重提醒读者，不要把我们在分析中提供的实际数值看得比我们的说明更重要。这些数值仅反映这样一种系统的一般数学特性，但它们并未可靠到足以能应用于结构计算之中。

因此：

使柱的加劲杆间距在底层达到最大，随着向上面的楼层越走越高，加劲杆的间距则越来越小。一幢特定楼房，其精确的柱间距将取决于高度、荷载和墙厚。下面表格中的数值只是为了说明情况、表示所需近似尺寸。

楼房层数	底层	二层	三层	四层
1	$2'\sim5'$			
2	$3'\sim6'$	$1'\sim3'$		
3	$4'\sim8'$	$3'\sim6'$	$1'\sim3'$	
4	$5'\sim\infty'$	$4'\sim8'$	$3'\sim6'$	$1'\sim3'$

将角柱之间的这些附加加强柱用点在为不同楼层所画的图纸上一一标出。调整好这些点，使它们在每一对角柱之间均匀分布；但在所有楼层都要保证做到，小房间墙的附加加强柱间距小，大房间墙的附加加强柱间距则大。

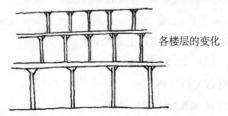

各楼层的变化

To the extent consistent with CEILING HEIGHT VARIETY(190),make walls and columns progressively shorter the higher you go in the building to keep slenderness ratios low.

And make wall thicknesses and column thicknesses vary with the height—see WALL MEMBRANE(218).Our calculations,for a typical lightweight concrete building of the kind we have been discussing,suggest the following orders of magnitude for wall thicknesses:Top story—2 inches thick;one below top story— 3 inches;two below top story—4 inches;three storys below top(ground floor on a four story building)—5 inches.Of course these numbers will change for different loads,or for different materials,but they show the type of variation you can expect.

Column thicknesses must be proportional to wall thicknesses, so that the thinnest walls have the thinnest columns. If they are very thin,it will be possible to make them simply by placing boards,or one thickness of material,outside the outer skins which form the wall membrane— see WALL MEMBRANE(218).If the walls are thick,they will need to be full columns,twice as thick as the walls,and roughly square in section,built before the walls,but made in such a way that they can be poured integrally with the walls—BOX COLUMNS(216)...

put stakes in the ground to mark the columns on the site,and start erecting the main frame according to the layout of these stakes;

214.ROOT FOUNDATIONS
215.GROUND FLOOR SLAB
216.BOX COLUMNS
217.PERIMETER BEAMS
218.WALL MEMBRANES
219.FLOOR-CEILING VAULTS
220.ROOF VAULTS

在可能范围内应符合**天花板高度变化**（190）。为了保持较小的细长比，往上面的楼层越走越高，墙和柱则越来越矮。

使墙的厚度和柱的粗细随高度而变化——参阅**墙体**（218）。我们的计算为正在讨论的这种典型的轻质混凝土楼房提出了如下的墙厚数量级：四层——墙厚 2in；三层——3in；二层——4in，底层——5in。当然，这些数值会因不同的荷载或不同的材料而起变化，但是它们表明这种变化正是你所期望的。

柱的粗细必须和墙的厚度成正比，所以最薄的墙就有最细的柱。如果墙很薄，就有可能将它们简化为板式墙，即厚度不变的单一材料，其外皮按墙体要求处理——参阅**墙体**（218）。如果墙较厚，就需要加设柱子。柱的厚度为墙的两倍，其截面大致呈正方形，在筑墙之前先立柱子，但它可以和墙连接成一整体灌筑而成——**箱形柱**（216）……

在地上立桩以示基地上柱的位置，并根据这些桩的位置，开始安装主框架；

214. 柱础

215. 底层地面

216. 箱形柱

217. 圈梁

218. 墙体

219. 楼面天花拱结构

220. 拱式屋顶

模式214　柱基

……一旦你有一幢楼房的粗略的柱平面图——**角柱**（212）、**柱的最后分布**（213）——你就具备在基地开始施工的条件了。首先，你在着手任何土方工程之前，要立桩标出底层柱的位置，以便有必要完整地保留原有岩石和植物，你能移动柱位——**基地修整**（104）、**与大地紧密相连**（168）。然后挖掘基础坑，并准备灌筑基础。

❧

各种基础中最佳的就是树根——树的整体结构直接延伸至地下，形成一个和地面完全结成一体的抗拉抗压系统。

当柱和基础是分离的构件时，它们应被连接起来。而这种连接就成为十分困难而又重要的了。正是在连接处弯曲应力和剪应力都十分巨大。如果采用第三种构件作为连接体，那么带来的麻烦会更多。而且，每一构件抵抗这些应力的效率都较低。

我们怀疑，只有把柱埋于柱础并与地面连成一体，才是建造基础和柱的较好方法。

为了实现我们在此说明的这一模式，柱础应具有简单的形状。因为柱开始是空心的，即**箱形柱**（216），我们可设计适于将空心柱插入的基础坑，然后在柱的下部和基础进行一次性整体浇筑，这样，基础就形成了。

至于谈到木质箱形柱方案，把木质箱形柱埋入地下，与地下潮湿的混凝土接触——这是一个严重的问题。为了防止柱木干裂和白蚁的破坏，可用压力向柱内注入五氯苯酚。

214 ROOT FOUNDATIONS

...once you have a rough column plan for the building—COLUMNS AT THE CORNERS(212),FINAL COLUMN DISTRIBUTION(213)—you are ready to start the site work itself.First,stake out the positions of the ground floor columns,before you do any other earthwork,so that you can move the columns whenever necessary to leave rocks or plants intact—SITE REPAIR(104),CONNECTION TO THE EARTH(168).Then dig the foundation pits and prepare to make the foundations.

₧₨

The best foundations of all are the kinds of foundations which a tree has—where the entire structure of the tree simply continues below ground level,and creates a system entirely integral with the ground,in tension and compression.

When the column and the foundations are separate elements which have to be connected,the connection becomes a difficult and critical joint.Both bending and shear stresses are extremely high just at the joint.If a connector is introduced as a third element,there are even more joints to worry about,and each member works less effectively to resist these stresses.

We suspect that it would be better to build the foundations and the columns in such a way that the columns get rooted in the foundation and become integral and continuous with the ground.

为每根柱挖一底
部大于顶部的坑

用混凝土灌筑柱和
基础并停止灌筑

轻质混凝土
40°/W·Pt

地下的木隔板用
五氯苯酚处理

1/4″钢筋

柱和地面之间的
2″间隙

我们已经做成的一种空心木质箱形柱的柱础方案

One version of a root foundation for a hollow wooden box column
which we have built

我们还认为，在木质箱形柱表面涂上一层厚沥青和防潮玛碲脂可能起作用，但问题并未真正解决。当然，砖石结构方案应当是可行的：柱可采用注入密实混凝土的陶瓦管或混凝土管。可是，即使在这些情况下，我们仍怀疑本模式的确切结构的有效程度。我们认为，某种与地面连成一体的结构应该是可以应用的：但是我们还没有十分的把握把它设计出来，我们说明这一模式，无非是想提出问题而已。

In the realization of this pattern which we illustrate,the root foundation takes a very simple form.Since columns start out hollow,BOX COLUMNS(216),we can form a root foundation by setting the hollow column into the foundation pit,and then pouring the lower part of the column and the foundation,integrally,in a single pour.

As far as the wood version is concerned,the problem of placing wood in contact with wet underground concrete is very serious.The wood of the column can be protected from dry rot and termites by pressure dipping in pentachlorophenol.We also believe that painting with thick asphalt or dampproof mastic might work;but the problem isn't really solved.Of course,masonry versions in which columns are made of terracotta pipe or concrete pipe and filled with dense concrete,ought to work alright.But even in these cases,we are doubtful about the exact structural validity of the pattern.We believe that some kind of structure which is continuous with the ground is needed:but we quite haven't been able to work it out.Meanwhile,we state this pattern as a kind of challenge.

Namely:

Try to find a way of making foundations in which the columns themselves go right into the earth,and spread out there—so that the footing is continuons with the material of the column,and the column,with its footing,like a tree root, can resist tension and horizontal shear as well as compression.

因此：

　　设法寻求一种建造基础的新方法：柱本身深埋地下，并在那里扩展——结果，柱础和柱连成一体，而且，柱和柱础就像一棵树的树根，既能抗拉力和水平剪力，又能抗压力。

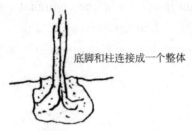

底脚和柱连接成一个整体

&Z&R

　　为了建造这种适于填充空心箱形混凝土柱的基础，首先要给每一柱础挖个坑，将空心柱放于坑内，并把柱和基础连成一体，一次连续灌筑而成——**箱形柱**（216）。最后，在灌筑底层地面时，把混凝土底层和基础连接起来——**底层地面**（215）。

∞

To make foundations like this for hollow concrete, filled box columns, start with a pit for each foundation, place the hollow column in the pit, and pour the column and the foundation integrally, in one continuous pour—BOX COLUMNS(216). Later, when you build the ground floor slab, tie the concrete into the foundations—GROUND FLOOR SLAB(215).

模式215　底层地面

……本模式有助于完善**与大地紧密相连**（168）、**有效结构**（206）、**角柱**（212）和**柱基**（214）。这是一简单的混凝土地面,形成房屋的底层平面,它将各柱基相互连接起来,同时,你也可以做成简单的条形基础以作为混凝土地面的一部分去支撑墙壁。

❧❧❧

平板是做底层地面可应用最容易、最便宜和最普通的方法。

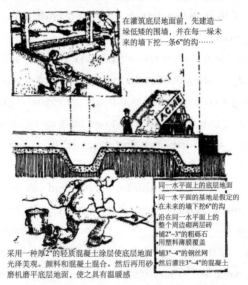

在灌筑底层地面前,先建造一垛低矮的围墙,并在每一垛未来的墙下挖一条6"的沟……

同一水平面上的底层地面
同一水平面上的基地是假定的
在未来的墙下挖6"的沟
沿在同一水平面上的整个周边放两层砖
铺2"~3"的粗砾石
用塑料薄膜覆盖
铺3"~4"的钢丝网
然后灌注3"~4"的混凝土

采用一种厚2"的轻质混凝土涂层使底层地面光泽美观。颜料和混凝土混合。然后再用砂磨机磨平底层地面,使之具有温暖感

四周为砖墙的抬高了的底层混凝土地板
A raised ground floor slab built inside a brick perimeter wall

当底层地面水平时,在素土上直接浇筑混凝土地板是建造房屋底层地面所采用的最普通和最便宜的方法。木质地板造价高,其下需有空气隔绝开,并需建造在连续的墙

215   GROUND FLOOR SLAB

...this pattern helps to complete CONNECTION TO THE EARTH(168),EFFICIENT STRUCTURE(206),COLUMNS AT THE CORNERS(212),and ROOT FOUNDATIONS(214).It is a simple slab,which forms the ground floor of the building,ties the root foundations to one another,and also allows you to form simple strip foundations as part of the slab,to support the walls.

༄༅༅

The slab is the easiest,cheapest,and most natural way to lay a ground floor.

When the ground is relatively level,a concrete slab which sits directly on the ground is the most natural and cheapest way of building a ground floor.Wood floors are expensive,need air space underneath them,and need to be built up on continuous foundation walls or beams.Prefabricated floor panels also need a structure of some sort to support them.A slab floor,on the other hand,uses the earth for support,and can supply the foundations which are needed to support walls,by simple thickening.

The one trouble with slabs is that they can easily feel cold and damp.We believe that this feeling is at least as much a psychological one as a physical one(given a well-made and insulated slab),and that the feeling is most pronounced with slabs that are on grade.We therefore propose that the slab be raised from the ground.This can be done by not excavating the ground at all,instead only leveling it,and placing the usual

基上或梁上。预制楼板也需要某种支承它们的结构。另一方面，混凝土地板可由素土支承，采用简单的增厚办法，就能提供支承墙壁所需的基础。

混凝土地板的缺点就是人们容易感到寒冷和潮湿。我们认为人们产生这种感觉，其心理的因素并不亚于物理上的因素（如混凝土地板施工质量好，且有保温层），而且，当混凝土地面和室外地面在同一高度时这种感觉最为明显。因此，我们建议，混凝土地面应比自然地面高一些。这是不必挖掘地面就可以办到的，只要把地面平整好，再铺上毛石和砾石作为常用的垫层，然后再浇筑混凝土即可。（在标准作业中，地面是要挖掘一定深度，以便填铺的毛石上皮略低于室外地面，而浇筑的混凝土的上皮恰好高出地面。）

因此：

浇筑底层混凝土地板应略高出室外地面 6in 或 9in：在房屋四周先筑一道交圈的矮墙，并使墙和柱础连接起来，然后用毛石、砾石和混凝土填铺浇筑成地面。

填充混凝土，砾石和毛石
混凝土
砾石
毛石
砖砌的边缘砖
高出地面

&OCB

公共区底面面层用砖、面砖或上蜡和抛光的轻质混凝土，甚至可用夯实的素土来做。至于那些更具私密性的区域，其地面比别处或高出一个台阶或降低一个台阶，用轻质混凝土饰面，以便铺设毛毡或地毯——**地面面层**（233）。

砌筑一道能限定底层混凝土地面边缘的矮砖墙，并将其和房屋周围的室外平台和小径直接连接起来——**与大地紧密相连**（168）和**软质面砖和软质砖**（248）。如果是在很陡的斜坡基地上建房，其中部分地面，不用挖土就可通过浇筑混凝土而形成拱结构楼面——**楼面天花拱结构**（219）……

bed of rubble and gravel on top of the ground.(In normal practice,the ground is excavated so that the top of the rubble is slightly below grade,and the top of the slab only just above the ground.)

Therefore:

Build a ground floor slab,raised slightly—six or nine inches above the ground—by first building a low perimeter wall around the building,tied into the column foundations, and then filling it with rubble,gravel,and concrete.

⊱✣⊰

Finish the public areas of the floor in brick,or tile,or waxed and polished lightweight concrete,or even beaten earth;as for those areas which will be more private,build them one step up or one step down,with a lightweight concrete finish that can be felted and carpeted—FLOOR SURFACE(233).

Build the low wall which forms the edge of the ground floor slab out of brick,and tie it directly into all the terraces and paths around the building—CONNECTION TO THE EARTH(168),SOFT TILE AND BRICK(248).If you are building on a steep sloped site,build part of the ground floor as a vaulted floor instead of excavating to form a slab—FLOOR-CEILING VAULTS(219)...

模式216　箱形柱**

216   BOX COLUMNS**

...if you use ROOT FOUNDATIONS(214),the columns must be made at the same time as the foundations,since the foundation and the columm are integral.The height,spacing,and thickness of the various columns in the building are given by FINAL COLUMN DISTRIBUTION(213).This pattern describes the details of construction for the individual columns.

ଽୠଔ

In all the world's traditional and historic buildings,the columns are expressive,beautiful,and treasured elements.Only in modern buildings have they become ugly and meaningless.

The fact is that no one any longer knows how to make a column which is at the same time beautiful and structurally efficient.We discuss the problem under seven separate headings:

1.Columns feel uncomfortable unless they are reasonably thick and solid.This feeling is rooted in structural reality. A long thin column,carrying a heavy load,is likely to fail by buckling:and our feelings,apparently,are particularly tuned in to this possibility.

We do not wish to exaggerate the need for thickness. Taken too far,it could easily become a mannerism of a rather ridiculous sort.But columns do need to be comfortable and solid,and only thin when they are short enough to be in no danger of buckling.When the column is a free-standing one,then the need for thickness becomes essential.This is fully discussed under COLUMN PLACE (226).

……如果采用**柱基**（214），柱和基础就必须同时施工。因为基础和柱是一个整体。建筑中柱的各种高度、间隔和厚度已在**柱的最后分布**（213）中给出。本模式描述单独柱的结构细节。

✽✽✽

全世界所有的传统古建筑中，柱是造价高、造型美的珍贵建筑构件。只是在现代建筑中，柱才变得丑陋难看而又毫无意义。

目前的事实是：没有人再想知道如何才能把柱造得既外形美观而又结构合理有效。我们现分 7 个小标题来论述这一问题。

1. 若柱粗细不合理，又不是实心的，那就会使人感到不舒服。这种感觉的根源在于结构的现实性。一根又细又高的柱子在承受重荷载时，很可能被压弯。我们的感觉，显然，是和这种可能性相一致的。

我们不愿意夸大柱粗的必要性。言过其实，往往就易出现滑稽可笑、矫揉造作的行为了。但是，柱必须是实心的、令人愉悦的，细到相应的柱高需应控制的没有压弯危险的限度之内。若柱是独立式的，必要的厚度就是十分重要的了。这一点将在**柱旁空间**（226）中加以充分论述。

2. 结构上的论证导致完全相同的结论。细的高强度材料，如钢管和预应力混凝土已被排除在**好材料**（207）之外。生态上合理的但强度较低的材料须相当粗壮才能应对荷载。

3. 柱必须是造价低廉的。一根 8in×8in 的实心木柱价格非常昂贵；而厚砖柱或石柱在今天的市场上几乎已不成问题了。

4. 柱应具有温暖的触感。混凝土柱和油漆钢柱的表面使人感到不快，而且也不容易进行饰面加工。

2.Structural arguments lead to exactly the same conclusion. Thin,high strength materials,like steel tubes and prestressed concrete,are ruled out by GOOD MATERIALS(207).Lower strength materials which are ecologically sound have to be relatively fat to cope with the loads.

3.The column must be cheap.An 8 by 8 solid wood column is too expensive;thick brick or stone columns are almost out of the question in today's market.

4.It must be warm to the touch.Concrete columns and painted steel columns have an unpleasant surface and are not very easy to face.

5.If the column takes bending,the highest strength materials should be concentrated toward the outside. Buckling and bending strength both depend on the moment of inertia,which is highest when the material is as far as possible from the neutral axis.A stalk of grass is the archetypal example.

6.The column must be easy to connect to foundations, beams,and walls.Precast concrete columns are very hard to connect.So are metal columns.Brick columns are easy to connect to brick walls—not to the lighter weight skin structures required by WALL MEMBRANE(218).

7.The column must be hand nailable,and hand cuttable to make on-site modification and later repair as easy as possible.Again,current materials do not easily meet this requirement.

A column which has all these features is a box column, where the hollow tube can be made as thick as is required,and then filled with a strong compressive material. Such a column can be made cheaper than comparable wood

5. 如果柱易发生弯曲，材料的最大应力将集中在外缘。抗压强度和抗弯强度两者都取决于惯性矩。材料离中和轴越远，抗压和抗弯强度的值就越大。草的主茎就是这一原型的实例。

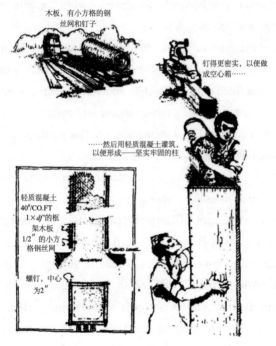

木板，有小方格的钢丝网和钉子

钉得更密实，以便做成空心箱……

……然后用轻质混凝土灌筑，以便形成——坚实牢固的柱

轻质混凝土 40#/CO.FT 1×df'的框架木板 1/2″的小方格钢丝网

螺钉，中心为2″

一种箱形柱方案：由1in厚的木板用有螺旋槽的钉子钉成箱形柱，其内紧贴木板放入钢丝网，并浇筑超轻质混凝土

A version of box columns made of 1 inch wood planks, nailed together with spiral groove nails, and filled with chicken wire and ultra-lightweight concrete

6. 柱必须和基础、梁、墙连接在一起。预制混凝土柱很难连接。金属柱也是如此。砖柱容易和砖墙连接，却不容易和**墙体**（218）所需的轻质墙面结构相连接。

7. 柱必须是可用手钉钉子的或在施工现场可进行手工切割的，以及今后尽可能容易修理的。而现在通用的材料

and steel columns;the outer skin can be made with a material that is beautiful,easy to repair,and soft to the touch;the column can be stiffened for bending,either by the skin itself,or by extra reinforcing;and,for structural integrity,the fill material can be made continuous with the column's footings and beams.

An example of a box column which we have built and tested is a wooden box column,made with 1 inch wooden planks and filled with lightweight concrete the same density as wood,so that it has the overall volume and mass of a heavy 8 inch solid column.The drawing opposite shows these wooden box columns being made.

Box columns can be made in many other ways. One kind is made by stacking 8 by 8 inch lightweight concrete blocks,and filling the cavity with a concrete of the same density.Some wire reinforcing inside the column is required to give the column tensile strength.A hollow brick column,filled with earth is another possibility. Concrete,vinyl,and terracotta sewer pipe filled with lightweight concrete and reinforced with mesh;a resin-impregnated cardboard tube filled with earth;or two concentric cardboard tubes with the outer ring filled with concrete and the inner ring filled with earth;still another is made from a tube of chicken wire wesh,filled with rubble,plastered and whitewashed on the outside.And still another can be made with self-aligning hollow tiles for the skin.The tiles can be molded by hand with a hand press—in concrete or tile;the soft tile will make beautiful rose red,soft warm columns.

无法满足这些要求。

具备上述特征的柱就是箱形柱，柱的壁厚可按需要而定，后用高强抗压材料灌筑。这样的柱比木柱或钢柱成本更低廉。它的外层可采用漂亮、易于修理、触感柔和的材料。为了提高抗弯强度，可加强外壁本身或另加附加钢筋；为了保证结构上的整体性，填料可与柱脚和梁浇筑成连续的整体。

我们已建成并进行了试验的箱形柱的一例就是木质箱形柱。它是由 1in 厚的木板制成的，其内浇筑和木料容重相同的轻质混凝土。柱的总体积和总重量相当于 8in 厚的沉重实心柱。上页插图说明建筑工人正在用超轻质混凝土浇筑木质箱形柱。

各种可能的箱形柱
Possible box columns

箱形柱还可以用其他方法来制作。一种方法是打桩：将 8in×8in 轻质混凝土空心柱打入地下，在柱的空腔内浇筑容重相同的混凝土。柱内用钢丝加固是必要的，以增加柱的抗拉强度。另一种可能的方法是采用空心砖柱：其空腔用土填实。还可用混凝土管、乙烯树脂管和陶瓦的污水管作柱，其内浇筑轻质混凝土，并用钢丝网加劲；用土充填的树脂浸渍的硬纸管；用混凝土充填外环、用土充填内环的同心双层硬纸管；钢丝网为管骨架，内填毛石，外部抹灰泥并用石灰刷白。还有一种柱外层采用自动调准空心面砖。这种面砖可用手压机压制混凝土或陶土模型而成。软质面砖将使柱呈现出美丽的玫瑰红色，从而使柱成为有

Therefore:

Make the columns in the form of filled hollow tubes, with a stiff tubular outer skin, and a solid core that is strong in compression. Give the skin of the column some tensile strength—preferably in the skin itself, but perhaps with reinforcing wires in the fill.

৪০১৫৪

As you already know, it is best to build the columns integral with ROOT FOUNDATIONS(214)on the ground floor, or integral with the FLOOR-CEILING VAULTS(219)on upper floors, and to fill them in one continuous pour. Once the columns are in position, put in the PERIMETER BEAMS(217), and fill the beams at the same time that you fill the upper part of the column. If the column is free standing, put in column braces or column capitals—COLUMN CONNECTION (227)—to brace the connection between the two. And make the columns especially thick, or build them in pairs, where they are free-standing, so that they form a COLUMN PLACE(226)...

柔和感的暖色柱。

由混凝土污水管内浇筑混凝土而制成的箱形柱
Box columns made from concrete sewer pipe,filled with concrete

因此：

把柱做成浇筑混凝土的空心管，它有坚硬的管外壁和抗压性能好的实心。使柱的外壁具有某种抗拉强度——加强外壁本身效果会更好些，但也可在填料时以钢丝加固。

填充

外壁

正如你早已知道的那样，在底层最好是将柱和**柱基**（214）浇筑成为一个整体，在上层最好是将柱和**楼面天花拱结构**（219）浇筑成为一个整体。一旦柱就位完毕，就要放置**圈梁**（217），同时要将梁和柱的顶部用混凝土浇筑好。如果是独立柱，那么就放置柱斜撑或柱帽——**柱的连接**（227）——在柱和柱帽之间应加强连接。尤其要使柱具有厚度，或做成独立式的双柱，以便它们能形成**柱旁空间**（226）……

模式217 圈梁*

　　……本模式有助于完善**箱形柱**（216）。一旦柱就位，就可将柱的端部连接起来。它也有助于形成**楼面天花拱结构**（219）边缘的支承面。为此，圈梁的位置必须和布置在**楼面和天花板布置**（210）内的边缘完全相符。

217   PERIMETER BEAMS*

...this pattern helps to complete BOX COLUMNS (216),by tying the tops of the columns together once they are in position.It also helps to form the bearing surface for the edge of the FLOOR-CEILING VAULTS (219).For this reason,the positions of the perimeter beams must correspond exactly to the edges of the vaults laid out in FLOOR AND CEILING LAYOUT(210).

⊰⊱

If you conceive and build a room by first placing columns at the corners,and then gradually weaving the walls and ceiling round them,the room needs a perimeter beam around its upper edge.

It is the beam,connecting the columns which creates a volume you can visualize,before it is complete;and when the columns are standing in the ground,you need the actual physical perimeter beam,to generate this volume before your eyes,to let you see the room as you are building it,and to tie the tops of the columns together,physically.

These reasons are conceptual.But of course,the conceptual simplicity and rightness of the beam around the room comes,in the end,from the more basic fact that this beam has a number of related structural functions,which make it an essential part of any room built as a natural structure.The perimeter beam has four structural functions:

8003

　　如果你构思并准备建造一个房间，首先要放置角柱，然后逐步砌墙和建墙所围的天花板，沿房间墙顶边缘需有一道圈梁。

　　正是连接柱的圈梁才产生出一个尚未形成而能想象出的体量；当柱竖立在地面时，你需要实际而有形的圈梁，以便在眼前产生出这种体量，帮助你联想到正在建造的房间，并从整体上把这些柱的端部连接在一起。

　　这些理由是概念性的。当然，关于房间四周的梁在概念上的简单明了和正确无误归根结底当然是源于更为基本的事实：圈梁具有若干相互有关的结构功能，使其本身成为任何一个按普通结构建造起来的房间的重要组成部分。圈梁有下列 4 种结构功能：

　　1. 圈梁在墙体和拱体之间形成在**有效结构**（206）中所描述的自然的加厚部分。

　　2. 凡在没有外扶壁和其他拱结构可支承和抵消水平推力的地方，圈梁都能抗衡天花拱的水平推力。

　　3. 凡在安放门和窗的墙体上打洞的地方，圈梁能起到门窗过梁的作用。

　　4. 圈梁能把来自上面楼层的柱荷载传递给下面的柱和墙体，并将这些荷载分散，均匀分布给柱和墙体。

　　圈梁的这些功能表明，梁应尽可能与上下的墙和柱连成一体，也应和楼面连成一体。如果我们遵循模式**好材料**（207），梁也应是易于制造，易于切割成不同长度。

　　现在可利用的梁材并不能满足上述要求。钢梁和预制梁或预应力梁都不易于和墙及楼面连接并和它们连成一体。更为重要的是，这些梁很难按不同房间的精确尺寸在施工现场进行切割。而在有机的平面图中，房间总是大小不同的。

CONSTRUCTION
构　造
1913

1.It forms the natural thickening between the wall membrane and vault membrane,described in EFFICIENT STRUCTURE(206).

2.It resists the horizontal thrust of the ceiling vault, wherever there are no outside external buttresses to do it,and no other vaults to lean against.

3.It functions as a lintel,wherever doors and windows pierce the wall membrane.

4.It transfers loads from columns in upper storys to the columns and the wall membrane below it,and spreads these loads out to distribute them evenly between the columns and the membrane.

These functions of the perimeter beam show that the beam must be as continuous as possible with walls and columns above,the walls and columns below,and with the floor.If we follow GOOD MATERIALS(207),the beam must also be easy to make,and easy to cut to different lengths.

Available beams do not meet these requirements.Steel beams and precast or prestressed beams cannot easily be tied into the wall and floor to become continuous with these membranes.Far more important,they cannot easily be cut on site to conform to the exact dimensions of the different rooms which will occur in an organic plan.

Of course,wood beams meet both requirements:they are easy to cut and can be tied along their lengths to wall and floor membranes.However,as we have said in GOOD MATERIALS(207),wood is unavailable in many places,and even where it is available,it is becoming scarce and terribly expensive,especially in the large sizes needed for beams.

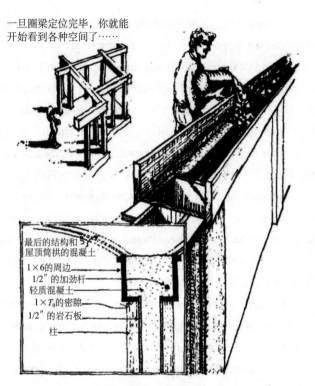

一旦圈梁定位完毕，你就能开始看到各种空间了……

最后的结构和
屋顶筒拱的混凝土
1×6的周边
1/2″的加劲杆
轻质混凝土
1×T_0的密隙
1/2″的岩石板
柱

一种和前一模式中所示的箱形柱相符的圈梁方案
*A version of the perimeter beam consistent with the box column
shown before*

当然，木梁能满足两种要求：木梁易于切割，并能沿其长度和墙体及楼面连接。可是，正如我们在**好材料（207）**中所说，木料在许多地方是不易得到的，即使能得到，木料也正日益成为稀罕之物，而且价格异常昂贵，需用大量木料作梁时尤其如此。

为了避免使用木料，我们已经设计出一种圈梁——如上图所示——它和我们的箱形柱是前后一致的，我们所设计的圈梁和箱形柱一并使用。这种圈梁的制作过程如下：在墙体建成前，先在柱上钉一木板槽；然后置入钢筋，在墙建成并用混凝土灌筑之后，再用每立方英尺受力60lbs的超轻质混

To avoid the use of wood,we have designed a perimeter beam-shown opposite—which is consistent with our box column,and designed to be used together with it.It is a beam made by first nailing up a channel made of wooden planks to the columns,before the wall membranes are made;then putting in reinforcing,and filling up with ultra-lightweight 60 pounds per cubic foot concrete,after the walls are made and filled.This beam is excellent for continuity.The wooden channel can first be made continuous with other skin elements by nailing,and the fill can then be made continuous by filling columns and beams and walls and vault in one continuous pour—see WALL MEMBRANES(218)and FLOOR-CEILING VAULTS(219).

Of course,there are many other ways of making a perimeter beam.First of all,there are several variants of our design:the U-shaped channel can be made of fiberboard,plywood,precast lightweight concrete,and,in every case,filled with lightweight concrete.Then there are various traditional perimeter beams— the Japanese version or the early American versions come to mind.And then there are a variety of structures which are not exactly even beams—but still act to spread vertical loads and counteract horizontal thrusts.A row of brick arches might function in this way,in a far fetched case so might a tension ring of jungle creeper.

凝土浇入槽内，于是圈梁就形成了。这种圈梁在连续性方面是非常有效的。也可以先把木槽和其他的表层构件钉在一起，然后用混凝土把柱、梁、墙和拱体一次连续浇筑成为一个整体——参阅**墙体**（218）和**楼面天花拱结构**（219）。

当然，还有制作圈梁的其他许多方法。首先，我们的设计有几种不同的方案：U形槽可用纤维板、胶合板、预制轻质混凝土来制成，并且，在上述各种情况下，均可用轻质混凝土浇筑。其次，还有几种传统用的圈梁，就我们回忆起来的有日本式的圈梁或早期美国式的圈梁。再次，还有各种不同的、甚至确切说不一定是梁的结构——但它们仍能起到分散垂直荷载并抗衡水平推力的作用。一排砖拱券可起着上述作用，更远的例子如林区定速运送器的拉力环。

因此：

在房间四周浇筑一道连续的圈梁，使其强度足以抗衡上面拱体的水平推力，它把上面楼层的荷载传递给柱，把柱连接在一起，同时它在墙洞上面又能起着过梁的作用。将这种圈梁和上面的柱、墙、地面及下面的柱和墙都浇筑成为一个整体。

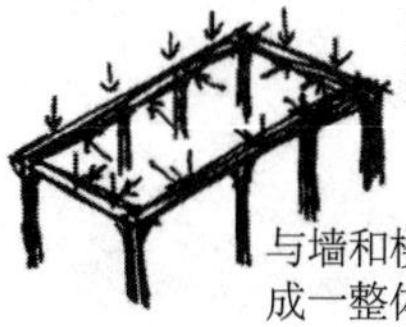
圈梁
与墙和楼层地面连接
成一整体

应记住钢筋的布置位置，使圈梁在水平方向和垂直方向都能受力。当圈梁形成**楼面天花拱结构**（219）的基座时，它应能起到抗衡拱结构未抵消的剩余水平外向推力的圈梁作用。用斜支撑来加强独立柱和圈梁之间的连接——**柱的连接**（227）……

Therefore:

Build a continuous perimeter beam around the room, strong enough to resist the horizontal thrust of the vault above, to spread the loads from upper stories onto columns, to tie the columns together,and to function as a lintel over openings in the wall.Make this beam continuous with columns,walls and floor above,and columns and walls below.

ഇൻ

Remember to place reinforcing in such a way that the perimeter beam acts in a horizontal direction as well as vertical. When it forms the base for a FLOOR-CEILING VAULT(219) it must be able to act as a ring beam to resist all those residual horizontal outward thrusts not contained by the vault.Strengthen the connection between the columns and the perimeter beam with diagonal braces where the columns are free standing—COLUMN CONNECTION(227)...

模式218　墙体*

……按照**有效结构**（206）和**柱的最后分布**（213），墙是抗压承载体，它在相邻的柱之间"被抻开"并和柱连接成一整体，而按一定间距布置的柱本身却起到加劲杆的作用。柱的间距随柱高而变化，墙体厚度也按类似方式变化。如果根据**箱形柱**（216），柱的加劲杆早已就位，本模式描述在柱与柱之间把墙体抻开而形成墙壁的方法。

❧❧❧

在有机的构造施工中，墙应承受它自己的那部分荷载。墙应和周围的其他结构共同连续工作。墙的功能是抗剪和抗弯，并承受压力荷载。

按上述要求工作的墙是结构的主要构件：墙是二维空间连续体；它和加劲杆、柱一起共同承受压力荷载；墙在柱、梁和上下楼面之间形成了连续不断的坚固的连接体，并有助于抗剪和抗弯。

相反，幕墙和真正的"填充"墙，则起不到墙体的作用。但是它们可以起到墙的其他方面的作用——如隔离、围合和限定空间——但却无助于增加建筑的总体结构的坚固性。这些墙是由它们的框架来起上述作用的，在结构上是浪费的。[对于这一论点的细节即结构的每一部分必须共同承受荷载，参阅**有效结构**（206）。]

另一方面，墙体使墙成为一个完整的构件，并和其周围的结构共同工作。我们应当如何来建造这样的墙体呢？

218   WALL MEMBRANE*

...according to EFFICIENT STRUCTURE(206)and
FINAL COLUMN DISTRIBUTION(213),the wall is a
compressive loadbearing membrane, "stretched" between
adjacent columns and continuous with them,the columns
themselves placed at frequent intervals to act as stiffeners.
The intervals vary from floor to floor,according to column
height;and the wall thickness(membrane thickness)varies in
a similar fashion.If the column stiffeners are already in place
according to BOX COLUMN(216),this pattern describes the
way to stretch the membrane from column to column to form
the walls.

෪෬

**In organic construction the walls must take their share of
the loads.They must work continuously with the structure on
all four of their sides;and act to resist shear and bending, and
take loads in compression.**

When walls are working like this,they are essentially
structural membranes:they are continuous in two dimensions;
together with stiffeners and columns they resist loads in
compression;and they create a continuous rigid connection
between columns,beams,and floors,both above and below,to
help resist shear and bending.

By contrast,curtain walls and walls which are essentially
"infill," do not act as membranes.They may function as walls
in other respects—they insulate,enclose,they define space—

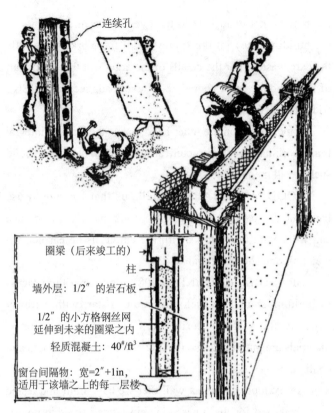

连续孔

圈梁（后来竣工的）

柱

墙外层：1/2″的岩石板

1/2″的小方格钢丝网
延伸到未来的圈梁之内

轻质混凝土：$40^{\#}/ft^3$

窗台间隔物：宽=2″+1in，
适用于该墙之上的每一层楼

一种内墙体方案：采用石膏板作为外层，并用超轻质混凝土填充
*A version of an interior wall membrane which uses gypsun board
as skin,and ultralightweight concrete for the fill*

　　模式**好材料**（207）告诉我们,应当使用能手工切割的、可钉钉子的、生态上合理的、人们可用家用工具进行加工的材料，以填土和板材为主。

　　模式**逐步加固**（208）告诉我们，施工过程应如下进行：人们可先从细薄结构开始，然后在施工过程中加固它；当各种材料一一放在该放的位置时，结构也就得到了加固。整个施工过程就能十分顺利和连续不断。

but they do not contribute to the overall structural solidity of the building.They let the frame do all the work;structurally they are wasted.[For the details of the argument that every part of the structure must cooperate to take loads,see EFFICIENT STRUCTURE(206).]

A membrane,on the other hand,makes the wall an integral thing,working with the structure around it.How should we build such a wall membrane?

GOOD MATERIALS(207)tells us that we should use hand cuttable,nailable,ecologically sound materials,which one can work with home tools,with the emphasis on earthen fill materials and sheet materials.

GRADUAL STIFFENING(208)tells us that the process of building should be such that one can start with a flimsy structure and stiffen it during the course of construction,as materials are put in place,so that the process can be smooth and continuous.

An example of such a wall that we have built and tested uses gypboard for the inner skin,ship-lapped wooden boards for the outer skin and ultra-lightweight concrete for the fill.The wall is built by fixing nailing blocks to the sides of columns. We nail the skin to the nailing blocks,put chickenwire into the cavity to reinforce the concrete against shrinkage,and then pour the lightweight concrete into the cavity.The wall needs to be braced during pouring,and you can't pour more than two or three feet at a time:the pressure gets too great.The last pour fills the perimeter beam and the top of the wall,and so makes them integral.The drawing opposite shows one way that we have made this particular kind of wall membrane.

现在介绍一个已经建造并测试的这种墙的例子。它采用石膏板作墙体的里表皮,鱼鳞板作外表皮,中间填充超轻质混凝土。建造这种墙的步骤是:将可钉钉子的板块先固定在柱的各个侧面,我们再把外层钉在这些板块上,继而再把钢丝网置入两层之间的空腔内,以加强混凝土的抗收缩能力,然后再把轻质混凝土灌注于腔内。灌注时,墙需要支撑,而且每次灌注不得超过 2ft 或 3ft,否则压力会太大。最后浇筑圈梁和墙的顶部,以便使圈梁和墙连接成一个整体。上图表示已建造这种特殊墙体的一种方法。

这种墙是实心的(容重大致与木材相当),它有很好的声学特性和热学特性,且适应那些平面自由和不规则的建筑,而且还可以钉钉子。因为这种墙有加劲杆,所以它的厚度足以保证坚实牢固。

本模式的其他方案有:(1)墙内外层可采用空心结构的面砖或混凝土砌块,内填混凝土或土;(2)外皮可采用砖,内皮可采用胶合板或石膏板。在这两种情况下,柱必须是空心瓦管或混凝土管,或其他砖砌箱形柱;(3)墙内外皮为钢丝网,其内逐步填充毛石和混凝土,墙外皮拉毛粉刷,内皮抹灰。在此情况下,柱可用同样的方法建造——在钢丝网制成的管内填充毛石和混凝土;(4)墙内外皮均可采用石膏板。石膏板的表面可贴上防潮纸,或镶上板条,或拉毛粉刷。

因此:

把墙建成为连接柱、门框和窗框的墙体,并且至少部分地和它们连成一体。建墙时,首先要放置墙体的内层和外层结构,它们同时又是墙的内外饰面;然后再把填充材料灌注于墙体内。

This wall is solid(about the density of wood),has good acoustic and thermal properties,can easily be built to conform to free and irregular plans,and can be nailed into.And because of its stiffeners,the wall is very strong for its thickness.

Other versions of this pattern:(1)The skin can be formed from hollow structural tiles or concrete blocks,with a concrete or earthen fill.(2)The exterior skin might be brick,the interior skin plywood or gypboard.In either case the columns would have to be hollow tile,or concrete pipe,or other masonry box columns.(3)The skin might be formed with wire mesh,gradually filled with concrete and rubble,and stuccoed on the outside,with plaster on the inside.The columns in this case can be built in the same way-out of a wire mesh tube filled with rubble and concrete.(4)It may also be possible to use gypboard for both skins,inside and out.The gypboard on the outer side could then be covered with building paper,lath,and stucco.

Therefore:

Build the wall as a membrane which connects the columns and door frames and windows frames and is,at least in part,continuous with them.To build the wall,first put up an inner and an outer membrane,which can function as a finished surface;then pour the fill into the wall.

෩෨

Remember that in a stiffened wall,the membranes can be much thinner than you might expect,because the stiffeners prevent buckling.In some cases they can be as thin as two inches in a one story building,three inches at the bottom of a two-story building and so on—see FINAL COLUMN DISTRIBUTION(213).

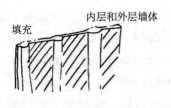

填充　　　　内层和外层墙体

❧❧❧

　　切记，加固墙体的厚度可以比你所想象的少，因为加劲杆能抗弯。在某些情况下，在单层建筑内它们可薄到 2in，在两层楼房的底部可薄到 3in，以此类推——参阅**柱的最后分布**（213）。

　　墙体的内外层可由空心砖、轻质混凝土砌块、胶合板、石膏板、木板或其他板材制成。这些材料应有良好的表面，容易钉钉子，且触感舒适，如此种种。如果墙的内层是石膏板，可用灰泥罩面——**有柔和感的墙内表面**（235）。其外层可采用 1in 厚的企口墙板；或室外用胶合板；或者挂面砖或挂木瓦的外墙板、或者抹上灰泥的墙板——**鱼鳞板墙**（234）。也可用普通砖或面砖作墙的外层：在这种情况下，柱应采用同一材料——**软质面砖和软质砖**（248）……

Membranes can be made from hollow tile,lightweight concrete block,plywood,gypboard,wood planks,or any other sheet type material which would make a nice surface,which is easy to nail into,comfortable to touch,and so on.If the inner sheet is gypsum board,it can be finished with a skim coat of plaster—SOFT INSIDE WALLS(235).The outer sheet can be made of 1 inch boards,tongue and grooved;or exterior grade plywood;or exterior board hung with tile,shingles,or plastered—LAPPED OUTSDE WALLS(234).It is also possible to build the outer skin of brick or tile:in this case,columns must be of the same material—SOFT TILE AND BRICK(248)...

模式219　楼面天花拱结构**

　　……我们已经论述了这样一个事实：普通梁式楼盖和板式楼盖是低效和不经济的，因为它们用相较于纯抗压材料更少的抗拉材料来抗弯曲——**有效结构**（206）和**好材料**（207），因此，比较理想的情况是，使用拱结构。本模式提出拱结构的形状和结构。拱结构将有助于完善**楼面和天花板布置**（210）和**圈梁**（217）；尤其重要的是拱结构将有助于创造各种不同的房间的**天花板高度变化**（190）。

219   FLOOR-CEILING VAULTS**

...we have already discussed the fact that ordinary joist floors and slab floors are inefficient and wasteful because the tension materials they use to resist bending are less common than pure compression materials—EFFICIENT STRUCTURE(206),GOOD MATERIALS(207),and that it is therefore desirable to use vaults wherever possible.This pattern gives the shape and construction of the vaults.The vaults will help to complete FLOOR AND CEILING LAYOUT(210),and PERIMETER BEAMS(217);and,most important of all,they will help to create the CEILING HEIGHT VARIETY(190)in different rooms.

೮೦Ӝಏ

We seek a ceiling vault shape which will support a live load on the floor above,form the ceiling of the room below, and generate as little bending and tension as possible so that compressive materials can be relied on.

The vault shape is governed by two constraints:the ceiling cannot be lower than about 6 feet at the edge of the room,except in occasional attic rooms;and the ceiling in the middle of the room should vary with the room size(8 to 12 feet for large rooms,7 to 9 feet for middle sized rooms,and 6 to 7 feet for the very smallest alcoves and corners—see CEILING HEIGHT VARIETY(190)).

　　我们现已找到一种天花拱结构的形状，它承受上面楼层的活荷载，形成下面房间的天花板，同时产生尽可能小的弯曲和拉力，从而使受压材料安全可靠。

　　拱结构的形状受下面两种情况制约：除非是正规阁楼房间，否则房间四周，天花板高度不能低于 6ft；房间中心的天花板高度应随房间的大小不同而变化（大房间的天花板高度变化范围为 8 ～ 12ft，中等房间为 7 ～ 9ft，最小的凹室和角隅为 6 ～ 7ft——参阅**天花板高度变化**（190））。

　　从结构考虑，我们知道，如果球形薄壳拱高超过其直径的 13％～ 20％，它实际上将不会产生任何弯矩（这一点我们已在薄壳结构的研究和试验中阐明，并由我们的计算机研究报告加以证实。）对于一个直径为 8ft 的房间来说，它所需的拱高约为 18in，在房间中心的总高度为 7 ～ 8in；对于一个直径为 15in 的房间来说，它需要的拱高为 2 ～ 3in，其中心的总高度为 8 ～ 10in。

　　幸运的是，这些拱结构的高度恰好和所需要的天花板高度相符合。因此，我们可以说，一个有人居住的空间其理想拱结构的边缘的高度为 6 ～ 7ft，并且拱高为短边长度的 13％～ 20％。

　　现在有各种不同的方法可以从一个正方形房间或一个长方形房间上建造起圆形或椭圆形拱结构。

　　1. 一种拱结构是由对角线发券形成拱形斜肋；然后由直线母线沿斜肋形成拱的空间。

　　2. 另一种是纯粹的圆形拱结构。它是由墙角支承上层

We know,from structural considerations,that a circular shell dome will generate virtually no bending moments when its rise is at least 13 to 20 per cent of its diameter.(This is established in studies and tests of shell structures,and is corroborated by our own computer studies.)For a room 8 feet across,this requires a rise of about 18 inches,making a total height of 7 to 8 feet in the middle;for a room 15 feet across,it requires a rise of 2-3 feet,making a height of 8 to 10 feet in the middle.

Luckily,these vault heights are just congruent with the needed ceiling heights.We may say,therefore,that the ideal vault for an inhabited space is one which springs from 6 to 7 feet at the edge,and rises 13 to 20 per cent of the smaller diameter.

There are various possible ways of making a circular or elliptical vault spring from a square or rectangular room.

1.One type of vault is made by arching diagonal ribs from corner to corner;and then spacing straight line elements across the ribs.

2.Another type is a pure dome supported on squinches.

3.Another is based on a rectangular grid of arched ribs.The edge ribs are entirely flat,and the center ribs have the greatest curvature.In the end,each part of the vault is curved in three dimensions,and the comers are slightly flattened.

Each of these three vaults makes sense in slightly different circumstances.The first is the easiest to conceive,but it has a slight structural disadvantage:its surface panels are curved in one direction only—because they are made of

结构的内角拱所构成的。

3. 还有一种拱结构是以矩形格网状的拱肋为基础的。其边缘肋完全是水平的，而肋的中心曲率最大。最后，这种拱结构的每一部分都是三维空间弯曲，而且各个角均略呈扁平。

在上述 3 种拱结构中，情况稍有不同时，每种拱结构都有其意义。第一种拱结构是最容易构想出来的，但在结构上稍有缺陷：它的板面只在一个方向上弯曲——因为它们是由直线母线制成——因此，达不到双曲拱结构的强度。第二种拱结构是最难设计的；可是，这种拱结构很自然地由一个球形结构和一个矩形结构相交而成的。假如有人利用气球外形来建造一个拱结构，并在圈梁之内将它推出，那么采用第二种拱结构就会是最容易的了。在非标准建筑的施工技术中我们一直采用的是第三种拱结构，使用它是最方便的，因为布置那些为支拱创造条件的拱肋是特别简单的。它的各个角都要取平，否则可能会产生弯矩，因而需要用抗拉材料。可是，在轻质混凝土中我们已经发现，第三种拱结构不需要消耗比任何其他形式更多的抗收缩配筋。

现在我们来说明建造拱结构的一种非常简便易行的方法。切记，我们认为重要的是：拱结构应逐步建造，以便它能毫无困难地适合于各种房间形状。这种技术不仅便宜简单，而且它也是我们在使拱结构适合于各种房间形状的

straight line elements—and cannot therefore achieve the strength of a doubly curved vault.The second is the hardest to conceive;however,it comes naturally from the intersection of a spherical shape and a rectangular one.If one were to make a vault by using a balloon as a form,pushed up within the perimeter beams,the second type would be the easiest to use. In the particular building technique we have been using,the third type is easiest to use,because it is particularly simple to lay out the arched ribs which provide the formwork.It flattens out at the corners,which could create bending moments and require tension materials.However,in lightweight concrete we have found that it does not require any more than the shrinkage reinforcement,which is needed anyway.

We shall now describe a very simple way of making a vault.Bear in mind that we considered it essential that the vault be built up gradually,and that it could be fitted to any room shape,without difficulty.This technique is not only cheap and simple.It is also one of the only ways we have found of fitting a vault to an arbitrary room shape.It works for rectangular rooms,rooms that are just off-rectangles, and odd-shaped rooms.It can be applied to rooms of any size. The height of the vault can be varied according to its position in the overall array of ceiling heights and floors—CEILING HEIGHT VARIETY(190),STRUCTURE FOLLOWS SOCIAL SPACES(205),FLOOR AND CEILING LAYOUT(210).

过程中所找到的仅有的几种方法之一。这种技术适用于矩形房间，也适用于非矩形房间或畸形房间。它能适用于各种尺寸的房间。这种拱结构的高度随其在天花高度和楼层的总序列中所处位置而变化——**天花板高度变化**（190）、**结构服从社会空间的需要**（205）以及**楼面和天花板布置**（210）。

首先，在间隔为 1ft 的各中心点上将网络结构的筋条，放在一方向上即相对的圈梁上，使每一筋条弯曲成合理的拱形状。再在另一个方向上，间距仍为 1ft 的各中心点上使筋条交织成网。可将筋条钉在房间四周圈梁的模板上。你会发现这种网状结构非常坚固和稳定。

已就位的网络筋条
Lattice strips in position

现将粗麻布绷紧覆盖在网络筋条上，并将它钉在筋条上，使它密实无缝。再在粗麻布上涂上一层厚厚的聚酯树脂来加固它。

覆盖在网络筋条构架上的粗麻布
Burlap over the lattice work

粗麻布树脂罩面非常坚固，足以支承 1 ~ 2in 厚的轻质混凝土。在准备过程中，先在加固了的粗麻布上放置一

First,place lattice strips at one foot centers,spanning in one direction,from one perimeter beam to the opposite perimeter beam,bending each strip to make a sensible vault shape.Now weave strips in the other direction,also at almost one foot centers,to form a basket.The strips can be nailed onto the form of the perimeter beam around the room.You will find that the basket is immensely strong and stable.

Now stretch burlap over the lattice strips,tacking it on the strips so it fits tightly.Paint the burlap with a heavy coat of polyester resin to stiffen it.

The burlap-resin skin is strong enough to support 1 to 2 inches of lightweight concrete.In preparation for this,put a layer of chickenwire,as shrinkage reinforcement,over the stiffened burlap.Then trowel on a 1-to-2 inch layer of lightweight concrete.Once again,use the ultra-lightweight 40-60 pound concrete described in GOOD MATERIALS(207).

The shell which forms is strong enough to support the rest of the vault,and the floor above.

The rest of the vault should not be poured until all edges are in,columns for the next floor are in position,and ducts are in—see BOX COLUMNS(216),DUCT SPACE(229).In order to keep the weight of the vault down,it is important that even the ultralightweight concrete be further lightened,by mixing it with 50 per cent voids and ducts.Any kind of voids can be used—empty beer cans,wine jugs,sono tubes,ducts,chunks of polyurethane.Or voids can be made very much like the vaults

层钢丝网，作为抗收缩配筋。然后，在它上面用泥刀涂抹 1～2in 厚的轻质混凝土层。最后，再涂上一层容重为每立方英尺 40～60lbs 的超轻质混凝土，详见**好材料**（207）。

涂抹在粗麻布上的树脂
Resin over burlap

　　这样形成的薄壳十分牢固，足以支承拱体其余部分和上层楼面的重量。

薄壳上的轻质混凝土
Lightweight concrete on

　　拱结构其余部分的浇筑应一直进行到所有的边缘都浇筑之后，上面一层楼的柱和管道都就位了为止——参阅**箱形柱**（216）、**管道空间**（229）。为了支承住拱结构的重量，重要的是甚至包括超轻质混凝土在内，应当进一步减轻重量，办法是在搅拌超轻质混凝土时再掺入 50％孔隙率的骨料。任何一种有孔隙的骨料都可以使用——空啤酒罐、葡萄酒罐、会发声的管子，其他有孔道的物体以及大小不等的聚氨酯泡沫块。或者孔隙体可以做得和拱体本身相似：在柱间用网络筋条做拱，并把这些拱铺设到圆顶位置。下

themselves by making arches with latticing between columns and then stretching burlap from these arches to the dome.The drawing opposite shows the sequence of construction.

A 16 by 20 foot vault similar to the one shown in our photographs has been analyzed by a computerized finite element analysis.The concrete was assumed to be 40 pounds perlite,with a test compressive strength of 600 psi.Tensile strength is taken as 34 psi,and bending as 25.5 inch pounds per inch.These figures are based on the assumption that the concrete is unreinforced.Dead loads were figured at 60 pounds per square foot assuming 50 percent voids in the spandrels of the vault.Live loads were taken to be 50 pounds per square foot.

According to the analysis,under such loading the largest compressive stress in this dome occurs near the base at mid points of all four sides and is 120 psi.Outward thrust is the greatest at quarter points along all four walls,and is 1769 pounds.The maximum tension of 32 psi occurs at the comers. Maximum bending is 10 inch pounds per inch.All of these are well within the capacity of the vault,and besides,shrinkage reinforcement in the vault will make it even stronger.

The analysis shows,then,that even though the vault is an impure form(it contains square panels which are actually sagging within the overall configuration of the vault shape),its structural behavior is still close enough to that of a pure vault to work essentially as a compression structure.There are small

图就表明这种孔隙体的施工顺序。

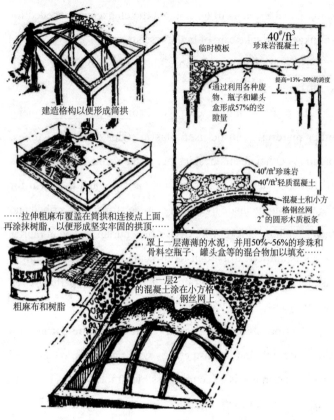

建造格构以便形成筒拱

……拉伸粗麻布覆盖在筒拱和连接点上面，再涂抹树脂，以便形成坚实牢固的拱顶……

粗麻布和树脂

临时模板

$40^{\#}/ft^3$ 珍珠岩混凝土

提高=13%~20%的跨度

通过利用各种废物、瓶子和罐头盒形成57%的空隙量

$40^{\#}/ft^3$珍珠岩
$40^{\#}/ft^3$轻质混凝土
混凝土和小方格钢丝网
$2''$的圆形木质板条

罩上一层薄薄的水泥，并用50%~56%的珍珠和骨料空瓶子、罐头盒等的混合物加以填充……

一层$2''$的混凝土涂在小方格钢丝网上

一种楼面天花拱结构方案：由薄的木筋条交织成网状物，
上面覆盖粗麻布、树脂、钢丝网和超轻质混凝土
One version of a floor-ceiling vault,made of thin wooden lattice
strips woven like a basket,burlap,resin,chickenwire and ultra-
lightweight concrete

　　一个 16ft×20ft 的拱结构和我们照片中的极为相似，已用计算机进行了有限元分析。混凝土被假定为容重每平方英尺 40lbs 的珍珠岩混凝土，试验抗压强度为 600lbs/in²。取抗拉强度为 34lbs/in²，抗弯强度为 25.5lbs/in²。这些数据是基于采用无筋混凝土。恒载值受力为 60lbs/ft²，并假定拱体

amounts of local bending;and the comer positions of the dome suffer small amounts of tension,but the chickenwire needed for shrinkage will take care of both these stresses.

Here are some other possible ways of building such a vault:

To begin with,instead of wood for the lattice work,many other materials can be used:plastic strips,thin metal tubes, bamboos.Other resins besides polyester resins can be used to stiffen the burlap.If resins are unavailable,then the form for the vault can be made by placing lattice strips as described,and then stretching chickenwire over it,then burlap soaked in mortar which is allowed to harden before concrete is placed.It might also be possible to use matting stiffened with glue,perhaps even papier mache.

It is possible that similar vaults could be formed by altogether different means:perhaps with pneumatic membranes or balloons.And it is of course possible to form vaults by using very traditional methods:bricks or stones,on centering,like the beautiful vaults used in renaissance churches,gothic cathedrals,and so on.

Therefore:

Build floors and ceilings in the form of elliptical vaults which rise between 13 and 20 per cent of the shorter span. Use a type of construction which makes it possible to fit the vault to any shaped room after the walls and columns are in position:on no account use a prefabricated vault.

的拱肩中有 50％的空隙量。活荷载受力则取值为 50lbs/ft^2。

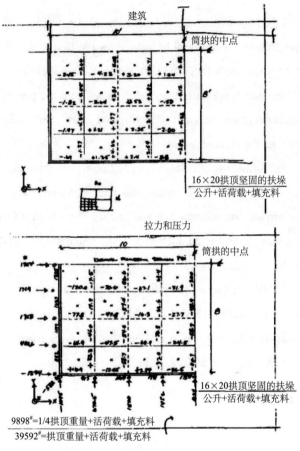

计算机分析结果
Results of computer analysis

　　根据分析，在这样的荷载下，该圆顶的最大抗压应力产生在四边的中点的底部，其值为 120lbs/in^2。沿四面墙的 1/4 各点上外推力最大，其值为 1769lbs。最大的拉应力为 32lbs/in^2 发生在各个角上。最大的弯曲应力为每 10lbs/in^2。所有这些数值都在拱结构的容许范围之内。此外，拱结构

When the main vault is finished,mark the positions of all those columns which will be placed on the floor above it—FINAL COLUMN DISTRIBUTION(213).Whenever there are columns which are more than 2 feet away from the perimeter beam,strengthen the vault with ribs and extra reinforcing to withstand the vertical forces.

Put all the upper columns in position before you pour the floor of the vault,so that when you pour it,the concrete will pour around the column feet,and anchor them firmly in the same way that they are anchored in the foundations—ROOT FOUNDATIONS(214).

To finish the under surface of the vault paint it or plaster it—SOFT INSDE WALLS(235).As for the floor surface above,either wax it and polish it or cover it with soft materials—FLOOR SURFACE (233)...

中的抗收缩配筋将使拱体更加牢固。

分析结果还表明，即使拱结构外形不纯（它所包含的矩形板面在拱体的总体结构外形之内，但实际上是下垂的），它的结构性能和纯拱结构的性能仍非常接近，实际上仍作为一种抗压结构而起作用。存在着少量的局部弯曲；圆顶的各角位承受小量的拉力，但是为抗收缩而设的钢丝网将会承担这两种应力。

下面介绍建造这种拱结构的其他的可能途径：首先，采用其他材料来代替木质筋条。这些材料有：塑料筋条、细金属管和竹子。除聚酯树脂之外，还可使用其他树脂来加强粗麻布。如果树脂不易获得，那么拱体可用如前所述的筋条来构成，并在其上覆盖一层绷紧的钢丝网，再铺上浸过灰浆的粗麻布。粗麻布在浇筑混凝土之前是来得及硬化的。还可以采用经胶合剂加强的粗编织物，甚至制型纸。

很可能，用完全不同的方法做成相似的拱结构：可能采用充气结构或气球。当然，利用各种传统方法也是可以建成拱结构：用砖或石砌成美丽的向心拱结构，见之于意大利文艺复兴时期的教堂、哥特式大教堂以及其他许多教堂。

因此：

将楼面和天花板建成椭圆形拱结构，其拱高为短跨的13%～20%。所采用的构造施工方式应能使拱体适合于在墙、柱就位后而形成的任何形状的房间，但绝不能采用预制拱。

非预制的

13%~20%的拱高

当主体拱完成后，一一标出上一楼层的全部柱的位置——**柱的最后分布**（213）。无论何时，只要柱和圈梁的间距大于 2ft，就要用肋或附加钢筋加固拱体，以抗衡各种垂直力。

在浇筑拱体的地面之前，务必使上面各柱一律就位，以便在浇筑混凝土时可绕柱脚浇筑，并以同样的方法将柱牢牢地固定在基础内——**柱础**（214）。

拱体下部面饰，可油漆或抹灰——**有柔和感的墙内表面**（235）。至于上面楼层的地面面饰，或上蜡抛光或覆盖有柔和感的材料均可——**地面面层**（233）……

模式220 拱式屋顶*

……如果屋顶是一个平的**屋顶花园**（118），它可采用任何一种**楼面天花拱结构**（219）。但是，当它是一个坡屋顶时，根据**带阁楼的坡屋顶**（117）的性质，就需采用一种新的构造，以满足某种体形的特殊需要。

∞∞

什么样的屋顶形状最佳？

出于某些原因，屋顶形状是一个最令人伤脑筋的、最令人感情激动的问题。当涉及有关房屋的构造施工时，这一问题是会被提出来的。我们在研究各种模式时没有发现

220   ROOF VAULTS*

...if the roof is a flat ROOF GARDEN(118),it can be built just like any FLOOR-CEILING VAULT(219).But when it is a sloping roof,according to the character of SHELTERING ROOF(117),it needs a new construction,specifically adapted to the shape which can enclose a volume.

⊰⊱

What is the best shape for a roof?

For some reason,this is the most loaded,the most emotional question,that can be asked about building construction.In all our investigations of patterns,we have not found any other pattern which generates so much discussion,so much disagreement,and so much emotion.Early childhood images play a vital role;so does cultural prejudice.It is hard to imagine an Arab building with a pitched roof;hard to imagine a New England farmhouse with a Russian onion roof over a tower;hard to imagine a person who has grown up among pitched steep wooden roofs,happy under the stone cones of the trulli.

For this reason,in this pattern we make our discussion as fundamental as we can.We shall do everything we can to obtain the necessary features which we can treat as invariant for all roofs,regardless of people or culture—yet deep enough to allow a rich assortment of cultural variations.

We approach the problem with the assumption that there are no constraints created by techniques or availability of

任何其他模式会引起如此多的争论、意见分歧和情绪激动。人在幼年时的印象常起到重大的作用；文化偏见也同样起重要作用。很难想象阿拉伯建筑会有斜屋顶；很难想象新英格兰的农舍会有高出塔楼的俄罗斯洋葱形屋顶；很难想象一个在陡斜的木屋顶下长大的人会在德鲁利人的锥形石砌小屋中感到愉快。

世界各地的屋顶
All over the world

为此，我们在本模式中尽量使论述有理有据。我们会尽可能得出我们处理各种屋顶所需要的那些不变特征，不管民族或文化的差异如何——并尽可能深入地考虑到各种文化的丰富多彩。

我们在处理这一问题时，先假设不存在技术上和材料有无的各种限制。我们只涉及最佳的屋顶形状和材料的分配。如果有一个粗略的矩形平面图，或由若干矩形连接组成的平面图，那么什么样的薄壳屋顶形状最佳呢？

影响屋顶形状的要求如下：

1. 遮蔽感——**带阁楼的坡屋顶**（117）。这要求屋顶遮盖建筑物某个翼的全部（即不只是逐个房间地遮盖）。这要求屋顶有使人看得见的高度，因此，屋顶就要有相当陡的斜面——并且还要求有一种平的屋顶，可作花园和露台。

materials.We are merely concerned with the optimum shape and distribution of materials.Given a roughly rectangular plan,or plan composed of rectangular pieces connected,what is the best shape for the shell of the roof which covers them?

The requirements influencing the shape are these:

1.The feeling of shelter—SHELTERING ROOF(117).This requires that the roof cover a whole wing(that is,not merely room by room).It requires that some of the roof be highly visible—hence,that it have a fairly steep slope—and that some of the roof be flat and usable for gardens or terraces.

2.The roof must definitely contain lived-in space that is,not just sit on top of the rooms which are all below—see SHELTERING ROOF(117).This means it needs rather a steep slope at the edge—because otherwise there is no headroom. This requires an elliptical section dome,or a barrel vault(which starts going up vertically at the edge),or a very steep slope.

3.In plan,each individual roof is a very rough rectangle,with occasional variations.This follows from the way the roofs of a building must,together,follow the social layout of the plan—ROOF LAYOUT(209).

4.The roof shape must be relaxed—that is,it can be used in any plan layout—and can be generated very simply from a few generating lines which follow automatically from the plan— that is,it must not be a tricky or contrived shape which needs a lot of fiddling around to define it—STRUCTURE FOLLOWS SOCIAL SPACES(205).

5.Structural considerations require a curved shell,dome or vault to eliminate as much bending as possible—see

2. 屋顶应明确地包含住人空间——即屋顶不应只是在它下面的所有房间的顶部——参阅**带阁楼的坡屋顶**（117）。这意味着屋顶边缘处需有一个相当陡的斜面——否则那里就不会有上部空间了。这需要有一个椭圆截面的拱顶或筒拱（在边缘处就开始垂直向上），或一个很陡的斜面。

3. 各单个屋顶的平面都是十分粗糙的矩形，而且时有变化。这是因为一幢房屋的所有屋顶都应共同服从于按社会要求安排的平面图——**屋顶布置**（209）。

4. 屋顶形状应是随意的——即它适用于任何平面布置，并且可以由若干母线根据平面图很简单而自然地产生出来——即它不应是一个错综复杂的或花了九牛二虎之力设计出来却毫无意义的——**结构服从社会空间的需要**（205）。

5. 结构上要求有带曲面的薄壳圆顶或拱结构，以便尽可能避免受弯——参阅**有效结构**（206）和**好材料**（207）。当然，木料、钢或其他受拉材料，在某种程度上是可以获得的，可不受这一要求的束缚。

6. 屋顶坡度应满足雨雪天雨水和积雪排泄的要求。很明显，从这一角度看，屋顶的斜度将因气候而异。

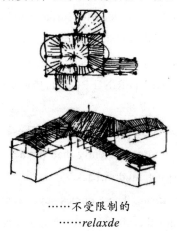

……不受限制的
……*relaxde*

这些要求会排除下面的几种屋顶：

EFFICIENT STRUCTURE(206)and GOOD MATERIALS (207).Of course,to the extent that wood or steel or other tension materials are available,this requirement can be relaxed.

6.The roof is steep enough to shed rain and snow in climates where they occur.Obviously,this aspect of the roof will vary from climate to climate.

These requirements eliminate the following kinds of roofs:

1.*Flat roofs.*Flat roofs,except ROOF GARDENS(118),are already eliminated by the psychological arguments of SHELTERING ROOFS(117)and,of course,by structural considerations.A flat roof is necessary where people are going to walk on it;but it is a very inefficient structural shape since it creates bending.

2.*Pitched Roofs.*Pitched roofs still require materials that can withstand bending moment.The most common material for pitched roofs—wood—is becoming scarce and expensive. As we have said in GOOD MATERIALS(207),we believe it is most sensible to keep wood for surfaces and not to use it as a structural material,except in wood rich areas.Pitched roofs also need to be very steep,indeed,to enclose habitable space as required by SHELTERING ROOF(117)—and hence rather inefficient.

3.*Dutch barn and mansard roofs.*These roofs enclose habitable space more efficiently than pitched roofs;but they have the same structural drawbacks.

4.*Geodesic domes.*These domes cover essentially circular areas,and are not therefore useful in their ordinary form— CASCADE OFROOFS(116),STRUCTURE FOLLOWS

1. 平屋顶。除了**屋顶花园**（118）之外，平屋顶早已遭到**带阁楼的坡屋顶**（117）的各种心理学论点的排斥，当然也会遭到有关结构考量方面的排斥。只要人们想在屋顶上散步，平屋顶就是必要的；但是，它是一种效率十分低的结构形式，因为它会产生弯曲。

2. 斜屋顶。斜屋顶仍需采用抗弯的材料。斜屋顶最常用的材料——木材——正越来越缺乏和昂贵。正如我们在**好材料**（207）中所说，我们认为，最合理的方法是利用木料作为饰面材料而不是作为结构材料，除了在木材丰富的地区以外。确实，斜屋顶也必须是很陡的，以便围合成可住人的空间，如**带阁楼的坡屋顶**（117）所要求的那样——因此，效率相当低。

3. 荷兰谷仓和折线形屋顶。这些屋顶在围合成可住人的空间方面比斜屋顶更为有效；但是它们有同样的结构缺陷。

4. 短程圆屋顶。这些圆屋顶主要是遮盖圆面积，因此，不用于常见形状的平面中——**重叠交错的屋顶**（116）和**结构服从社会空间的需要**（205）。当你将底面积捅拉成近似的矩形时，这种修正后的短程圆屋顶就会变得有些符合于本模式所说明的这类拱结构了。

5. 钢缆网和帐篷。这些屋顶采用受拉材料以取代受压材料——它们不符合**好材料**（207）的要求。它们的效率也是十分低的，而且在围合成可住人空间方面，也不能满足**结构服从社会空间的需要**（205）的要求。

凡满足上述要求的那些屋顶都是矩形筒拱或薄壳筒拱，其中有的有屋脊，有的没有屋脊；有两坡形的，也有四坡形的，还有各种各样的横截面。几乎所有薄壳筒拱都在拱结构的纵向，做成波浪形使结构进一步得到加固。各种各样的横截面的例子如下图所示。[要记住，薄壳筒拱并不包括平屋顶的**屋顶花园**（118），这种屋顶花园是建造在**楼面**

SOCIAL SPACES(205).In the modified form,which comes when you stretch the base into a rough rectangle,they become more or less congruent with the class of vaults defined by this pattern.

5.*Cable nets and tents.*These roofs use tensile materials instead of compressive ones—they do not conform to the requirements of GOOD MATERIALS(207).They are also very inefficient when it comes to enclosing habitable space—and thus fail to meet the requirements of STRUCTURE FOLLOWS SOCIAL SPACES(205).

The roofs which satisfy the requirements are all types of rectangular barrel vaults or shells,with or without a peak,gabled or hipped,and with a variety of possible cross sections. Almost any one of these shells will be further strengthened by additional undulations in the direction of the vault.Examples of possible cross sections are given below.[Remember that this does not include those flat ROOF GARDENS(118)built over FLOOR-CEILING VAULTS(219).]

We have developed a range of roof vaults which are rather similar to a pitched roof—but with a convex curve great enough to eliminate bending,in some cases actually approaching barrel vaults.One is shown in the drawing opposite;another is shown below.

We build the roof vault very much like the floor vaults:

1.First span the wing to be roofed with pairs of lattice strips which are securely nailed at their ends to the perimeter beam,and weighted at their apex so that the two pieces become slightly curved.

天花拱结构（219）之上的。]

　　我们已研究了一系列拱结构屋顶，它们和坡屋顶极为相似——但呈凸形曲线，其曲率大到足以抵消弯曲。在某些情况下，它们是真正的筒拱。一种拱结构如右图；另一种如下图所示。

各种常见的拱屋顶
Possible roof vaults

另一种拱式屋顶由鲍勃·哈里斯建于俄勒冈
Another version of a roof vault, built by Bob Harris in Oregon

　　我们建造的拱式屋顶酷似楼面拱结构。

　　1. 首先将建造屋顶用的成对的筋条从一侧跨越至另一侧，并将它们的下端钉在圈梁上，在顶尖下挂上重量，以便两侧的成对筋条稍呈弯曲。

　　2. 同时根据**楼面天花拱结构**（219），在屋顶构架下面建造天花构架。

　　3. 每隔18in，重复建造一个这样的构架，直到整个翼的构架都建完为止。外构架将按上述方法建造，而天花板的内构架可随其下面房间不同而变化。

2.Make the frame for the ceiling under the roof frame at the same time according to FLOOR-CEILING VAULTS(219).

3.Repeat this frame every 18 inches,until the entire wing is framed.The outer one will be the same,while the inner frame for the ceiling may change according to the rooms under it.

4.Now lay burlap over the ceiling frame,then resin,then $1\frac{1}{2}$ inches of ultra-lightweight concrete—as for FLOOR-CEILING VAULTS(219).

5.Now lay burlap over the roof frame,tacking it onto the lattice strips so that there is a 3-inch scallop in between the ribs-to form structural undulations in the skin.Again,paint the burlap with resin;lay chickenwire and put a layer of lightweight concrete over the entire roof.

We have analyzed a 48-foot roof of this type by means of a computerized finite element analysis similar to the one described for FLOOR-CEILING VAULTS(219).The analysis shows that the maximum membrane compressive stress in the roof is 39.6 psi;the maximum membrane tensile stress is 2.5 psi,and the maximum diagonal membrane stress which develops from the maximum shear of 41.7 psi is 15.2 psi.These stresses are within the capacity of the material [See allowable stresses given in FLOOR-CEILING VAULT(219)].The maximum membrane bending moment is 46 inch pounds per inch which is higher than the capacity of the unreinforced section,but extrapolations from our data show that this will be comfortably taken care of by the reinforcing which is needed anyway for shrinkage.Roofs with smaller spans,for a typical WING OF LIGHT(106),will

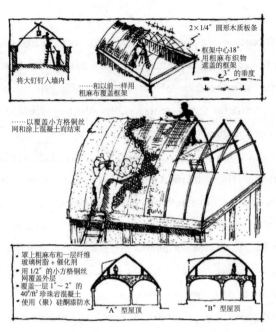

将大钉钉入墙内

……和以前一样用粗麻布覆盖框架

2×1/4″圆形木质板条

框架中心18″用粗麻布织物遮盖的框架

c3″的垂度

……以覆盖小方格铜丝网和涂上混凝土而结束

罩上粗麻布和一层纤维玻璃树脂＋催化剂

用1/2″的小方格钢丝网覆盖外层

覆盖一层1″~2″的40#/ft²珍珠岩混凝土

使用(聚)硅酮漆防水

"A"型屋顶

"B"型屋顶

一种和楼面天花拱结构相似的拱屋顶是由网络筋条、粗麻布、钢丝网和超轻质混凝土所构成。它有屋脊尖，有坡屋面和为加固而做成波形起伏

A type of roof vault, similar to the floor-ceiling vault, made from Lattice strips, burlap, chicken-wire and ultralightweight concrete, but with an apex, and a pitch, and undulations for strength

4. 将粗麻布铺在天花构架上，然后用树脂涂抹，再浇筑 1.5in 厚的超轻质混凝土——参阅**楼面天花拱结构**（219）。

5. 将粗麻布铺在屋顶框架上，再把它钉在筋条上，钉时使其肋间形成 3in 高差的扇形凹面，屋面就形成结构上的波形起伏。再用树脂涂粗麻布；在整个屋顶上铺钢丝网，再在其上浇筑一层轻质混凝土。我们通过计算机的有限元分析方法——和**楼面天花拱结构**（219）描写的分析法相似——已经分析了一个面积为 48ft 的这类屋顶。分析结果表明，这类屋顶的最大压应力为 39.6 lbs/in²；最大拉应力

be even stronger.

Of course there are dozens of other ways to make a roof vault.Other versions include ordinary barrel vaults,lamella structures in the form of barrel vaults,elongated geodesic domes(built up from struts),vaults built up from plastic sheets,or fiberglass,or corrugated metal.

But,in one way or another,build your roofs according to the invariant defined below,remembering that it lies somewhere in between the Crystal Palace,the stone vaults of Alberobello,mud huts of the Congo,grass structures of the South Pacific,and the corrugated iron huts of our own time.This shape is required whenever you are working with materials which are in pure compression.

Obviously,if you have access to wood or steel and want to use it,you can modify this shape by adding tension members. However,we believe that these tension materials will become more and more rare as time goes on and that the pure compression shape will gradually become a universal.

Therefore:

Build the roof vault either as a cylindrical barrel vault, or like a pitched roof with a slight convex curve in each of the two sloping sides.Put in undulations along the vault, to make the shell more effective.The curvature of the main shell,and of the undulations,can vary with the span; the bigger the span,the deeper the curvature and undulations need to be.

为 2.5 lbs/in^2；由最大剪力 41.7 lbs/in^2 而产生的、最大的斜分应力为 15.2 lbs/in^2。这些应力都在材料的容许范围之内［参阅**楼面天花拱结构**（219）中所给出的容许应力］。最大弯矩为 46 lbs/in^2，此值高于无筋截面的容许范围。但从我们的数据的演绎表明，这种屋顶，在配置抗收缩所必需的钢筋时，将受到充分的关注。对于典型的**有天然采光的翼楼**（107）来说，跨度小的屋顶将更加牢固。

当然，建造拱结构屋顶还有其他数十种方法，包括：普通的筒拱屋顶、筒拱形式的网状结构、加长的短程圆屋顶（由压杆建成）以及用塑料板、玻璃纤维和波纹金属板制成的各种拱屋顶。

但是，不管用哪种方法来建造屋顶，都需符合下面所说的不变原则并牢记你的屋顶总是介于"水晶宫"、阿尔贝罗贝洛的石拱、刚果简陋的土房、南太平洋的茅舍和我们这个时代的波纹金属的棚屋之间的某种混合型的形状。无论何时只要你用纯受压材料建造屋顶时，就都应采用这种形状。

很明显，如果你能取得木料或钢材，并想使用它们，就可用附加受拉构件来修改这种形状。然而，我们认为这些受拉材料随着时间的推移将变得越来越稀缺，纯压缩形状的屋顶将逐步得到普及。

实验性的拱式屋顶
Experimental roof vaults

Leave space for dormers at intervals along the vault—DORMER WINDOWS(231),and build them integral with it.Finish the roof with ROOF CAPS(232).And once the vault is complete,it needs a waterproof paint or skin applied to its outer surface—LAPPED OUTSIDEWALLS(234).It can be painted white to protect it against the sun;the undulations will carry the rainwater...

within the main frame of the building,fix the exact positions for openings—the doors and the windows-and frame these openings.

221.NATURAL DOORS AND WINDOWS

222.LOW SILL

223.DEEP REVEALS

224.LOW DOORWAY

225.FRAMES AS THICKENED EDGES

因此：

将拱屋顶建成圆形筒拱屋顶，或建成近似斜屋顶，其倾斜的两侧均呈现为轻微凸出的曲线。沿筒拱纵向做成波形起伏结构，使薄壳筒拱更加有效。主薄壳筒拱的曲率和起伏结构的曲率可随跨度而变化；跨度越大，曲率和起伏就越大。

筒拱

起伏结构

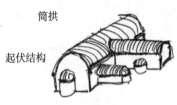

沿拱体按一定间距留出天窗的空间——**老虎窗**（231），并把老虎窗和拱体建成一体。用**屋顶顶尖**（232）对屋顶进行装饰。一旦拱体建成，在其表面就应上防水漆或铺防水层材料——**鱼鳞板墙**（234）。为了保护拱体不受阳光破坏，可在拱体表面刷白；波形起伏结构可将雨水排出……

在房屋的主框架之内，为开口——门和窗——规定精确的位置，并为这些开口安装框架。

221. 借景的门窗

222. 矮窗台

223. 深窗洞

224. 低门道

225. 门窗边缘加厚

模式221　借景的门窗**

221   NATURAL DOORS AND WINDOWS**

...imagine that you are now standing in the built-up frame of a partly constructed building,with the columns and beams in place—BOX COLUMNS(216),PERIMETER BEAMS(217). You know roughly where you want doors and windows from ZEN VIEW(134),STREET WINDOWS(164),WINDOW PLACE(180),WINDOWS OVERLOOKING LIFE (192), CORNER DOORS(196).Now you can settle on the exact positions of the frames.

❧❧❧

Finding the right position for a window or a door is a subtle matter.But there are very few ways of building which take this into consideration.

In our current ways of building,the delicacy of placing a window or a door has nearly vanished.But it is just this refinement,down to the last foot,even to the last inch or two,which makes an immense difference.Windows and doors which are just right are always like this.Find a beautiful window.Study it.See how different it would be if its dimensions varied a few inches in either direction.

Now look at the windows and doors in most buildings made during the last 20 years.Assume that these openings are in roughly the right place,but notice how they could be improved if they were free to shift around,a few inches here and there,each one taking advantage of its own special circumstances—the space immediately inside and the view outside.

It is almost always a rigid construction system,combined with a formal aesthetic,which holds these windows in such a death

……想象一下，你现在正站在一幢楼房已部分施工的、由已就位的柱和梁构成的框架内——**箱形柱**（216）和**圈梁**（217）。根据**禅宗观景**（134）、**临街窗户**（164）、**窗前空间**（180）、**俯视外界生活之窗**（192）和靠近**墙角的房门**（196），你大概就会知道要在什么地方开门窗洞。现在你就可以正确安排门窗框架的位置了。

☯☯☯

为门窗选定正确的位置是十分讲究的。但是只有很少几种施工方法考虑到这一点。

在我们流行的施工方法中，细致考虑门窗位置的情况已不多见。可是，正是这种细微的尺寸，哪怕是一英尺，甚至一英寸或两英寸就可能造成差别很大的结果。门窗恰恰需要这样精确。去发现生活中美观雅致的窗户吧！去认真研究窗户吧！去看一看，要是窗户的尺寸在各个方向上变化几英寸，会有何等的不同。

现在来考虑一下近二十年来大部分建筑的门窗吧。假定这些开口大致都有正确的位置，但要注意，假如门窗都能自由地来回移动几英寸，并都能利用各自的特殊环境——就近的内部空间和外部景观，那么门窗怎样才能是最优方案呢？

使窗户借景恰到好处
Getting it just right

grip.There is nothing else to this regularity, for it is possible to relax the regularity without losing structural integrity.

It is also important to realize that this final placing of windows and doors can only be done on site,with the rough frame of the building in position.It is impossible to do it on paper. But on the site it is quite straightforward and natural: mock up the openings with scraps of lumber or string and move them around until they feel right;pay careful attention to the organization of the view and the kind of space that is created inside.

As we shall see in a later pattern—SMALL PANES(239),it is not necessary to make the windows any special dimensions,or to try and make them multiples of any standard pane size.Whatever dimensions this pattern gives each window,it will then be possible to divide it up,to form small panes,which will be different in their exact shape and size,according to the window they are in.

However,although there is no constraint on the exact dimension of the windows,there is a general rule of thumb, which will make window sizes vary:Windows,as a rule, should become smaller as you get higher up in the building.

1.The area of windows needed for light and ventilation depends on the size of rooms,and rooms are generally smaller on upper stories of the building—the communal rooms are generally on the ground floor and more private rooms upstairs.

2.The amount of daylight coming through a window depends on the area of open sky visible through the window. The higher the window,the more open sky is visible(because nearby trees and buildings obscure less)—so less window area is needed to get sufficient daylight in.

3.To feel safe on the upper stories of a building,one wants more enclosure,smaller windows,higher sills—and the higher

通常采用的一种和形式美相结合的、死板的构造施工系统把窗户都固定死了。这种规定是不足取的，因为不按这一规定处理窗户而又不致失去结构上的完整性，还是有可能做到的。

同样重要的是，要承认这种最后放置门窗的做法只有在楼房的粗糙的框架已经就位，才能在施工现场来实现。这在图纸上是无法办到的。但在施工现场这是十分简单又自然的：用废木料或线绳模拟开口，将它们来回移动，直至合适为止；要倍加注意组织好景观和内部空间的形状。

正如我们在后面的一个模式——**小窗格**（239）中将会看到的那样，没有必要使窗户具有任何特殊尺寸，并竭力使窗户比任何标准窗格尺寸大许多倍。无论本模式给每一窗户提供什么尺寸，都可以把它划分为小窗格，而这些小窗格的确切尺寸和形状将按窗户的大小而有所不同。

虽然对窗户的确切尺寸没有任何限制，可是存在一个惯用的经验法则，它可定出窗户的不同尺寸：通常情况，在楼房内越往高处走，窗户应变得越小。

1. 光线和通风所需要的窗户面积取决于房间的尺寸，一般说来，位于楼房上层的房间较小——公用房间通常位于底层，而较私密性的房间位于上层。

2. 窗户的采光量取决于通过窗户能看到的开阔天空的面积。窗户越高，可见的开阔的天空就越大（因为附近的树木和楼房的遮挡少）——结果，小窗户也可以有充足的采光量。

3. 人们为了在楼层上有安全感，希望围合的空间大些，窗户小些，窗台高些——人们离地越高，就越需要这些心理上的安全保障。

因此：

绝对不要采用标准门窗。使每一窗户按其不同的位置而大小各异。在房间的构架基本造好以前，不要固定门框

of f the ground one is,the more one needs these psychological protections.

Therefore:

On no account use standard doors or windows.Make each window a different size,according to its place.

Do not fix the exact position or size of the door and window frames until the rough framing of the room has actually been built,and you can really stand inside the room and judge,by eye,exactly where you want to put them,and how big you want them.When you decide,mark the openings with strings.

Make the windows smaller and smaller,as you go higher in the building.

☙❧

Fine tune the exact position of each edge,and mullion,and sill,according to your comfort in the room,and the view that the window looks onto—LOW SILL(222),DEEP REVEALS(223). As a result,each window will have a different size and shape,according to its position in the building.This means that it is obviously impossible to use standard windows and even impossible to make each window a simple multiple of standard panes.But it will still be possible to glaze each window,since the procedure for building the panes makes them divisions of the whole,instead of making up the whole as a multiple of standard panes—SMALL PANFS(239)...

和窗框的精确位置和尺寸，并且你实际上能站在房间内利用目测来判断，在什么地方可分毫不差地安置门窗，你想要的门窗究竟有多大。当你作出决定之后，用线绳标出开口的位置和大小。你在楼房内越往高处走，窗户就应越小。

窗户尺寸的变化

"被感知的"门窗位置

❧❧

根据你在房间内是否感到舒适以及窗外的景色是否合意，精心调整每一边缘、窗的竖框和窗台的准确位置——**矮窗台**（222）和**深窗洞**（223）。结果，每一窗户，根据其在楼房中的位置，都将具有不同的尺寸和形状。这意味着，十分明显，使用标准窗户是不可能的，简单地加大标准窗格也是不可能的。但是给每一处窗户配玻璃仍然是可能的，因为制造窗格时就是对整樘窗进行分割，而不是把标准窗格放大许多倍——**小窗格**（239）……

模式222　矮窗台

222   LOW SILL

...this pattern helps to complete NATURAL DOORS AND WINDOWS (221),and the special love for the view,and for the earth outside,which ZEN VIEW(134),WINDOW PLACE(180) and WINDOWS OVERLOOKING LIFE(192)all need.

ಬಂಞ

One of a window's most important functions is to put you in touch with the outdoors.If the sill is too high,it cuts you off.

The "right" height for a ground floor window sill is astonishingly low.Our experiments show that sills which are 13 or 14 inches from the floor are perfect.This is much lower than the window sills which people most often build:a standard window sill is about 24 to 36 inches from the ground.And it is higher than French doors and windows which usually have a bottom rail of 8 to 10 inches.The best height,then,happens to be a rather uncommon one.

We first give the detailed explanation for this phenomenon, and we then explain the modifications which are necessary on upper floors.

People are drawn to windows because of the light and the view outside—they are natural places to sit by when reading, talking,sewing,and so on,yet most windows have sill heights of 30 inches or so,so that when you sit down by them you cannot see the ground right near the window.This is unusually frustrating—you almost have to stand up to get a complete view.

In "The Function of Windows:A Reappraisal" (*Building Science*,Vol.2,*Pergamon Press*,1967,pp.97-121.),Thomas Markus

……本模式有助于完善**借景的门窗**（221），并能满足人们对窗外景色和田园风光的特殊爱好。它也是**禅宗观景**（134）、**窗前空间**（180）和**俯视外界生活的窗**（192）等模式所需要的。

❧

窗户的最重要功能之一是使你和室外世界保持接触。如果窗台太高，就会切断你同外界的联系。

一层楼窗台“合适的”高度矮得惊人。我们的实验表明，离地面 13in 或 14in 的窗台是合适的。这一高度是比人们一般建造的窗台矮得多：标准窗台离地面约 24 ～ 36in。但它比法国式的玻璃落地门窗高，通常，这种落地门窗都有离地面 8 ～ 10in 的下槛。由此可见，这一最佳高度目前尚未普遍采用。

我们首先要详细说明这一点，然后再解释上面楼层的窗台高度必须修正的原因。

人们到窗口去是因为有光线和外界景观的吸引——窗口是读书、谈话、做针线活等十分合理的可以坐人的地方，可是现在大多数窗台的高度为 30in 左右，所以，当你坐在这么高的窗台旁时，却看不到窗下的室外地面。这会使人感到十分扫兴——你几乎不得不站在窗台旁才能看见窗外的全景。

在《窗户的功能：重新估价》(“The Function of Windows : A Reappraisal”，*Building Science*，Vol.2，*Pergamon Press*，1967，pp.97~121）一文中，托马斯·马库斯指明，窗户的第一功能不是提供光线；而是提供与外界的联系，而且，这种联系意义非常重大，当人们能看到窗外的大地和地平线上的景物时更是如此。高窗台的窗户隔绝了人们与大地景观的联系。

shows that the primary function of windows is not to provide light but to provide a link to the outside and, furthermore,that this link is most meaningful when it contains a view of the ground and the horizon.Windows with high sills cut out the view of the ground.

On the other hand,glass all the way down to the floor is undesirable.It is disturbing because it seems contradictory and even dangerous.It feels more like a door than a window;you have the feeling that you ought to be able to walk through it.If the sill is 12 to 14 inches high,you can comfortably see the ground,even if you are a foot or two away from the window,and it still feels like a window rather than a door.

On upper stories the sill height needs to be slightly higher. The sill still needs to be low to see the ground,but it is unsafe if it is too low.A sill height of about 20 inches allows you to see most of the ground,from a chair nearby,and still feel safe.

Therefore:

When determining exact location of windows also decide which windows should have low sills.On the first floor, make the sills of windows which you plan to sit by between 12 and 14 inches high.On the upper stories,make them higher, around 20 inches.

೮೦೦೮

Make the sill part of the frame,and make it wide enough to put things on—WAIST-HIGH SHELF(201),FRAMES AS THICKENED EDGES (225),WINDOWS WHICH OPEN WIDE(236).Make the window open outward,so that you can use the sill as a shelf,and so that you can lean out and tend the flowers.If you can,put flowers right outside the window,on the ground or raised a little,too,so that you can always see the flowers from inside the room—RAISED FLOWERS (245)...

另一方面，窗玻璃做到直达地面总是不理想的。这会令人烦恼，因为似乎是自相矛盾的，甚至是危险的。这种落地玻璃窗与其说是窗，倒不如说是门。当你感到它是门时，就想能穿越它。如果窗台高 12 ~ 14in，你就可以舒舒服服地看到大地，即使你离开窗户有 1ft 或 2ft 远，你仍然会感到它是窗而不是门。

上面各层窗台的高度需要略微高些。窗台仍应矮到能让人看见窗外的大地，但如果太矮，就不安全了。大约 20in 高的窗台，如果坐在窗台附近的椅子上，人们就能看见大部分地面而仍有安全感。

因此：

在决定窗户的精确位置时，也应决定哪些窗户应有矮窗台。底层窗台要矮，高 12 ~ 14in，你可以坐在那里远眺窗外景物。上面楼层的窗台则可高些，约 20in。

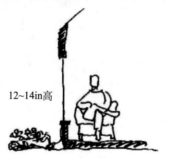

12~14in高

8003

使窗台成为窗框的一部分，并使其有足够的宽度以便搁放东西——**半人高的搁架**（201）、**门窗边缘加厚**（225）和**大敞口窗户**（236）。向外开窗，这样你才能利用窗台放置物品，并且你可以探身窗外照料鲜花。如若可能，将鲜花栽种在窗外的地面上或略高一些的台地上，这样你就可以随时从室内看到窗外的鲜花——**高花台**（245）……

模式223 深窗洞

……本模式有助于完善每室**两面采光**（159），进一步减少眩光，并有助于形成**门窗边缘加厚**（225）。

❧❧❧

在窗框和墙的会合处转角突然的窗户会造成刺目的眩光，并使开窗的房间变得不舒适了。

窗户与一辆正在驶近的汽车的明亮前灯具有相同的效果：眩光使你看不见公路上任何别的东西，因为你的眼睛不能在一瞬间适应明亮的前灯和公路的黑暗。窗户也是如

223   DEEP REVEALS

...this pattern helps to complete the work of LIGHT ON TWO SIDES OF EVERY ROOM(159),by going even further to reduce glare;and it helps to shape the FRAMES AS THICKENED EDGES(225).

৪০৫৪

Windows with a sharp edge where the frame meets the wall create harsh,blinding glare,and make the rooms they serve uncomfortable.

They have the same effect as the bright headlights of an oncoming car:the glare prevents you from seeing anything else on the road because your eye cannot simultaneously adapt to the bright headlights and to the darkness of the roadway.Just so,a window is always much brighter than an interior wall;and the walls tend to be darkest next to the window's edge.The difference in brightness between the bright window and the dark wall around it also causes glare.

To solve this problem,the edge of the window must be splayed,by making a reveal between the window and the wall. The splayed reveal then creates a transition area—a zone of intermediate brightness—between the brightness of the window and the darkness of the wall.If the reveal is deep enough and the angle just right,the glare will vanish altogether.

But the reveal must be quite deep,and the angle of the splay quite marked.In empirical studies of glare,Hopkinson and Petherbridge have found:(1)that the larger the reveal is,the less

此，它总比内墙亮得多；而靠近窗户边缘的墙往往是最暗的。在明亮的窗户和其四周的暗墙之间，亮度的差别也会引起眩光。

有眩光
Glare…

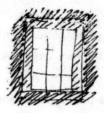

无眩光
and no glare

　　为了解决眩光这一问题，窗户的两侧应呈八字形向外展宽，在窗和墙之间形成一个深窗洞。呈八字形展宽的窗洞就会形成一个过渡区，即介于窗亮度和墙暗度之间的中间亮度区。如果窗洞有足够的深度，而且展宽的角度正确，则眩光将可完全避免。

　　但是，窗洞必须很深，呈八字形展宽的角度要十分明显。霍普金森和佩西布里奇从研究眩光的切身经验中已经发现：(1) 窗洞侧面越宽，眩光越小；(2) 当窗洞侧面的亮度恰好是窗亮度和墙亮度的平均值时，它的效果最佳。("Discomfort Glare and the Lighting of Buildings," *Transactions of the Illuminating Engineering Society*, Vol.XV, No.2, 1950, pp.58~59.)

　　我们自己的实验表明，当窗洞侧面和窗户平面成50°～60°夹角时，结果与上述效果就非常接近。可是，当然，这一角度将随局部条件的不同而变化。而且我们还发现，为了满足"宽"窗洞侧面的需要，窗洞侧面本身应有一合适的宽度：10～12in。

glare there is;(2)the reveal functions best,when its brightness is just halfway between the brightness of the window and the brightness of the wall.("Discomfort Glare and the Lighting of Buildings," *Transactions of the Illuminating Engineering Society*,Vol.XV,No.2,1950,pp.58-59.)

Our own experiments show that this happens most nearly,when the reveal lies at between 50 and 60 degrees to the plane of the window;though,of course,the angle will vary with local conditions.And,to satisfy the need for a "large" reveal,we have found that the reveal itself must be a good 10 to 12 inches wide.

Therefore:

Make the window frame a deep,splayed edge:about a foot wide and splayed at about 50 to 6o degrees to the plane of the window,so that the gentle gradient of daylight gives a smooth transition between the light of the window and the dark of the inner all.

⊱✺⊰

Build the depth of the frame so that it is continuous with the structure of the walls—FRAMES AS THICKENED EDGES(225);if the wall is thin,make up the necessary depth for the reveal on the inside face of the wall,with bookshelves,closets or other THICK WALLS (197);embellish the edge of the window even further,to make light even softer,with lace work,tracery,and climbing plants—FILTERED LIGHT(238),HALF-INCH TRIM(240),CLIMBING PLANTS (246)...

因此：

窗框一定要具有一个深的呈八字形展宽的斜侧面：约1ft宽，与窗户平面的倾斜角度为50°～60°，这样日光逐渐变化，在明亮的窗户和较暗的内墙之间形成一平稳的过渡区。

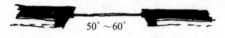

50°～60°

∞∞

窗框应具有一定深度，以便它和墙的结构连成一体——**门窗边缘加厚**（225）；如果墙薄，仍按窗洞所需厚度布置，并利用书架、壁橱或其他的**厚墙**（197）来弥补墙内面与窗洞深度的不足；进一步装饰窗户的边缘，利用镂空花饰、窗花格和攀援植物使光线变得比较柔和——**过滤光线**（238）、**半英寸宽的压缝条**（240）和**攀援植物**（246）……

模式224　低门道

224   LOW DOORWAY

...some of the doors in a building play a special role in creating transitions and maintaining privacy:it may be any of the doors governed by FAMILY OF ENTRANCES(102),or MAIN ENTRANCE(110),or THE FLOW THROUGH ROOMS(131)or CORNER DOORS(196),or NATURAL DOORS AND WINDOWS(221).This pattern helps to complete these doors by giving them a special height and sape.

❧❧❧

High doorways are simple and convenient.But a lower door is often more profound.

The 6′8″rectangular door is such a standard pattern,and is so taken for granted,that it is hard to imagine how strongly it dominates the experience of transition.There have been times,however,when people were more sensitive to the moment of passage,and made the shape of their doors convey the feeling of transition.

An extreme case is the Japanese tea house,where a person entering must literally kneel down and crawl in through a low hole in the wall.Once inside,shoes off,the guest is entirely a guest,in the world of his host.

Among architects,Frank Lloyd Wright used the pattern many times.There is a beautifully low trellised walk behind Taliesin West,marking the transition out of the main house, along the path to the studios.

If you are going to try this pattern,test it first by pinning cardboard up to effectively lower the frame.Make the doorway low enough so that it appears "lower than usual" —then people will

……房屋的某些门在形成过渡区和保持私密性方面起到特殊的作用。这些门可能是受下列模式所支配的各种门的任何一种：**各种入口**（102）、**主入口**（110），或**穿越空间**（131）或**靠近墙角的房门**（196），或**借景的门窗**（221）。本模式通过确定这些门的特殊高度和形状使门得到改进和完善。

୨୦୯୧

高门道简单又方便。但是低的门往往更富于深奥感。

6.8ft 的矩形门理所当然地被认为是一种标准模式，但很难想象它是如何强有力地支配着人经过门时的心理感受的。可是，从古至今，人们在穿过门的一瞬间总是更加敏感的，他们常借门的形状表达出他们通过门时的心情。

一个极端的例子就是日本的茶馆，在那里正要进门的人，毫不夸张地说，要双膝跪下，爬过墙壁上的一个矮洞。客人一旦入内，就要脱鞋，在主人的眼里，就完全是他的一位客人了。

在建筑师当中，弗兰克·劳埃德·赖特曾一再利用本模式。试举一例。在塔利埃新威斯特后面，有一条十分美丽的低矮的棚下小径，他在主宅之外布置了一个过渡区，沿着小径通往工作室。

如果你想试用本模式，首先把硬纸板钉在明显低于门框的地方进行试验。门道应低得足以使人感到"比平常的还要低"——然后人们会立即适应它，而且高个子也不会碰着自己的脑袋。

immediately adapt to it,and tall people will not hit their heads.

Therefore:

Instead of taking it for granted that your doors are simply 6′8″rectangular openings to pass through,make at least some of your doorways low enough so that the act of going through the door is a deliberate thoughtful passage from one place to another.Especially at the entrance to a house,at the entrance to a private room,or a fire corner—make the doorway lower than usual,perhaps even as low a 5′8″.

৪০৫৪

Test the height before you build it,in place—NATURAL DOORS AND WINDOWS(221).Build the door frame as part of the structure-FRAMES AS THICKENED EDGES(225),and make it beautiful with ORNAMENT(249)around the frame. If there is a door,glaze it,at least partially—SOLID DOORS WITH GLASS(237)...

因此：

门就是 6ft8in 矩形通过口这样一种理所当然的看法已经过时。至少你的若干门道要低得足以使你的过门的行为变成一种从一处通往另一处的深思熟虑的富有创见的活动。尤其在进入住宅的入口处，在进入私人房间或炉火熊熊之隅的入口处——务必使门道低于常见的，也许甚至低到 5.8ft。

（突然受痛时的叫声）哎哟

⌘

在建造门道之前，就地试验它的高度——**借景的门窗**（221）。把门框建成结构的一部分——**门窗边缘加厚**（225），并以**装饰**（249）美化门窗的边缘。如果有门，则至少部分镶上玻璃——**镶玻璃板门**（237）⋯⋯

模式225 门窗边缘加厚**

……假定柱和梁已经就位，并且你已经用线绳或铅笔标出了门窗的精确位置——**借景的门窗**（221）。在一切准备就绪后，就可以开始建造门窗的框架了。切记，造得好的门窗框架应和周围的墙连成一体，以便它在结构上加强房屋的牢固性——**有效结构**（206）和**逐步加固**（208）。

❧❧❧

凡有洞的均匀墙体，如果洞的边缘不增厚加固，都容易在有洞处形成破坏。

225   FRAMES AS THICKENED EDGES**

...assume that columns and beams are in and that you
have marked the exact positions of the doors and windows
with string or pencil marks—NATURAL DOORS AND
WINDOWS(221).You are ready to build the frames.
Remember that a well made frame needs to be continuous
with the surrounding wall,so that it helps the building
structurally—EFFICIENT STRUCTURE(206),GRADUAL
STIFFENING(208).

❧

**Any homogeneous membrane which has holes in it will
tend to rupture at the holes,unless the edges of the holes are
reinforced by thickening.**

The most familiar example of this principle at work is in
the human face itself.Both eyes and mouth are surrounded by
extra bone and flesh.It is this thickening,around the eyes and
mouth,which gives them their character and helps to make them
such important parts of human physiognomy.

A building also has its eyes and mouth:the windows
and the doors.And following the principle which we observe
in nature,almost every building has its windows and doors
elaborated,made more special,by just the kind of thickening we
see in eyes and mouths.

The fact that openings in naturally occurring membranes
are invariably thickened can be easily explained by considering
how the lines of force in the membrane must flow around

说明这一原理的最熟悉的例子就是人脸。两只眼睛和一张嘴都被外围的骨肉包着。正是眼和口的边缘增厚了才使它们各具特征，并有助于使它们成为人体外貌的重要器官。

房屋同样也有自己的眼和口：窗和门。根据我们所遵循的这一原理，实际上，几乎每幢房屋都有它自己精致而又独具风格的门窗。这一点正如我们在眼和口处所具有的边缘增厚一样。

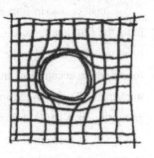

力线密度表示增加应力集中
The density of the lines represents increasing stress concentrations

自然界生成的薄膜上的开口处，其边缘总是加厚的；这一事实是极易解释的，因为薄膜中的传力线必须绕洞流动。洞周边传力线的密度增加了，所以要求在这里多放材料，加厚边缘，以防止被拉断。

请考虑一下肥皂泡的薄膜吧。当你戳它时，张力就把薄膜拉开，结果它就破裂了。但是，如果你把一个线环放入肥皂泡中，环孔可以在薄膜上固定住，

起加固作用的门框
A door frame as a thickening

因为环孔周围所积聚的张力能被这一较粗的线环承担了。此时，该环受拉。对于压曲和压缩也同样如此。当一有洞的薄板受压时，其洞的边缘必须加固。重要的是要认识到这种加固不仅使洞本身不被破坏，而且还要注意到墙体缺掉一块的那部分的应力重新分布。船和汽车上窗口凸边就是通俗的例子。

房屋的门窗也是如此。木板和轻质混凝土填料筑成的墙——参阅**墙体**（218）——可用同样的木板做成加厚框的

the hole.The increasing density of lines of force around the perimeter of the hole requires that additional material be generated there to prevent tearing.

Consider a soap film.When you prick the film,the tension pulls the film apart,and it disintegrates.But if you insert a ring of string into the film,the hole will hold,because the tensile forces which accumulate around the opening can be held by the thicker ring.This is in tension.The same is true for buckling and compression.When a thin plate is functioning in compression and a hole is made in it,the hole needs stiffening.It is important to recognize that this stiffening is not only supporting the opening itself against collapse,but it is taking care of the stresses in the membrane which would normally be distributed in that part of the membrane which is removed.Familiar examples of such stiffening in plates are the lips of steel around the portholes in a ship or in a locomotive cab.

The same is true for doors and windows in a building. Where the walls are made of wood planks and lightweight concrete fill—see WALL MEMBRANES(218)—the thickened frames can be made from the same wood planks,placed to form a bulge,and then filled to be continuous with the wall.If other types of skin are used in the wall membranes,there will be other kinds of thickening:edges formed with chicken wire,burlap,and resin,filled with concrete;edges formed with chicken wire filled with rubble,and then mortar,plaster;edges formed with brick,filled,then plastered.

More general examples of frames as thickened edges exist all over the world.They include the thickening of the mud around the windows of a mud hut,the use of stone edges to the opening in a brick wall because the stone is stronger,the use of

模板，然后再浇筑混凝土和墙连成一体。如果采用其他外层墙材，则加厚洞口边缘的方法有所不同：有的使用钢丝网、粗麻布和树脂，并浇筑混凝土；有的使用钢丝网、填充毛石，再作砂浆粉刷；有的使用砖，浇筑混凝土，然后再粉刷。

框架作为门窗加厚的边缘是比较普遍的，这种例子在世界各地都有。其中还包括：土屋的窗边缘加上厚厚的泥框；砖墙的门窗边缘用石砌，因石头的强度比砖大；在灰板墙筋结构中门窗边缘采用双板墙筋；在哥特式教堂的窗户周缘采用附加的石块；在筐式结构的茅屋洞口四周要做成波形特别加固。

因此：

不要把门窗的框架视为分立的、插入墙洞的刚性结构，而要将它们视为墙体本身真正结构的加厚部分，以防门窗边缘处的应力集中。

根据本模式的概念，要把门窗的框架作为墙的增厚部分来建造，并与墙连成一体。增厚的方法是：采用与墙相同的材料，浇筑或是砌筑，但都必须和墙建成一体。

墙的增厚

❀

窗户的边缘呈八字形展宽加厚，以便形成**深窗洞**（223）：装入框架的门窗形状取决于下列模式：**深窗洞**（223）、**镶玻璃板门**（237）和**小窗格**（239）……

当你建造主框架及其门窗开口时，采用如下相应的辅助模式：

double studs around an opening in stud construction,the extra stone around the windows in a gothic church,the extra weaving round the hole in any basket hut.

Therefore:

Do not consider door and window frames as separate rigid structures which are inserted into holes in walls.Think of them instead as thickenings of the very fabric of the wall itself, made to protect the wall against the concentrations of stress which develop around openings.

In line with this conception,build the frames as thickenings of the wall material,continuous with the wall itself, made of the same materials,and poured,or built up, in a manner which is continuous with the structure of th wall.

♏♏

In windows,splay the thickening,to create DEEP REVEALS(223);the form of doors and windows which will fill the frame,is given by the later patterns—WINDOWS WHICH OPEN WIDE(236),SOLID DOORS WITH GLASS (237),SMALL PANES(239)...

as you build the main frame and its openings,put in the following subsidiary patterns where they are appropriate;

226.COLUMN PLACE

227.COLUMN CONNECTIONS

228.STAIR VAULT

229.DUCT SPACE

230.RADIANT HEAT

231.DORMER WINDOWS

232.ROOF CAPS

模式226 柱旁空间*

……某些柱，尤其是独立柱，不仅起着**角柱**（212）的结构作用，而且还起着重要的社会作用。正是柱特别有助于形成拱廊、长廊、门廊、人行道和有围合的户外空间——**户外亭榭**（69）、**拱廊**（119）、**有围合的户外小空间**（163）、**回廊**（166）、**六英尺深的阳台**（67）、**棚下小径**（174）。本模式说明柱在发挥社会功能方面所必须具有的特征。

❧∞❧

细柱即细长柱，其形状仅由结构功能而定，绝不能形成舒适的环境。

226   COLUMN PLACE*

...certain columns,especially those which are free standing,play an important social role,beyond their structural role as COLUMNS AT THE CORNERS(212).These are,especially,the columns which help to form arcades,galleries,porches,walkways,and outdoor rooms—PUBLIC OUTDOOR ROOM(69),ARCADES(119),OUTDOOR ROOM(163),GALLERY SURROUND(166),SIX-FOOT BALCONY(167),TRELLISED WALK (174).This pattern defines the character these columns need to make them function soially.

❧❧

Thin columns,spindly columns,columns which take their shape from structural arguments alone,will never make a com fortable environment.

The fact is,that a free-standing column plays a role in shaping human space.It marks a point.Two or more together define a wall or an enclosure.The main function of the columns,from a human point of view,is to create a space for human activity.

In ancient times,the structural arguments for columns coincided in their implications with the social arguments. Columns made of brick,or stone,or timber were always large and thick.It was easy to make useful space around them.

事实是独立柱在形成人类活动空间方面起着作用。独立柱可用点表示。两个或更多的点连在一起就限定一片墙或一个围合的空间。从人文的观点来看，柱的主要功能是创造人类活动的空间。

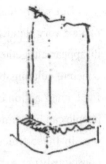

一根粗大的柱
A big thick column

在古代，关于柱的结构功能及其社会功能的论点是完全吻合一致的。由砖、石或木料构筑成的柱总是粗大而结实的。柱在其周围极易形成有用的空间。

塑料世界的细长柱
Thin columns of the plastic world

但是，现在由于使用钢材和钢筋混凝土，很可能使柱变得十分细长，甚至会细长到完全丧失其社会功能的地步。4in粗的钢管或6in厚的钢筋混凝土柱把空间分割开来，就会破坏人类活动的空间，因为这样一些柱创造不出令人舒适的"场所"。因此，当前，就有必要有意识地重新提出：柱不仅具有社会功能，而且具有结构功能。现在我们力求确切地说明柱的社会功能。

柱影响其周围空间的体量因情况而异。这种空间具有一个近似圆的面积，其半径约为5ft。

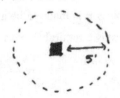

柱周围的空间
The space around the column

But with steel and reinforced concrete,it is possible to make a very slender column;so slender that its social properties disappear altogether.Four inch steel pipes or 6 inch reinforced concrete columns break up space,but they destroy it as a place for human action,because they do not create "spots" where people can be comfortable.

In these times,it is therefore necessary to reintroduce, consciously,the social purposes which columns have,alongside theirstructural functions.Let us try to define these social purposes exactly.

A column affects a volume of space around it,according to the situation.The space has an area that is roughly circular,perhaps 5 feet in radius.

When the column is too thin,or lacks a top or bottom, this entire volume—an area of perhaps 75 square feet— is lost.It cannot be a satisfactory place in its own right:the column is too thin to lean against,there is no way to build a seat up against it,there is no natural way to place a table or a chair against the column.On the other hand,the column still breaks up the space.It subtly prevents people from walking directly through that area:we notice that people tend to give these thin columns a wide berth;and it prevents people from forming groups.

In short,if the column has to be there,it will destroy a considerable area unless it is made to be a place where people feel comfortable to stay,a natural focus,a place to sit down,a place to lean.

如果柱太细，或柱没有顶部和底部，这一全部体量——面积约为75ft²——就失掉了。本来柱本身就不能成为令人满意的空间：柱太细而无法令人或依或靠，无法建造背靠柱的坐位，也无法挨着柱摆放桌椅。另一方面，柱还分割了这一空间。细柱巧妙地不让人直接穿越柱区：我们注意到人们倾向于回避细柱，因为细柱阻止人们集合在一起。

　　总之，凡有细柱的地方，如果那里未能成为人们乐意逗留的地方、一个天然的集合点、一个可坐可倚的休息场所，那么一片相当大的面积将会遭到破坏。

　　因此：

　　当柱独立时，务必使它有人体那样粗——至少有12in粗，最好16in：在柱的周围形成一些人们可以坐下或可以背靠柱舒舒服服休息的地方：有凭靠着柱建造的台阶、小坐位或由一对柱形成的空间。

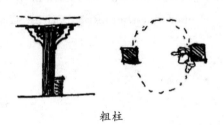

粗柱

∞◌◌◌

　　如果你把柱做成**箱形柱**（216），则不必花多少钱就可以获得很粗的柱；给柱所形成的"场所"，加上一个柱头式的"屋顶"或从柱顶上凸起的拱顶，或在梁和柱间支斜撑——**柱的连接**（227）。只有当柱下部变成为**可坐矮墙**（243），或种花的地方——**高花台**（245），或摆放桌椅的地方——**各式坐椅**（251）之时，柱才具有意义……

Therefore:

When a column is free standing,make it as thick as a man—at least 12 inches,preferably 16 inches:and form places around it where people can sit and lean comfortably: a step, a small seat built up against the column,or a space formed by a pair of columns.

ֆꙮƈ

You can get the extra thickness quite cheaply if you build the column as a BOX COLUMN(216);complete the "place" the column forms,by giving it a "roof" in the form of a column capital,or vault which springs from the column,or by bracing the column against the beams—COLUMN CONNECTION(227). And when it makes sense,make the column base a SITTING WALL(243),a place for flowers—RAISED FLOWERS(245),or a place for a chair or table-DIFFERENT CHAIRS(251)...

模式227 柱的连接**

227   COLUMN CONNECTIONS**

...the columns are in position,and have been tied together by a perimeter beam—BOX COLUMNS(216),PERIMETER BEAMS(217).According to the principles of continuity which govern the basic structure—EFFICIENT STRUCTURE(206), the connections need stiffening to lead the forces smoothly from the beams into the columns,especially when the columns are free standing as they are in an arcade or balcony—ARCADE (119),GALLERY SURROUND(166),SIX-FOOT BALCONY (167),COLUMN PLACE(226).You may also do the same in the upper corners of your door and window frames—FRAMES AS THICKENEDEDGES(225)—making arched openings.

∞⁂

The strength of a structure depends on the strength of its connections;and these connections are most critical of all at corners,especially at the corners where the columns meet the beams.

There are two entirely different ways of looking at a connection:

1.As a source of rigidity,which can be strengthened by triangulation,to prevent racking of the frame.This is a moment connection:a brace.See the upper picture.

2.As a source of continuity,which helps the forces to flow easily around the comer in the process of transferring loads by changing the direction of the force.This is a continuity connection:a capital.See the lower picture.

……柱逐一就位后，就由圈梁紧紧连接在一起——**箱形柱**（216）、**圈梁**（217）。根据支配基本结构的连续性原理——**有效结构**（206），连接处必须加固，以便使力平稳地从梁传到柱，尤其当它们是拱廊或阳台的独立柱时——**拱廊**（119）、**回廊**（166）、**六英尺深的阳台**（167）和**柱旁空间**（226）。你也可以在你的门窗框的上部角隅处进行加固——**门窗边缘加厚**（225）——建造拱形开口。

හ෴෴ය

结构的强度取决于其连接处的强度；这些连接处的角隅处最重要，特别是在梁和柱相连接的角隅处。

看待连接处的两种态度截然不同：

（1）把连接处视为刚性连接，它可被三角形的连接处所加强，从而使框架避免断裂。这是一种力矩连接：斜撑。参阅前页图。

（2）把连接处视为连续性连接，有助于在传递荷载的过程中改变力的方向，使力绕过转角。这是一种连续性连接：柱帽。参阅下图。

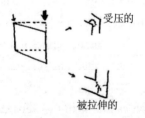

受压的

被拉伸的

不均匀应力对框架的影响
Effect of uneven stresses on a frame

1.斜撑式柱连接。当房屋安装完毕后，在整个使用期间，它要沉降，其结构内部会产生极小的应力。正如常见的那样，如果下沉不均匀，应力就会失去平衡；不管应力是否被设计成承受应变并把各力向下传到地面，其在房屋的每一部

1.A column connection as a brace.

As a building is erected,and throughout its life,it settles,creating tiny stresses within the structure.When the settling is uneven,as it most always is,the stresses are out of balance;there is strain in every part of the building,whether or not that part of the building was designed to accept strain and transmit the forces on down to the ground.The parts of the building that are not designed to carry these forces become the weak points of the building subject to fracture and rupture.

Rectangular frames,especially,have these cracks at the corners because the transmission of the load is discontinuous there.To solve this problem the frame must be braced—made into a rigid frame that transmits the forces around it as a whole without distorting.The bracing is required at any right-angled corner between columns and beams or in the corners of door and window frames.

2.A column connection as a capital.

This happens most effectively in an arch.The arch creates a continuous body of compressive material,which transfers vertical forces from one vertical axis to another.It works effectively because the line of action of a vertical force in a continuous compressive medium spreads out downward at about 45 degrees.

And a column capital is,in this sense,acting as a small, under-developed arch.It reduces the length of the beam—and so reduces bending stress.And it begins to provide the path for the forces as they move from one vertical axis to another,through the medium of the beam.The larger the capital,the better.

A column connection will work best when it acts both as a column capital and as a column brace.This means that it

分都会有应变。未被设计成承受这些力的部分就会成为房屋产生裂缝和受破坏的薄弱点。

矩形框架尤其在角隅处有这些裂缝，因为在那里荷载的传递是不连续的。为了解决这一问题，框架必须有斜撑——必须成为刚性框架。而这种刚性框架作为一个整体可传递其周围的各种力而不发生变形。在柱和梁之间的任何一个呈直角的角隅处或在门窗框的角隅处都必须有斜撑。

2. 柱帽式柱连接。这种柱帽连接在拱结构中非常有效。拱形成抗压材料的连续体，而这种连续体把垂直力从一垂直轴线传递到另一垂直轴线。柱帽十分有效，因为在连续抗压介质中垂直力的作用线大约在 45° 处向下扩散。

在此意义上，柱帽起着一个小的、近似拱的作用。它缩短梁的长度，从而也就减少弯曲应力，而且柱帽开始为力从一垂直轴线通过梁介质传递到另一垂直轴线提供通路。柱帽越大越好。

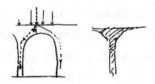

一个和拱起着同样作用的柱帽
A capital that acts the same way as an arch

柱的连接处同时起到柱帽和斜撑两者的作用时效果最佳。这意味着柱的连接必须像柱帽一样，是厚实的，结果就有许多材料可让力通过；而材料必须是结实而又坚固的，并且像斜撑一样，和柱、圈梁完全连成一体，结果柱的连接就能支承住剪力和弯曲。

如下图所示为骨头内部结构，同时符合上述两原理，以便把抗压应力从一个支柱传到另一个支柱，并且使它连续不断地通过支柱的三维空间框架。这一结构的截面在力改变方向的各连接点上最大。

needs to be thick and solid,like a capital,so that there is a lot of material for the forces to travel through,and stiff and strong and completely continuous with the column and perimeter beam,like the brace,so that it can work against shear and bending.

The bone structure,shown below uses both principles, to transfer compressive stress from one strut to another, continuously, throughout a three-dimensional space frame of struts. The structure is most massive at the connections,where the forces change direction.

A similar column connection can be made integral with poured hollow columns and beams.The forms for the connection are gussets made of skin material:then fill the column and the gussets and the beam in a continuous concrete pour.

Of all the patterns in the book,this is one of the most widespread and has taken the greatest variety of outward forms throughout the course of history.A solid wood capital on a wood column,or a continuously poured column top,and arches of stone,brick,or poured concrete are all examples. And,of course,typical column capitals-a larger stone on a stone column or typical gusset plate or brace-even if weak in some ways,also help a great deal.But only relatively few of the historical column connections succeed fully in acting both as braces and as capitals.

Therefore:

Build connections where the columns meet the beams. Any distribution of material which fills the corner up will do:fillets,gussets,column capitals,mushroom column,and most general of all,the arch,which connects column and beam in a continuous curve.

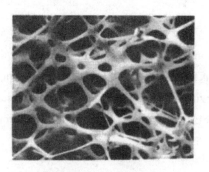

骨头内部结构的各连接点
Connections inside a bone

　　有一种相似的柱连接，就是把空心柱和梁用混凝土浇筑在一起。连接的形式是先用外层材料做成角撑板；然后用混凝土向柱、角撑板和梁进行连续浇筑。

　　本模式是此书所有模式中应用最广泛的，在历史上有过丰富多彩的外形。木柱的实心木柱帽，或混凝土连续浇筑的柱顶，以及石拱、砖拱或浇筑混凝土的拱都是实例。而且，当然，典型的柱帽——石柱之上扩大了的石块或典型的角撑板或斜撑——即使在某些方面有弱点，仍有很大益处。但是，在历史上柱的连接处同时起到斜撑和柱帽两者作用的、完全成功的例子真是寥若晨星。

　　因此：

　　在柱和梁的连接处建造各种连接。浇筑角隅的材料不管如何分配都应做到：圆抹角、角撑板、柱帽、蘑菇柱以及最普遍的拱都要把柱和梁以连续的曲线形式连接在一起。

45°连接

The connection is one of the most natural places for ORNAMENT (249):there is a wide variety of possible connections,carvings,fretwork,painting,for this critical position. In certain cases,the connection may act as an umbrella for a COLUMN PLACE(226)...

∽❀∾

　　连接处是**装饰**（249）最合理的地方之一：在这一关键
性的位置上，有各种不同的连接方式，雕刻、雕花和油漆。
在某些情况下，连接处可起到**柱旁空间**（226）一把伞的作
用……

模式228　楼梯拱*

228   STAIR VAULT*

...this pattern helps complete the rough shape and location of stairs given by STAIRCASE AS A STAGE(133) and by STAIRCASE VOLUME (195).If you want to build a conventional stair,you can find what you need in any handbook. But how to build a stair in a way which is consistent with the compressive structure of EFFICIENT STRUCTURE (206),without using wood or steel or concrete—GOOD MATERIALS (207)?

❧✥❧

Within a building technology which uses compressive materials as much as possible,and excludes the use of wood,it is natural to build stairs over a vaulted void,simply to save weight and materials.

A concrete stair is usually made from precast pieces supported by steel stringers;or it is formed in place,and then stripped of its forms.But for the reasons already given in GOOD MATERIALS(207),precast concrete and steel are undesirable materials to use-they call for modular planning;they are unpleasant materials to touch,look at,and walk on;they are hard to work with and modify in any relaxed way,since they call for special tools.

Given the principles of EFFICIENT STRUCTURE(206), GOOD MATERIALS(207),and GRADUAL STIFFENING(208), we suggest that stairs be made like FLOOE-CEILING VAULTS(219)-by making a half-vault (to the slope of the stair),with lattice strips,burlap,resin, chickenwire,and

......本模式有助于完善由**有舞台感的楼梯**（133）和**楼梯体量**（195）所决定的楼梯的大致形状和位置。如果你想建造一座常规的楼梯，在任何一本手册中都能找到你所需要的一切资料。但是，怎样才能建成一座与**有效结构**（206）的抗压结构完全一致的楼梯而又不使用木料、钢材或混凝土呢？——参阅**好材料**（207）。

◊◊◊◊◊

在尽量使用抗压材料而不用木料的施工技术中，显而易见的事实是，把楼梯建成拱状空间结构将会明显减轻楼梯的重量和节省用料。

混凝土楼梯通常由楼梯钢斜梁支承的预制踏步板组成；或楼梯就地用混凝土浇筑而成，然后再拆去模板。但是，由于在**好材料**（207）中早已阐明的理由，预制构件和钢材不是理想的材料——它们要求按模数布置平面；它们是使人手感、观感和行走都不愉快的材料；对它们很难进行加工也不便修改，因为只有使用特种工具才行。

根据**有效结构**（206）、**好材料**（207）和**逐步加固**（208）等原理，我们建议楼梯应当造得和**楼面天花拱结构**（219）一样——利用格构筋条、粗麻布、树脂、钢丝网和轻质混凝土把楼梯建成半拱形的（楼梯的斜跑做成拱形）。然后，楼梯踏步则以木板或软质面砖作竖板，再填以混凝土并抹光。

当我们开始撰写本模式时，我们以为它是一个应用效果尚存疑的模式——并且将其主要部分同楼面和拱屋顶保持连贯一致。后来我们建成了一座拱形的楼梯。这是一次巨大的成功——这一楼梯拱非常漂亮出色——现在我们由衷地向大家推荐。

lightweight concrete.The steps themselves can then be formed by using wood planks,or tiles,as risers,and filling in the steps with trowelled concrete.

When we first wrote this pattern,we thought it was very doubtful—and put it in mainly to be consistent with floor and roof vaults.Since then we have built a vaulted stair.It is a great success—beautiful—and we recommend it heartily.

Therefore:

Build a curved diagonal vault in the same way that you build your FLOOR-CEILING VAULTS(219).Once the vault hardens,cover it with steps of lightweight concrete, trowelformed into position.

❧∽∾❧

A lightweight concrete tread,colored,waxed,and polished can be quite beautiful and soft enough to be comfortable— see FLOOR SURFACE(233)—and will eventually take on the patina of wear called for in SOFT TILE AND BRICK(248).

The vaulted space under the stair can be used as an ALCOVE (179)a CHILD CAVE(203),or CLOSETS BETWEEN ROOMS(198).If it is plastered,like a regular ceiling—see FLOOR-CEILING VAULTS(219),it makes a much more pleasant and useful space than the space under an ordinary stair.

因此：

**以与建造楼面天花拱结构（219）相同的方式建造呈
对角曲线的楼梯拱。一旦楼梯拱硬化，就在其上做成轻质
混凝土踏步，并抹灰修整。**

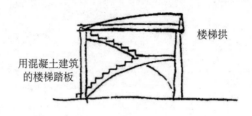

着色的、上蜡的或抛光的轻质混凝土可达成美观并柔
和舒适的效果——参阅**地面面层（233）**——最后可以罩上
软质面砖和软质砖（248）中所介绍的面层。

楼梯下的拱结构空间可用作**凹室（179）、儿童猫耳洞**
（203）或**居室间的壁橱**（198）。如果楼梯拱抹灰后，就宛
如一个规则的天花板——参阅**楼面天花拱结构（219）**，它
就会比普通楼梯下所形成的空间更为有用和更令人愉快。

模式229　管道空间

……在根据**有效结构**（206）原理或以拱形楼面——**楼面天花拱结构**（219）所建成的房屋中，每一房间四周的边缘，都有一个未被使用的三角形体量。这就是放置管道最理想的地方。

❧❧❧

你绝不会知道各种管道埋设在什么地方；它们都埋在墙内的某一处；但是它们究竟在什么地方呢？

在大多数的房屋内，电线管路、上下水管道、煤气管道、电话线管路等总以一种完全不协调和无组织的方式埋设在墙内。这就使新建房屋的施工复杂化，因为随着房屋各部分施工的开展，就很难去协调各种不同的服务设施。一旦房屋建成，因你不知道各种管线埋设在什么地方，所以就难以考虑对它们进行某种修改或补充。这就使我们在对周围环境的认识上出现了一块空白：我们对所居住的房屋内各种服务设施的安排会有一种神秘感。

我们建议，所有这些服务设施都应集中布置，并应布置在每个房间的天花四周的拱形天花和楼面之间的拱肩内——**楼面天花拱结构**（219）。

一般供热和电线管道布满整幢房屋，并按上述方式分布在每一房间的四周。但上下水和煤气管道仅通到某些房间。所有管线都应垂直集中于各个房间的角隅处。各种管线就这样形成分出水平环行管道的垂直干管。这样安排各种管线是不难理解的，并且极易分出支管。

229   DUCT SPACE

...in a building built according to the principles of
EFFICIENT STRUCTURE(206)and built with vaulted floors—
FLOOR-CEILING VAULTS(219),there is a triangular volume,
unused, around the edge of every room.This is the most natural
place to put the ducts.

❧❧❧

**You never know where pipes and conduits are;they are
buried somewhere in the walls;but where exactly are they?**

In most buildings electric conduits,plumbing,drains,gas
pipes,telephone wires,and so on,are buried in the walls,in a
completely uncoordinated and disorganized way.This makes
the initial construction of the building complicated since it is
difficult to coordinate the installation of the various services
with the building of various parts of the building.It makes it
difficult to think about making any changes or additions to
the building once it is built since you don't know where the
service lines are.And it leaves a gap in our understanding of
our surroundings:the organization of utilities and services in the
buildings we live in are a mystery to us.

We propose that all the services be located together and
run around the ceiling of each room in the spandrel between
the vaulted ceiling and the floor above—FLOOR-CEILING
VAULTS(219).

Heating and electrical conduits will be universal throughout
the building and should thus be run around every room.
Plumbing and gas lines will be around some rooms only.

各种管道都汇集在一处
All in one place

因此：

将热风管道、上下水管道、煤气管道和其他服务设施的管线一律集中在拱体所包含的三角形空间内，环绕在每个房间的上部边缘。将各种不同房间的管道用敷设在管槽内的垂直干管在房间的角隅处连接起来。沿着干管，每隔一定距离装有接口和接线板，以便接出支管。

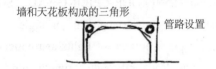

墙和天花板构成的三角形　　　　　　　　　管路设置

❧❦

一旦管道设在房间角隅处的三角形空间内，就可用轻质混凝土填实它——**楼面天花拱结构**（219）。沿该三角形空间表面放置散热板——**辐射热**（230）；在干管下面，每隔一定距离，应安装照明电线引出口，导线和管路顺着窗框敷线槽口板向下引出——**投光区域**（252）。

All lines will also be concentrated vertically at the corners of rooms.Thus the lines form vertical trunks from which horizontal loops spring.This configuration of pipes and conduits is easy to understand and plug into.

Therefore:

Make ducts to carry hot air conduit,plumbing,gas,and other services in the triangular space,within the vault, around the upper edge of every room.Connect the ducts for different rooms by vertical ducts,in special chases,in the corners of rooms.Build outlets and panels at intervals along the duct for access to the conduits.

࿇

Once the duct is in,you can fill up the triangle with lightweight concrete—FLOOR-CEILING VAULTS(219).Place heating panels along the surface of the triangle—RADIANT HEAT(230);and place outlets for lights at frequent intervals below the duct,with leads and conduits running down in rebates along the window frames—POOLS OF LIGHT(252)...

模式230 辐射热*

……为了完善**墙体**（218）、**楼面天花拱结构**（219）和
管道空间（229），应利用一种符合生物学的供暖系统。

∞⋙∞

本模式从生物学角度而言，是直觉的正确陈述，即阳
光和炽烈燃烧之火是最佳的热源。

热量传递的方式有三：辐射（热波穿过空间）；对流
（因分子混合和热空气的上升而引起的空气和液体的流
动）；传导（流经固体）。

在大部分地方，我们就是以这3种方式从周围环境获
取热量：我们接触的固体所传导的热量；我们周围空气中对
流的热量；我们视线范围内的那些辐射源所辐射的热量。

其中，传导热是微不足道的，因为把热量直接传导到
我们的任何一种表面都发烫得令人不舒服。至于其他两种
热——对流热和辐射热——我们可以问一下，在它们对人
类的影响方面是否有任何生物学上的差别。事实上也是
有的。

业已证明，当人们在接受比周围空气温度略高的辐射
热时，就会感到十分舒服。有两个最简单的例子可供说明：
（1）在春光明媚、和风拂面的室外；（2）在有凉意的夜晚，
围坐在火炉旁。

大多数人都会直觉地承认这是两种不同寻常的舒服环
境。鉴于我们是在野外受到充足的阳光照射而进化成的有
机体这一事实，所以，我们只要置身其间，就会感到格外
舒服，这是不足为奇的。从生物学观点考虑，这一点已构
成我们体系的一部分了。

230 RADIANT HEAT*

...to complete WALL MEMBRANES(218),FLOOR-CEILING VAULTS(219) and DUCT SPACE(229),use a biologically sensible heating system.

৪০>৫৪

This pattern is a biologically precise formulation of the intuition that sunlight and a hot blazing fire are the best kinds of heat.

Heat can be transmitted by radiation(heat waves across empty space),convection(flow in air or liquids by mixing of molecules and hot air rising),and conduction(flow through a solid).

In most places,we get heat in all three ways from our environment:conducted heat from the solids we touch, convected heat in the air around us,and radiated heat from those sources of radiation in our line of sight.

Of the three,conducted heat is trivial,since any surface hot enough to conduct heat to us directly is too hot for comfort.As far as the other two are concerned—convected heat and radiant heat—we may ask whether there is any biological difference in their effects on human beings.In fact there is.

It turns out that people are most comfortable when they receive radiant heat at a slightly higher temperature than the temperature of the air around them.The two most primitive examples of this situation are:(1)Outdoors,on a spring day when the air is not too hot but the sun is shining.(2)Around an open fire,on a cool evening.

很不幸，许多最为广泛使用的供暖系统往往忽略了这一基本事实。

热风系统、埋设在墙内的供热管道和所谓的热水"散热器"，通过辐射，把它们的部分热量传送给我们，但我们从它们所获得的大部分热量却来自对流。空气逐渐变热并在我们周围旋转时，我们就会感到温暖。但是，当热空气正常旋转之时，就会形成一种令人最不愉快的、气闷的、过热的、干燥的环境。当对流加热器给我们提供热量时，我们也会感到热得喘不过气来。如果我们减小供热量，空气就会变得太冷。

人们感到非常舒服的那种环境必须有对流热和辐射热之间的微妙平衡。实验已经表明，两者之间令人最感舒服的平衡是平均辐射温度高于周围环境温度约2℃。为了获得室内的平均辐射温度，我们测量了室内所有可见的表面的温度，并将每一表面面积乘上它的温度，再把所有这些乘积加在一起，除以总面积。为了舒适起见，这一平均辐射温度必须比空气温度高2℃左右。

因为一个房间的某些表面（窗户和外墙）通常总比室内空气温度低，所以，这意味着，至少一些表面必须相当暖和才能获得这一平均值。

一个散热面积小而温度高的明火在一个微冷的室内能创造出这一条件。奥地利和瑞典的砖炉美观精致，同样具有良好的热效应。它们炉体大，由黏土砖或面砖构成，其内有一个小炉膛。在炉膛内添加一把小树枝，炉火就能加热炉体的黏土，而黏土像泥土一样，能保存热量，并缓慢地、在长达数小时之内，将热辐射出来。

奥地利砖炉
Austrian tiled stove

Most people will recognize intuitively that these are two unusually comfortable situations.And in view of the fact that we evolved as organisms in the open air,with plenty of sun,it is not surprising that this condition happens to be so comfortable for us.It is built into our systems,biologically.

Unfortunately,it happens that many of the most widely used heating systems ignore this basic fact.

Hot air systems,and buried pipes,and the so-called hot water "radiators" do transmit some of their heat to us by means of radiation,but most of the heat we get from them comes from convection.The air gets heated and warms us as it swirls around us.But,as it does so it creates that very uncomfortable stuffy,over-heated,dry sensation.When convection heaters are warm enough to heat us we feel stifled.If we turn the heat down,it gets too cold.

The conditions in which people feel most comfortable require a subtle balance of convected heat and radiant heat. Experiments have established that the most comfortable balance between the two,occurs when the average radiant temperature is about two degrees higher than the ambient temperature.To get the average radiant temperature in a room,we measure the temperature of all the visible surfaces in a room,multiply the area of each surface by its temperature,add these up,and divide by the total area.For comfort,this average radiant temperature needs to be about two degrees higher than the air temperature.

Since some of the surfaces in a room(windows and outside walls),will usually be cooler than the indoor air temperature,this means that at least some surfaces must be considerably warmer to get the average up.

An open fire,which has a small area of very high

由单个房间控制的辐射散热板和悬挂在墙壁和天花板上的红外加热器都是可用的高技术辐射源。很可能，低水平的辐射热源——像一只热水箱——同样具有良好的热效应。这种用不着隔热保温的热水箱放在建筑的正中间，很可能是效果非常出色的辐射热源。

　　因此：

　　选择一种加热方法来为你的房间供热——尤其要供热给天冷时人们常常聚首会晤的那些房间——供热过程基本上应是辐射热大于对流热。

比空气温度略高的表面

❈❈

　　如果你已按前面的一些模式行事，也许你已经建成一个拱形天花板的房间了，在近墙处有一很陡的斜面，在斜面后埋设着主要管道——**楼面天花拱结构**（219）、**管道空间**（229）。在此情况下，在斜面上放置辐射散热板就是合理的了。

　　但是，至少部分辐射面的位置要很低，以便围绕着或对着这些散热面布置坐位，这也是一件令人愉悦的乐事；在寒冷的日子里，没有什么东西能比得上一个挨近暖融融火炉的坐位了——**嵌墙坐位**（202）……

temperature, creates this condition in a cool room.The beautiful Austrian and Swedish tiled stoves also do it very well.They are massive stoves,made of clay bricks or tiles,with a tiny furnace in the middle.A handful of twigs in the furnace give all their heat to the clay of the stove itself,and this clay,like the earth, keeps this heat and radiates it slowly over a period of many hours.

Radiant panels,with individual room control,and infrared heaters hung from walls and ceilings,are possible high technology sources of radiant heat.It is possible that sources of low-grade radiant heat—like a hot water tank—might also work to very much the same effect.Instead of insulating the tank,it might be an excellent source of radiant heat,right in the center of the house.

Therefore:

Choose a way of heating your space—especially those rooms where people are going to gather when it is cold—that is essentially a radiative process,where the heat comes more from radiation than convectio.

⁂

If you have followed earlier patterns,you may have rooms which have a vaulted ceiling,with a steeply sloping surface close to the wall,and with the major ducts behind that surface—FLOOR-CEILING VAULTS(219),DUCT SPACE(229).In this case, it is natural to put the radiant heating panels on that sloping surface.

But it is also very wonderful to make at least some part of the radiant surfaces low enough so that seats can be built round them and against them;on a cold day there is nothing better than a seat against a warm stove—BUILT-IN SEATS(202)...

模式231　老虎窗*

231   DORMER WINDOWS*

...this pattern helps to complete SHELTERING ROOF(117).If you have followed sheltering roof,your roof has living space within it:and it must therefore have windows in it,to bring light into the roof.This pattern is a special kind of WINDOW PLACE(180),which completes the ROOF VAULTS(220),in these situation.

🙰

We know from our discussion of SHELTERING ROOF(117) that the top story of the building should be right inside the roof,surrounded by it.

Obviously,if there is habitable space inside the roof,it must have some kind of windows;skylights are not satisfactory as windows—except in studios or workshops—because they do not create a connection between the inside and the outside world—WINDOWS OVERLOOKING LIFE(192).

It is therefore natural to pierce the roof with windows;in short,to build dormer windows.This simple,fundamental fact would hardly need mentioning if it were not for the fact that dormer windows have come to seem archaic and romantic.It is important to emphasize how sensible and ordinary they are—simply because people may not build them if they believe that they are old fashioned and out of date.

Dormers make the roof livable.Aside from bringing in light and air and the connection to the outside,they relieve the low ceilings along the edge of the roofs and create alcoves and window places.

……本模式有助于完善**带阁楼的坡屋顶**（117）。如果你已经采用带阁楼的坡屋顶，那么屋顶之内已具有居住空间：因此，应设置有光线可照进屋顶的窗户。本模式是一种特殊的**窗前空间**（180），并在这些情况下完善**拱式屋顶**（220）。

∞∞∞

我们从讨论模式带阁楼的坡屋顶（117）中知道，楼房的顶层应当恰好在屋顶之内，并由屋顶所包围。

显而易见，如果在屋顶之内有一居住空间，就必须具有某种窗户；天窗作为窗户已不能满足要求——除了工作室或车间以外——因为天窗不能形成内部世界和外部世界之间的联系——**俯视外界生活之窗**（192）。

因此，必然要剖开屋顶装上窗户；一言以蔽之，就是要建造老虎窗。若非老虎窗看来具有古典风格和浪漫色彩，这一简单的基本事实就几乎无须提及。强调老虎窗的合理性和正常性是极其重要的——因为只要人们认定老虎窗是旧式的或已过时的设施，就不会再建造老虎窗了。

老虎窗使屋顶成为可居住的空间。老虎窗不仅让光线和空气进入屋顶的居住空间，并与外部世界保持联系，而且还减轻了沿屋顶边缘的低天花板的压抑感，同时还能形成凹室和窗前空间。

怎样来构筑老虎窗呢？在我们已经描述过的拱屋顶之内，在形成拱的筐式结构时，在柱和圈梁的框架开口的上面，就可简单地连续形成老虎窗了。

建老虎窗的其他方法取决于将要采用的施工体系。不论门窗过梁、柱和墙采用何种材料，你都能简单地进行修改，并能使用组合方法来建造老虎窗。

How should the dormers be constructed? Within the roof vault we have described,the basket which forms the vault can simply be continued to form the roof of the dormer,over a frame of columns and perimeter beams which form the opening.

The other ways of building dormer windows depend on the construction system you are using.Whatever you are using for lintels,columns,and walls,can simply be modified and used in combination to build the dormer.

Therefore:

Wherever you have windows in the roof,make dormer windows which are high enough to stand in,and frame them like any other alcoves in the buildin.

୫୦୧୪

Frame them like ALCOVES(179) and WINDOW PLACE (180)with GRADUAL STIFFENING(208),COLUMNS AT THE CORNERS(212),BOX COLUMNS(216), PERIMETER BEAMS(217),WALL MEMBRANES(218), FLOORCEILING VAULTS(219),ROOF VAULTS(220) and FRAMES AS THICKENED EDGES(225).

Put WINDOWS WHICH OPEN WIDE(236) in them,and make SMALL PANES(239)...

因此：

不管你在屋顶的什么地方开窗洞，都要建老虎窗。它要具有足够的高度以便可让人站得直，并和楼内的任何其他凹室一样，给老虎窗安装上边框。

老虎窗

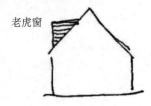

୫୦୦୫

采用**逐步加固**（208）、**角柱**（212）、**箱形柱**（216）、**圈梁**（217）、**墙体**（218）、**楼面天花拱结构**（219）、**拱式屋顶**（220）和**门窗边缘加厚**（225）等模式，和**凹室**（179）和**窗前空间**（180）一样，给老虎窗安装上框架。

老虎窗要采用**大敞口窗户**（236）和**小窗格**（239）……

模式232　屋顶顶尖

　　……本模式对**屋顶花园**（118）或**拱式屋顶**（220）进行修饰。假定你已建造了拱式屋顶——或至少已开始做好支承拱结构的模板；或假定你已开始建造屋顶花园，并开始修筑栏杆将它围住。在上述两种情况下——屋顶应怎样修饰才好呢？

୨୦ଓ

　　在传统建筑中，建造者几乎都利用屋顶细部给建筑加上装饰性的屋顶顶尖。
　　下面是一些例子：希腊建筑的三角形山花、阿尔贝罗贝洛的德罗利人的石砌拱的顶尖、日本庙堂的顶部、谷仓

232 ROOF CAPS

...and this pattern finishes the ROOF GARDENS(118)or the ROOF VAULTS(220).Assume that you have built the roof vaults—or at least that you have started to build up the splines which will support the cloth which forms the vault.Or assume that you have begun to build a roof garden,and have begun to fence it or surround it.In either case—how shall the roof be finishe?

ଚ୪ଏ୪

There are few cases in traditional architecture where builders have not used some roof detail to cap the building with an ornament.

The pediments on Greek buildings;the caps on the trulli of A1berobello;the top of Japanese shrines;the venting caps on barns.In each of these examples there seems to be some issue of the building system that needs resolution,and the builder takes the opportunity to make a "cap."

We suspect there is a reason for this which should be taken seriously.The roof cap helps to finish the building;it tops the buildin gwith a human touch.Yet,the power of the cap,its overall effect on the feeling of the building,is of much greater proportions than one would expect.Look at these sketches of a building,with and without a roof cap.They look like different buildings.The difference is enormous.

Why is it that these caps are so important and have such a powerful effect on the building as a whole?

的透气帽顶。其中每一例似乎都是由于建筑体系所需要解决的某个问题，而建造者就借机造了一顶"帽子"。

我们认为不必对这一说法过于较真。屋顶顶尖有助于修饰建筑物；它以富有人情味的艺术风格为建筑物增添光彩。而且，顶尖的力量以及它使人们对建筑的感受所造成的全部影响都远远出人意料。请看下面两张草图：一张有屋顶顶尖，另一张无屋顶顶尖。两者看上去就是两所不同的房屋。一所房屋有无屋顶顶尖的差别何等悬殊！

有屋顶顶尖和无屋顶顶尖的两张草图
With and without a roof cape

为什么这些顶尖如此重要，而且对整幢建筑具有如此巨大的影响呢？

下面是一些可能的理由：

1. 顶尖为屋顶加冕。顶尖赋予屋顶应有的地位。屋顶是重要的，而顶尖则强调了这一点。

2. 顶尖增加了细部处理。顶尖使屋顶显得不那么千篇一律，并使屋顶不致成为孤立单调之物。门、窗和阳台衬托墙壁，而墙则增加比例和特征；当屋顶有许多老虎窗时，似乎屋顶顶尖就无须那么多了。

3. 顶尖表明了和天际的联系，它在某一时期内曾具有宗教色彩。正如建筑必须有与大地相连的感觉一样——参阅**与大地紧密相连**（168）——也许屋顶也必须和天空相连。

我们建议，在建筑体系中，屋顶顶尖就是我们放在屋脊上的重物，以便使屋顶的两个斜面略呈曲线。这些顶尖以等间距放置在带有贝壳饰的屋脊上。这些重物未必是大体积的——一小袋砂子或一块石头就行，再抹上混凝土，

Here are some possible reasons.

1. They crown the roof. They give the roof the status that it deserves. The roof is important, and the caps emphasize this fact.

2. They add detail. They make the roof less homogeneous, and they relieve the roof from being a single uninterrupted thing. The walls get this relief from windows, doors, balconies, which add scale and character; when a roof has many dormers, it seems to need the caps less.

3. The caps provide a connection to the sky, in a way that might have had religious overtones at one time. Just as the building needs a sense of connection to the earth—see CONNECTION TO THE EARTH(168)—perhaps the roof needs a connection to the sky.

In the building system we propose, the roof caps are weights we use at the ridge of the roof to make the slight curve in the pitched sides of the roof. They happen at regular intervals, at the ridges of the scallops. They need not be large— a small bag of sand or a stone will do, plastered with concrete and shaped so the bulge is obvious. It may be nice to paint them a different color from the roof.

Of course, there are hundreds of other possible kinds of roof caps. They can be brick chimneys, statues, vents, structural details, the pinnacles on a gothic buttress, weather vanes, or even windmills.

使之具有一定的形状,于是,一个鼓包就呈现得清清楚楚了。给顶尖涂上与屋顶颜色不同的油漆是十分精美的。

当然,屋顶顶尖的形状千差万别,不可胜数。它们可能是砖砌的烟囱、雕塑像、透气洞、结构细部、哥特式扶垛的尖顶、风向标,甚至于风车。

因此:

选择一种合理的方法为屋顶加顶尖——即某种与构造施工和建筑物的意义相呼应的方法。顶尖可能是结构需要;但其主要功能是装饰性的——标志着屋顶与天空相连。

和天空相连接

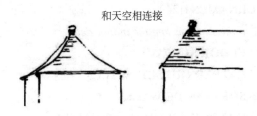

以你所希望的任何方式来修饰屋顶顶尖,但不要忘却它们——装饰(249)……

修饰表面和室内细部;

233. 地面面层

234. 鱼鳞板墙

235. 有柔和感的墙内表面

236. 大敞口窗户

237. 镶玻璃板门

238. 过滤光线

239. 小窗格

240. 半英寸宽的压缝条

Therefore:

Choose a natural way to cap the roof-some way which is in keeping with the kind of construction,and the meaning of the building.The caps may be structural;but their main function is decorative—they mark the top—they mark the place where the roof penetrates the sky.

𝄢𝄡

Finish the roof caps any way you want,but don't forget them—ORNAMENT(249)...

put in the surfaces and the indoor details;

233.FLOOR SURFACE

234.LAPPED OUTSIDE WALLS

235.SOFT INSIDE WALLS

236.WINDOWS WHICH OPEN WIDE

237.SOLID DOORS WITH GLASS

238.FILTERED LIGHT

239.SMALL PANES

240.HALF-INCH TRIM

模式233　地面面层**

　　……本模式告诉你如何去铺砌地面面层以及如何去修饰**底层地面**（215）和**楼面天花拱结构**（219）。如果地面修饰得好，也将有助于强化建筑的私密性层次——**私密性层次**（127）。

୫୦୯ଓ

　　我们希望地面舒适、触感温暖、引人注目，但我们也希望地面有足够的耐磨硬度，且容易擦洗。

233 FLOOR SURFACE**

...this pattern tells you how to put the surface on the floors,to finish the GROUND FLOOR SLAB(215) and FLOOR-CEILING VAULTS(219).When properly made,the floor surfaces will also help intensify the gradient of intimacy in the building—INTIMACY GRADIENT(127).

❧✣❧

We want the floor to be comfortable,warm to the touch,inviting.But we also want it to be hard enough to resist wear,and easy to clean.

When we think of floors,we think of wood floors. We hope,if we can afford it,to have a wooden floor.Even in hot countries,where tiles are beautiful,many people want hardwood floors whenever they can afford them.But the wood floor,though it seems so beautiful,does little to solve the fundamental problem of floors.The fact is that a room in which there is a bare wood floor,seems rather barren,forbidding,makes the room sound hollow and unfurnished.To make the wooden floor nice,we put down carpets.But then it is not really a wood floor at all.This confusion makes it clear that the fundamental problem of "the floor" has not been properly stated.

When we look at the problem honestly,we realize that the wooden floor,*and* the wooden floor with a carpet on it,are both rather uneven compromises.The bare wooden floor is too bare,too hard to be comfortable;but not in fact hard enough to resist wear particularly well if it is left uncovered—it scratches

当我们考虑地面时，我们就会想到木质地面。如果我们的预算可以承受，我们就使用木质地面。甚至在气候炎热的国家，虽地砖色彩绚丽，但只要付得起钱，许多人仍希望要硬木地面。尽管木质地面纹细美丽，但解决不了地面的基本问题。事实上，如果室内只铺木质地面，会使声音空洞沉重，令人难亲近而感到索然无味。为了使木质地面美观雅致，我们可铺上地毯。但这样一来，地面就不是名副其实的木质的了。这种名实不符的混淆表明"地面"这一基本问题还没有被恰如其分地说清楚。

　　当我们正视这一问题时，我们就会认识到木质地面和铺上地毯的木质地面不能相提并论。木质地面太裸露太硬会使人感到不舒服；事实上，如果不铺上地毯，它的耐磨度并不很好——会被磨损、凹陷或碎裂。而当木质地面铺上了地毯，木质所具备的全部的美就看不见了。除了地毯的边缘以外，再多一点也看不到了。铺上的地毯确实硬度很小，经不起磨损。而且，最漂亮的地毯、手工做的小地毯和地毡是如此的精致，以致很不耐磨。穿着户外用鞋在波斯地毯上行走是一种野蛮的习惯，而制作这些地毯的工人和知道如何对待它们的人绝不会这样做——他们总是脱掉鞋走上去。但是现代的尼龙地毯、丙烯地毯和机织耐磨地毯都失去了地毯的全部华丽和令人愉悦之感：它们仿佛是软质混凝土。

　　这一问题是无法解决的。冲突是根本性的。这一问题只能回避了之，解决办法是在住宅内划分两个区：人来人往较频繁的区域采用容易清洗的硬质耐磨地面；人员进出较少的区域就可以轻松铺上各种丰富多彩的、质地柔软的地毯、垫子和地毡，在此处人人必须脱鞋。

　　传统的日本住宅和俄国住宅就是完全这样来解决问题的：日本人和俄罗斯人把地面分成两个区域——耐磨区和舒适区。他们在舒适区使用非常干净的且往往很珍贵的材料，而使耐磨区成为街道的延伸——即肮脏不堪的、铺砌

and dents and splinters.And when the floor is covered with a carpet,the whole point of the beauty of the wood is lost. You cannot see it any more,except round the edges of the carpet;and the carpet on the floor is certainly not hard enough to resist any substantial wear.Furthermore,the most beautiful carpets,handmade rugs and tapestries,are so delicate that they cannot take very rough wear.The practice of walking on a Persian rug with outdoor shoes on is a barbarian habit,never practiced by the people who make those rugs,and know how to treat them—they always take their shoes off.But the modern nylon and acrylic rugs,machine—made for hard wear,lose all the sumptuousness and pleasure of the carpet:they are,as it were,soft kinds of concrete.

The problem cannot be solved.The conflict is fundamental. The problem can only be *avoided* by making a clear distinction in the house between those areas which have heavy traffic and so need hard wearing surfaces which are easy to clean,and those other areas which have only very light traffic,where people can take off their shoes,and where lush,soft,beautiful rugs,pillows,and tapestries can easily be spread.

Traditional Japanese houses and Russian houses solve the problem in exactly this way:they divide the floor into two zones—serviceable and comfortable.They use very clean,and often precious materials in the comfortable zone,and often make the serviceable zone an extension of the street—that is,dirt,paving,and so on.People take their shoes off,or put them on,when they pass from one zone to the other.

We are not sure whether taking shoes off and on could become a natural habit in our culture.But it still makes sense to zone the house so that the floor material changes

路面的，如此等等。当他们从这一区到那一区时，人人都要脱鞋或穿鞋。

硬区和软区之间的门槛
The threshold between hard and soft

脱鞋和穿鞋是否能成为我们文化中合理的习惯，我们对此没有把握。但是，把住宅分成两个区域仍然是有意义的，所以地面材料随人们进入住宅中私密程度不同而变化。**私密性层次**（127）中需要一种公用房间、半公用房间和私人房间的层次。由此得出结论：当人们进入宅内越深，他们所要的地面就越柔软——入口和厨房的地面是较好的硬质耐磨地面；餐厅、家庭公用室、儿童游戏室需要耐磨地面，但要有一些舒适的部位；卧室、书房和私人房间需要柔软舒适的地面，以供人们坐、卧或光脚行走。

材料应当是什么样的？在硬质材料和软质材料之间，硬质材料的问题更多。因为儿童们和软质材料与硬质材料都形影不离，硬质材料应有温暖感——同时还应是容易清洗的。有一种"软质的"混凝土可用于硬质地面。如果这种地面使用一种轻质的、孔洞较多的混凝土并经质感处理，兼具耐磨性和愉悦感。如果这种地面在做成后用某种防水和着色混合物进行罩面并且上蜡抛光，就兼具耐磨性和防水性了。不管怎么说，如果地面是混凝土的，造价就相当低，因此是有意义的。

可以制成硬质地面的其他材料有泥土、橡胶或软木面砖和在秘鲁称为着色地砖的未焙制的软质面砖——参阅**软质**

as one gets deeper into the house.The pattern INTIMACY GRADIENT(127) calls for a gradient of public,semi-public,and private rooms.It follows that one wants the floor to get softer as one goes deeper into the house—that is,the entrance and the kitchen are better floored with a hard,serviceable surface,while the dining,family room,and children's playrooms need a serviceable floor but with comfortable spots,and the bedrooms,studies,rooms of one's own need soft comfortable floors,on which people can sit,lie,and walk barefoot.

What should the materials be? Of the hard and soft materials,the hard is more of a problem.Since children are close to these floors,as well as the soft ones,they must be warm to the touch, —and at the same time they must be easy to clean. For these hard floors,a "soft" concrete might work.It can be made serviceable and pleasant at the same time if it is finished off with a lightweight textured floor finish,which is relatively porous.It can be made to wear and repel water by making the color integral with the mix and by waxing and polishing after it is set.It is fairly cheap and makes sense if the floor is a concrete floor anyway.Other materials which would work as hard floors are earth,rubber or cork tile,soft unbaked tile known as pastelleros in Peru—see SOFT TILE AND BRICK (248)—and wood planks,but these materials are more expensive.

For soft materials,carpet is the most satisfactory— for sitting,lying,and being close to the ground.We doubt that an improvement can be made on it—in fact we guess that if a substitute is used instead,it will eventually get carpeted over,anyway.This means that the areas which are going to be carpeted might as well have a cheap subfloor with matting laid wall to wall.

面砖和软质砖（248）——和木板，但这些材料比较昂贵。

地毯是最令人满意的软质材料——地毯紧贴地面，可供人们或坐或卧。我们怀疑对地毯还能作出什么改进——事实上，我们猜测，如果使用地毯的替代品，这种替代品最终将像地毯那样覆盖地面。这意味着在要铺地毯的那些区域同样可以满铺造价相对较低的编织物作为地面底层。

为了强调房间内的这两个区域，为了改善从一个区域到另一个区域的脱鞋和穿鞋的条件，我们建议在两区域之间有一个升高或下降台阶。这将极大地有助于保持每个区域的"纯净度"，并必定对各个区域的活动有利。

因此：

将住宅或建筑分成两个区域：公用区和私人区或私密性较强的区域。在公用区采用硬质地面材料，诸如上蜡的、红色抛光的混凝土、地面面砖或硬木。在私密性较强的区域，在地面底层上铺质地柔软的材料，诸如毛毡、廉价的尼龙地毯或有布套的草垫、枕头、地毯和地毡等。两区域之间界线分明——或甚至有一个台阶——以便人们从公用区进入私密区时能够脱鞋。

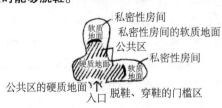

私密性房间
私密性房间的软质地面
软质地面
公共区
私密性房间
硬质地面
软质地面
公共区的硬质地面
入口　脱鞋、穿鞋的门槛区

৪০৫৩

在硬质地面上，可采用和铺砌室外小径和露台的相同地面材料——手工焙制砖和面砖——**软质面砖和软质砖**（248）。在私密房间的软质地面上，采用富有装饰性的、色泽鲜明的材料和织物——**装饰**（249）和**暖色**（250）……

To emphasize the two zones,and to promote the taking off and on of shoes from one zone to the next,we suggest that there be a step up or a step down between the zones.This will help tremendously in keeping each zone "pure," and it is sure to help the activities in each zone.

Therefore:

Zone the house,or building,into two kinds of zones: public zones,and private or more intimate zones.Use hard materials like waxed,red polished concrete,tiles,or hardwood in the public zones.In the more intimate zone, use an underfloor of soft materials,like felt,cheap nylon carpet, or straw matting,and cover it with cloths,and pillows, and carpets,and tapestries.Make a clearly marked edge between the two—perhaps even a step—so that people can take their shoes off when they pass from the public to the intimte.

೮೦೦೮

On the hard floor,you can use the same floor as you use on outdoor paths and terraces—hand fired brick and tile—SOFT TILE AND BRICK(248).On the soft intimate floors,use materials and cloths that are rich in ornament and color—ORNAMENT(249),WARM COLORS (250)...

模式234　鱼鳞板墙*

　　……本模式修饰**墙体**（218）和**拱式屋顶**（220）。它说明上述两模式的外表特征。

∞⊂⊱

　　房屋外墙的主要功能是防风雨、避寒暑。如果建筑材料的搭接处能做成防渗接缝，上述功能才能实现。

234   LAPPED OUTSIDE WALLS*

...this pattern finishes the WALL MEMBRANES(218),and ROOF VAULTS(220).It defines the character of their outside surfaes.

❧❧❧

The main function of a building's outside wall is to keep weather out.It can only do this if the materials are joined in such a way that they cooperate to make impervious joints.

At the same time,the wall must be easy to maintain;and give the people outside some chance of relating to it.

None of these functions can be very well managed by great sheets of impervious material.These sheets,always in the same plane,have tremendous problems at the joints.They require highly complex,sophisticated gaskets and seals,and,in the end,it is these seals and joints which fail.

Consider a variety of natural organisms:trees,fish,animals. Broadly speaking,their outside coats are rough,and made of large numbers of similar but not identical elements.And these elements are placed so that they often overlap:the scales of a fish,the fur of an animal,the crinkling of natural skin,the bark of a tree.All these coats are made to be impervious and easy to repair.

In simple technologies,buildings follow suit.Lapped boards, shingles,hung tiles,thatch,are all examples.Even stone and brick though in one plane,are still in a sense lapped internally to prevent cracks which run all the way through.And all of these walls are made of many small elements,so that individual pieces can be replaced as they are damaged or wear out.

同时，墙必须是容易维护的；并必须给室外的人与墙发生联系的某种机遇。

如果采用大块的防渗板材作墙，上述功能全部实现不了。这些板材总是在同一平面上，所以在接缝处将会存在很大问题。这些接缝需要高度复杂的衬垫和密封。而最终出毛病的还是这些密封和接缝处。

请考虑一下自然界的各种生物：树、鱼、飞禽走兽。一般说来，它们的外表是粗糙的，由大量相似的但并非同一的素材构成。它们的排列往往是重叠的：如鱼的鳞片、飞禽走兽的软毛、天然的皮肤的皱纹和树皮等。所有这些外皮都是防渗透的，并且容易再生。

房屋也应采用简单工艺手段仿效自然，重叠的木板、墙面板、挂瓦和茅草都是例子。甚至石头和砖，即使处于同一平面，在某种意义上，它们仍然是内部重叠的，以便防止在任何方向上发生裂缝。所有这些墙都由许多小型建材筑成，所以单独的小块建材一旦遭到破坏或磨损时，均可一一更换。

务必记住，当你为了避风雨、防寒暑选择某种外墙饰面时，选用的材料应当是容易重叠的、便于局部修理的。因此，外墙就可以不定期、一部分一部分地得到维修。当然，你无论选择何种饰面材料，都应使外墙表面能吸引你去抚摸它和倚傍它。

在我们浇筑轻质混凝土结构的施工过程中，已经采用重叠的木板作为轻质混凝土填料的外模板。如果人们能获得或买得起也可采用其他材料作为外墙覆面材料。石板、波纹铁皮和瓷砖都是极好的护墙覆面材料，它们都可以作为混凝土墙的外模板。还可以设想（虽然我们对此还没有证据），科学家或许能创出一种有向材料，它内部的晶体结构或纤维结构基本上是"重叠的"，因为所有的接缝线均呈

Bear in mind then,as you choose an exterior wall finish, that it should be a material which can be easily lapped against the weather,which is made of elements that are easy to repair locally,and which therefore can be maintained piecemeal,indefinitely.And of course,whatever you choose, make it a surface which invites you to touch it and lean up against it.

In making our filled lightweight concrete structures,we have used lapped boards as the exterior formwork for the lightweight concrete fill.And it is,of course,possible to use many other kinds of external cladding if they are available and if one can afford them.Slate,corrugated iron,ceramic tiles will produce excellent shingled wall claddings,and can all be placed in such a way as to provide exterior formwork for the pouring of a wall.It is also conceivable (though we have no evidence for it),that scientists might be able to create an oriented material whose internal crystal or fiber structure is in effect "lapped," because all the split lines run diagonally outward and downward.

Therefore:

Build up the exterior wall surface with materials that are lapped against the weather:either"internally lapped," like exterior plaster,or more literally lapped,like shingles and boards and tiles.In either case,choose a material that is easy to repair in little patches, inexpensively,so that little by little, the wall can be maintained in good condition indefnitely.

对角向外和向下。

鱼翅状　　非鱼翅状

一种想象的叠加材料的内部结构
The internal structure of an imaginary lapped material

因此：

以重叠材料的方式来建造外墙表面，以便抵风雨、御
寒暑；或采用"内部重叠的"方式，如外部抹灰浆，或真
正的重叠方式，如木瓦板、板材和瓷砖。在任何一种情况下，
选择某种可以小块修补的廉价材料，以便使墙能够不定期
地在良好的条件下一部分一部分地得到维护保养。

墙体的重叠材料

容易修理

模式235 有柔和感的墙内表面*

......本模式修饰**墙体**（218）的内表面和**楼面天花拱结构**（219）的下部表面。如有可能，墙体内表面采用质感柔和的材料，则墙一开始就将具有合理的特性。

❧✦❧

墙太硬或太冷或太实心都会使人有不快的触感；这种墙不可能形成装饰，并会产生沉闷的回声。

质感柔软的白色石膏灰泥是一种极佳的材料。它是暖色的（甚至是白的），触感温暖，质地松软，可钉大小钉子和钩子，易于修理，且吸音能力相当高，室内音响圆润。

235   SOFT INSIDE WALLS*

...and this pattern finishes the inner surface of the WALL MEMBRANES(218),and the under surface of FLOOR-CEILING VAULTS(219).If it is possible to use a soft material for the inner sheet of the wall membrane,then the wall will have the right character built in from the beinning.

ଽଽ୦ଔଔ

A wall which is too hard or too cold or too solid is unpleasant to touch;it makes decoration impossible,and creates hollow echoes.

A very good material is soft white gypsum plaster.It is warm in color(even though white),warm to the touch,soft enough to take tacks and nails and hooks,easy to repair,and makes a mellow sound,because its sound absorption capacity is reasonably high.

However,cement plaster,though only slightly different— and even confused with gypsum plaster—is opposite in all of these respects.It is too hard to nail into comfortably;it is cold and hard and rough to the touch;it has very low absorption acoustically—that is,very high reflectance—which creates a harsh,hollow sound;and it is relatively hard to repair,because once a crack forms in it,it is hard to make a repair that is homogeneous with the original.

In general,we have found that modern construction has gone more and more toward materials for inside walls that are hard and smooth.This is partly an effort to make buildings clean

可是，水泥灰泥，即使只有稍许不同——甚至和石膏灰泥极易混淆——其实在各方面都和石膏灰泥相对立。水泥灰泥硬得无法钉钉子，触感寒冷，坚硬而又粗糙；它的吸音能力较差——有很高的声音反射率——这就造成一种刺耳的沉闷回声；而且它难于修理，一旦出现裂缝，就很难修补成和原来一样的匀质。

一般说来，我们已经发现现代构造施工中越来越倾向于采用坚硬平滑的墙内层材料。这种情况部分源于人们努力使房屋保持干净和避免人为磨损所致，但部分也由于目前所采用的各种材料都是机制的——每一构件都是完整的、一模一样的。

由这些完整、坚硬、平滑的表面所构成的许多房屋，总体说来，我们和它们不发生任何关系。我们宁可远远离开它们，不仅因为它们使我们在心理上感到奇怪，而且事实上我们一旦倚傍上它们，就会感觉到身体不舒服：它们没有弹性，它们对我们没有反应。

这一问题的解决办法如下：

1. 石膏灰泥和水泥灰泥相对立，焙制的软质面砖和烧制的硬质面砖相对立。如果材料是多孔的和容重低的，通常它们就会更具有柔和感和温暖感。

2. 采用带颗粒状的、具有天然质感的材料，将它们做成小件，或以同样的小件重复使用。用木材装饰的墙面所具有的品质就是木材本身的纹理质感，墙面板会在较大的面积上重现这些纹理。如采用手工修饰墙面时，灰泥就具有这种特征。首先，墙面有灰泥的颗粒状特点；其次，有抹灰泥时人手操作所造成的较明显的质感纹理。

本模式最美丽的范例之一就是印地安人村舍中所采用的墙面。其内墙面有一层用手涂抹的牛粪和泥浆的混合物，干燥后就成为一种美丽且有柔和感的饰面，而涂抹者的手的五指形状在全部的墙上被逐一显示出来。

and impervious to human wear.But it is also because the kinds of materials used today are machine made—each piece perfect and exactly the Same.

Buildings made of these flawless,hard and smooth surfaces leave us totally unrelated to them.We tend to stay away from them not only because they are psychologically strange,but because in fact they are physically uncomfortable to lean against;they have no give;they don't respond to us.

The solution to the problem lies in the following:

1.Gypsum plaster as opposed to cement plaster.Soft baked tiles as opposed to hard fired ones.When materials are porous and low in density they are generally softer and warmer to the touch.

2.Use materials which are granular and have natural texture,and which can be used in small pieces,or in such a way that there is repetition of the same small element.Walls finished in wood have the quality—the wood itself has texture;boards repeat it at a larger scale.Plaster has this character when it is hand finished.First there is the granular quality of the plaster and then the larger texture created by the motion of the human hand.

One of the most beautiful versions of this pattern is the one used in Indian village houses.The walls are plastered,by hand,with a mixture of cow dung and mud,which dries to a beautiful soft finish and shows the five fingers of the plasterer's hand all over the walls.

印第安人村舍的内墙面上有一层掺牛粪的灰泥
Cow dung plaster in an Indian village house

因此:

应使墙的每一内表面都具有温暖的触感、柔和感和某种轻微的"弹性",并可钉入大小钉子。软质灰泥是极佳的内墙面材料;纺织品帷幔、藤编工艺品以及其他编织品也都具有这种质感柔和的特性。如果你的预算可以承受,木材是内墙面的好材料。

触感柔和

对钉入的钉子有足够的"弹性"

ഇൠ

在我们的施工体系中,我们发现,在**墙体**(218)和**楼面天花拱结构**(219)的内表面涂上一层薄薄的灰膏灰泥是值得的。在各个地方,饰面的灰泥遇上柱、梁和门窗框时,都要用半英寸宽的木质饰边盖住接缝处——**半英寸宽的压缝条**(240)……

Therefore:

Make every inside surface warm to the touch,soft enough to take small nails and tacks,and with a certain slight "give"to the touch.Soft plaster is very good;textile hangings,canework, weavings,also have this character.And wood is fine,where you can afford it.

கூ௸

In our own building system,we find it is worth putting on a light skim coat of plaster over the inner surfaces of the WALL MEMBRANE(218) and FLOOR-CEILING VAULTS(219). Wherever finish plaster meets columns,and beams,and doors and window frames,cover the joint with half—inch wooden trim—HALF-INCH TRIM (240)...

模式236 大敞口窗户*

236   WINDOWS WHICH OPEN WIDE*

...this pattern helps to complete WINDOW PLACE(180), WINDOWS OVERLOOKING LIFE(192),and NATURAL DOORS AND WIDOWS(221).

৪৩৫

Many buildings nowadays have no opening windows at all; and many of the opening windows that people do build, don't do the job that opening windows ought to do.

It is becoming the rule in modern design to seal up windows and create "perfect" indoor climates with mechanical air conditioning systems.This is crazy.

A window is your connection to the outside.It is a source of fresh air;a simple way of changing the temperature, quickly,when the room gets too hot or too cold;a place to hang out and smell the air and trees and flowers and the weather;and a hole through which people can talk to each other.

What is the best kind of window?

Double-hung windows cannot be fully opened—only half of the total window area can ever be opened at once. And they often get stuck—sometimes because they have been painted,sometimes because their concealed operating system of cords,counter-weights,and pulleys gets broken;it becomes such an effort to open them that no one bothers.

Sliding windows have much of the same problem— only part of the window area can be open,since one panel goes behind another;and they often get stuck too.

……本模式有助于完善**窗前空间**（180）、**俯视外界生活的窗**（192）和**借景的门窗**（221）。

∞✧

现在许多住宅根本没有大敞口窗户；而人们所造的许多大敞口窗户又不得其所。

密封窗户并利用机械的空调系统来创造出"完美无缺的"室内气候——这正在成为现代设计的一条规则。但这是不切实际的做法。

窗户是居者同外界联系的媒介。它是新鲜空气的源泉，当室内太热或太冷时，它是迅速改变室内温度的简单手段。凭窗处是人们晾晒东西、观察天气、呼吸新鲜空气、观赏树木和花卉的地方，也是人们能够相互谈话的一个出口。

什么样的窗户最佳呢？

上下推拉窗是不能完全打开的——只有窗户总面积的一半能够立即被打开。而且，它们往往相互卡住无法移动——有时因为它们已被刷上的油漆粘在一起；有时因为它们的绳索、平衡重锤和滑轮的隐蔽操作系统损坏了，要打开上下推拉窗真要花一番功夫，结果，就无人敢问津了。

横向推拉窗也有同样的弊病——只有一部分窗户的面积可以打开，因为一扇窗在另一扇的后面；而且它们也会常常卡住动不了。

而平开窗易开易关。开窗范围最大，从而能够最大限度地调节空气和温度，并且它的窗口敞得足以让里面居者的头和肩膀可探出窗外，它也是供人爬进爬出的最简易的窗户。

The side hung casement is easy to open and close.It gives the greatest range of openings,and so creates the greatest degree of control over air and temperature;and it makes an opening which is large enough to put your head and shoulders through. It is the easiest window to climb in and out of too.

The old time French windows are a stunning example of this pattern.They are narrow,full length upstairs windows,which swing out onto a tiny balcony,large enough only to contain the open windows.When you open them you fill the frame,and can stand drinking in the air:they put you intensely close to the outside—yet in a perfectly urban sense,as much in Paris or Madrid as in the open countryside.

Therefore:

Decide which of the windows will be opening windows. Pick those which are easy to get to,and choose the ones which open onto flowers you want to smell,paths where you might want to talk,and natural breezes.Then put in side-hung casements that open outward.Here and there,go all the way and build full Frnch windows.

❧❧❧

Complete the subframe of the casement with SMALL PANES (239)...

往昔法国的落地窗是本模式的一个出色例子。这是一种狭窄的楼上的落地窗，窗向外开到小眺台。眺台不大，只能容纳敞开的落地窗。当你打开落地窗时，你就置身于窗框之内，并能站在窗外眺台上饮酒作乐：落地窗会使你明显地感到与大自然更加亲近——而就理想城市这层意义上来说，人在巴黎或马德里却宛如在辽阔的乡村。

因此：

决定哪些窗户是大敞口的。选择人们乐于接近的窗户并选好窗户的位置。让窗户面向鲜花敞开，你就可以闻到扑鼻的芬芳；让窗户向小路曲径敞开，你就可以和行人攀谈；且时有微风拂面，沁人心脾。继而，安装向外开的平开窗。无论何处，都要尽量做通长的落地窗。

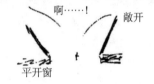

用**小窗格**（239）完善平开窗的窗框……

模式237　镶玻璃板门

　　……本模式完善由**靠近墙角的房门**（196）和**低门道**（224）所规定的门。它也有助于修饰**明暗交织**（135）和**内窗**（194），因为它要求门上镶上玻璃，因而使室内一些较暗的地方能射进天然光。

ഔരു

　　不透光的门在庞大的屋宇或宫殿内才具有意义，因为在那里每个房间本身就大得像一个世界似的；但在含有许多小房间的小建筑内，不透光的门用处甚微。

　　我们所需要的一种门必须兼备两种功能：既能使你看清门外的事物，又能隔音，你能透过门看清东西，但听不到声音。

　　在某些时期内，镶玻璃板门，早已成为一种传统——这种门美丽透光，扩大了联系意识，使室内生活同样具有与室外联系的感觉，但仍能满足人们必不可少的私密性条件。镶玻璃板门会使房间入口处增添美感，也会使迎接宾客的过程更加多彩多姿，因为它能使宾主双方早作相互施礼的准备。它也允许不同程度的私密性：你可以敞开大门，或为了隔音可以紧闭大门，但你仍可以和外界保持视觉联系；或者可以在这种门的玻璃处挂上门帘，以便与外界断绝一切视听联系。但是，最重要的一点是：镶玻璃板门会使宅内的每个人都感到，他虽在自己的私人房间里仍是和外界有联系的，而不是隔绝的。

237   SOLID DOORS WITH GLASS

...this pattern finishes the doors defined by CORNER DOORS(196) and LOWDORWAY(224).It also helps to finish TAPESTRY OF LIGHT ANDDARK(135)and INTERIOR WINDOWS(194),since it requires glazing in the doors,and can help to create daylight in the darker parts of idoor places.

❧❦

An opaque door makes sense in a vast house or palace, where every room is large enough to be a world unto itself;but in a small building,with small rooms,the opaque door is only very rarely useful.

What is needed is a kind of door which gives some sense of visual connection together with the possibility of acoustic isolation:a door which you can see through but can't hear through.

Glazed doors have been traditional in certain periods— they are beautiful,and enlarge the sense of connection and make the life in the house one,but still leave people the possibility of privacy they need.A glazed door allows for a more graceful entrance into a room and for a more graceful reception by people in the room,because it allows both parties to get ready for each other.It also allows for different degrees of privacy:You can leave the door open,or you can shut it for acoustical privacy but maintain the visual connection;or you can curtain the window for visual and acoustic privacy.And, most important,it gives the feeling that everyone in the building is connected—not isolated in private rooms.

因此：

制作门时尽可能镶上玻璃，这样，至少你可以从门的上半部分向外观看。同时要把门造得足够厚实，使它能隔声，而且在关门时能起到一种令人舒服的"形式和实在转换程序"。

厚实的镶玻璃板门

"形式和实在的转换程序"

⊷⊶

用小块玻璃镶门——**小窗格**（239）；并把门造得比较厚实，就像**建造墙体**（218）一般……

Therefore:

As often as possible build doors with glazing in them,so that the upper half at least,allows you to see through them. At the same time,build the doors solid enough,so that they give acoustic isolation and make a comfortable"thunk"when tey are closed.

⊷⊶

Glaze the door with small panes of glass—SMALL PANES(239); and make the doors more solid,by building them like WALL MEMBRANES(218)...

模式238　过滤光线*

　　……即使窗户的位置选得非常合适，但刺目的强光仍然是一个问题——**借景的门窗**（221）。窗内和窗周围的光线会对室内的明暗效果造成巨大的差异。窗框的形状能够起一部分作用——**深窗洞**（223）——但还需有附加的措施与之配合。

238   FILTERED LIGHT*

...even if the windows are beautifully placed,glare can still be a problem—NATURAL DOORS AND WINDOWS(221). The softness of the light,in and around the window,makes an enormous difference to the room inside.The shape of the frames can do a part of it—DEEP REVEALS(223)—but it still needs aditional help.

❧❦

Light filtered through leaves,or tracery,is wonderful. But why?

We know that light filtering through a leafy tree is very pleasant—it lends excitement,cheerfulness,gaiety;and we know that areas of uniform lighting create dull,uninteresting spaces. But why?

1.The most obvious reason:direct light coming from a point source casts strong shadows,resulting in harsh images with strong contrasts.And people have an optical habit which makes this contrast worse:our eye automatically reinforces boundaries so that they read sharper than they are.For example,a color chart with strips of different colors set next to each other will appear as though there are dark lines between the strips.These contrasts and hard boundaries are unpleasant objects appear to have a hard character,and our eyes,unable to adjust to the contrast,cannot pick up the details.

穿过叶子或窗花格过滤的光线是奇妙的。这是为什么呢?

我们知道,经过枝繁叶茂的树木过滤的光线是令人惬意的——它使我们感到赏心悦目,而且我们也知道,均匀的光线会使室内空间显得沉闷而又索然无味。但是原因何在?

1. 最明显的理由是:来自点光源的直射光线会投下浓重的阴影,强烈的明暗对比而造成刺目的效果。而人们有一种使这种对比更恶化的视觉习惯:我们的眼睛会自动强化影子的边缘,结果影子的边缘更加清晰可辨。例如,一张具有不同颜色的条纹排列在一起的彩色图,在条纹间仿佛出现暗线条。这些明暗对比和硬线条是令人感到不愉快的——物体呈现出硬的特征,而我们的眼睛,由于不能适应这种对比,就无法辨认物体的细部了。

鉴于上述理由,我们自然会有一种愿望:利用灯罩或非直接照射来扩散光线,以便光线所投下的影子"比较柔和",即我们所感知的影子的边缘不那么鲜明,明暗对比减弱,阴影较少,而物体的细部较易看清。这也就是为什么摄影者在拍摄物体时使用反射光而不使用直射光;他们这样才能拍出物体的细部,不然的话,物体的细部就会在阴影中消失。

2. 减少窗户周围的眩光。当明亮的光线透过窗户照进室内时,由于窗户四周墙壁上暗区的衬托,就会形成眩光——参阅**深窗洞**(223)。特别在窗户的边缘,如光线经过过滤而射入室内的光线较少,就会减弱眩光。

3. 第三个理由是纯粹的臆测。这种臆测可能是如此简单,即一束光造成的图案般的影子在某物上雀跃移动就会使我们感到兴奋愉快,从生物学角度而言,就是光对我们造成刺激。某些影片摄影师宣称,光对人的视网膜的作用自然会使人的身体受到刺激。

For all these reasons,we have a natural desire to diffuse light with lamp shades or indirect lighting,so that the images created by the light will be "softer," that is,that the boundaries perceived are not sharp,there is less contrast,fewer shadows,and the details are easier to see.This is also why photographers use reflected light instead of direct light when photographing objects;they pick up details which otherwise would be lost in shadow.

2.The second reason:to reduce the glare around the window.When there is bright light coming in through the window,it creates glare against the darkness of the wall around the window—see DEEP REVEALS(223).Filtering the light especially at the edges of the window cuts down the glare by letting in less light.

3.A third reason which is pure conjecture:it may simply be that an object which has small scale patterns of light dancing on it is sensually pleasing,and stimulates us biologically.Some filmmakers claim the play of light upon the retina is naturally sensuous,all by itself.

To create filtered light,partially cover those windows which get direct sunlight,with vines and lattices.Leaves are special because they move.And the edge of the window can have fine tracery—that is,the edge of the glass itself,not the frame,so that the light coming in is gradually stronger from the edge to the center of the window;the tracery is best toward the top of the window where the light is strongest.Many old windows combine these ideas.

为了形成过滤光线，可以利用藤蔓植物和窗格部分地阻挡阳光直接射入窗内。窗前藤蔓的绿叶婆娑，婀娜多姿，独具特色。窗户的边缘有十分精致雅观的窗花格——即玻璃的边缘本身，而不是窗框，结果，射入的光线从窗户边缘到中心渐渐变强；窗花格最好靠近光线最强烈的窗户上部。许多旧式窗户的设置都包含上述理念。

因此：

窗户的边缘或屋顶挑檐的边缘和天空相接处呈现的轮廓线形成一幅丰富多彩、错综复杂的明暗交织图，从而使光线散射，变得柔和。

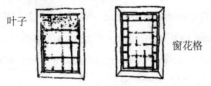

叶子　　　　　　　　　　　　　　窗花格

৪০০৪

你能轻而易举地营造出过滤光线，办法是栽种攀援植物，可以有意让它们在窗外的周围攀爬——**攀援植物**（246）。如果什么植物也没有，你可以非常出色地利用简易的帆布凉篷来形成过滤光线——**帆布顶篷**（244），也可利用色彩——**暖色**（250）。你还可以在光线强烈的窗户上部采用更小、更精致雅观的窗格来过滤光线——**小窗格**（239）……

Therefore:

Where the edge of a window or the overhanging eave of a roof is silhouetted against the sky,make a rich,detailed taspestry of light and dark,to break up the light and soften it.

ಬಿಂದ

You can do this,most easily,with climbing plants trained to climb around the outside of the window—CLIMBING PLANTS(246).If there are no plants,you can also do it beautifully with simple canvas awnings—CANVAS ROOFS(244), perhaps colored—WARM COLORS (250).You can also help to filter light by making the panes smaller,more delicate, and more elaborate high in the window where the light is strong—SMALL PANES(239)...

模式239 小窗格**

239   SMALL PANES**

...this pattern gives the glazing for the windows in INTERIOR WIN-DOWS(194),NATURAL DOORS AND WINDOWS(221),WINDOWS WHICH OPEN WIDE(236),and SOLID DOORS WITH GLASS(237).In most cases,the glazing can be built as a continuation of the FRAMES AS THICKENED EDGES(225).

❧❧❧

When plate glass windows became possible,people thought that they would put us more directly in touch with nature. In fact,they do the opposite.

They alienate us from the view.The smaller the windows are,and the smaller the panes are,the more intensely windows help connect us with what is on the other side.

This is an important paradox.The clear plate window seems as though it ought to bring nature closer to us,just because it seems to be more like an opening,more like the air.But,in fact,our contact with the view,our contact with the things we see through windows is affected by the way the window frames them.When we consider a window as an eye through which to see a view,we must recognize that it is the extent to which the window frames the view,that increases the view,increases its intensity,increases its variety,even increases the number of views we seem to see—and it is because of this that windows which are broken into smaller windows,and windows which are filled with tiny panes,put us so intimately

……本模式说明**内窗**（194）、**借景的门窗**（221）、**大敞口窗户**（236）和**镶玻璃板门**（237）要如何镶玻璃。在大多数情况下，镶玻璃是**门窗边缘加厚**（225）工作的延续。

∞∞

　　人们普遍认为，如有建筑物可以应用大块的玻璃窗这种窗户将使我们更直接地和大自然接触。事实上，却恰恰相反。

　　平板玻璃窗使我们与外部的景物疏远了。窗户尺寸越小，窗格越小，窗户就越强烈地促使我们与窗外的东西保持联系。

　　这是一个重要的反证。透明的大块玻璃窗看起来似乎应当使大自然和我们更接近，因为它更像一个大敞口，更像空的。但是，事实上，我们同外界的景物接触，同我们透过窗户所见的东西接触，无不受到窗框的限制。当我们把窗户看作观看外界景物的眼睛时，我们必须承认，景色是被窗框所限定了，同时也必须承认，这样会强化景物，增强了景物的鲜明性和多样性，甚至增加我们似乎看到的景色的数量——正因为如此，那些分隔成较小的窗户和窗格分小的窗户会使我们在和窗外的景物接触时感到格外亲切。这是因为它们创造出多得多的窗框：正是窗框倍增才使框景千变万化。

　　托马斯·马库斯曾广泛地研究了各种窗户，现已得出相同的结论：被分隔的窗户会造成更加有趣的景观（"The Function of Windows—A Reappraisal," *Building Science*, Vol.2, 1967, pp.101~104）。他指出，透过狭小窗户从室内的不同位置观看，会出现不同景物，而透过大窗户或水平窗户观看，景物总是一模一样的。

in touch with what is on the other side.It is because they create far more frames:and it is the multitude of frames which makes the view.

Thomas Markus,who has studied windows extensively, has arrived at the same conclusion:windows which are broken up make for more interesting views.("The Function of Windows-A Reappraisal," *Building Science*, Vol.2,1967,pp.101-104.).He points out that small and narrow windows afford different views from different positions in the room,while the view tends to be the same through large windows or horizontal ones.

We believe that the same thing,almost exactly,happens within the window frame itself.The following picture shows a simple landscape,broken up as it might be by six panes.Instead of one view,we see six views.The view becomes alive because the small panes make it so.

Another argument for small panes:Modern architecture and building have deliberately tried to make windows less like windows and more as though there was nothing between you and the outdoors.Yet this entirely contradicts the nature of windows.It is the function of windows to offer a view and provide a relationship to the outside,true.But this does not mean that they should not at the same time,like the walls and roof,give you a sense of protection and shelter from the outside. It is uncomfortable to feel that there is nothing between you and the outside,when in fact you are *inside* a building.It is the nature of windows to give you a relationship to the outside *and* at the same time give a sense of enclosure.

Not only that.Big areas of clear glass are sometimes even dangerous.People walk into plate glass windows,because they

我们相信，窗框也会有同样的情况，下图表示一幅简单的景色。如该图所示，这幅画被 6 个小窗格所分隔。我们看到的是 6 幅风景画，而不是一幅。因为小窗格所具有这种神奇的功效，使得该景物栩栩如生。

六幅风景画
Six views

　　采用小窗格的另一论据是：现代建筑总是故意使居者和室外之间的窗户与其说像窗户，倒不如说仿佛一无所有。可是，这完全和窗户的性质相矛盾了。确实如此，窗户的功能是向人们提供外界的景物，并使之与外界发生关系。但这并不意味着，窗户不应该同时如墙和屋顶那样，给人一种庇护感，使你免遭风雨寒暑的袭击。当你实际上在房屋内，如果你和外界之间一无所有，你会感到不舒服的。窗户的本质就是使你和外界发生关系，并且同时使你有一种闭合感。

加州曼多细诺角的小窗格
Small panes in Mendocino

look like air.By comparison,windows with small panes give a clear functional message—the frames of the panes definitely tell you that something is there separating you from the outside. And they help to create FILTERED LIGHT(238).

Therefore:

Divide each window into small panes.These panes can be very small indeed,and should hardly ever be more than a foot square.To get the exact size of the panes,divide the width and height of the window by the number of panes. Then each window will have different sized panes according to its height and width.

හ⊃෴

In certain cases you may want to make the small panes even finer near the window edge,to filter the light around the upper edge of windows which stand out against the sky— FILTERED LIGHT(238).As for the muntins,they can be made from the same materials as trim—HALF-INCH TRIM(240)...

不仅如此，大尺寸的透明玻璃有时甚至是危险的。人们会朝大片玻璃窗里走，因为这种窗宛如空无一物一样透明。比较而言，小窗格窗户会提供清晰的功能信息——小窗格窗户的窗框肯定会告诉你：有某种东西把你和外界分隔开来。而且，小窗格有助于形成**过滤光线**（238）。

因此：

应把每一窗户划分成许多小窗格。这些窗格可能做得很小，而且几乎不超过 1ft². 为了求得窗格的准确尺寸，可将窗户的宽度和高度除以窗格数。而后，每一窗户根据自身的高度和宽度就会有大小不等的窗格了。

小窗格

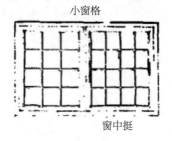

窗中挺

ೞೞೞ

在某些情况下，你兴许想使接近窗户边缘的小窗格更精致雅观，以便在和天空相连的窗户上部的边缘周围过滤光线——**过滤光线**（238）。至于窗中挺，则可用和压缝条相同的材料来制作——**半英寸宽的压缝条**（240）……

模式240　半英寸宽的压缝条**

……本模式修饰在**有柔和感的墙内表面**（235）、**鱼鳞板墙**（234）和与墙连成一体的各种地面、拱结构、框架、加劲杆、装饰之间的接缝：**箱形柱**（216）、**圈梁**（217）、**楼面天花拱结构**（219）、**门窗边缘加厚**（225）和**装饰**（249）。

❧❧❦❧

整体化的机械化施工不需要压缝条，因为它的精密程度很高。但它的精密度是以高昂的代价换取的：在施工计划中没有丝毫的灵活性。

240 HALF-INCH TRIM**

...and this pattern finishes the joints between SOFT INSIDE WALLS (235),or LAPPED OUTSIDE WALLS(234) and the various floors and vaults and frames and stiffeners and ornaments which are set into the walls:BOX COLUMNS(216),PERIMETER BEAMS(217),FLOOR-CEILING VAULTS(219),FRAMES AS THICKENED EDGES(225),and ORNAMENT (249).

❧❦

Totalitarian,machine buildings do not require trim because they are precise enough to do without.But they buy their precision at a dreadful price:by killing the possibility of freedom in the building plan.

A free and natural building cannot be conceived without the possibility of finishing it with trim,to cover up the minor variations which have arisen in the plan,and during its construction.

For example,when nailing a piece of gypsum board to a column—if the board is cut on site—it is essential that the cut can be inaccurate within a half-inch or so.If it has to be more accurate,there will be a great waste of material,and on-site cutting time and labor will increase,and,finally,the very possibility of adapting each part of the building to the exact subtleties of the plan and site will be in jeopardy.

一幢不受约束的、合规的房屋若无法利用压缝条来修饰以掩盖其施工过程和平面图中所产生的各种不重要的误差，那将是无法想象的。

　　例如，当把一块石膏板钉到柱上时——如果这块石膏板是在工地上被切割的——重要的是这种切割可能不精确，误差 1in 左右。如果切割需更精确，则将会大大浪费材料，还将增加在工地切割的时间和用工量。最后，房屋的各部分与平面图和工地现场的精确尺寸完全一致的可能性是微乎其微的。

　　正是在解决这一困难的过程中，建筑施工的现代化体系才应运而生。这种施工法的容许误差的确很小——1/8in甚至更小——结果就用不着压缝条来掩饰误差了。但是，构件的精密度只有靠对平面图的最严格的控制才能获得。但仅从施工单方面着想的观点就有损于建造者去盖一幢合理、有机、适应工地现场的房屋的能力。

　　正如我们所建议的，如果施工程序不太严格，允许较大误差——甚至可有半英寸或更大的误差，则使用压缝条来掩盖材料之间的接缝就变得十分重要了。确实，在这样一个误差值之内施工，压缝条作为一种修饰技巧不是无足轻重的装饰，而是施工的一个重要阶段。于是我们就会明白，压缝条往往和古老的建筑有密切联系，并被人们看成怀旧的象征。事实上，压缝条在使房屋造得合理的施工过程中起着重要作用。

　　最后，关于压缝条的实际尺寸值得再提一笔。最近 25 年中建成的房屋往往具有想象力丰富的优点，并有一种采用大压缝条取代小压缝条的倾向。如果按此原理，使用 2in 或 3in 厚的压缝条，就其功能和重量而言，似乎是合适的。而我们认为上述观点是错误的：因为太大或太厚的压缝条是不起什么作用的。这不是一个建筑风格问题。为了符合人们的心理要求，务必要使房屋的每一构件至少都有一些

It is in response to difficulties of this sort that modern system building has arisen.Here tolerances are very low indeed—1/8 inch and even lower-and there is no need for trim to cover up inaccuracies.However,the precision of the components can only be obtained by the most tyrannical control over the plan.This one aspect of construction has by itself destroyed the builder's capacity to make a building which is natural,organic,and adapted to the site.

If,as we suggest,the building procedure is looser and allows much larger tolerance-even mistakes on the order of half an inch or morethen the use of trim to cover the connection between materials becomes essential.Indeed,within this attitude to building,the trim is not a trivial decoration added as a finishing touch,but an essential phase of the construction.We see,then,that trim,so often associated with older buildings,and treated as an emblem of nostalgia,is in fact a vital part of the process of making buildings natural.

Finally,it is worth adding a note about the actual size of the trim pieces.Buildings built in the last 25 years often make a virtue out of boldness,and there is a tendency to use very large oversized pieces of trim instead of small pieces.Within the framework of this philosophy,it might seem right to use pieces of trim 2 or 3 inches thick for their effect and heaviness. We believe that this is wrong: Trim which is too large,or too thick,doesn't do its job.This is not a matter of style.There is a psychological reason for making sure that every component in the building has at least some pieces of trim which are of the order of half an inch or an inch thick,*and no more.*

压缝条，其厚度约为 0.5in 或 1in，而不能再厚了。

试比较下面两个压缝条的例子：左右两图相比，由于某种原因，我们感到右图的压缝条更精致、更亲切、更适合我们的感情。

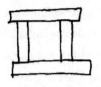

厚实的压缝条
Chunky trim...

精致的压缝条
...fine scale trim

对此作如下简要解释：我们自己的肉体和我们赖以进化的自然环境都包含着各种细部的连续系列，其范围包括从分子的精细结构直到（我们人体的）臂、腿和我们的自然环境的树干、树叉等。

我们从认识心理学的研究成果中知道，在这种系列（如果我们认为它是自然系列）中的任何一级的细部比例都可能不大于 1：5、1：7 或 1：10。我们还无法理解这种系列中存在细部比例的飞跃现象：1：20 或更大。对于我们的环境，甚至是人为的环境来说，正是这一事实说明有必要显示出细部的相似连续性。

大多数材料都具有某种天然纤维结构或晶状结构，它们的细部比例约为 1/20in。但是如果最小的建筑细部尺寸约为 2in 或 3in，那么这就会使这些细部和材料的细微结构之间造成飞跃的比例：1：40 或 1：60。

为了让我们能感知精心的建筑施工和材料的细小结构之间的联系，重要的是最小的建筑细部应为半英寸左右，以使其不大于材料的颗粒结构和纤维结构的细小尺寸的 10 倍左右。

Compare the following two examples of trim.For some reason the right-hand one,in which the trim is finer,is closer and better adapted to our feelings than the left-hand one.

The reason for this seems to be the following.Our own bodies and the natural surroundings in which we evolved contain a continuous hierarchy of details,ranging all the way from the molecular fine structure to gross features like arms and legs(in our own bodies)and trunks and branches(in our natural surroundings).

We know from results in cognitive psychology that any one step in this hierarchy can be no more than 1:5,1:7,or 1:10 if we are to perceive it as a natural hierarchy.We cannot understand a hierarchy in which there is a jump in scale of 1:20 or more.It is this fact which makes it necessary for our surroundings,even when manmade,to display a similar continuum of detail.

Most materials have some kind of natural fibrous or crystalline structure at the scale of about 1/20 inch.But if the smallest building detail dimensions are of the order of 2 or 3 inches,this leaves a jump of 1:40 or 1:60 between these details and the fine structure of the material.

In order to allow us to perceive a connection between the fine building construction and the fine structure of the materials,it is essential that the smallest building details be of the order of a half inch or so,so that it is no more than about 10 times the size of the granular and fibrous texture of the materials.

因此：

不论何处，两种材料相接，就应将一压缝条置于它们连接的缝隙处。选择压缝条时，为使最小的压缝条能够应用于每一构件中，始终保持约半英寸的宽度。压缝条可能就是木头、灰泥、琉璃砖等。

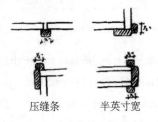

压缝条　　　半英寸宽

❧❧

在许多情况下，你都可以利用压缝条作为装饰——**装饰**（249）；而且压缝条有时也可以是彩色的：甚至少量的色彩也有助于室内的光线具有温暖感——**暖色**（250）……

建造户外细部，以便像对待室内空间那样来充分修饰户外空间；

241. 户外设座位置
242. 大门外的条凳
243. 可坐矮墙
244. 帆布顶篷
245. 高花台
246. 攀援植物
247. 留缝的石铺地
248. 软质面砖和软质砖

Therefore:

Wherever two materials meet,place a piece of trim over the edge of the connection.Choose the pieces of trim so that the smallest piece,in each component,is always of the order of 1/2 inch wide.The trim can be wood,plaster,terracotta...

⊃⃝⊂

In many cases,you may be able to use the trim to form the or-naments—ORNAMENT(249);and trims may occasionally be colored:even tiny amounts can help to make the light in a room warm—WARM COLORS(250)...

build outdoor details to finish the outdoors as fully as the indoor spaces;

241.SEAT SPOTS

242.FRONT DOOR BENCH

243.SITTING WALL

244.CANVAS ROOFS

245.RAISED FLOWERS

246.CLIMBING PLANTS

247.PAVING WITH CRACKS BETWEEN THE STONES

248.SOFT TILE AND BRICK

模式241　户外设座位置**

　　……假设建筑的主结构已经完成。为了将它进一步完善，你应在它周围建造花园和露台的细部。在某些情况下，你还要筑围墙、修花台和建坐位，至少使户外细部初具轮廓。但是，关于这些户外细部的最后决定最好在建筑真正竣工之后作出——以求这些细部与你的建筑相适应，并有助于建筑和其周围环境保持联系——**小路的形状**（121）、**袋形活动场地**（124）、**私家的沿街露台**（140）、**建筑物边**

241 SEAT SPOTS**

...assume that the main structure of the building is complete.To make it perfectly complete you need to build in the details of the gardens and the terraces around the building. In some cases,you will probably have laid out the walls and flowers and seats,at least in rough outline;but it is usually best to make the fihal decisions about them after the building is really there—so that you can make them fit the building and help to tie it into its surroundings—PATH SHAPE (121), ACTIVITY POCKETS(124),PRIVATE TERRACE ON THE STREET (140),BUILDING EDGE(160),SUNNY PLACE(161), OUTDOOR ROOM (163),CONNECTION TO THE EARTH(168), TRELLISED WALK(174),GARDEN SEAT(176), etc.First,the outdoor seats,public and private.

୫୦୦ଓ

Where outdoor seats are set down without regard for view and climate,they will almost certainly be useless.

We made random spot checks on selected benches in Berkeley,California,and recorded these facts about each bench:Was it occupied or empty? Did it give a view of current activity or not? Was it in the sun or not? What was the current wind velocity? Three of the eleven benches were occupied;eight were empty.

At the moment of observation,all three occupied benches looked onto activity,were in the sun,and had a wind velocity of less than 1.5 feet per second.At the moment of observation,none of the eight erupty benches had all three of these characteristics. Three of them had shelter and activity but no sun;three of them had activity

缘（160）、有阳光的地方（161）、有围合的户外小空间（163）、与大地紧密相连（168）、棚下小径（174）、园中坐椅（176）等。首先我们叙述公用的和私人的户外设座位置。

ಬಂಲ

凡不考虑景观和气候的户外设座位置，几乎可以肯定，毫无用处。

我们在加利福尼亚的伯克利在户外设座中任意选择了一些长椅进行了抽样调查，并对每一长椅的使用率作了如实的记录：它一直有人坐着还是空着？它是否能让人看到行人当时的活动情景？它是否有阳光？经常的风速有多大？11 条长椅中 3 条是有人坐的；8 条是空着的。

我们在调查时发现，那 3 条一直有人坐的长椅位置好，处于人来人往的热闹地方；那里有阳光；那里的风速小，每秒不到 1.5ft。而那 8 条一直空着的长椅都缺少这 3 个特点。8 条长椅中的 3 条有遮阴，也颇热闹，但无阳光；还有 3 条长椅处于闹区，但无阳光，而且风速大于每秒 3ft；最后的 2 条长椅有太阳和遮阴，但所在的地方太冷清。

第二项调查是对下午 3 时坐在联合广场上的老人人数进行比较：在阳光灿烂的一天有 65 名，但是在满天乌云的一天却只有 21 名，即使这两天的气温是一样的。

当然这很明显——但下面一点是关键——当你在设计图中用点标出户外设座位置、能坐的矮墙、能坐的台阶、园中坐椅时，宜选择具有如下特征的地方：

1. 长椅直接面向行人的活动。

2. 冬天长椅朝南向阳。

3. 在冬天寒风吹得到的地方筑矮墙挡风。

4. 在酷暑盛夏季节——日当中天之时，要有遮阳设施，长椅应面向清风徐徐吹来的方向。

but no sun,and wind greater than 3 feet per second;two of them had sun and shelter but no activity.

A second series of observations compared the numbers of old people sitting in Union Square at 3:00 P.M.on a sunny day with the number at 3:00 P.M.on a cloudy day:65 people on the sunny day and 21 on the cloudy day,even though the air temperature was the same on both days.

It's obvious,of course—but the point is this—when you are going to mark in spots in your project for the location of outdoor seats,sitting walls,stair seats,garden seats,look for places with these characteristics:

1.Benches facing directly onto pedestrian activity.

2.Benches open to the south for sun exposure during winter months.

3.A wall on those sides where the winter wind comes down.

4.In hot climates—cover to give sun protection during the midday hours of summer months,and the bench open to the direction of the summer breeze.

Therefore:

Choosing good spots for outdoor seats is far more important than building fancy benches.Indeed,if the spot is right,the most simple kind of seat is perfect.

In cool climates,choose them to face the sun,and to be protected from the wind;in hot climates,put them in shade and open to summer breezes.In both cases,place them to face activities.

ଇଷଓଓ

If these seats can be made continuous with stairs or building entrances or low walls or ballustrades,so much the better—STAIR SEATS(125),FRONT DOOR BENCH (242),SITTING WALL(243)...

新英格兰的长椅
New England benches

因此：

为户外设座选择合适的地方远比制作精致的长椅重要。确实，如果地方选得好，就连最简单的坐位也是完美的。

冷天，长椅要朝南向阳，并要有防风的屏障；热天，长椅要位于荫凉处，对着微风吹来的方向。在上述两种情况下，长椅都应面对着人们活动的方向。

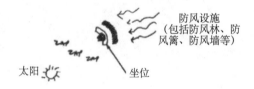

如果这些坐位能和楼梯、建筑的入口、矮墙或栏杆建成一体，那就更好了——**能坐的台阶**（125）、**大门外的条凳**（242）和**可坐矮墙**（243）……

模式242 大门外的条凳*

......在几个较大的模式中起着作用的**户外设座位置**
（241），在建筑物边缘的周围形成一种使人留连忘返的气
氛——**拱廊**（119）、**建筑物边缘**（160）、**有阳光的地方**
（161）、**与大地紧密相连**（168）；入口附近是十分显眼而
又重要的地方——**入口空间**（130）。本模式说明一种特殊
的**户外设座位置**（241）：条凳有助于形成入口空间及入口
周围的建筑边缘。大门外的条凳在任何情况下都是重要的；
但或许**老人住所**（155）门口的条凳将是最重要的。

242 FRONT DOOR BENCH*

...SEAT SPOTS(241),acting within several larger patterns,creates an atmosphere around the edge of the building which invites lingering—ARCADES(119),BUILDING EDGE(160),SUNNY PLACE(161),CONNECTION TO THE EARTH(168);it is most marked and most important near the entrance—ENTRANCE ROOM(130).This pattern defines a special SEAT SPOT(241):a bench which helps to form the entrance room and the building edge around the entrance.It is always important;but perhaps most important of all,at the door of an OLD AGE COTTAGE(155).

☙◊❧

People like to watch the street.

But they do not always want a great deal of involvement with the street.The process of hanging out requires a continuum of degrees of involvement with the street,ranging all the way from the most private kind to the most public kind.A young girl watching the street may want to be able to withdraw the moment anyone looks at her too intently.At other times people may want to be watching the street,near enough to it to talk to someone who comes past,yet still protected enough so that they can withdraw into their own domain at a moment's notice.

The most public kind of involvement with the street is sitting out.Many people,especially older people,pull chairs out to the front door or lean against the front of their houses,either while they are working at something or just for the pleasure

人们喜欢注视街道。

　　但是，他们并不总是希望大量卷入街道的活动。探身户外的过程就是连续卷入街道活动的过程，包括从最隐蔽的活动到最公开的活动。一位年轻的姑娘在注视街道时，也许希望能躲开别人向她投去的含情的目光。在另外一些场合，当人们注视街道时，兴许想尽量接近街道，以便同某一路人攀谈几句。但是，他们又是有所防范的，一旦有人盯住他们，就可立即退回到自己的领域。

　　最公开的卷入街道活动就是坐在户外。许多人，尤其是上了年纪的人，总是把椅子搬到门外坐下或背靠着自己房屋的正面墙，不是在于正在干什么活，而是兴高采烈地在注视着街道的生活。但是他们不愿太公开露面，所以他们都是自备板凳或坐位，即使在公共场所也是如此。把条凳放在他们的私人土地的边缘是最好的办法——但是，这样摆放条凳的结果，这样形成的个人空间就和法定的公共空间有一定重合。

秘鲁的大门外的条凳
Front door benches in Peru.

of watching street life.But since there is some reluctance to be too public,this activity requires a bench or seat which is clearly private,even though in the public world.It is best of all when the bench is placed so that people are sitting on the edge of their world on private land—yet so placed that the personal space it creates overlaps with land that is legally public.

Therefore:

Build a special bench outside the front door where people from inside can sit comfortably for hours on end and watch the world go by.Place the bench to define a half-private domain in front of the house.A low wall,planting,a tree,can help to create the same domain.

ക്കരു

The bench may help to make the entrance visible— MAIN ENTRANCE(110);it can be part of a wall—SITTING WALL(243),with flowers in the sunshine next to it—RAISED FLOWERS(245).Place it with care,according to the rules given in SEAT SPOTS(241)...

因此：

在**大门外建造一种特殊的条凳，人们从室内出来就能够在条凳上舒适地连续坐上几小时，并能注视来去匆匆的行人。建造大门外的条凳是为了规定住宅正面的半私密性区域。矮墙、花草树木都有助于形成半私密性区域。**

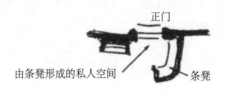

正门

由条凳形成的私人空间

条凳

❧❧❧

条凳能使入口更加清晰可见——**主入口**（110）；它可能是墙的一部分——**可坐矮墙**（243），在它附近有阳光和鲜花——**高花台**（245）。根据**户外设座位置**（241）的规则，建造大门外的条凳须谨慎行事……

模式243　可坐矮墙**

……如果一切都顺利，那么户外空间大多由正空间构成——**户外正空间**（106）；你已以某种方式在花园和街道之间、露台和花园之间、有围合的户外小空间和露台之间、游戏区和花园之间都一一标出边界——**绿茵街道**（51）、**步行街**（100）、**半隐蔽花园**（111）、**外部空间的层次**（114）、**小路的形状**（121）、**袋形活动场地**（124）、**私人的沿街露台**（140）、**有围合的户外小空间**（163）、**向街道的开敞**（165）、**回廊**（166）和**花园野趣**（172）。

你可利用本模式使这些天然边界具有各自的特点，并修筑矮到能坐人、高到能标界的墙。

243   SITTING WALL**

...if all is well,the outdoor areas are largely made up of positive spaces—POSITIVE OUTDOOR SPACES(106);in some fashion you have marked boundaries between gardens and streets,between terraces and gardens,between outdoor rooms and terraces,between play areas and gardens—GREEN STREETS(51),PEDESTRIAN STREET(100),HALF-HIDDEN GARDEN(111),HIERARCHY OF OPEN SPACE(114),PATH SHAPE(121),ACTIVITY POCKETS(124),PRIVATE TERRACE ON THE STREET(140),OUTDOOR ROOM(163),OPENING TO THE STREET(165),GALLERY SURROUND(166),GARDEN GROWING WILD(172).With this pattern,you can help these natural boundaries take on their proper character,by building walls,just low enough to sit on,and high enough to mark the boundaries.

If you have also marked the places where it makes sense to build seats—SEAT SPOTS(241),FRONT DOOR BENCH(242)—you can kill two birds with one stone by using the walls as seats which help enclose the outdoor space wherever its positive character is weakest.

❧❧❧

In many places walls and fences between outdoor spaces are too high;but no boundary at all does injustice to the subtlety of the divisions between the spaces.

Consider,for example,a garden on a quiet street.At least somewhere along the edge between the two there is a need

如果你也已标出修筑坐位的理想地方——**户外设座位置（241）、大门外的条凳（242）**——你就可以利用作为坐位的墙一箭双雕，这些可坐矮墙有助于围合任何一处"正"特征最弱的户外空间。

∞⋘∞

**　　在许多地方，各户外空间之间的墙和篱笆总是太高；但根本没有边界对于细微地划分空间是不合理的。**

　　例如，考虑一条幽静的街道上的一座花园吧。至少在花园的某处，沿两个空间之间的边缘应有一条接合线。这条接合线把两个空间连接起来，但又没有破坏两个空间又是相互分离的这样一个事实。如果筑一高墙或篱笆，那么花园中的人就无法与街道联系；而街道上的人也无法与花园发生联系。但如果根本没有物障——那么街道和花园两者之间的分界就难以保持。迷路的狗就会随意进进出出闲逛；坐在花园里的人们甚至会感到不舒服，因为就好像是坐在大街上似的。

　　只有利用这样一种物障，即它既是起分隔作用的屏障，同时又是起连接作用的接合线，这一问题才能迎刃而解。

　　高度适于坐人的矮墙或栏杆是理想的。矮墙是形成隔离的物障。因为矮墙能吸引人们去坐——他们首先把腿放到一侧而坐上矮墙，然后再把腿搁在矮墙的顶部，接着又转过身把腿放到另一侧，或跨骑着矮墙——矮墙也具有接合线的功能，从而在两个空间之间形成一个正连接。

　　下面是几种矮墙的例子：一种是一侧有儿童的沙箱，另一侧有环行小路的矮墙；另一种是把住宅和公用小路连接起来的花园前面的矮墙；还有一种是能坐人的挡土墙，在它的一侧栽有花草树木，人们可以坐在芬芳的鲜花丛中享用午餐。

　　拉斯金对他所体验过的一种能坐的矮墙作了如下描述：

for a seam,a place which unites the two,but does so without breaking down the fact that they are separate places.If there is a high wall or a hedge,then the people in the garden have no way of being connected to the street;the people in the street have no way of being connected to the garden.But if there is no barrier at all—then the division between the two is hard to maintain.Stray dogs can wander in and out at will;it is even uncomfortable to sit in the garden,because it is essentially like sitting in the street.

The problem can only be solved by a kind of barrier which functions as a barrier which separates,and as a seam which joins,at the same time.

A low wall or balustrade,just at the right height for sitting,is perfect.It creates a barrier which separates.But because it invites people to sit on it—invites them to sit first with their legs on one side,then with their legs on top,then to swivel round still further to the other side,or to sit astride it—it also functions as a seam,which makes a positive connection between the two places.

Examples:A low wall with the children's sandbox on one side,circulation path on the other;low wall at the front of the garden,connecting the house to the public path;a sitting wall that is a retaining wall,with plants on one side,where people can sit close to the flowers and eat their lunch.

Ruskin describes a sitting wall he experienced:

Last summer I was lodging for a little while in a cottage in the country,and in front of my low window there were,first,some beds of daisies,then a row of gooseberry and currant bushes,and then a low wall about three feet above the ground,covered with stonecress. Outside,a corn-field,with its green ears glistening in the sun,and a field path through it,just past the garden gate.From my window I could

去年夏天我在乡村别墅做了短暂歇息。在我的矮窗前首先映入眼帘的是几个种有雏菊的花台，其次是一排醋栗树和红醋栗灌木错落相间，最后是一垛矮墙，高出地面约 3ft，上面长满了石水芹。矮墙外是一片玉米田，青青的玉米穗在阳光下闪闪发亮。一条田间小路恰好经花园的大门穿过玉米田。我从窗口向外眺望，可以看到一个个的村民走过那条小路，他们臂上挂着篮子到市场去或肩扛铁铲下田干活。当我想了解那里的社会人情时，就能俯身在矮墙上同任何人谈话；当我想学点科学时，就可沿矮墙的顶部研究植物的生态——矮墙上有 4 种石水芹独自生长着；当我想锻炼身体时，还可进行向前向后的跳墙运动。这是在基督教国度里才有的一种围墙。这既不是你把自己当作野兽才能入内的围墙，又不是在清晨你透过窗户能看见且不希望夜里有人被墙头上的尖物刺伤的那种围墙。(John Ruskin, *The Two Paths*, New York : Everyman's Library, 1907, p.203.)

　　因此：

将任何一种天然户外空间围合起来，用矮墙在户外空间之间形成次要边界，矮墙高约 16in，宽以能坐为限，至少 12in。

宽阔的顶部　　坐位高度

两用或多用边界

⊱⊰

　　修筑可坐矮墙要和天然坐位位置相一致，这样就不需要附加的条凳了——**户外设座位置**（241）；如果可能，则用砖或面砖修筑——**软质面砖和软质砖**（248）；如果矮墙隔离两个高度略有不同的区域，可将它筑成镂空栏杆——**装饰**（249）。将可坐矮墙修建在有阳光的地方，面积要大得足以在矮墙上或在矮墙前可栽种鲜花——**高花台**（245）……

see every peasant of the village who passed that way,with basket on arm for market,or spade on shoulder for field.When I was indined for society,I could lean over my wall,and talk to anybody;when I was inclined for science,I could botanize all along the top of my wall— there were four species of stone—cress alone growing on it;and when I was inclined for exercise,I could jump over my wall,backwards and forwards.That's the sort of fence to have in a Christian country;not a thing which you can't walk inside of without making yourself look like a wild beast,nor look at out of your window in the morning without expecting to see somebody impaled upon it in the night.(John Ruskin,*The Two Paths*,New York:Everyman's Library,1907,p.203.)

Therefore:

Surround any natural outdoor area,and make minor boundaries between outdoor areas with low walls,about 16 inches high,and wide enough to sit on,at least 12 inches wide.

∞CR

Place the walls to coincide with natural seat spots,so that extra benches are not necessary—SEAT SPOTS(241);make them of brick or tile,if possible—SOFT TILE AND BRICK(248);if they separate two areas of slightly different height,pierce them with holes to make them balustrades—ORNAMENT(249). Where they are in the sun,and can be large enough,plant flowers in them or against them—RAISED FLOWERS(245)...

模式244　帆布顶篷*

　　……在每幢建筑的周围有**屋顶花园**（118）、**拱廊**（119）、**私人的沿街露台**（140）、**有围合的户外小空间**（163）、**回廊**（166）、**棚下小径**（174）和**窗前空间**（180），甚至**小停车场**（103）。所有这些模式因使用帆布顶篷和遮阳篷都变得更加精致和美丽了。而遮阳篷总是有助于形成**过滤光线**（238）。

⁜

　　帐篷和帆布遮阳篷具有非常特殊的美。帆布具有柔软和舒适感，它和风、光线和阳光和谐协调。用某种帆布建成的住宅或任何其他建筑比只用普通硬质材料建成的建筑和全部构件的关系更为密切。

244   CANVAS ROOFS*

...around every building there are ROOF GARDENS(118), ARCADES (119),PRIVATE TERRACES ON THE STREET(140), OUTDOOR ROOMS (163),GALLERY SURROUNDS(166), TRELLISED WALKS(174),and WINDOW PLACES(180), even SMALL PARKING LOTS (103),which all become more subtle and more beautiful with canvas roofs and awnings.And the awnings always help to create FILTERED LIGHT(238).

⋅⋅⋅⋅

There is a very special beauty about tents and canvas awnings.The canvas has a softness,a suppleness,which is in harmony with wind and light and sun.A house or any building built with some canvas will touch all the elements more nearly than it can when it is made only with hard conventional materials.

In conventional building,it is easy to think that walls and roofs must either be solid,or missing altogether.But cloth and canvas lie just exactly halfway in between.They are translucent,let a little breeze pass through,and they are very cheap,and easy to roll up and easy to pull down.

We can identify three kinds of places that need these properties:

1.Awnings—sunshades over windows,retractable,and used to filter very bright hot sunlight.

2.Curtains—moveable,half-open walls on outdoor rooms, balconies,and galleries-places that are occupied mainly during

在常规建筑中，人们很容易认为墙和屋顶要么是厚实且密不透风，要么完全没有。但布和帆布恰好介于这两者之间。它们是半透明的，微风可从其中穿透过去，而且价格低廉，易于卷起和拆除。

我们能认定有三种地方需有上述特点。

1. 遮阳篷——覆罩窗户的布幔伸缩自如，可用于过滤非常强烈而又灼热的阳光。

2. 窗帘——可挂在活动的户外空间的半空墙上、阳台和走廊内，主要在白天挂，有滤光和通风的良好功能。

3. 户外空间的帐篷式屋顶——帐篷可防蒙蒙细雨，并使户外空间、拱廊或庭院在春秋两季和夜晚成为可住人的地方。

下面是弗兰克·劳埃德·赖特对他在塔利埃辛·威斯特地区的最早期的结构中使用帆布顶篷情况的描述：

……塔利埃辛联谊会是在幅员辽阔的亚利桑那州大台地上的一个沙漠营地。为了冬季的工作和生活，我和一群男孩现在正在建造它。其中许多单元都是帆布顶篷。这种帆布顶篷被紧绷在红木框架上，而这些框架支在厚重的石墙上。石墙是由木箱构筑的：先把沙漠里的扁平石头放入木箱内，然后再放入石子和混凝土。大多数帆布框架都能开启或关闭……帆布顶篷呈半透明。我们在帆布顶篷下工作和生活，感到光线非常美丽悦目。我除了在日本时住在有纸糊的推拉墙或"纸门"的日式房屋内有此同感外，在别处我均未有此感受。（*The Future of Architecture*，London：The Architectural Press，1955，pp.255~256.）

另外一个例子是：在意大利，帆布遮阳篷作为朝南和朝西的窗户的简易遮阳篷已被广泛使用。帆布遮阳篷宛如一个个明亮而又美丽的橘子，给街道增添了色彩，也给内部空间增添了温暖。

现在我们把在秘鲁利马建房工程中利用本模式的情况作为最后一个例子提供给大家。我们用可移动的帆布材料做内庭院的顶篷。在热天，帆布顶篷可卷起，让阵阵微风

the day,but might benefit from extra wind protection.

3.Tent-like roofs on outdoor rooms—a tent which can hold off a drizzle and make outdoor rooms,or trellises,or courtyards habitable in the spring and autumn and at night.

Here is Frank Lloyd Wright describing his use of the canvas roof in the very early structures at Taliesin West:

...the Taliesin Fellowship(is a)desert camp on a great Arizona mesa which the boys,together with myself,are now building to work and live in during the winter-time.Many of the building units have canvas tops carried by red-wood framing resting on massive stone walls made by placing the flat desert stones into wood boxes and throwing in stones and concrete behind them.Most of the canvas frames may be opened or kept closed... The canvas overhead being translucent,there is a very beautiful light to live and work in;I have experienced nothing like it elsewhere except in Japan somewhat,in their houses with sliding paper walls or "shoji." (*The Future of Architecture*,London:The Architectural Press,1955,pp.255-256.)

Another example:In Italy,the canvas awning is used quite commonly as a simple awning over south and west windows. The canvas is often a bright and beautiful orange,giving color to the street and a warm glow to the interior rooms.

As a final example,we report on our own use of this pattern in the housing project in Lima.We roofed interior patios with movable canvas material.In hot weather the covers are rolled back,and a breeze blows through the house.In cold weather,the canvas is rolled out,sealing the house,and the patio is still useful.In Lima,there is a winter dew which normally makes patio floors damp and cold for eight months in the year. The cover on the patios keeps them dry and warm and triples their useful life.They eliminate the need for glass windows

吹进室内。在冷天，帆布顶篷要全打开，密密地盖住房屋，这样内庭院仍然是可使用的。在利马，冬天的露水在正常情况下会使内庭院的地面潮湿和寒冷，一年之内长达8个月。内庭院的帆布顶篷会使地面保持干燥和温暖，并使地面的使用寿命增至为原来的3倍。内庭院几乎完全不需要玻璃窗。光线透过内庭院的窗户射入房间，并可挂窗帘来进行光线的视觉控制——但是，因为内庭院的帆布屋顶能消除寒冷和潮湿，窗户就既不需要玻璃也不需要昂贵的活动零件了。

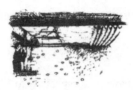

我们在秘鲁建造的内庭院的帆布顶篷
Our patio covers in Peru

因此：

凡在夏天需要有比较柔和的光线和部分荫凉的地方，或在秋冬两季必须防止雾气和露水的地方，都应建造帆布顶篷、帆布墙和遮阳篷。把帆布顶篷建成折叠式的、可用绳索或钢丝拉开的，以便帆布顶篷能够很容易地启开。

柔和的光线

帆布顶篷

৪৩৫৪

尤其是朝西和朝南的窗户要使用帆布遮阳篷来过滤光线并消除眩光，因为这些窗户对着天空——**过滤光线**（238）。五颜六色的帆布美化生活，增添特色——**装饰**（249）、**暖色**（250）……

almost entirely. The windows which look into patios give light to rooms and may be curtained for visual control—but since the cold and damp are kept out by the patio canvas there need be no glass in the windows and no expensive moving parts.

Therefore:

Build canvas roofs and walls and awnings wherever there are spaces which need softer light or partial shade in summer, or partial protection from mist and dew in autumn and winter. Build them to fold away, with ropes or wires to pull them, so that they can easily be opened.

⊰⊱

Use the canvas awnings, especially, to filter light over those windows which face west and south and glare because they face the sky—FILTERED LIGHT(238). Colored canvas will add special life—ORNAMENT(249), WARM COLORS(250)...

模式245 高花台*

245   RAISED FLOWERS*

...outdoors there are various low walls at sitting height—
SITTING WALL(243);terraced gardens,if the garden has
a natural slope in it—TERRACED SLOPE(169);and paths
and steps and crinkled building edges—PATHS AND
GOALS(120),STAIR SEATS(125),BUILDING EDGE
(160),GARDEN WALL(173).These are the best spots for
flowers,and flowers help to make them beautiful.

❧❧

**Flowers are beautiful along the edges of paths, buildings,
outdoor rooms—but it is just in these places that they need
the most protection from traffic.Without some protection
they cannot easily survive.**

Look at the positions that wildflowers take in nature.
They are as a rule in protected places when they occur in
massive quantities:places away from traffic—often on grassy
banks,on corners of fields,against a wall.It is not natural
for flowers to grow in bundles like flower beds;they need a
place to nestle.

What are the issues?

1.The sun—they need plenty of sun.

2.A position where people can smell and touch them.

3.Protection from stray animals.

4.A position where people see them,either from inside
ahouse or along the paths which they naturally pass coming and
going.

……户外有各种可坐人的矮墙——**可坐矮墙**（243）；如果花园中有一天然斜坡，就会有露台花园——**梯形台地**（169）；还有小路、台阶和波纹状的建筑物边缘——**小路和标志物**（120）、**能坐的台阶**（125）、**建筑物边缘**（160）、**花园围墙**（173）。这些都是布置花卉的最佳地点，而花卉又可美化这些地点。

❀❀❀

花卉点缀着小路、建筑物、户外空间的边缘，分外美丽——但正是在这些地方，花卉需备加爱护，以避免来往行人和车辆对其造成损坏。若不采取保护措施，花卉是无法生存的。

观察一下自然界野花生长的地方。通常野花在受到保护的地方都大量密集：它们生长在远离交通的地方——往往生长在杂草丛生的河岸边、田野的偏僻角落或靠近墙根。花卉成堆地长在花台里是不自然的；花卉需要有合适的生长环境。

哪些是要解决的问题？

1. 阳光——花卉需要充足的阳光。

2. 位置合适——人们能接触并闻到花卉的芬芳。

3. 防止迷路的小动物践踏花卉。

4. 人们或从室内或从经常走过的小路上都能看到花卉。

典型的狭长花台往往太深和太暴露。花台很低，花卉可望而不可及。为保护花卉而采用的混凝土栽培箱往往陷入另一极端。花卉得到了保护，但人们无法直接接触，除非和它们保持一定距离。这种栽培方法几乎无用。花卉应是可接近的，这样才能接触它们，并闻到其芳香。

Typical flower borders are often too deep and too exposed. And they are so low the flowers are out of reach.Concrete planter boxes made to protect flowers often go to the other extreme.They are so protected that people have no contact with them,except from a distance.This is next to useless.The flowers need to be close,where you can touch them,smell them.

Therefore,instead of putting the flowers in low borders,on the ground,where people walk,or in massive concrete tubs,build them up in low beds,with sitting walls beside them,along the sides of paths,around entrances and edges.Make quite certain that the flowers are placed in positions where people really can enjoy them—and not simply as ornament:outside favorite windows,along traveled paths,near entrances and round doorways,by outdoor seats.

Therefore:

Soften the edges of buildings,paths,and outdoor areas with flowers.Raise the flower beds so that people can touch the flowers,bend to smell them,and sit by them.And build the flower beds with solid edges,so that people can sit on them,among the flowers too.

因此，不要把花栽在低而狭长的花台里、人们步行的地面上或大量的混凝土花盆内，而应把花栽在四周有可坐矮墙的低花坛里，或在小路的两旁，入口和边缘的周围。务必把花栽在人们能真正欣赏的地方——而不是简单地作为装饰品：栽在你最喜爱的窗外、散步小径的两侧、靠近入口的地方、门道的周围以及户外设座位置的附近。

高花台
Raised flowers

因此：

栽种花卉可使建筑物、小径和户外空间等的边缘有柔和感。把花台加高，以便人们能够接触朵朵鲜花，俯身去闻花香，或坐在花旁赏花。花台的边缘要造得很厚实，以便人们身处花丛中也能坐在花台边缘上赏花。

高花台   1~3ft高

模式246 攀援植物

……建筑周围的攀援植物能有助于前面的两个模式：**棚下小径**（174）和**过滤光线**（238）。

246   CLIMBING PLANTS

...two earlier patterns can be helped by climbing plants around the building:TRELLISED WALK(174)and FILTERED LIGHT(238).

∝

A building finally becomes a part of its surroundings when the plants grow over parts of it as freely as they grow along the ground.

There is no doubt that buildings with roses or vines or honeysuckle growing on them mean much more to us than buildings whose walls are blank and bare.That is reason enough to plant wild clematis around the outside of a building,to make boxes to encourage plants to grow at higher storys,and to make frames and trellises for them to climb on.

We can think of four ways to ground this intuition in function.

1.One argument,consistent with others in the book,is that climbing plants effect a smooth transition between the built and the natural.A sort of blurring of the edges.

2.The quality of light.When the plants grow around the openings of buildings,they create a special kind of filtered light inside.This light is soft,reduces glare,and stark shadows— FILTERED LIGHT(238).

3.The sense of touch.Climbing and hanging plants also give the outside walls a close and subtle texture.The same kind of texture can be achieved in the building materials,but it is uniquely beautiful when it comes from a vine growing across a

当建筑被攀援植物自然生长爬满外墙时，它终于成为周围环境的一部分。

毫无疑问，在住宅的外墙上长满玫瑰花、葡萄树或金银花，对于我们来说，总比光秃而呆板的外墙更有意思。这就是我们提倡在住宅四周、房前屋后栽种野生女萝，在较高的楼层上多种箱栽攀援植物，并让它们在门窗框和花格墙上蔓延攀援的充分理由。

我们想从以下四个方面来说明这种直观的功能。

1. 有一种论据是和本书中的其他论据前后一贯的，即攀援植物在已建成的住宅和自然界之间起到一种平稳的过渡作用。它会使两者之间的界线模糊起来。

2. 光线的性质。当在建筑的门窗之外生长着花草树木时，它们就会在室内形成一种特殊的过滤光线。这种光线柔和，会减少眩光和轮廓反差大的阴影——**过滤光线**（238）。

3. 触感。攀援植物和悬挂植物还赋予外墙一种密实而又微妙的质感。利用建筑材料也可以获得同样质感，但由攀爬过墙的或缠绕在拱廊檐口的藤蔓枝叶形成的质感所具有的美是无与伦比的。那时，它会吸引你去触摸青藤绿蔓，去闻它的芬芳，或摘下它的一片翠叶。或许最重要的是攀援植物的质感总是因时而异；当风和阳光嬉弄它的枝叶时，它天天都会呈现出细微的变化；而从一个季节到另一个季节，其质感有极明显的差异。

4. 爱护花草树木。如果精心管理，茁壮生长在窗前的或栽种在楼上阳台上花箱中向外探出的花草树木都能美化街道，使街道变得更加宜人、赏心悦目。花草树木展示了建筑内某种宁静的社会秩序，因此，人们在街道上会感到如同在家里一样安逸舒适。仿佛花草树木就是室内居民送给街上行人的一件礼物。

wall or winding around the eaves of an arcade.Then,the texture invites you to touch and smell it,to pick off a leaf.Perhaps most important,the texture of climbing plants is ever different;it is subtly different from day to day,as the wind and sun play upon it;and it is greatly different from season to season.

4.Tending the plants.When they are well-tended,healthy plants and flowers growing around the windows and out of flower boxes in the upper storys,make the street feel more comfortable.They bespeak a social order of some repose within the buildings,and therefore it is comfortable to be on the streets—one feels at home.It is as if the plants were a gift from the people inside to people on the street.

Therefore:

On sunny walls,train climbing plants to grow up round the openings in the wall—the windows,doors,porches, arcades,and trellises.

奉献给街道的礼物
The contribution to the street

因此:

在向阳的墙上,让攀援植物环绕墙的敞口——窗、门、门廊、拱廊和花格墙生长。

攀援植物

花木箱

棚架

模式247 留缝的石铺地**

247 PAVING WITH CRACKS
BETWEEN THE STONES**

...many patterns call for paths and terraces and places where the outdoor areas around a building feel connected to the earth—GREEN STREETS(51),PATH SHAPE(121),PRIVATE TERRACE ON THE STREET (140),OUTDOOR ROOM(163), CONNECTION TO THE EARTH(168), TERRACED SLOPE (169).This pattern provides a way of building the ground surface that makes these larger patterns come to life.

৪৩৫৪

Asphalt and concrete surfaces outdoors are easy to wash down,but they do nothing for us,nothing for the paths, and nothing for the rainwater and plants.

Look at a simple path,made by laying bricks or paving stones directly in the earth,with ample cracks between the stones. It is good to walk on,good for the plants,good for the passage of time,good for the rain.You walk from stone to stone,and feel the earth directly under foot.It does not crack,because as the earth settles,the stones move with the earth and gradually take on a rich uneven character.As time goes by,the very age and history of all the moments on that path are almost recorded in its slight unevenness.Plants and mosses and small flowers grow between the cracks.The cracks also help preserve the delicate ecology of worms and insects and beetles and the variety of plant species. And when it rains,the water goes directly to the ground;there is no concentrated run-off, no danger of erosion,no loss of water in the

……许多模式需要小路和露台，并且人们在建筑周围的户外区都会感到和大地紧密相连——**绿茵街道**（51）、**小路的形状**（121）、**私家的沿街露台**（140）、**有围合的户外小空间**（163）、**与大地紧密相连**（168）和**梯形台地**（169）。本模式提供一种铺砌地面的方法，以便上述较大的模式得以实现。

ՍᏦᏽᏦᏽ

户外的沥青地面和混凝土地面容易冲洗，但这些地面并没有为我们人类、小路、雨水和植物做什么事。

来看一条简单的小路吧。它是在泥地上用铺路砖或铺路石直接铺砌而成，路面石块之间有着大量的缝隙。这种小路适于人们步行、栽花种草、消磨时光和吸收雨水。你从一块石头走到另一块石头，就会直接感到大地在自己的脚下。它不会出现裂缝，因为泥土塌陷时，石头会随着移动，并渐渐呈现出坎坷不平的特征。随着时间的推移，这种道路的年代和各个时期的历史几乎都会记录在它轻微的高低不平之中。草木、苔藓和小花夹杂生长在缝隙间。缝隙也有助于维持蠕虫、甲虫和各种植物品种微妙的生态平衡；下雨时，雨水会直接渗入地面；不会有集中的雨水径流，不会有被侵蚀的危险，小路四周的地下水也不会流失。

所有这一切都是铺砌石地面要留缝隙的充分理由。至于谈到平坦、光滑、坚硬的沥青路面和混凝土路面，几乎没有什么要介绍的。当人们忘却留缝的石铺地面的这些小优点时，就会修筑沥青路面和混凝土路面。

ground around the path.

All these are good reasons to set paving stones loosely.As for the flat,smooth,hard concrete and asphalt surfaces,they have almost nothing to recommend them.They are built when people forget these small advantages that come about when paving is made out of individual stones with cracks between the stones.

Therefore:

On paths and terraces,lay paving stones with a 1 inch crack between the stones,so that grass and mosses and small flowers can grow between the stones.Lay the stones directly into earth,not into mortar,and,of course,use no cement or mortar in between the stones.

ଷଠଓଷ

Use paving with cracks,to help make paths and terraces which change and show the passage of time and so help people feel the earth beneath their feet—CONNECTION TO THE EARTH(168);the stones themselves are best if they are simple soft baked tiles—SOFT TILE AND BRICK(248)...

因此:

　　小路和露台要铺砌成留缝的石铺地面，缝隙宽 1in，以便青草、苔藓和小花可以杂生其间。把石块直接铺入土中，而不是铺入砂浆中。当然，在石块之间的缝隙内也不要浇筑水泥和砂浆。

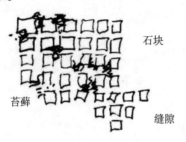

石块

苔藓

缝隙

ଈୠ୯ଔ

　　使小路和露台成为留缝的石铺地面。小路和露台将会发生变化并体现出流逝的岁月，因此有助于人们感觉到大地就在他们的脚下——**与大地紧密相连**（168）；如果砖块是简单的软质焙制砖则最佳——**软质面砖和软质砖**（248）……

模式248　软质面砖和软质砖

……下面几个模式需用面砖和砖——**与大地紧密相连**（168）、**好材料**（207）、**地面面层**（233）、**可坐矮墙**（243）和**留缝的石铺地**（247）。

⚜

当一个人行走在坚硬的、便于机械冲洗的混凝土路面、沥青路面、建筑用的铺路过火砖路面或人工配制的混合物如水磨石路面时，他怎样才能感觉到与大地、与时代、与他的周围环境的联系呢？

最重要的是我们赖以行走的大地的水平面——然而在我们住宅周围或在室内，比如过道和厨房的地面，却不得不全部是坚硬的——至少应柔软得足以体现出在缓缓的起伏和高低不平之中所流逝的岁月，诉说着成千上万人走过时的情景，并清楚表明，建筑物和人们一样——不是无动于衷的，不是异己的，而是活生生的，随时间而变化的，并且记录下人们所踩出的足印。

没有什么东西能像软质的、焙制的或文火烧制的砖和面砖那样出色地体现出流逝的岁月。这些面砖最便宜，是由普通黏土焙制而成，是可以进行分解和还原处理的，总是能够给人以美感，而且能从人们的使用过程所造成的高低不平之中反映出流逝的岁月痕迹。

除此之外，为了**与大地紧密相连**（168），住宅周围铺砖的地方起着特殊的作用。这些地方恰好位于采用人造材料的建筑和完全是自然的大地两者之间。为了使人感受到这种联系，材料本身也应具有这种介于建筑和大地两者之间的性质。因此重申，软质的、初步定型的面砖是最合适的材料。

CONSTRUCTION
构　造

248 SOFT TILE AND BRICK

...several patterns call for the use of tiles and bricks—
CONNECTION TOTHE EARTH(168),GOOD MATERIALS
(207),FLOOR SURFACE(233),SITTING WALL(243),
PAVING WITH CRACKS BETWEEN THE STONES(247).

❧❧❧

**How can a person feel the earth,or time,or any
connection with his surroundings,when he is walking on the
hard mechanical wash-easy surfaces of concrete, asphalt,
hard-fired architectural paving bricks,or artificially
concocted mixes like terrazo.**

It is essential,above all,that the ground level surfaces
we walk on—both around our buildings and indoors in those
places like passages and kitchens where the floor has to be
hard—be soft enough,at least,to show the passage of time,in
gradual undulations and unevenness,that tell the story of a
thousand passing feet,and make it clear that buildings are like
people—not impervious and alien,but alive,changing with
time,remembering the paths which people tread.

Nothing shows the passage of time so well as very
soft,baked or lightly fired,bricks and tiles.They are among
the cheapest tiles that can be made;they use ordinary clay,are
biodegradable,and always develop a beautiful sense of wear
and time in the undulations made by people walking over them.

In addition,those paved areas around a building required
by CONNECTION TO THE EARTH(168)play a special role.

我们认为这一点非常重要，所以我们特别提倡，凡正在建房的人应为铺砌他们的底层地面和室外地面准备大量面砖，这些面砖应当利用当地的黏土堆放成垛，就在基地上用文火烧制而成。

这是容易办到的。现在我们对如何制作面砖和建造户外简易火坑进行详细介绍。

我们从黏土谈起：怎样摆弄黏土最好还是从头说起。

黏土是已风化了的长石质的火成岩。这种火成岩地球上到处都有，储量丰富。如果幸运，人们就能在自己的后院找到它。

为了验明它是否为黏土，只需挑出一些用水浸湿。如果它是塑性的，而且它的黏度足以形成一个光滑的球，那么它就是黏土了……

处理黏土的步骤如下：

1. 首先，除去黏土中的杂质，如细枝、树叶、草根和石子。

2. 其次，让黏土块在阳光下晒干。

3. 粉碎这些黏土块，并尽量把它们磨成精细的粉末。

4. 把这种磨成精细粉末的黏土放入水中，以便形成一个高出水面的土堆。

5. 让这种混合物浸泡一天，然后搅拌，并用筛子进行筛选。

6. 再把筛选后的黏土搁置一天，去掉多余的水分。

7. 然后将黏土放入灰泥容器中；灰泥吸收水分，从而使混合物变稠而成为可加工砖的黏土了。

8. 取黏土少许进行试验。如它出现裂纹，则说明它"缺乏"某些成分。

如发生这种情况，就将7%的膨润土掺入混合物。如果黏土太黏，就掺入"耐火黏土"。

在黏土中掺入硬石粉或耐火黏土会减少收缩。耐火黏土是将黏土经坯窑烧制再粉碎而成的。一些人是利用坯窑制品的碎片自己加工耐火黏土。任何一个供销公司都出售各种细度不同的耐火黏土，而且价格低廉。黏土内加入的耐火黏土的颗粒越粗，则烧

They are the places which are halfway between the building-with its artificial materials—and the earth—which is entirely natural.To make this connection felt,the materials themselves must also be halfway,in character,between the building and the earth.Again,soft,lightly fixed tiles are most appropriate.

We consider this so important,that we advocate, specifically, that the people who are making the building,make the quantity of bricks and tiles they need for ground floor and outdoor surfaces—and that these be made in local clay and soft fired, in stacks,right on the site.

It is easy to do.We shall now give detailed instructions for making the tiles themselves and for making a rudimentary outdoor firing pit.

We start with the clay:it would be best to make one's own clay from scratch.

Clay is decomposed feldspathic rock.There is an abundance of it all over the earth.One may be fortunate enough to find it in one's back yard.

To test whether it is clay,pick up a bit of it and wet it.If it is plastic and sticky enough to form a smooth ball,it is clay....

Process the clay as follows:

1.First,remove impurities such as twigs,leaves,roots and stones.

2.Then,let the chunks dry in the sun.

3.Break up these chunks and grind them up as finely as possible.

4.Put this ground-up clay in water so that there is a mound above water.

5.Let this mixture sonk for one day,then stir it,and sieve it through a screen.

6.Let stand again for another day,and remove excess water.

7.Then put the clay in a plaster container;plaster absorbs water,thus stiffening the mixture into workable clay.

制的黏土砖的结构就越粗。

耐火黏土使黏土成为多孔的，并被用于不保持水分的黏土砖。耐火黏土还可防止翘曲，因此，它对制造面砖和雕塑是非常有用的。黏土混合物内含 20% 的耐火黏土，则比例适当。（Muriel ParghTuroff, *How to Make Pottery and Other Ceramic Ware*, New York：Crown Publishers，1949，p.13.）

当你有了黏土后就能够制作面砖了。

以此法制作地砖，必须使用与成品砖尺寸相同的木模。这种木模由 4 块木板条一起钉到一块平整的木板上而成。木板条应宽 1in，厚 3/8～3/4in，可随你所想造的砖的厚度而定。在把 4 块木板条钉到底板上之前，在底板上铺一块油布是个好办法。这可以防止底板翘曲……

将一大块黏土辗平……然后根据模子的大小，从大块辗平的黏土上切下一块装入模子，再用滚筒把它弄平。滚筒不必在该小块黏土的整个表面滚动，只要在其中心处向四侧滚动即可……让面砖干燥，直至它的表面变硬为止；然后再用一把小刀沿面砖的四周边缘刮一刮，使它和模子脱离……

让黏土面砖很缓慢地干燥，为此，应当把它们放在阴凉处。如果它们因受热而干燥得过快，就容易出现裂纹或翘曲不平。面砖的边缘总是比中心处干燥得快，所以，通常要不时地在它们的边上洒点水，以防发生上述情况。（Joseph Leeming, *Fun With Clay*, Philadelphia and New York：J.B.Lippincott Company.）

为了烧制软质面砖和砖，无须建造正规的砖窑，可以在露天的坑中烧制，很像早期的陶工在露天坑内烧制他们的陶器。对于这种烧制面砖的露天坑，丹尼尔·罗兹（Danid Rhodes）在《砖窑》（*Kilns:Design Construction and Operation*, Philadelphia:Chilton Book Company）一文中作了详尽无遗的描写。摘要如下：

8.Work the clay a little to test it.If cracks appear,it is "short";when that happens, add to the mixture,up to 7%bentonite.If clay is too plastic,add "grog." ...

Shrinkage may be decreased by adding flint or grog to the clay. Grog is clay that has been biscuit-fired and then crushed.Some people prepare their own grog from broken biscuit-fired pieces.It can be bought at very little cost at any supply company in varying degrees of fineness.The coarser the particles of grog added to the clay,the coarser the texture of the fired object will be.

Grog makes clay porous and is used for objects which are not intended to hold water.Grog also prevents warpage and is,therefore,very useful for tile making and for sculpture.20% is a good proportion of grog in a clay mixture.

(Muriel Pargh Turoff,*How to Make Pottery and Other Ceramic Ware*,New York:Crown Publishers,1949,p.13.)

Once you have the clay,you can make the tiles.

In this method of tile making,a wooden form is used that has the dimensions desired for the finished tiles.It is put together by nailing four strips of wood to a smooth piece of board.The strips should be 1 inch wide and their height may vary from 3/8 inch to 3/4 inch,depending on how thick you wish the finished tiles to be.It is a good plan to put a piece of oilcloth on the base board before nailing down the strips.This will keep the board from warping....

Roll out a slab of day....Then cut from the slab a piece that will fit comfortably into the form and roll it down with a rolling pin.Do not roll the pin all the way across the surface of the clay,but work from the center outwards to all four sides....Let the tile dry until it is leather-hard;then separate it from the form by running a knife around its edges....

Clay tiles should be allowed to dry very slowly,and for this reason should be put in a cool place.If they dry too quickly under heat,they

挖掘一个浅坑，深 14 ~ 20in，面积为数个平方英尺。把粗细大小不等的树枝、树杈和芦苇等整齐地码放在坑内（坑底和坑的四周）。把要烧制的面砖坯和砖坯放在衬垫上，码放密实，坯间留稍许空隙——（这些坯能堆放成十字交叉）……如果你将旧面砖整齐地铺在坑底，则坑的保温性能更好。坑的一端气孔低有助于燃烧……在坯垛之间以及坯垛的上面堆放一些燃料。然后点燃坑内的燃料，并让其慢慢燃烧——燃料只好这样徐徐开始燃烧，因为空气供应不足。当火苗冒出坑的某一水平面时，就要添放更多的燃料。在整个坑内以及坑内的全部砖坯达到赤热温度后，就要把火熄灭，并用叶子、牛马粪便或灰烬盖住火苗，以便保持坑温。在火完全熄灭以及余烬冷却之后，面砖即可取出。

一座简易的砖窑
A simple kiln

因此：

利用文火烧制的砖和面砖，它们将随着岁月的流逝而逐渐磨损，并显示出使用的痕迹。在工地上你可利用黏土在简易的模子里制造砖坯；把细枝小杈和木柴放在坯垛的周围；点燃柴火，把砖坯直烧到有柔和感的粉红色为止，这样烧制成的软质面砖将随时间的消逝而受磨损。

文火烧制的
黏土地砖坯

are apt to crack or warp.The edges have a tendency to dry more rapidly than the center and usually should be dampened from time to time to prevent this.(Joseph Leeming,*Fun With Clay*,Philadelphia and New York:J.B.Lippincott Company.)

To fire soft tiles and bricks,it is not necessary to build real kilns.They can be fired in open pits much like those which primitive potters used to fire their pottery.This type of open pit firing is described in detail by Daniel Rhodes,in *Kilns:Design, Construction and Operation*, Philadelphia:Chilton Book Company.Briefly:

Diga shallow pit about 14 to 20 inches deep,and several square feet in area.Line this pit(bottom and sides)with branches,reeds,twigs,etc.Place the tiles and bricks to be fired on the lining,so that they are compactly piled with just a tiny bit of airspace between them—(they can be criss-crossed)....If you use old tiles to line the pit,it will keep the heat in even better;and air holes low down at one end will help combustion....Put some fuel in between stacks and over them.Then light the fuel in the pit,and allow it to burn slowly— which it will to begin with because not much air can get to it.Pile more fuel on as the fire burns up to a level above the pit.After the entire pit and its contents reach red heat,allow the fire to die down,and cover the top of the fire with wet leaves,dung or ashes to retain the heat.After the fire has died down,and the embers cooled,the tiles can be removed.

Therefore:

Use bricks and tiles which are soft baked,low fired—so that they will wear with time,and show the marks of use.

You can make them in a simple mold from local clay, right on the site;surround the stack with twigs and firewood; and fire them,to a soft pink color which will leave them soft enough to wear with time.

有柔和感的粉红色有助于形成**暖色**（250）。在烧制前，可按自己的要求，对面砖加以某种**装饰**（249）……

以装饰物、光线颜色和生活中的纪念品来使你的住宅建筑更加完美。

249. 装饰

250. 暖色

251. 各式坐椅

252. 投光区域

253. 生活中的纪念品

The soft pink color helps to create WARM COLORS(250). Before firing,you may want to give the tiles some ORNAMENT(249)...

complete the building with ornament and light and color and your own things.

249.ORNAMENT

250.WARM COLORS

251.DIFFERENT CHAIRS

252.POOLS OF LIGHT

253.THINGS FROM YOUR LIFE

模式249 装饰**

　　……一旦建筑主体和花园建成；墙、柱、窗、门和各种表面都已竣工；各种边界、边缘和过渡区都已确定——主入口（110）、建筑物边缘（160）、与大地紧密相连（168）、花园围墙（173）、窗前空间（180）、靠近墙角的房门（196）、门窗边缘加厚（225）、柱旁空间（226）、柱的连接（227）、屋顶顶尖（232）、有柔和感的墙内表面（235）、可坐矮墙（243）等——那么接着要做的就是完成各种修饰，以装饰来填缝和标志边界。

249 ORNAMENT**

...once buildings and gardens are finished; walls, columns, windows,doors,and surfaces are in place; boundaries and edges and transitions are defined—MAIN ENTRANCE(110),BUILDING EDGE(160),CONNECTION TO THE EARTH(168),GARDEN WALL(173),WINDOW PLACE (180),CORNER DOORS(196),FRAMES AS THICKENED EDGES(225),COLUMN PLACE(226), COLUMN CONNECTION(227),ROOF CAPS(232),SOFT INSIDE WALLS(235),SITTING WALL(243),and so on—it is time to put in the finishing touches,to fill the gaps,to mark the boundaries,by making ornamen.

᠀ᢆᡄᡝᢆᠻᡝ

All people have the instinct to decorate their surroundings.

But decorations and ornaments will only work when they are properly made:for ornaments and decorations are not only born from the natural exuberance and love for something happy in a building;they also have a function,which is as clear,and definite as any other function in a building.The joy and exuberance of carvings and color will only work,if they are made in harmony with this function.And,further,the function is a necessary one—the ornaments are not just optional additions which may,or may not be added to a building,according as the spirit moves you—a building needs them,just as much as it needs doors and windows.

In order to understand the function of ornament,we must begin by understanding the nature of space in general.

装饰美化周围的环境是人所共有的天性。

但是，只有当花纹和装饰恰到好处时，它们才能发挥效用。这是因为花纹和装饰不仅来源于绚丽多彩的自然界和人们喜欢住宅内有某种令人愉悦的气氛，而且具有与建筑的任何其他功能一样鲜明而又明确的功能。雕刻和色彩的丰富多样和欢乐气氛只有与此功能和谐一致时才会发挥作用。进而言之，这一装饰功能是必不可少的——根据精神会使你感动这一原则，装饰恰恰不是建筑中可有可无的任意添加物——建筑需要装饰，正如它需要门窗一样。

为了理解装饰的这种作用，我们必须从理解普遍的空间性质开始。空间具有恰当的形状时就是一个整体。空间的每一部分，即一个城镇、一个邻里、一幢房屋、一座花园或一个房间的每一部分统统都是一个整体。从这种意义上讲，空间本身既是一个完整的统一体，同时又与其他统一体相结合，从而形成更大的整体。这一过程在很大程度上取决于边界。

本书中有如此众多的模式涉及各种事物之间边界的重要性，一点不亚于事物本身的重要性，这绝非偶然。诸如**亚文化区边界**（13）、**邻里边界**（15）、**拱廊**（119）、**建筑物边缘**（160）、**回廊**（166）、**与大地紧密相连**（168）、**半敞开墙**（193）、**厚墙**（197）、**门窗边缘加厚**（225）、**半英寸宽的压缝条**（240）和**可坐矮墙**（243）。

某一事物只有当它本身是完整的，并和它外部的另一事物结合在一起而形成更大的统一体时才是一个整体。但是，这种情况只能在两物之间的边界非常宽泛且又模糊不清时才发生，两者区分不明显，但能够或作为分离的统一体，或作为一个更大的、没有内部裂缝的整体而发挥效能。

Space, when properly formed,is whole.Every part of it,every part of a town,a neighborhood,a building,a garden,or a room, is whole,in the sense that it is both an integral entity,in itself, and at the same time,joined to some other entities to form a larger whole.This process hinges largely on the boundaries.It is no accident that so many of the patterns in this pattern language concern the importance of the boundaries between things,as places that are as important as the things themselves—for example,SUBCULTURE BOUNDARY (13), NEIGHBORHOOD BOUNDARY(15), ARCADES(119), BUILDING EDGE (160),GALLERY SURROUND(166), CONNECTION TO THE EARTH(168),HALF-OPEN WALLS(193), THICK WALLS(197), FRAMES AS THICKENED EDGES(225),HALF-INCH TRIM(240), SITTING WALL(243).

A thing is whole only when it is itself entire and also joined to its outside to form a larger entity.But this can only happen when the boundary between the two is so thick,so fleshy,so ambiguous,that the two are not sharply separated,but can function either as separate entities or as one larger whole which has no inner cleavage in it.

In the left-hand diagram where there is a cleavage that is sharp,the thing and its outside are distinct entities—they function individually as wholes—but they do not function together as a larger whole.In this case the world is split.In the right-hand diagram where there is ambiguous space between them,the two entities are individually entire,as before,but they are also entire together as a larger whole.In this case the world is whole.

This principle extends throughout the material universe, from the largest organic structures in our surroundings,to the

左图有一条很明显的裂缝。此物与其外部的另一物是完全不同的统一体，它们各自都以整体发挥功能——但它们并未合在一起作为更大的整体而发挥功

分裂的……和完整的
Split...and whole

能。在此情况下，空间是分裂的。右图中两物之间的空间是模棱两可的。两个统一体如前一样，各自都是完整的，但同时，它们也是作为一个更大的整体而完整地结合在一起。

这一原理普遍适用于物质世界，从我们周围环境中最大的有机结构直到原子和分子这样的微粒。

这一有效原理在人造物品中的极端例子就是从所谓"（中世纪）黑暗时期"以来产品的无穷无尽的面以及土耳其和波斯的地毯和砖地面。现在且把这些"装饰"的深远意义搁置一旁，事实上它们的主要功能是创造出了各种表面。而表面的每一部分同时既是图案又是边界，表面的设计同时在几个不同的层次上既是图案又是边界。

一幅完整的装饰品，因为它不能被分成若干部分
A decoration which is whole,because it cannot be broken into parts

这种古代的地毯没有一部分能和它的相邻各部分分离开来，因为每一部分在不同的层次上同时既是图案又是边界，所以它在极大程度上是完整的。

very atoms and molecules.

Extreme examples of this principle at work in manmade objects are in the endless surfaces of objects from the so-called "dark ages" and in the carpets and tilework of Turkey and Persia.Leaving aside the profound meaning of these "ornaments," it is a fact that they function mainly by creating surfaces in which each part is simultaneously figure and boundary and in which the design acts as boundary and figure at several different levels simultaneously.

Since none of the parts can be separated from their surroundings,because each part acts as figure and as boundary,at several levels,this ancient carpet is whole,to an extraordinary degree.

The main purpose of ornament in the environment—in buildings,rooms,and public spaces—is to make the world more whole by knitting it together in precisely the same way this carpet does it.

If the patterns in this language are used correctly,then these unifying boundaries will already come into existence without ornament at almost all the scales where they are necessary in spaces and materials.It will happen in the large spaces,like the entrance transition or the building edge.And,of course,it happens of its own accord,in those smaller structures which occur within the materials themselves—in the fibers of wood,in the grain of brick and stone.But there is an intermediate range of scales,a twilight zone,where it will not happen of its own accord.*It is in this range of scales that ornament fills the gap.*

As far as specific ways of doing it are concerned,there are hundreds,of course.In this balustrade the ornament is made entirely of the boundary,of the space between the boards.The boards are cut in such a way,that when they are joined together

环境——建筑、房间和公共空间——中的装饰主要目的在于像精心编织这种地毯一样使世界变得更加完整。

如果本书中的许多模式能够被正确利用，这些合成一体的边界就会早已存在，而几乎不需要空间和材料上所必要的各种尺度的装饰了。这种情况将发生在较大的空间，如入口过渡区或建筑物边缘。当然，这种情况在材料本身内部某些较小的结构中——在木材的纤维中，在砖和石头的颗粒中——会自行发生。但存在中间的尺度范围，处于难于明确划界的过渡区，这种情况不会自行发生。正是在这一尺度范围内才需装饰来填补空隙。

至于说到进行装饰的特殊方法，当然有成百上千种。在下图的栏杆中，装饰完全由木板之间的空间边缘构成。木板要锯成能连结成栅栏的材料，以便形成某种空间。

一种栏杆
...A balustrade

下图是一种更加复杂的情况——罗马式建筑风格的教堂入口。

一种门道
A doorway

in the fence,they make something of the space between them.

Here is a more complicated case—the entrance to a Romanesque church.

The ornament is built up around the edge of the entrance. It creates a unifying seam between the entrance space and the stone.Without the ornament,there would be a gap between the arch of the entry and the passage itself:the ornament works on the seam,between the two,and holds them together.It is especially lavish and developed in this place,because just this seam—the boundary of the entrance to the church—is so important,symbolically,to the people who worship there.

In fact,doors and windows are always important for ornament,because they are places of connection between the elements of buildings and the life in and around them.It is very likely that we shall find a concentration of ornament at the edges of doors and windows,as people try to tie together these edges with the space around them.

And exactly the same happens at hundreds of other places in the environment;in rooms,around our houses,in the kitchen,on a wall,along the surface of a path,on tops of roofs,around a column—in fact,anywhere at all where there are edges between things which are imperfectly knit together,where materials or objects meet,and where they change.

Most generally of all,the thing that makes the difference in the use of ornament is the eye for the significant gap in the continuum:the place where the continuous fabric of interlock and connectivity is broken.When ornament is applied badly it is always put in—to some place where these connections are not really missing,so it is superfluous,frivolous.When it is well used,it is always applied in a place where there is a genuine

沿入口的边缘有装饰。装饰在入口空间和石块之间形成一条合成一体的缝。如果没有装饰，在入口的拱门和过道本身之间就会造成一空隙，而装饰在两者之间的空隙上发挥功能，使两者浑然一体。在这种存在间隙的地方，尤其要多用装饰，绝不可吝惜。因为正是这条缝——教堂入口处的边界——对那些去教堂顶礼膜拜的信徒具有非常重要的象征意义。

　　事实上，门窗始终是重要的装饰，因为它们是建筑构件的连接处，在门窗的内外都有着生活。很可能，当人们竭力将这些门窗边缘及其周围空间连接在一起时，我们就会发现，门窗的边缘是装饰的集中点。

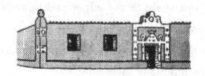

努比亚人的门
Nubian door

　　上述装饰的情况同样会发生在周围环境中的其他许多地方；在房间内、在墙壁上、沿小路的表面、在屋顶的顶部和柱的四周——事实上，各物之间有边缘但未完美结合成一体的任何一处，以及材料或物体相接或发生变化的任何一处都会存在这种情况。

美国早期的镂花型板喷刷
Early American stencilling

gap,a need for a little more structure,a need for what we may call metaphorically "some extra binding energy," to knit the stuff together where it is too much apart.

Therfore:

Search around the building,and find those edges and transitions which need emphasis or extra binding energy. Corners,places where materials meet,door frames, windows,main entrances,the place where one wall meets another, the garden gate,a fence—all these are natural places which call out for ornament.

Now find simple themes and apply the elements of the theme over and again to the edges and boundaries which you decide to mark.Make the ornaments work as seams along the boundaries and edges so that they knit the two sides together and make them one.

ଧର୍ଷ

Whenever it is possible,make the ornament while you are building—not after—from the planks and boards and tiles and surfaces of which the building is actually made— WALL MEMBRANE(218),FRAMES AS THICKENED EDGES(225),LAPPED OUTSIDE WALLS(234),SOFT INSIDE WALLS(235),SOFT TILE AND BRICK(248). Use color for ornament—WARM COLORS(250);use the smaller trims which cover joints as ornament—HALF-INCH TRIM(240);and embellish the rooms themselves with parts of your life which become the natural ornaments around you— THINGS FROM YOUR LIFE(253)...

最常见的是，能区别被采用的装饰的就是我们的眼睛，它们会看出连续统一体中有意义的空隙，即联锁和连续结构的中断处。如果装饰使用不当，它总是在并未真正失掉这些联系的某处出现，那么它是多余的、毫无意义的。如果装饰使用得当，它总是可以应用于真正有空隙的、需要多一些结构的、我们可比喻为需要"某种附加的结合能量"的地方，那么它就可以把分离得太远的部分连结得浑然一体。

因此：

在建筑物四周寻找并发现那些需要强调或需要附加结合能量的边缘。角隅、材料相接的地方、门框、窗、主入口、一片墙和另一片墙相接的地方、花园大门、栅栏——所有这些都是需要装饰的合理位置。

请找出简单的主题思想，并把它的诸要素反复应用于决定要标志出的各种边缘和边界。务必要使装饰为沿边界和边缘的缝发挥功能，以便把两条边联结起来，形成一个整体。

边界

重复　　　　　　　　　　主题

☘☙

当你正在建造房屋时——不是在房屋建成后——利用木板条、木板、面砖这类材料来实际构成房屋的各种表面，并在任何时候都能对它们进行装饰——**墙体（218）、门窗边缘加厚（225）、鱼鳞板墙（234）、有柔和感的墙内表面（235）**和**软质面砖和软质砖（248）**。利用色彩进行装饰——**暖色（250）**；利用覆盖接缝的较小的压缝条做装饰——**半英寸宽的压缝条（240）**；用生活中的部分纪念品装饰房间，使它们成为你周围的天然装饰——**生活中的纪念品（253）**……

CONSTRUCTION
构 造

2151

模式250　暖色**

……本模式有助于创造和生产**好材料**（207）、**地面面层**（233）、**有柔和感的墙内表面**（235）。凡是可以使用的地方，都要使材料保持其自然状态。只要有足够的装饰色彩，就会使室内的光线充满生气和温暖。

❧❧❧

医院和办公室走廊的绿色和灰色是令人压抑的、寒冷的。天然的树木、阳光和鲜明的色彩是温暖的。室内各种颜色的温暖感会以某种方式造成舒适和不舒适的巨大差别。

但是，暖色和冷色都指何种颜色？简明扼要地说，赤、黄、橙、褐是暖色；蓝、绿、灰是冷色。但是，很明显，如下说法是不正确的：具有红黄两色的房间会使人感觉良好；而具有蓝、灰两色的房间会使人感到冷冰冰。存在某些表面实情可供说明：红、褐、黄三色有助于形成室内的舒适气氛；而白、蓝、绿三色也都会使人感到舒适。毕竟天是蓝的、草是绿的。显然，我们在蓝天之下、绿草丛中，都会感到惬意舒适。

解释既简单而又有诱惑力。绝不是物品和表面的颜色使某处变得温暖或寒冷，而恰恰是光的颜色造成的。这究竟意味着什么？我们能测定出在空间某一特定点上光的颜色，只要让其显示在一块纯白色的表面上。如果光是暖色的，该表面将会被轻微地染成黄红色。如果光是冷色的，该表面将会轻微地染成蓝绿色。这种色泽上的变化是很轻微的：的确，在一小块白色表面上，可能很难看到这一点，所以你应备一台分光仪。

但是，当你认识到空间内的每一物体——人们的脸面、

250   WARM COLORS**

...this pattern helps to create and generate the right kind of GOOD MATERIALS(207),FLOOR SURFACE(233),SOFT INSIDE WALLS(235).Where possible leave the materials in their natural state.Just add enough color for decoration,and to make the light inside alive and warm.

⁊⸙

The greens and greys of hospitals and office corridors are depressing and cold.Natural wood,sunlight,bright colors are warm.In some way,the warmth of the colors in a room makes a great deal of difference between comfort and discomfort.

But just what are warm colors and cold colors?In a very simple minded sense,red and yellow and orange and brown are warm;blue and green and grey are cold.But,obviously,it is not true that rooms with red and yellow feel good;while rooms with blue and grey feel cold.There is some superficial truth to this simple statement:it is true that reds and browns and yellows help to make rooms comfortable;but it is also true that white and blue and green can all make people comfortable too.After all,the sky is blue,and grass is green.Obviously,we feel comfortable out in the green grass of a meadow,under the blue sky.

The explanation is simple and fascinating.It is not the color of the things,the surfaces,which make a place warm or cold,*but the color of the light.*What exactly does this mean?We can estimate the color of the light at a particular point in space by holding a perfectly white surface there.If the light is warm,this surface will be slightly tinted toward the yellow-red.If the light

手、男衬衫、女装、童装、食物、纸张等——都有轻度的色泽变化时，那就不难看出这种色泽变化了。所以这一认识对正在那里体验色泽变化的人们的情绪会产生巨大的影响。

再则，某一空间内光的颜色不是简单地取决于表面的颜色，而是取决于光源的颜色和许多表面间反复反射的光两者复杂的相互作用。在绿草地上，在春光明媚的日子里，从绿草反射回来的阳光依然是温暖的——在淡黄和粉红两色的波长范围之内。医院走廊内，荧光灯发出的并从绿色墙壁上反射回来的光是冷光——在绿蓝两色的波长范围内。在自然光线充足的室内，全部光都是温暖的。某一房间的窗户朝向马路对面的灰色楼房，它内部的光线也许是寒冷的，除非布置了强烈的黄红两色的纺织品。

如果你对室内光线的客观存在特性有怀疑，身边又没有分光仪，那么你唯一的办法只有使用彩色胶卷了。如果光是暖色的，而且胶卷曝光很好，白墙会略呈粉红色。如果光是冷色的，白墙会略呈蓝色。

所以，为了使房间舒适宜人，应汇集与光源和户外反射面有关的各种颜色，使其混合，以求房间中央的反射光成为暖光，即黄红色光。黄色和红色始终是暖色。而蓝色、绿色、白色只有在和其他颜色相抵消的适当地方和光源起良好作用之时才能成为暖色。

为了完善我们的论述，现在我们利用色度边界来使暖光的概念精确化。设光线射到一房间中央任何一个给定的表面上。该光线包含许多不同的波长。它的特性可以精确地由光谱能量 $p(\lambda)$ 的某种分布来说明，而光谱能量分布却为此光线中的不同波长提供相对的比例。

我们知道，不论何种光线——简言之，任何 $p(\lambda)$ ——都能在色彩三角形上被标绘成一个单一的点——比较正式的名称是二维色度图——是根据冈特·怀赞基和斯泰尔

is cold,this surface will be slightly tinted toward the blue-green. This tinting will be very slight:indeed,on a small white surface it may be so hard to see that you need a spectrometer to do it.But when you realize that everything in that space is lightly tinted—people's faces,hands,shirts,dresses,food,paper,everyth ing—it is not so hard to see that this can have a huge effect on the emotional quality that people experience there.

Now,the color of the light in a space does not depend in any simple way on the color of the surface.It depends on a complex interaction between the color of the light sources and the way this light then bounces on and off the many surfaces.In a meadow,on a spring day,the sunlight bouncing off the green grass is still warm light—that is,in the yellowish reddish range. The light in a hospital corridor,lit by fluorescent tubes,bouncing off green walls is cold light—in the green-blue range.In a room with lots of natural light,the overall light is warm.In a room whose windows face onto a grey building across the street,the light may be cold,unless there is a very strong concentration of yellow and red fabrics.

If you are in any doubt about the objective character of the light in the room and you don't have a spectrometer,all you need to do is to try to use color film.If the light is warm and the film is properly exposed,white walls will come out slightly pink.If the light is cold,white walls will come out slightly blue.

So,in order to make a room comfortable,you must use a collection of colors which together with the sources of light and the reflecting surfaces outside the room,combine to make the reflected light which exists in the middle of the room warm,that is,toward the yellow-red.Yellow and red colors will always do it.Blues and greens and whites will only do it in the proper places,balanced

斯所提出的标准色彩匹配功能理论（*Color Science*，New York，1967，pp.228~317）而绘制的。此色彩三角形图中的坐标标明任何一种已给出的能量分布的色度。

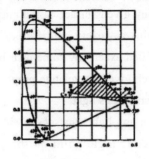

色度图
Chromaticity diagram

现在我们可以来鉴定此色度图上我们称之为暖区的一个区。这在图上由阴影线示出。

该阴影线区是以若干经验数据为基础的。例如，我们知道，人们对于不同空间中相对的温暖或寒冷都有一个清晰的主观印象。(See, for instance, Committee on Colorimetry of the Optical Society of America, *The Science of Color*，New York，1953，p.168.) 有一项研究是试图鉴定已感知的"温暖"的客观相关事物。(S.M.Newhall, "Warmth and Coolness of Colors," *Psychological Record*, 4，1941，pp.198~212.) 它揭示出"最温暖的区"就是判定值在主波长610nm（纳米）处，即在橙色波长范围的中心。个别研究人员所作的判定值的稳定性很高。因此，有一项研究提出温暖的可靠性系数为0.95，寒冷的可靠性系数为0.82。(N.Collins, "The Appropriateness of Certain Color Combinations in Advertising," M.A.thesis,Columbia University，New York，1924.)

最后，最为重要的是要记住本模式所需要的只是光线——室内中央的全部光线，来源于太阳光、各种灯光、

with other colors,and when the light sources are helping.

To complete the discussion we now make the concept of warm light precise in terms of chromaticity.Consider the light falling on any given surface in the middle of the room. This light contains a variety of different wavelengths.Its character is specified,exactly,by some distribution of spectral energies $p(\lambda)$,which gives the relative proportions of different wavelengths present in this light.

We know that any light whatsoever—in short,any $p(\lambda)$—can be plotted as a single point on the color triangle—more formally known as the two-dimensional chromaticity diagram—by means of the standard color matching functions given in Gunter Wyszecki and W.S.Stiles,*Color Science*,New York,1967,pp.228-317.The coordinates of a plot in this color triangle define the *chromaticity* of any given energy distribution.

We may now identify a region on the chromaticity diagram which we shall call the *warm region*.It is shown hatched on the drawing.

This hatched area is based on a number of empirical results. For example,we know that people have a clear subjective impression of the relative warmth,or coldness, of different spaces. See,for instance,Committee on Colorimetry of the Optical Society of America,*The Science of Color*, New York,1953,p.168.One study which attempts to identify the objective correlates of perceived "warmth" is S.M.Newhall, "Warmth and Coolness of Colors," *Psychological Record*,4,1941,pp.198-212.This study revealed a maximum for "warmest" judgments at dominant wave-length 610 millimicrons,which is in the middle of the orange range.And individual observer stability in such judgments is high. Thus,one study gives reliability coefficients of 0.95 for warmth and

墙壁的反射光、户外的反射光和地毯的反射光——全部光线位于我们称之为色彩三角形的"暖区"之内。这并不要求室内任何单独的彩色表面必须是红色、橙色或黄色的，而仅仅要求室内所有表面和光线的综合效应在室中央所形成的光线处于色彩三角形的暖区。

因此：

选择表面的颜色要同自然光、反射光和各种灯光的颜色协调一致，从而形成室内的暖光。

这意味着为了挑选压缝条、灯罩和少数的细部，黄色、红色和橙色是必不可少的——**半英寸宽的压缝条**（240）、**装饰**（249）、**投光区域**（252）。彩色的**帆布顶篷**（244）和**软质面砖和软质砖**（248）也都有助于形成暖色光。采用蓝色、绿色和灰色要难得多；尤其在北侧，那里的光线既寒冷又灰暗，但蓝、绿、灰三色始终能用于装饰，因为它们有助于衬托出较温暖的颜色——**装饰**（249）……

0.82 for coolness—N.Collins, "The Appropriateness of Certain Color Combinations in Advertising," M.A.thesis,Columbia University,New York,1924.

Finally,it is vital to remember that this pattern requires only that the light—the total light in the middle of a room, coming from sunlight,artificial lights,reflections from walls, reflections from outside,from carpets—the total light,lies in that part of the color triangle we call "warm." It does not require that any individual color surfaces in the room should be red or orange or yellow—only that the combined effect of all the surfaces and lights together,creates light in the middle of the room which lies in the warm part of the color triangle.

Therefore:

Choose surface colors which,together with the color of the natural light,reflected light,and artificial lights,create a warm light in therooms.

∞○∞

This means that yellows,reds,and oranges will often be needed to pick out trim and lampshades and occasional details—HALF-INCH TRIM(240),ORNAMENT(249), POOLS OF LIGHT(252).Colored CANVAS ROOFS(244)and SOFT TILE AND BRICK(248)also help to make warm colored light. Blues and greens and greys are much harder to use;especially on the north side where the light is cold and grey,but they can always be used for ornament,where they help to set off the warmer colors—ORNAMENT(249)...

模式251　各式坐椅

　　……当你准备用家具布置房间时，你选择家具就应像建房过程那样谨慎从事，以便你所购置的每件家具，非嵌墙的或嵌墙的，都如房间和凹室一样，具有独特的、有机的个性——每件不同的家具要摆放在不同的位置上——**起居空间的序列**（142）、**坐位圈**（185）、**嵌墙坐位**（202）。

❧❦

　　人们的身材大小高矮各不相同；他们坐的姿势也各不相同。可是，现代有一种倾向：所有的椅子都造得一模一样。

251 DIFFERENT CHAIRS

...when you are ready to furnish rooms,choose the variety of furniture as carefully as you have made the building,so that each piece of furniture,loose or built in,has the same unique and organic individuality as the rooms and alcoves have—each different,according to the place it occupies—SEQUENCE OF SITTING SPACES(142),SITTING CIRCLE(185),BUILT-IN SEAT(202).

※※※

People are different sizes;they sit in different ways.And yet there is a tendency in modern times to make all chairs alike.

Of course,this tendency to make all chairs alike is fueled by the demands of prefabrication and the supposed economies of scale.Designers have for years been creating "perfect chairs" — chairs that can be manufactured cheaply in mass.These chairs are made to be comfortable for the average person.And the institutions that buy chairs have been persuaded that buying these chairs in bulk meets all their needs.

But what it means is that some people are chronically uncomfortable;and the variety of moods among people sitting gets entirely stifled.

Obviously,the "average chair" is good for some,but not for everyone.Short and tall people are likely to be uncomfortable.And although situations are roughly uniform—in a restaurant everyone is eating,in an office everyone is working at a table—even so,there are important distinctions:people

当然，所有椅子造得雷同的这种倾向是和产品工业生产的种种要求分不开的，也是和推测的经济效益分不开的。多少年来，设计人员一直在创造着"完美无缺的椅子"——即可以大批量生产的、成本低廉的椅子。这些椅子对于中等身材的人来说是舒适的。购买椅子的机关团体都被逐一说服：大量购买这些椅子就能满足它们的需要。

可是，这意味着有一批人会长期不舒服；坐椅子的人的情绪变化完全受到抑制。

很明显，"平平常常的椅子"对一些人是合适的，但并不是对每个人都合适。矮个子和高个子很可能会感到不舒服。虽然情况大致相同——在饭馆里每人都在用餐，在办公室里每人都在桌旁工作——即使如此，仍然存在着重要的区别：有的坐的时间长，有的短；有的靠椅背坐着，在那里冥思苦想；有的探身向前，在那里进行热烈辩论；有的正襟危坐，在那里稍事等候。如果所有的椅子全都一模一样，上述人们神态上的差异就会被抑制，而一些人的心情就不会舒畅。

下面一点虽不太明显，然而也许却是最重要的：我们将使自己的心境和个性在我们所坐的椅子中形象地体现出来。在第一种心境中，我们坐大而宽的椅子正合适；在第二种心境中，就要坐摇椅；在第三种心境中，坐挺直的靠背椅；而在第四种心境中，坐凳子和沙发。当然，这不单是我们喜欢随自己的心绪而变换坐椅；有的椅子使我们非常喜爱，有的椅子使我们感到非常安全和舒服；但对此每个人的感受又是各不相同的。摆满千差万别的各式坐椅的环境会立即创造出一种使人感受丰富的气氛；而摆放着一模一样的坐椅的环境会造成一种微妙的直接限制人们感受的气氛。

sitting for different lengths of time;people sitting back and musing; people sitting aggressively forward in a hot discussion;people sitting formally,waiting for a few minutes.If the chairs are all the same,these differences are repressed,and some people are uncomfortable.

What is less obvious,and yet perhaps most important of all,is this:we project our moods and personalities into the chairs we sit in.In one mood a big fat chair is just right;in another mood,a rocking chair;for another,a stiff upright;and yet again,a stool or sofa.And,of course,it isn't only that we like to switch according to our mood;one of them is our favorite chair,the one that makes us most secure and comfortable;and that again is different for each person.A setting that is full of chairs,all slightly different,immediately creates an amosphere which supports rich experience;a setting which contains chairs that are all alike puts a subtle straight jacket on experience.

Therefore:

Never furnish any place with chairs that are identically the same.Choose a variety of different chairs,some big,some small,some softer than others,some rockers,some very old,some new,with arms,without arms,some wicker,some wood,some cloth.

ഇൗരു

Where chairs are placed alone and where chairs are gathered,reinforce the character of the places which the chairs create with POOLS OF LIGHT(252),each local to the group of chairs it marks...

因此：

无论是在哪里，千万不要陈设完全相同的坐椅，要选择各种不同的坐椅：有的大，有的小，有的较柔软，有的是摇椅，有的旧，有的新，有的有扶手，有的无扶手，有的是柳条编的，有的是木制的，有的是布做的。

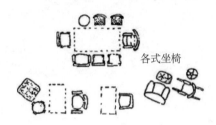

各式坐椅

ഔൟ

凡只摆放椅子或椅子密集的地方，其特点就会因椅子和**投光区域**（252）而显得突出。投光区域标志一部分椅子或一批椅子……

模式252 投光区域**

……本模式有助于修饰小的社会空间，如**凹室**（179）、**工作空间的围隔**（183）；较大的空间，如**中心公共区**（129）、**入口空间**（130）、**灵活办公室空间**（146）；室内摆设，如**进餐气氛**（182）、**坐位圈**（185），还有各式坐椅（251）。本模式甚至有助于产生**暖色**（250）。

❧❦❧

均匀照明——照明工程师的佳作——无论如何使用效果还是差的。事实上，均匀照明会破坏空间的社会特性，使人们感到晕头转向和漫无边际。

请看下面这幅照片。这是一个带方框的板条天花，在天花上有数十盏间距相等的荧光灯。这意味着尽可能使光线成为无明暗差别的均匀光线，不过这种模拟天光的努力是徒劳无益的。

无明暗差别的均匀光线
Flat, even light

但是，均匀照明所依据的是两个错误观念。首先，户外光线几乎永远是不均匀的。大多数天然地区，尤其是人类有机体不断进化的那些环境，都有持续不断、每时每刻地从一处向另一处变化着的有圆形斑点的光线。

更为严重的是：我们按社会空间使用的空间，部分地

252 POOLS OF LIGHT**

...this pattern helps to finish small social spaces like
ALCOVES(179)and WORKSPACE ENCLOSURE(183),
larger places like COMMON AREAS AT THE HEART(129),
ENTRANCE ROOM(130),and FLEXIBLE OFFICE
SPACE(146), and the furnishing of rooms like EATING
ATMOSPHERE (182),SITTING CIRCLE(185),and DIFFERENT
CHAIRS(251).It even helps to generate WAM COLORS(250).

❧❧❧

**Uniform illumination—the sweetheart of the lighting
engineers—serves no useful purpose whatsoever.In fact,it
destroys the social nature of space,and makes people feel
disoriented and umbounded.**

Look at this picture.It is an egg-crate ceiling,with dozens
of evenly spaced fluorescent lights above it.It is meant to make
the light as feat and even as possible,in a mistaken effort to
imitate the sky.

But it is based on two mistakes.First of all,the light
outdoors is almost never even.Most natural places,and
especially the conditions under which the human organism
evolved, have dappled light which varies continuously from
minute to minute,and from place to place.

More serious,it is a fact of human nature that the space we use
as social space is in part defined by light.When the light is perfectly
even,the social function of the space gets utterly destroyed:it
even becomes difficult for people to form natural human groups.
If a group is in an area of uniform illumination,there are no light

是由光线限定，这是人类天性的实情。当光线格外均匀时，空间的社会功能就会遭到彻底破坏：人们形成自然群体就变得越发困难。如果一个群体处于均匀照明区内，没有对应于群体边界的光线梯度，则群体的分界线、内聚性及其"存在"都将被削弱。如果该群体处于"投光区域"内，而且其大小和边界相应于群体的大小和边界，这就会增强群体的界线、内聚性以及它的存在（从现象学角度而言）。

霍普金森和朗莫尔根据实验结果提出了一种可能的解释。他们指出，面积又小又亮的光源分散注意力的程度小于较不明亮的大面积的光源。这两位首创者得出结论说，工作台上的局部照明比均匀的背景照明更能使工人把自己的注意力集中到工作上去。这似乎有理由来推断：如果群体有局部照明而无均匀的背景照明，就更有可能保持社会群体的内聚性所需的人对人的高度注意力。(R.G.Hopkinson and J.Longmore，"Attentionand Distraction in the Lighting of Workplaces," *Ergonomics*，2，1959,p.321ff.Also reprinted in R.G.Hopkinson, *Lighting*, London : HMSO, 1963, pp.261~268.)

实地考察的结果证实这种推测。在伯克利的加利福尼亚大学国际部，有一个大房间，在这里人们通常可以懒散地坐着等候客人和其他居民。室内有坐位 42 个，其中 12 个紧靠着电灯。在两次考察中，我们清点了坐在室内的总人数为 21 名；其中 13 人选择的坐位紧靠着电灯（X^2=11.4，有效数字在 0.1%的范围内）。可是，室内光线的总亮度只够供阅读之用。我们终于得出如下结论：人们在寻找"投光区域"。

日常经验所证明的和千百次观察中得出的结果相同。每一个设施完善的餐馆都会使每一餐桌成为一个单独的投光区域，因为，众所周知，这样有助于餐桌的私密的和亲切的气氛。在宅内，一张"你自己的"、坐上去非常舒适的旧椅，在比较昏暗的环境里有一盏电灯照明——结果你就会摆脱家庭中的喧闹而能够去安静地读书了。再如，居室

gradients corresponding to the boundary of the group,so the definition,cohesiveness,and "existence" of the group will be weakened.If the group is within a "pool" of light,whose size and boundaries correspond to those of the group,this enchances the definition,cohesiveness,and even the phenomenological existence of the group.

One possible explanation is suggested by the experiments of Hopkinson and Longrnore,who showed that small bright light sources distract the attention less than large areas which are less bright.These authors conclude that local lighting over a work table allows the worker to pay more attention to his work than uniform background lighting does.It seems reasonable to infer that the high degree of person to person attention required to maintain the cohesiveness of a social group is more likely to be sustained if the group has local lighting,than if it has uniform background lighting.(See R.G.Hopkinson and J.Longrnore, "Attention and Distraction in the Lighting of Wor kplaces," *Ergonomics*,2,1959,p.321 ff.Also reprinted in R.G.Ho pkinson,*Lighting*,London:HMSO,1963,pp.261-268.)

On-the-spot observation supports this conjecture.At the International House,University of California,Berkeley,there is a large room which is a general waiting and sitting lounge for guests and residents.There are 42 seats in the room,12 of them are next to lamps. At the two times of observation we counted a total of 21people sitting in the room;13 of them chose to sit next to lamps.These figures show that people prefer sitting near lights (X^2=11.4,significant at the 0.1%level).Yet the overall light level in the room was high enough for reading.We conclude that people do seek "pools of light."

Everyday experience bears out the same observation in

内常有一盏孤灯悬吊在餐桌之上——灯光似乎和坐在餐桌周围的人形影不离。在较大的空间，情况也属如此。试想一下在公园中的孤灯长椅为一对情侣所形成的幽静的环境吧；或试想一下在卡车停车站上，一群休戚相关的人坐在灯光明亮的咖啡座周围啜饮着咖啡的情景吧。

　　尚有一言提醒。本模式是容易被理解的；或许也是容易被赞同的。但是，在周围环境中实际创造出效能高的投光区域是一个相当微妙的问题。我们知道许多失败的例子：例如，一些小灯破坏均匀照明的地方无论如何不适于做人们聚集的场所。

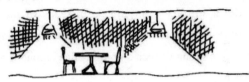

与社会空间不协调的投光区域
Light pools at odds with social space

因此：

把灯位放低并相互分开，以便形成单独的投光区域。它像水泡似地环绕桌椅，以便加强其所形成的空间的社会性质。务必记住，如果没有较暗的空间夹在中间，就无法形成投光区域。

投光区域

❧

　　让灯罩和靠近灯的帷幔带有色彩，以便使从灯罩和帷幔上反射回来的光成为暖色光——**暖色**（250）……

hundreds of cases.Every good restaurant keeps each table as a separate pool of light,knowing that this contributes to its privalte and intimate ambience.In a house a truly comfortable old chair "yours," has its own light in dimmer surroundings— so that you retreat from the bustle of the family to read the paper in peace.Again,house dining tables often have a single lamp suspended over the table—the light seems almost to act like glue for all the people sitting round the table.In larger situations the same thing seems to be true.Think of the park bench,under a solitary light,and the privacy of the world which it creates for a pair of lovers.Or,in a trucking depot,the solidarity of the group of men sipping coffee around a brightly lit coffee stand.

One word of caution.This pattern is easy to understand;and perhaps it is easy to agree with.But it is quite a subtle matter to actually create functioning pools of light in the environment.We know of many failures:for example,places where small lights do break down even illumination,but do not correspond in any real way with the places where people tend to gather in the space.

Therefore:

Place the lights low,and apart,to form individual pools of light which encompass chairs and tables like bubbles to reinforce the social character of the spaces which they form. Remember that you can't have pools of light without the darker places in between.

ᚼᚾᚼ

Color the lampshades and the hangings near the lights to make the light which bounces off them warm in color—WARM COLORS (250)...

模式253　生活中的纪念品*

253 THINGS FROM YOUR LIFE*

...lastly,when you have taken care of everything,and you start living in the places you have made,you may wonder what kinds of things to in up on the walls.

❧❧❧

"Decor"and the conception of"interior design"have spread so widely,that very often people forget their instinct for the things they really want to keep around them.

There are two ways of looking at this simple fact.We may look at it from the point of view of the person who owns the space,and from the point of view of the people who come to it.From the owner's point of view,it is obvious that the things around you should be the things which mean most to you,which have the power to play a part in the continuous process of self—transformation,which is your life.That much is clear.

But this function has been eroded,gradually,in modem times because people have begun to look outward,to others,and over their shoulders,at the people who are coming to visit them,and have replaced their natural instinctive decorations with the things which they believe will please and impress their visitors.This is the motive behind all the interior design and decor in the women's magazines. And designers play on these anxieties by making total designs,telling people they have no right to move anything,paint the walls,or add a plant,because they are not party to the mysteries of Good Design.

But the irony is,that the visitors who come into a room don't want this nonsense any more than the people who live there.It is far more fascinating to come into a room which is the living

……最后，当你对每件事物都已注意到，并开始住进已布置好家具的住宅，你或许会想到要往墙上钉些什么。

&OCR

"装饰"和"室内设计"的概念早已广泛传播，以致人们往往忘却了在他们的周围环境中布置本是他们天生爱好的东西。

现在对于这一简单的事实有两种看法。我们可以持住宅空间占有者的观点，我们也可以持外来者的观点。从占有者的观点看，很明显，周围的一切东西应当是对你最有意义，并在你的连续不断的自我改造过程中起一定作用。不言而喻，这些东西就代表着你的生命。

但是，这种功能在现代社会中已被逐步削弱，因为人们现在已开始向外看，眼睛盯着别人，并越过别人的肩膀注视行将来访的客人，他们用自己确信会使来客高兴并留下深刻印象的物件来取代他们天生喜爱的装饰品。这就是在妇女杂志上连篇累牍刊登室内设计和装饰的背后的动机。设计者利用人们热衷于装饰的急切心情，进行总体设计，并不断表明，居民无权挪动任何东西，无权漆墙，无权增加一草一木，因为这些做法都是和尽善尽美的设计的奥秘相悖的。

但是，具有讽刺意味的是：登堂入室的宾客除了想看望一下住在那里的人，并不想看这种无聊的东西。当你跨进一个能显示出某人或某一群体的活生生的思想感情的房间时，一定会感到这样的房间更具有吸引力。你能看到他们的生活、经历和倾向，他们的一切都会在墙壁的四周、在家具中、在书架上清清楚楚地显示出来。和这种平凡得像青草一样的感受相比，矫揉造作的、舞台布景式的"现代装饰"会彻底破产。

expression of a person,or a group of people,so that you can see their lives,their histories,their inclinations,displayed in manifest form around the walls,in the furniture,on the shelves.Beside such experience-and it is as ordinary as the grass—the artificial scene—making of "modern decor" is totally bankrupt.

Jung describes the room that was his study,how he filled the stone walls with paintings that he made each day directly on the stones—mandalas,dream images,preoccupations—and he tells us that the room came gradually to be a living thing to him—the outward counterpart to his unconscious.

Examples we know:A motel run by a Frenchman, mementos of the Resistance all around the lounge,the letter from Charles de Gaulle.An outdoor market on the highway, where the proprietor has mounted his collection of old bottles all over the walls;hundreds of bottles,all shapes and colors;some of them are down for cleaning;there is an especially beautiful one up at the counter by the cash register.An anarchist runs the hot dog stand,he plasters the walls with literature, proclamations,manifestoes against the State.

A hunting glove,a blind man's cane,the collar of a favorite dog,a panel of pressed flowers from the time when we were children,oval pictures of grandma,a candlestick,the dust from a volcano carefully kept in a bottle,a picture from the news of prison convicts at Attica in charge of the prison,not knowing that they were about to die,an old photo,the wind blowing in the grass and a church steeple in the distance,spiked sea shells with the hum of the sea still in them.

Therefore:

Do not be tricked into believing that modern decor must be slick or psychedelic,or"natural"or"modemrn art,"or"plants" or anything else that current taste-makers claim.It is must beautiful when it comes straight from your life—the things you care for,the things that tell your story.

荣格在描述他的书斋时写道，他是如何日复一日地直接在石墙上画满密密麻麻的画——曼陀罗（佛教中菩萨形象之画——译注）、梦境的幻影和神志的恍惚——他告诉我们，对他来说，他的书斋已渐渐成为栩栩如生的东西，即他的潜在意识向外表露的体现物。

我们知道的类似例子还有：法国人经营的汽车旅馆，客厅四周放满了反法西斯战争的纪念品，查理·戴高乐的信。在公路旁的户外集市上，业主把收集到的旧瓶子都堆放在墙头上；旧瓶成百上千，各种形状和颜色应有尽有；有些旧瓶排列着正准备清洗；有一个特别漂亮的旧瓶子竖放在收银机旁的柜台上。一个无政府主义者经营着热狗摊，他把广告、传单、声明、宣言等糊在墙上，以表示反对他的政府。

可以作装饰的纪念品比比皆是：诸如打猎用的五指分开的手套，盲人的拐棍，爱犬的项圈，我们童年时代的一束压平的花，老祖母的蛋壳画，烛台，小心翼翼地珍藏于瓶中的火山灰，从管理犯人的古希腊雅典城邦传出的有关犯人的罪行、但并不知道犯人即将被处决的消息而作的画，具有诗情画意的旧照片——如风吹草动、远处天际的教堂尖顶，依然散发着海腥味的长而尖的贝壳等。

因此：

请不要上当受骗，认为现代装饰必须是第一流的或颜色鲜艳的，或"自然的"，或"现代艺术"，或"花草树木"，或当前被时髦风尚带头人所宣称的其他所有东西。如果装饰品直接来自你自己的生活，那才是最美的——因为这些东西是自己所珍惜的，是自己的历史的见证者。

收藏品　　全家合影和其他照片

过去的历险纪念品　　纪念品

致 谢

我们在八年的构思和创作中，得到了许多友人的大力支持和协助。在此我们谨向他们表示由衷的感谢。

我们的"环境结构中心"（以下称"中心"）是一个工作小组。根据工作的需要，人员经常变动，3～8人不等。自1967年"中心"成立以来，一些同行已和我们合作共事，虽然时间长短不同，但在许多方面我们均受益匪浅。丹尼·艾布拉姆斯曾任"中心"财务经理三年。他在"中心"的初创时期起了重要的作用，并协助过我们确定小组的性质。他对本书初稿的编排和摄影实验工作也都提供了帮助，并和我们一起从事过俄勒冈的试验工作。罗恩·沃尔基在"中心"度过了两个春秋，他大力协助过我们提高改进了本书第一部分中所描绘的模式和城市总体概念。上述两位从一开始，就和本模式语言的发展结下了不解之缘；尤其是，每当午餐后，他们悦耳动听的音乐使我们一起度过了难以忘怀的美好时光。

我们到处有朋友。西姆·范德·赖恩和罗斯林·林海姆两位远在我们动手写本书的前几年就一直帮助和鼓励我们。克里斯蒂·科芬、吉姆·琼斯和巴巴拉·施赖纳都帮助过我们丰富本书最初几个模式的内容。

吉姆·阿克斯莱对本书最后部分最难写好的结构模式所给予的帮助比谁都大。更早一些时候，桑迪·赫希恩就和我们在秘鲁同心协力地进行过施工实验，并开始完善我们关于施工技术的新观点。

ACKNOWLEDGMENTS

We have had a great deal of help and support over the eight years it has taken us to conceive and create this work. And we should here like to express our feelings of gratitude to everyone who helped us.

The Center has always been a small workgroup,fluctuating in size from 3 to 8,according to the demands of the work. Since the Center was incorporated in 1967,a number of people have worked with us,for different lengths of time,and helped in many ways.Denny Abrams was financial manager of the Center for three years.He played a critical role in the early days of the Center,helping to shape our nature as a work group. He also helped with layout and photographic experiments in the early drafts of the book and worked with us on the Oregon experiment.Ron Walkey spent two years at the Center,and helped especially to develop the patterns and the overall conception of the city portrayed in the first section of the book.The two of them were very close to the development of the pattern language,from the beginning;and above all,their music,after lunch,made unforgettable times together for all of us.

In more general terms,both Sim Van der Ryn and Roslyn Lindheim gave us help and encouragement when we first began the project,years ago.Christie Coffin,Jim Jones,and Barbara Schreiner all helped us develop the contents of the earliest versions of the language.

Jim Axley helped more than anyone on the very difficult development of the structural patterns,in the last part of the language.And earlier,Sandy Hirshen,collaborating with us during the Peru project,had begun to develop our attitude to construction techniques.

Harlean Richardson has worked tremendously hard on the

哈里恩·里查森在详细设计本书方面作了大量的艰苦工作。海伦·格林8年的文书工作成绩斐然，她打印了浩繁的文稿；玛丽·露易丝·罗杰斯在协调工作、提供支援方面成效卓著。

我们得到的另一种宝贵的支持就是那些信赖我们正在尝试我们的事业的人们为我们提供试验的机会，并让我们为他们建造体现本书思想的房屋。肯·西蒙斯允许我们在他的一个建筑工程中完善我们的第一个建筑模式语言。约翰尼斯·奥利夫格伦，约翰·埃伯哈德，鲍勃·哈里斯，唐·康韦，弗里德·威特曼，休伊特·赖恩和埃德加·考夫曼都是如此竭诚相助。他们给我们以信心、精神支持、友谊和金钱，以此来支持我们的事业。所有这一切支援都是无法估量的。

我们要特别感谢全国精神康复研究所的迪克·韦克菲尔德和克莱德·多塞特。这本《建筑模式语言》之所以得以完善是因为在最重要的四年内得到了全国精神康复研究所附设城市问题研究中心的一系列赞助。如果没有这些赞助，我们是无法进行这项创作的。

最后，我们要衷心感谢牛津大学出版社，尤其是我们的编辑詹姆斯·赖姆斯，他首先同意出版这套丛书（共三卷），我们也同样衷心感谢吉姆斯·赫胡斯—戴维斯和拜伦·霍林斯海德。他们三位全都支持本书的出版和其他书籍的出版，甚至在他们过目本书之前就欣然同意：这表明，在完成这部书的过程中，他们一再给予我们巨大的动力，在我们缺乏信心时，他们及时对我们进行鼓励。在本书的排印过程中，我们给牛津大学出版社增添了许多麻烦和困难。但出版社的友人却和我们站在一起，通力合作。

本书最终问世完全仰赖于朋友们的全力支持。

detailed design of the book itself.And we have had wonderful secretarial help over the years from Helen Green,who typed many many versions of the patterns,and from Mary Louise Rogers who helped in many ways coordinating the work and providing support.

Another invaluable kind of help we have had was that given by people who believed in what we were trying to do,gave us an opportunity to work on it,and to do projects for them which incorporated these ideas.Ken Simmons,who allowed us to develop our very first pattern language in a professional job,Johannes Olivegren,John Eberhard,Bob Harris,Don Conway,Fried Wittman,Hewitt Ryan,and Edgar Kaufmann all helped us in this way.What they gave us in confidence,and emotional support,and friendship,and,often,in money that supported the work,cannot be counted.

Even more specifically,we want to thank Dick Wakefield,Coryl Jones,and Clyde Dorsett at the National Institute for Mental Health. The evolution of the pattern language was supported for the four most important years by a sequence of grants from the Center for the Study of Metropolitan Problems of the National Institute for Mental Health—and it would have been quite impossible for us to do the work if it had not been for those grants.

Finally,we owe a great deal to Oxford University Press, especially to James Raimes,our editor,who first agreed to try and publish all three books,in a series,and also to James Huws-Davies and Byron Hollinshead.All three of them supported the publication of this book,and the other books,before they had even seen them: and once again,gave us enormous energy to do the work,by putting their confidence in us at a time when we badly needed it.During the production of the book,we have often created severe difficulties for Oxford;but they have stood by us throughout.

It is only because all of our friends have helped us as they did that it has actually been possible.

向提供照片的友人致谢

我们为本书选用的许多照片取自第二或第三来源。在每一情况下我们都力图确定原拍摄者是谁，并向他表示诚挚的谢意。可是在某些情况下，来源含混不清，我们简直无法追根溯源。所以，我们感到遗憾的是我们无法一一都感谢到，如有冒犯不周之处，希请见谅。名单如下：

77	H.Armstrong Roberts	531	Martin Hurlimann
91	Andreas Feininger	589	Anne-Marie Rubin
113	Emil Egli and	609	Marc Foucalt
	Hans Richard Muller	617	Ivy De Wolfe
117	Clifford Yeich	627	R.Blijstra
123	Gutkind	655	Henri Cartier-Bresson
129	Claude Monet	661	Andre George
161	Henri Cartier-Bresson	669	V.S.Pritehett
175	Walter Sanders	693	Henri Cartier-Bresson
203	Herman Kreider	701	Henri Cartier-Bresson
219	Sam Falk	709	Iain Macmillan
257	Joanne Leonard	771	Martin Hurlimann
327	Edward Weston	815	Ken Heyman
333	Iain Macmillan	835	Robert Doisneau
339	Fred Plaut	841	Edwin Smith
401	Bernard Rudofsky	885	Alfred Eisenstaedt
415	Gilbert H.Grosvenor	931	Andre'Kertesz
447	Edwin Smith	947	Ralph Crane
501	Martin Hurlimann	959	Eugene Atget
521	Andre George	965	V.S.Pritchett

PHOTO ACKNOWLEDGMENTS

Many of the pictures we have selected for this book come from secondary and tertiary sources.In every case we have tried to locate the original photographer and make the appropriate acknowledgment.In some cases,however,the sources are too obscure,and we have simply been unable to track them down.In these cases,we regret that our acknowledgments are incomplete and hope that we have not offended anyone.

77	H.Armstrong Roberts	501	Martin Hurlimann
91	Andreas Feininger	521	Andre George
113	Emil Egli and	531	Martin Hurlimann
	Hans Richard Muller	589	Anne-Marie Rubin
117	Clifford Yeich	609	Marc Foucalt
123	Gutkind	617	Ivy De Wolfe
129	Claude Monet	627	R.Blijstra
161	Henri Cartier-Bresson	655	Henri Cartier-Bresson
175	Walter Sanders	661	Andre George
203	Herman Kreider	669	V.S.Pritchett
219	Sam Falk	693	Henri Cartier-Bresson
257	Joanne Leonard	701	Henri Cartier-Bresson
327	Edward Weston	709	Iain Macmillan
333	Iain Macmillan	771	Martin Hurlimann
339	Fred Plaut	841	Edwin Smith
401	Bernard Rudofsky	885	Alfred Eisenstaedt
415	Gilbert H.Grosvenor	931	André Kertesz
447	Edwin Smith	947	Ralph Crane

959	Eugene Atget	1559	Henri Cartier-Bresson
965	V.S.Pritchett	1607	Berthe Morisot
971	André Kertesz	1631	A.F.Sieveking
1003	Charles E.Rotkin	1661	R.Rodale
1043	Bernard Rudofsky	1727	C.H.Baer
1075	Wu Pin	1757	Pierre Bonnard
1105	Tonk Schneiders	1767	G.Nagel
1119	Eugene Atget	1791	Henri Matisse
1135	Erik Lundberg	1807	Dorothy and Richard Pratt
1191	Martin Hurlimann		
1211	Bernard Rudofsky	1929	Alan Fletcher
1219	Francois Enaud	1945	Erik Lundberg
1227	Bernard Rudofsky	1979	Clifford Yeich
1237	Herbert Hagemann	2045	Erik Lundberg
1243	Lazzardo Donati	2081	Carl Anthony
1327	Pierre Bonnard	2089	Winslow Homer
1349	Russell Lee	2097	Edwin Smith
1357	Joanne Leonard	2103	Avraham Wachman
1373	Joanne Leonard	2117	Ivy De Wolfe
1433	Ken Heyman	2163	Bruno Taut
1451	Dorien Leigh	2193	Izis Bidermanas
1493	Ernest Rathnau	2217	André Kertesz
1499	Aniela Jaffe	2223	Pfister
1505	Orhan Ozguner	2235	Roderick Cameron
1511	Marian O.Hooker	2247	Marc Foucault
1523	Erik Lundberg	2305	J.Szarhouski